Turtles of the World

DISCARD

Turtles
of the World

Carl H. Ernst and Roger W. Barbour

Smithsonian Institution Press

Washington, D.C., and London

Designer: Janice Wheeler
Editor: Nancy Dutro

Library of Congress Cataloging-in-Publication Data
Ernst, Carl H.
   Turtles of the world.

   Bibliography: p.
   Includes index.
   1. Turtles.   I. Barbour, Roger William, 1919–
II. Title.
QL666.C5E77   1989   597.92   88-29727
ISBN 0-87474-414-8 (cloth)
ISBN 1-56098-212-8 (paper)
British Library Cataloging-in-Publication Data
available

∞ The paper used in this publication meets the
minimum requirements of the American National
Standard for Performance of Paper for Printed
Library Materials Z39.48-1984.

Color illustrations printed by South China Printing
Company, Hong Kong
All other illustrations and text printed in the United States
of America
Manufactured in the United States of America
96   5

Contents

Illustrations

All photographs not otherwise credited are by Roger W. Barbour. Specimens marked with an asterisk were provided by William P. McCord from his collection of living turtles.

Color Plates

The following species are illustrated in the color plate section. The page number given references the text discussion of the species.

Preface

After publication of our *Turtles of the United States* in 1972, we discussed the possibility of preparing a companion volume on the identification and relationships of all the world's turtles. We recognized there was a definite need for such a volume, for many of the 257 species of living turtles have been poorly described and are difficult to identify.

The first serious attempt at a monograph covering the turtles, with descriptions and illustrations for identification, was by Gray (1831b) in which he summarized the then-known species. That volume spurred research on turtles which resulted in a number of new species being described. In 1835 and 1851, A. M. C. Duméril and his colleagues published lists and descriptions of the turtles in the Muséum d'Histoire Naturelle, Paris.

In 1855, Gray published a second volume based on the turtles at the British Museum of Natural History, which was the most extensive collection at that time. A supplement was published by Gray in 1870(b), and a further appendix in 1872. After Gray's death Boulenger (1889) published a new edition which included excellent identification keys and illustrations of the turtles in the British Museum. Although Siebenrock (1909) published a later list, Boulenger's (1889) catalogue is still the most scholarly identification reference for the classification and identification of turtles. Unfortunately, only about 70 percent of the now-recognized species are included in Boulenger's work, and most of the taxonomic names have changed; one must have a previous comprehensive knowledge to use it.

More recent general books on turtles by Wermuth and Mertens (1961) and Pritchard (1967, 1979) contain erroneous keys and are of limited use in identifying turtles. Therefore, it is clear that there is a need for a modern volume summarizing the classification of tur-

tles with clear descriptions and keys to their identification. After a long delay we have attempted to do this, endeavoring to present the material in such a form as to be useful to both scientists and interested laypersons. We also indicate existing problems within turtle systematics, and hope this will spur research to solve these questions.

Our experience with turtles spans over 50 years. CHE has studied turtles in the eastern United States, Central America, and Europe for over 25 years and has given special attention to the ecology and behavior of freshwater species; his studies have resulted in more than 100 scientific papers on turtles. As a Research Associate of the Division of Reptiles and Amphibians, National Museum of Natural History, Washington, D.C., CHE has valuable experience with the almost 200 species in its collection. RWB has done research on aspects of the natural history of vertebrates, including turtles, since his college days; his experience as a wildlife photographer covers over 45 years. We have made several collecting trips to parts of the United States and Europe, seeking acquaintance with species not previously encountered in the field and obtaining photographs. RWB had previously spent two years researching reptiles and mammals in Indonesia and Southeast Asia. Several trips were taken by us to zoological gardens and museums in Europe and the United States to examine and photograph additional species of turtles.

Several persons have aided us in various ways during the preparation of this volume. W. Ronald Heyer, Roy W. McDiarmid, George R. Zug, and Ronald I. Crombie of the National Museum of Natural History provided library and research facilities, advice, and encouragement. C. J. McCoy, Carnegie Museum of Natural History, and Edward O. Moll, Eastern Illinois University, critically read the first draft of the manuscript and offered many helpful suggestions. Charles C. Crumly, John B. Iverson, William W. Lamar, and Michael E. Seidel also critically reviewed parts of the manuscript. Jens and Larry Wishmeyer, California Academy of Sciences, provided measurements on various tortoises of the *Geochelone elephantopus* complex. Fred Medem sent Colombian specimens, and Charles R. Crumly, J. Whitfield Gibbons, C. J. McCoy and Peter C. H. Pritchard offered advice and encouragement. Photographs or specimens were provided by the Natal Parks, Game and Fish Preservation Board and by John L. Behler, James F. Berry, Charles M. Bogert, O. Bourquin, Donald G. Broadley, Christopher W. Brown, Stephen D. Busack, John Cann, Kenneth Darnell, Richard A. Davis, James L. Dobie, Llewelyn M. Ehrhart, Michael A. Ewert, Wayne Frair, Dale Fuller, Avital Gasith, Michael Goode, Steve W. Gotte, George H. Grall, Frank Groves, John Groves, James H. Harding, Bryon C. Harper, John B. Iverson, Karl

Kranz, Arndt F. Laemmerzahl, William W. Lamar, John M. Legler, Klaus M. Lehmann, Colin J. Limpus, Jeffrey E. Lovich, C. J. McCoy, Roy W. McDiarmid, Sean McKeown, Federico Medem, K. E. Methner, Sherman A. Minton, Jr., Ardell Mitchell, Joseph C. Mitchell, Russell A. Mittermeier, Edward O. Moll, Kenneth T. Nemuras, Ellen B. Nicol, Gregg W. North, Reiner Praschag, Peter C. H. Pritchard, William H. Randall, Anders G. J. Rhodin, David A. Ross, Michael E. Seidel, Jay C. Shaffer, Ian J. Smales, John H. Tashjian, Ramon A. Velez, J. Vijaya, Laurie J. Vitt, and Robert G. Webb. William P. McCord was particularly helpful, allowing us to examine and photograph his large collection of living turtles. The line drawings were prepared by Evelyn M. Ernst and photographed by Jan F. Endlich. Various parts of the text were typed by Bonnie Contos, Evelyn M. Ernst, Gale McCune, Mary Roper, and the staff of the George Mason University Word Processing Center.

The staffs of the following institutions allowed us to examine or photograph specimens in their collections. *Zoological gardens:* Atlanta Zoo, Baltimore Zoo, Brookfield Zoo, Cincinnati Zoological Society, Columbus Zoo, Dallas Zoo, Detroit Zoological Park, Fort Worth Zoological Park, Gladys Porter Zoo, Honolulu Zoo, Houston Zoological Gardens, Knoxville Zoo, Nairobi Snake Park (Nairobi, Kenya), National Zoological Park (Washington, D.C.), Philadelphia Zoological Garden, Regents Park Zoo (London), San Antonio Zoo, San Diego Zoo, Steinhardt Aquarium, St. Louis Zoological Park, and Woodland Park Zoo. *Museums and other institutions:* Academy of Natural Sciences, Philadelphia; American Museum of Natural History; British Museum (Natural History); California Academy of Sciences; Carnegie Museum of Natural History; Cincinnati Museum of Natural History; Cornell University; Field Museum of Natural History; Florida State Museum of Biological Sciences; George Mason University; Los Angeles County Museum; Louisiana State University; Miami Seaquarium; Museo de Zoología, Universidade de São Paulo; Museum of Comparative Zoology, Harvard; Museum of Vertebrate Zoology, Berkeley; Museum of Zoology, University of Michigan; Muséum National d'Histoire Naturelle, Paris; Naturhistorisches Museum, Vienna; North Carolina Museum of Natural History; North Museum, Franklin & Marshall College; Rijksmuseum van Natuuralijke Historie, Leiden; Senckenberg-Museum, Frankfurt; Texas Cooperative Wildlife Collection, Texas A & M University; Tulane University; United States National Museum of Natural History; University of Illinois Museum of Natural History; University of Kansas Museum of Natural History; University of Kentucky; and the Zoologische Sammlung des Bayerischen Staats, Munich.

Research and travel for this project were subsidized in part by grants from the American Philosophical Society, George Mason University Foundation, and the Kentucky Research Foundation. Travel monies were also provided to CHE by the Department of Biology, George Mason University.

Introduction

Turtles are ancient animals that evolved into a shelled form over 200 million years ago. Together with lizards, amphisbaenids, snakes, crocodilians, and the tuatara, they constitute the vertebrate class Reptilia. Reptiles are ectotherms that, in part, have evolved walking limbs and a dry, scaly skin. They evolved from the amphibians during the Pennsylvanian Period, toward the close of the Paleozoic Era. During the following Mesozoic Era the reptiles underwent rapid adaptive radiation and became the dominant animals on earth. Reptiles, along with the amphibians, represent a transitional group in vertebrate evolution, occurring between the aquatic fishes and the terrestrial birds and mammals. Reptiles were the first vertebrates adapted to life in dry places. While it is true that many amphibians spend much time on land, their eggs must be laid in water or damp places. Reptiles were able to lay a specialized egg, which has a calcareous or parchmentlike shell that retards moisture loss. Hence, reptiles could take advantage of the great expanses of dry land not previously available to their ancestors. Their eggs also have embryonic membranes (amnion, chorion, and allantois) that are not found in amphibian eggs, as well as a yolk sac that contains nutrients. The amnion forms a fluid-filled compartment surrounding the embryo, thus bringing the aquatic environment within the egg.

Other characteristics also evolved that contributed to the freeing of reptiles from the water requirements of their amphibian predecessors. The scaly skin has few surface glands, and very little moisture is lost cutaneously. The well-developed lungs provided ample oxygen for the heart, now a three-chambered organ in most reptiles, but four-chambered in the crocodilians. The partitioning of the ventricle into two chambers was more efficient, produced higher blood pressure,

and led indirectly to the development of more efficient kidneys. Other major developments include the appearance of claws on the toes, a palate separating the oral and nasal passages, and the evolution of a male copulatory organ (absent in the tuatara) which facilitates internal fertilization.

Turtles are the only living members of the subclass Anapsida, which is characterized by a primitive skull with a solid cranium and no temporal openings (anapsid). These shelled reptiles constitute the order Testudines. All living shelled reptiles are turtles, but the terms tortoise and terrapin have also been applied, and these have different meanings in various parts of the world. Tortoise is best applied to terrestrial turtles. Terrapin is usually applied to edible, more or less aquatic, hard-shelled turtles.

In this book we have attempted to present data on turtles in a logical, systematic order. The first section, on the order Testudines, includes a description and illustrations of turtle anatomy and a discussion of their origins. Also included are keys to the identification of the various families of turtles. Each family is treated in a separate section, which begins with a description of the family and an identification key to the various genera which constitute that family. If the genera have been placed in subfamilies or complexes, these are compared in the text. As with the various families, each genus is described and a key to its species is included. If pertinent, species groups are discussed. A description of the karyotype (chromosome number and morphology) is also given, if known. Finally, each species is treated in a format which begins with a detailed description (RECOGNITION) of the turtle. The geographic range of the species is given under DISTRIBUTION (see Iverson, 1986, for maps of the distribution of individual species). If subspecies have been described, their characteristics and areas of occurrence are presented under GEOGRAPHIC VARIATION. The turtle's ecological needs are described under HABITAT, and reproductive and feeding habits are discussed under NATURAL HISTORY.

Turtle populations have been declining at an alarming rate throughout the world in recent years. A number of factors have contributed to this decline. Overcollecting is certainly a major problem. The volume of the pet trade has resulted in the removal of many adults from the populations, and the gathering of eggs and juveniles reduces the rate of replacement of those adults left to die of natural causes. Adults of many species are harvested for food. With their relatively slow rate of maturation turtles cannot withstand heavy cropping and still maintain their populations. Fortunately, recent salmonella scares and conservation legislation have reduced the pet trade in turtles. Insecticides and herbicides probably contribute to the decrease in the numbers of turtles. Large quantities of chlorinated hydrocarbons (ingredients in many pesticides) may be stored in the body fat and be released later to poison the turtle while it fasts. In urbanized countries the automobile has had a detrimental effect: thousands of turtles die on the highways each year; in all too many cases they are struck deliberately.

The general deterioration and disappearance of the natural environment has eliminated entire turtle populations, and in many of the underdeveloped areas turtles and their eggs constitute a major protein source for native peoples.

If turtles are to remain a conspicuous part of our fauna we must initiate conservation measures. Although we do not yet know enough about turtle biology to formulate adequate conservation plans for all species, certain needs are obvious. The waterways and lands harboring important populations should be protected from undue human disturbance and pollution. The trend away from the use of the dangerous residual pesticides must be continued. Countries must pass and enforce legislation controlling the capture of these creatures in the wild.

Equally important is the fact that more people need to become acquainted with the many fascinating aspects of turtle biology. Such awareness will surely make people more interested in the protection of these shy creatures. The creation of such an attitude—not only toward turtles but also toward our dwindling wildlife—is a major purpose of this book.

Order Testudines
The Turtles

The order Testudines is composed of shelled reptiles with solid, anapsid skulls. It is the most primitive order of living reptiles, and is apparently close to the ancestral reptilian lineage (as indicated by its unspecialized skull). Living turtles are found on all continents except Antarctica. The marine species occur predominantly in tropical waters of the Atlantic, Pacific, and Indian oceans, but several also range far poleward in these waters. In maximum size, adult turtles range from 11.5 cm *(Clemmys muhlenbergii)* to 180 cm and 590 kg *(Dermochelys coriacea)*.

The most exceptional feature of the turtle is its shell; this extremely conservative character has remained little changed for about 200 million years. The shell is divided into two halves: an upper part, the carapace, and lower part, the plastron. The two parts are joined on each side by a bridge.

The typical carapace usually consists of about 50 bones. The nuchal is the most anterior bone along the midline, behind it are eight neurals (absent in some chelid turtles), two suprapygals, and a pygal, in that order. Occasionally a preneural may be found between the nuchal and the 1st neural, and the total number of neural bones may vary. The neurals are attached to the neural arches of the dorsal vertebrae, but the other bones of the series are free from the vertebrae. On each side of the neurals are usually eight costal bones; in some species a precostal is also present. Outside the costals and extending along each side from the nuchal to the pygal is a series of about 11 peripherals. Each carapacial bone articulates with the adjacent bones along a suture.

Members of the tortoise genus *Kinixys* have a movable hinge in their carapaces that is located between the 4th and 5th costal bones and 7th and 8th peripheral bones, which allows the posterior part of the carapace to be lowered over the hind limbs and tail, thus affording protection of these parts.

The forepart of the typical plastron is composed of a median bone, the entoplastron (absent in Kinosternidae), surrounded anteriorly by two epiplastra and posteriorly by two hyoplastra. Behind these are a pair each of hypoplastra and xiphiplastra. In some primitive species a pair of mesoplastra occurs between the hyoplastra and hypoplastra. Between the forelimbs and the hind limbs the hyoplastra and hypoplastra articulate with the 3d to 7th peripherals. The forelimb emerges from the axillary notch, the hind limb from the inguinal notch. Just behind the axillary notches the axillary buttresses solidly attach the hyoplastra to the 1st costals, and in front of the inguinal notches the inguinal buttresses solidly attach the hypoplastra to the 5th costals. In some turtles, such as *Terrapene*, there is no bridge connecting the carapace to the plastron, and in others, such as *Claudius*, the bridge may be narrow or poorly buttressed.

In the families Dermochelyidae and Trionychidae and in the tortoise *Malacochersus tornieri*, there has been a reduction in the size and/or number of bones in the shell, so the structures of their carapaces and plastra are quite aberrant.

The bones of the shell are covered with horny scutes. The division between adjacent scutes is called the seam. A seam often leaves an impression, termed a sulcus, on the underlying bones. We follow the terminology for the various scutes proposed by Zangerl (1969), but other systems of scute nomenclature have been proposed (Table 1).

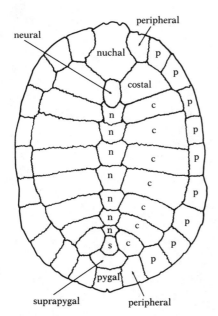

Carapacial bones of turtles

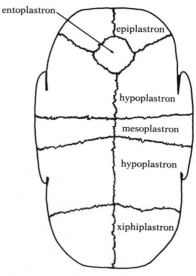

Plastral bones of turtles

Table 1. Synonymy of the various names applied to the epidermal scutes of the turtle shell

Zangerl (1969)	Carr (1952)	Boulenger (1889) Loveridge & Williams (1957)	Pritchard (1979)	Hutchison & Bramble (1981)
cervical	precentral	nuchal	nuchal	—
vertebral	central	vertebral	central	—
pleural	lateral	costal	costal	—
marginal	marginal	marginal	marginal	—
12th marginal	postcentral	supracaudal	supracaudal	—
intergular	—	intergular	intergular	gular (Chelidae)
gular	gular	gular (Dermatemydidae) extragular (Chelidae)	gular	intergular (Kinosternidae)
humeral	humeral	humeral	humeral	gular (Kinosternidae)
pectoral	pectoral	pectoral humeral (Dermatemydidae) pectoral (Chelidae)	pectoral (Kinosternidae)	anterior humeral
abdominal	abdominal	abdominal	abdominal	posterior humeral (Kinosternidae)
femoral	femoral	femoral	femoral	femoral
anal	anal	anal	anal	anal

Along the anterior midline of the carapace is a single cervical scute. This is followed posteriorly by a series of five vertebral scutes. Along each side and touching the vertebrals is a series of four pleurals. Outside the pleurals and extending along each side from the cervical are 12 marginal scutes (11 in Kinosternidae). In the alligator snapping turtle (*Macroclemys*) a series of small scutes, called supramarginals, appears between the posterior pleurals and the marginals.

The scutes of the plastron are divided into pairs by a median longitudinal seam. Anteriorly there is a pair of gular scutes (except that in the family Kinosternidae a single gular is usual); in some families an intergular is also present. Paired humerals, pectorals, abdominals, femorals, and anals follow, respectively, and in the

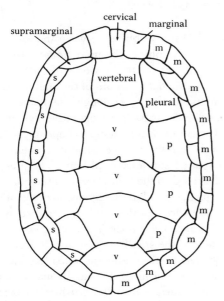

Carapacial scutes of turtles

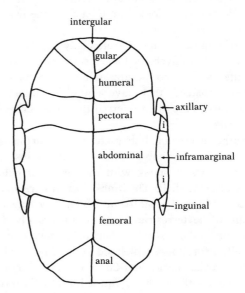

Plastral scutes of turtles

Cheloniidae an interanal is sometimes present. At the posterior edge of each axillary notch there may be an axillary scute, and at the front edge of each inguinal notch there may be an inguinal scute. Inframarginals, a series of small scutes lying between the carapacial marginals and the sides of the adjacent plastral scutes, are present in the families Cheloniidae, Chelydridae, Dermatemydidae, and Platysternidae.

Included in the description of each hard-shelled species is a plastral formula that denotes the relationship of the length of each plastral scute, as measured along its midseam, to those of the other scutes on the plastron.

Recently, Hutchison and Bramble (1981) analyzed the scutes of fossil and living specimens of the Kinosternidae and related Dermatemydidae, and compared these scutes to those of testudinoid turtles. They found that the gular scute of kinosternids is really a new scale of different origin, which they term an intergular. They also found that the humeral scute of extant kinosternids is divided into two scutes by the development of the anterior plastral hinge between the epiplastron and hyoplastron; thus, kinosternids are unique in not having true pectoral or abdominal scutes. Hutchison and Bramble have prepared the alternate plastral scute terminology for the Kinosternidae, Dermatemydidae, and the chelid genus *Chelodina* shown in Table 1. Although we feel their conclusions are accurate we have not incorporated them in this book, but instead call the attention of the reader to their paper.

Turtles of the genera *Pelusios, Emys, Emydoidea, Terrapene, Cuora, Cyclemys, Pyxidea, Notochelys, Pyxis,* and *Testudo* have a tranverse hinge, more or less developed, located on the anterior edge of the abdominal scutes, and in most species of *Kinosternon* a pair of hinges borders the abdominals. These hinges allow the plastron to be folded up to enclose the head and limbs if it is large enough.

Turtles of the families Carettochelyidae, Dermochelyidae, and Trionychidae have lost the horny covering of scutes, and the bony material in their shells is much reduced. Instead, they have a tough, leathery skin. These turtles are often referred to as leatherbacks or softshells.

Another exceptional characteristic of turtles is the migration of the limb girdles to positions inside the rib cage (which, along with the vertebral column, helps form the shell). The limbs of turtles are adapted to the medium through or on which they travel. For instance, the terrestrial tortoises have evolved elephantine hind limbs, which help to support them, and shovellike forelimbs, which aid in digging. The limbs of the marine species and *Carettochelys* are modified as flippers with which they quite literally fly through the water. The semiaquatic turtles have developed

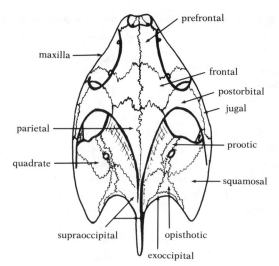

Dorsal view of turtle skull

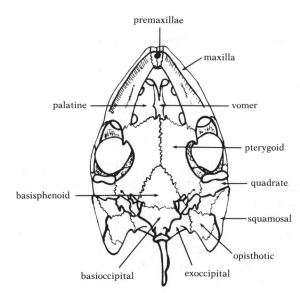

Ventral view of turtle skull

various degrees of webbing between the toes; generally, the more webbing, the more aquatic the turtle.

We refer the reader to the works of Gaffney (1972, 1979b) for detailed discussions of the anatomical variations in turtle skulls. Although the turtle skull is highly modified it is basically primitive, evidenced by the solid cranium with no temporal openings. Posteriorly the cranium has a large otic notch on either side of an elongated supraoccipital bone. The pterygoids are solidly fused to the braincase, and the quadrate is closely attached to a lateral expansion of the otic capsule. The jaws are toothless and are modified as sharp shearing beaks. The upper jaw is composed of two pairs of bones: the small premaxillae, in front (fused in soft-shelled turtles), and the maxillae, along the sides. The lower border of the maxilla may be either a sharp cutting surface or a flat crushing surface. The maxilla forms the lateral border of the nasal cavity

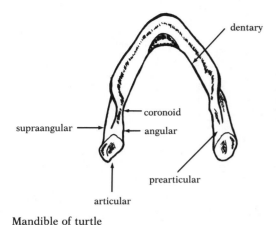

Mandible of turtle

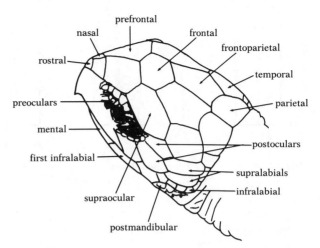

Head scalation of turtles

and the lower border of the orbit. Posteriorly it articulates with the jugal. The posterior rim of the orbit is formed by the jugal ventrally and the postorbital dorsally. The paired bones along the dorsal midline of the skull are the prefrontals, the frontals, and the parietals. The parietals form the roof and much of the lateral walls of the braincase. Laterally the parietal has a strong process projecting downward and joining the pterygoid. The lower border of the parietal also articulates with the prootic and the supraoccipital. The elongated supraoccipital extends backward and forms the upper border of the foramen magnum. Paired exoccipitals (= paroccipitals) form the lateral borders, and a basioccipital forms the lower border; parts of these three bones constitute the occipital condyle. The cheek bones are the large posterior squamosal and the smaller anterior quadratojugal (a small bone lying between the quadrate and jugal bones). The squamosals are connected to the supraoccipital by the exoccipitals. The palatal complex is composed, from front to rear, of the single vomer and palatine bones and the paired pterygoids. These are often covered with a secondary palate. Anteriorly the vomer touches the premaxillae. The palatines form part of the roof of the nasal passage. Between each palatine and the adjacent maxilla is an opening, the posterior palatine foramen. The pterygoids are separated posteriorly by the basisphenoid. The free lateral projections of the pterygoids are the ectopterygoid processes. The pair of quadrate bones, which lie posterior to the pterygoids, forms the lower-jaw articulation. Each half of the lower jaw is composed of six bones. The most anterior is the dentary, which forms the crushing or cutting surface; the two dentaries are firmly united at the symphysis. On the lower border of the jaw the dentary extends backward nearly to the articulation with the quadrate. On the upper side behind the crushing surface are the coronoid and supraangular bones. At the inner posterior end of the lower jaw are, ventrally, the angular bone and, dorsally, the prearticular. The articular bone forms the articulation with the quadrate.

Head scalation illustrated in the figure is typical.

We follow, with some modification, the higher classification of turtles proposed by Gaffney (1984). In his taxonomic scheme, living turtles belong to the gigaorder Casichelydia, and either the megaorder Pleurodira (side-necked turtles, see below) or the megaorder Cryptodira (hidden-necked turtles, see below). There are 12 living families and 257 living species. See Hunt (1958) for a discussion of the proper ordinal name for turtles.

The oldest fossil turtles have been found in the Triassic deposits of Germany. These fossils have been assigned to the genus *Proganochelys* (= *Triassochelys*), which, along with the fossil genera *Proterocheris*, *Saurischiocomes*, and *Chelytherium*, belong to the gigaorder Proganochelydia (Romer, 1956; Gaffney, 1975b). Only some skeletal features of these are known completely. The skull is solidly roofed, with the external parts sculptured. The postfrontal, lacrimal, and supratemporal are (or may be) present, and the external nares are divided by a bony bar. The quadrate lacks the strong curvature around the stapes that is characteristic of most turtles. There are small teeth on bones of the palate and rudimentary teeth on the jaw margins. Proganochelydians have seven cervicals—the 8th vertebra is a dorsal rather than a cervical, its arch fused to the nuchal—and the cervicals are amphicoelous, with two-headed ribs. The bones of the pectoral girdle, although partially fused with the plastron, can still be identified individually as clavicle, interclavicle, and cleithrum. The pelvis probably was more or less firmly attached, even sometimes fused, to the plastron. The peripherals of the carapace apparently were very numerous; the neurals are long and narrow; and there are nine costals. The plastron contains several extra bones, in the form of two pairs of

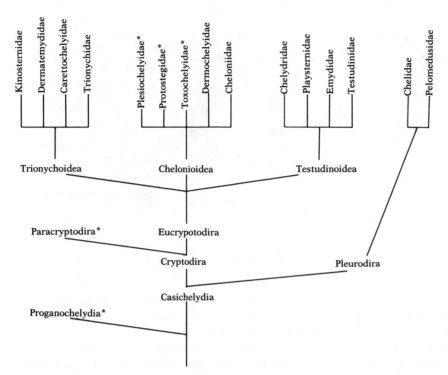

Relationships within the order Testudines (* = extinct taxa)

mesoplastra, which are absent in most modern turtles. The buttress elements of the plastron do not reach the carapacial costals. There are dermal tubercles on the neck and tail, and presumably the head and limbs could not be pulled into the shell. Likely these primitive turtles were amphibious (Romer, 1956).

All other turtles are assigned to the gigaorder Casichelydia. Gaffney (1975b, 1984) compared the skulls of proganochelydians and casichelydians, and found the floor of the middle ear (acoustic-jugular cavity) and the interpterygoid vacuity to be open in proganochelydians but closed in casichelydians. The quadrate does not form the lateral wall of the middle ear in proganochelydians, but does so in casichelydians, and the following features are present in Proganochelydia but not in Casichelydia: a bladelike (cultiform) process on the parasphenoid, palatal teeth, a well-developed median tubercle on the basioccipital, and the external opening of the nostrils divided by a dorsal premaxillary process. In contrast, casichelydians have both an expanded supraoccipital crista and an antrum postoticum which are absent in proganochelydians. The Casichelydia became dominant during the Jurassic and remained so ever since. They were mainly amphibious, but some species have become marine (Cheloniidae, Dermochelyidae) or terrestrial (Emydidae, Testudinidae). Two modern megaorders of turtles have arisen from early casichelydians: the Cryptodira, during the Jurassic, and the Pleurodira, during the Cretaceous. The Cryptodira continued as the main evolutionary line of turtles, with the Pleurodira appearing as an aberrant but structurally conservative side branch. Table 2 compares the Cryptodira and Pleurodira.

The earliest cryptodirans are from the Late Jurassic. By the Early Cretaceous they had become the dominant turtles of northern regions, and some species had invaded the oceans. Their heads can be withdrawn in a vertical flexure, because the cervical vertebrae can be bent into a sigmoid curve. The Cryptodira is considered to be the more advanced group of turtles. According to Gaffney (1984) the cryptodiran line is composed of two subdivisions called hyperorders. The extinct hyperorder Pleurosternoidea is comprised of the family Pleurosternidae (= Glyptopsidae, Jurassic). The second hyperorder, Daiocryptodira, includes the parvorder Baenoidea (family Baenidae, Cretaceous–Eocene), and the parvorder Eucryptodira, which includes living cryptodirans. Pleurosternids and baenoids retain several primitive characters. They retain nasal bones and thus have the prefrontals separated and not in contact at the midline of the skull. The foramen for the internal posterior cortical canal lies midway along the length of the basisphenoid-pterygoid suture. They also have a well-developed stapedial artery and reduced orbital and palatine arteries. The neck vertebrae lacked mecha-

Table 2. A comparison of pleurodiran and cryptodiran turtles (modified from Romer, 1956, and Gaffney, 1975b)

Character	Pleurodira	Cryptodira
Skull:		
Temporal roof	Emarginated posteriorly and variably emarginated ventrally	Emarginated posteriorly, often not emarginated ventrally
Nasal	Rarely present	Present or absent
Prefrontal	No descending process	Descending process meets vomer ventromedially
Pterygoid	With upturned lateral expansions	Without upturned lateral expansions
Quadrate	Separates pterygoid from basioccipital	Does not separate pterygoid and basioccipital
Epiterygoid	Absent	Present (except in *Dermochelys*)
Hyomandibular branch of facial nerve (*VIII*)	Travels outside the *canalis cavernosus* in its own canal for most of the distance	Traverses the *canalis cavernosus*
Mandibular artery	Does not exit from the trigeminal nerve foramen, does not travel through the *canalis cavernosus*	Usually exits from the trigeminal nerve foramen, usually travels through the *canalis cavernosus*
Posterior palatine foramen	Behind orbit	In floor of orbit (absent in cheloniids)
Anterior exit of vidians canal	In ventral surface of pterygoid in pelomedusids, absent in chelids	In posterior wall of the posterior palatine foramen
Lower jaw articulation	Hemispherical articular	No hemispherical articular
Vertebrae:		
Head retraction	Laterally	Vertically
Posterior cervical spines	High	Low
Postzygapophyses	Close or fused	Widely separated
Cervical central articulations	Single, well developed	Double on posterior cervicals, broad and well developed
Transverse processes	Well developed, in middle of vertebrae	Rudimentary, in front of vertebrae
Sacral (posterior-dorsal) ribs	Rudimentary, arise from neural arch	Well developed, only connect with neural arch
Caudal vertebrae	Procoelous	Usually procoelous, sometimes opisthocoelous
Atlas	More or less solidly fused to odontoid	Rarely fused to odontoid
Limbs and Girdles:		
Pelvis	Fused to plastron	Not fused to plastron
Limbs	Aquatic, toes webbed	Aquatic with webbed toes or paddlelike, or terrestrial with no toe webbing
Carapace:		
Neurals	Usually reduced	Usually reduced only in highly derived families
Plastron:		
Mesoplastra	May be present	Absent
Buttresses	Usually strong (absent in *Pelusios*), extending dorsally and medially to the costals	Variable in development
Habitat:	Freshwater	Freshwater, marine, terrestrial

nisms to retract the head, and no formed central articulations were present in the early species. Mesoplastral bones and paired intergular scutes were also present. The parvorder Eucryptodira is composed of several superfamilies: the Trionychoidea (Cretaceous–Recent), including the living families Kinosternidae, Dermatemydidae, Carettochelyidae, and Trionychidae; the marine Chelonioidea (Jurassic–Recent), including the extinct families Plesiochelyidae (Jurassic), Protostegidae (Cretaceous), and Toxochelyidae (Cretaceous–Eocene) and the living families Cheloniidae and Dermochelyidae; the Testudinoidea (Paleocene–Recent), including the living families Chelydridae, Platysternidae, Emydidae, and Testudinidae. Gaffney (1984) places the Chelydridae in the separate infraorder Chelydroidea.

The Pleurodira are considered by many experts to be more primitive than the Cryptodira. Surprisingly, these turtles appear in the fossil record an entire period (about 50 million years) later than the cryptodirans: the oldest pleurodirans are from the Upper Cretaceous. The pleurodirans withdraw their necks laterally; that is, they bend the neck sideways to tuck the head under the rim of the shell. This has given rise to their common name: side-necked turtles. Two families are included, Pelomedusidae and Chelidae.

Key to the families of Cryptodira:

1a. Forelimbs modified as oarlike flippers   2
 b. Forelimbs not modified as oarlike flippers   4
2a. Surface of carapace covered with horny scutes: **Cheloniidae**
 b. Surface of carapace covered with smooth or granulated skin; scutes absent   3
3a. Limbs with claws; piglike snout; no keels on carapace: **Carettochelyidae**
 b. No claws on limbs; snout not elongated and piglike; seven longitudinal keels on carapace: **Dermochelyidae**
4a. Shell bony and complete; snout not elongated and tubelike   5
 b. Bony shell reduced and incomplete; snout elongated and tubelike: **Trionychidae**
5a. 12 plastral scutes present; anterior plastral hinge, if present, between hyo- and hypoplastra   6
 b. Less than 12 plastral scutes present; anterior plastral hinge present between epi- and hyoplastra: **Kinosternidae**
6a. Inframarginal scutes present   7
 b. Inframarginal scutes absent   9
7a. Upper jaw not medially hooked; jaw margins serrated; tail of moderate length; seams separating carapacial scutes disappear with age leaving a smooth, undivided surface: **Dermatemydidae**
 b. Upper jaw medially hooked; jaw margins not serrated; tail long; seams separate carapacial scutes   8
8a. Plastron large; head cannot be withdrawn into shell: **Platysternidae**
 b. Plastron reduced and cruciform; head can usually be entirely withdrawn into shell: **Chelydridae**
9a. Hind limbs elephantine (columnar), two phalanges in digits of each hind foot; toes not webbed: **Testudinidae**
 b. Hind limbs not elephantine, three or more phalanges in digits 2 and 3 of hind foot, toes usually with some degree of webbing: **Emydidae**

Key to the families of Pleurodira:

1a. Mesoplastral bones present; intergular scute usually reaches anterior plastral rim; quadratojugal present; splenial bone absent from lower jaw: **Pelomedusidae**
 b. Mesoplastral bones absent; intergular scute usually separated from anterior plastral rim by marginals; quadratojugal absent; splenial bone present in lower jaw: **Chelidae**

Pelomedusidae

graphicus, which attained a carapace length of 230 cm (Wood, 1976).

The skulls are only moderately posteriorly emarginated and the ventral emargination is variable. A quadratojugal is present and the bony temporal arch is complete. This bone may contribute to the temporal roof lying above and in front of the quadrate and touching the parietal. Jugal-parietal contact may occur; however, there is a loss of contact between the parietals and squamosals. The prefrontals touch dorsally, but nasal bones are absent. In the roof of the mouth, the palatines meet, but the vomer is small or absent, and the premaxillae are unfused. In the lower jaw a single fused dentary bone extends posteriorly almost to the articular region between the angular and well-developed supraangular bones. There is no splenial bone. The carapace may be arched or depressed; it contains six to eight neural bones, and one or more pairs of the posterior costal bones meet dorsally. There is no cervical scute. On the plastron, a single intergular scute lies between the paired gulars. A pair of mesoplastral bones is also present on the plastron. These are usually positioned laterally, but do meet medially in the genus *Pelusios*. A movable transverse hinge is present between the pectoral and abdominal scutes in *Pelusios;* all other genera have rigid plastra. The 2d cervical vertebra is convex; the others are procoelous or amphicoelous. The hind toes are heavily webbed. The five genera and 23 species are placed in two subfamilies. The subfamily Podocnemidinae includes the genera *Podocnemis*, *Peltocephalus*, and *Erymnochelys*, those turtles with

These primitive, semiaquatic side-necked turtles are today found naturally only in South America, Africa, Madagascar, and on some of the Seychelles Islands in the Indian Ocean. Their fossil record extends back to the Upper Cretaceous of North America and Europe, indicating they probably evolved in the Northern Hemisphere and were formerly more widespread. The oldest fossil pelomedusids are North American species assigned to the genus *Bothremys* (Gaffney and Zangerl, 1968). Among the numerous species of fossil pelomedusids is included what may be the largest freshwater turtle that ever lived, *Stupendemys geo-*

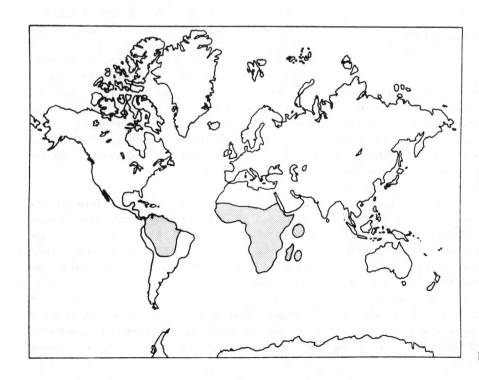

Distribution of Pelomedusidae

hingeless plastra and four claws on their hind feet. The subfamily Pelomedusinae includes the genera *Pelomedusa* and *Pelusios*, which have five claws on the hind feet.

Key to the genera of Pelomedusidae

1a. Hind feet with five claws   2
  b. Hind feet with four claws   3
2a. Plastron with movable hinge between pectoral and abdominal scutes; mesoplastra touching at plastral midline: ***Pelusios***
  b. Plastron rigid, lacking a movable hinge between pectoral and abdominal scutes; mesoplastra separated widely: ***Pelomedusa***
3a. Long intergular scute completely separates adjacent gulars   4
  b. Short intergular scute does not completely separate gulars: ***Erymnochelys***
4a. Interorbital groove present; upper jaw not hooked: ***Podocnemis***
  b. Interorbital groove absent; upper jaw hooked: ***Peltocephalus***

Pelomedusa Wagler, 1830

The genus *Pelomedusa* is composed of only one species of sideneck.

Pelomedusa subrufa (Lacépède, 1788)
African helmeted turtle

Recognition: The brown to olive carapace is oval, broad, and rather flattened dorsally. It may be smooth or slightly serrated posteriorly, and may have a slight keel on the 2d to 4th vertebrals. Vertebral 1 is broader than long and the largest of the five. The 4th is also broader than long. No cervical scute is present. The plastron is usually yellow to cream colored, but may be brownish or have dark seams. It is broadest anteriorly but narrows and is deeply notched posteriorly. No hinge is present. A large plastral fenestra remains open until adulthood. The plastral formula is usually: hum > fem > intergul > an > abd > gul > pect. The yellow to brown bridge is well developed, but somewhat narrow. On it are small wedge-shaped mesoplastral bones between the hyo- and hypoplastral bones. Ventrally, the marginal scutes are usually plain yellow, but in one race dark triangles are pres-

Pelomedusa subrufa

ent along the seams. The skull lacks a supratemporal roof, and the quadratojugal is widely separated from the parietal. A very small, almost indistinguishable ridge occurs on the median alveolar surface of the upper jaw. The snout protrudes. The head is brown to olive dorsally, and dark or light mottlings may be present. Laterally it becomes yellow to cream colored at the level of the dorsal edge of the tympanum. Ventrally it is unicolored yellow to cream, and two yellow barbels are present on the chin. Dorsally there are two supraorbital scales separated by a longitudinal seam. These are followed by a large frontal scale laterally bordered by two temporal scales. Neck, limbs, and tail are gray brown to olive dorsally or anteriorly, but yellowish ventrally or posteriorly. The webbed toes are composed of only two phalanges.

Most individuals are under 20 cm in carapace length, but Pritchard (1979) recorded a South African male approximately 33 cm long.

Stock (1972), Kiester and Childress (in Gorman, 1973), and Killebrew (1975a) reported a karyotype of 2n = 34 for *Pelomedusa*, but Bull and Legler (1980) reported it as 2n = 36 with 5 pairs of macrochromosomes and 13 pairs of microchromosomes.

Adult males have concave plastra and long, thick tails. Females have somewhat broader carapaces, flat plastra, and short tails.

Distribution: *Pelomedusa subrufa* inhabits subtropical and tropical Africa from Ethiopia and the Sudan westward to Ghana, Senegal, Mali, Nigeria, and the Cameroons, and southward to Cape Province, South Africa. It is also known from many localities on Madagascar.

Geographic Variation: Three subspecies have been described, but more study is needed to determine their validity. The nominate race, *Pelomedusa subrufa*

subrufa (Lacépède, 1788), has its pectoral scutes meeting at the midline of the plastron. It occurs from Somalia and the Sudan west to Ghana and southward to the Cape of Africa and on Madagascar. *P. s. olivacea* (Schweigger, 1812) is a more northern race, ranging from Ethiopia and the Sudan westward to Nigeria and the Cameroons. Its pectoral scutes are widely separated. *P. s. nigra* Gray, 1863c is found only in Natal and South Africa from Orange State and east of Cape Province to the line joining Kuruman, Kimberley, Graaf Reinet, and Grahamstown (Bour, 1986). Its pectoral scutes meet at the midline of the black to dark-brown plastron, and it has dark triangles on the ventral surface of the marginals and dark dashes on the dorsal surface of the head. Wide zones of intergradation occur where the races meet.

Habitat: The helmeted turtle is semiaquatic, living in temporary marshes, creeks, and rain holes in the open country of Africa, generally south of the Sahara Desert. It is seldom found in heavily forested areas, but is known to occur from the coastal plain to upland savannas of 3100 m elevation. Large populations sometimes inhabit the rain pools used as watering holes by African ungulates. The habitat requirements of *Pelomedusa* are apparently too dry for *Pelusios*, and for this reason Wood (1973) feels *Pelomedusa* has not been preserved as fossils as often as *Pelusios*, which inhabits permanent water bodies. When the temporary pools dry up, *Pelomedusa* buries itself in the mud bottoms until the next rainy season. Loveridge (1941) reported that South African individuals may hibernate out of water in soft soil or beneath leaves from May to August.

Natural History: Courtship and mating occur in the spring. The male pursues the female with extended head and neck, often touching her hindquarters and vent with his snout. If she is nonreceptive, the male often snaps at her tail and hind limbs. He then mounts her carapace from the rear and hooks the claws of all four feet under her marginals. His head and neck are then extended forward and downward and he sways these back and forth in front of her face while expelling a stream of water through his nostrils (Ernst, 1981a). Eventually coitus is achieved. Nesting occurs in late spring or early summer. Nests are flask shaped and about 10–17 cm deep. Apparently only one clutch of up to 42 eggs (Jaques, 1966; Pritchard, 1979) is laid each season; however, the normal clutch seems to be about 13–16 eggs. The white elongated (38 × 22 mm) eggs have a membraneous shell and are covered with slime when laid. Hatching occurs in 75–90 days. Hatchlings are olive to black, with 25–30-mm carapaces.

Pelomedusa is carnivorous, feeding on a variety of insects, earthworms, crustaceans, snails, fish, amphibians, small reptiles, birds, and mammals. It seizes prey in its mouth and then uses the forefoot claws to tear it into shreds. Where population densities are high, several turtles may attack larger prey such as a small aquatic bird in consort, drag it underwater, and tear it apart while alive. They also feed on carrion. Rochat et al. (1962) have reported that they may act as cleaners and pick off ectoparasites from the hides of rhinoceroses that enter their water holes.

Basking is common in subtropical wild individuals, but apparently the sun is too hot for them to bask in the more tropical parts of their range. During the rainy season they wander from rain pool to rain pool, thus becoming widely distributed.

According to Loveridge (1941), when first captured, helmeted turtles emit an offensive musky odor from glands opening opposite the 4th and 8th marginals; however, they soon tame and become excellent pets.

Pelusios Wagler, 1830

The tropical African freshwater pelomedusid turtles with hinged plastra belong to the genus *Pelusios*. The carapace is elongated, oval, and somewhat flattened in most species. A middorsal keel represented by raised protuberances on the posterior parts of the vertebrals may be present, and the posterior carapacial rim is serrated in one species. Neural bones vary in number from five to eight (rarely four) and they may be in contact with or separated from the nuchal and single suprapygal bones, or both, depending on the species. A hinge is present on the large plastron, which also contains a pair of mesoplastral bones that extend to the midline. The hinge lies between the hypoplastra and mesoplastra, permitting the anterior lobe to close in adults (but rather rigid in *P. broadleyi*). Bramble and Hutchison (1981) studied the hinge kinesis and found that a modified axillary buttress serves as a lever for attachment of a large closing muscle, the *M. levator plastralis*. This muscle is apparently derived from the exhalent respiratory muscle *M. diaphragmaticus*. A supratemporal roof is lacking on the skull, and the quadratojugal is separated from the parietal by the postorbital and from the squamosal by the quadrate. The jugal is also prevented from touching the parietal by the postorbital. Posterior emargination of the skull prevents contact between the parietal and the squamosal bones. The prefrontals are well developed and in contact, but nasals are absent. The palatines are in wide contact in the roof of the mouth, but a vomer is usually absent. Only an indistinct medial ridge occurs on the triturating surface of the upper jaw. Between the orbits is a pair of supraorbital scales separated by a longitudinal seam; behind these a large frontal scale is bordered by temporal scales. All toes are webbed, and the middle toe contains three phalanges.

Killebrew (1975a) and Bull and Legler (1980) reported the karyotype is 2n = 34 (22 macrochromosomes, 12 microchromosomes).

Williams (1954b) and Auffenberg (1981) proposed that the living species of *Pelusios* can be divided into two apparently valid groups. The *P. adansonii* group, which also includes *P. broadleyi*, *P. gabonensis*, and *P. nanus*, is characterized by having the anterior lobe of the plastron relatively long and the abdominal scutes relatively short, so that the anterior lobe is twice as long or longer than the length of the interabdominal seam, and a short bridge. Also, the mesoplastra are more or less tapered toward the plastral midseam. The *P. subniger* group includes the remaining species. In these turtles the plastral anterior lobe is shorter and, correspondingly, the abdominal scutes are longer, so that the anterior lobe is less than twice as long as the length of the interabdominal seam (usually it is about 1.5 times as long), and the bridge of moderate length; their mesoplastra are not tapered medially, having instead parallel transverse contacts with the hyo- and hypoplastra.

This genus is in dire need of revision. There is disagreement on the status of several taxa between Broadley (1981a) and Bour (1983, 1984), who have most studied this group, and Obst (1986) and ourselves.

Key to the species of *Pelusios*:

1a. Plastral forelobe at least twice as long as length of interabdominal seam — 2
 b. Plastral forelobe about 1.5 times as long as length of interabdominal seam — 5
2a. Head dark with light vermiculations, carapace lacking a black medial longitudinal stripe — 3
 b. Head buff colored with a broad black Y-shaped stripe connecting orbits and continuing backward to neck, carapace with a distinct black medial longitudinal stripe: *P. gabonensis*
3a. Adult carapace length greater than 13 cm, a vertebral keel present — 4
 b. Adult carapace length to 12 cm, no vertebral keel present: *P. nanus*
4a. Plastron and bridge yellow, medial abdominal lacking: *P. adansonii*
 b. Plastron and bridge dark brown (some yellow pigment may occur medially on plastron), medial abdominal fontanelle present: *P. broadleyi*
5a. Posterior rim of carapace strongly serrated — 6
 b. Posterior rim of carapace smooth or only slightly serrated — 7
6a. Black border of plastron confined to forelobe, axillary scute absent from bridge: *P. carinatus*
 b. Black border on both lobes of plastron, axillary scute present on bridge: *P. sinuatus*
7a. Plastron with a well-developed constriction at level of abdominal-femoral seam — 8
 b. Plastron without or with only a slight constriction at level of abdominal-femoral seam — 9
8a. Plastron yellow with dark seams or a dark border: *P. subniger*
 b. Plastron totally black or with a lighter midseam: *P. bechuanicus*
9a. Upper jaw slightly hooked, but not bicuspid — 10
 b. Upper jaw bicuspid — 11
10a. Eight to ten enlarged transverse scales on anterior surface of each foreleg; Seychelles Islands: *P. seychellensis*
 b. Three to four enlarged transverse scales on anterior surface of each foreleg; West Africa: *P. niger*
11a. Plastron totally black, without any constriction at level of abdominal-femoral seam; eight neural bones, the 1st in contact with nuchal bone: *P. rhodesianus*
 b. Plastron not totally black but instead with some yellow pigment, and with a slight constriction or no constriction at level of abdominal-femoral seam; five to eight neural bones in carapace, the 1st may be in contact with or well separated from nuchal bone — 12
12a. Plastron black with a yellow rim: *P. williamsi* (in part)
 b. Plastron mostly yellow with dark pigment only along seams, as a medial blotch, or as scattered spots — 13
13a. Dark pigment on plastron reduced to some small spots: *P. williamsi* (in part)
 b. Dark pigment on plastron found only along seams, on anterior plastral rim, or as a large medial blotch — 14
14a. Plastral hinge level with the posterior of 5th marginal or with the seam separating marginals 5 and 6; neural bones widely separated from nuchal bone; dark pigment restricted to anterior rim of plastron or only along anterior seams: *P. castanoides*
 b. Plastral hinge level with the center of 5th marginal; 1st neural bone in contact with nuchal bone; plastron with a large medial dark blotch, or with dark pigment along all seams: *P. castaneus*

Pelusios gabonensis (Duméril, 1856)

African forest turtle

Plate 1

Recognition: The oval carapace (to 30 cm) is flattened dorsally and bears a low medial keel which may disappear with age. All five vertebrals are broader than long, with the 1st the broadest and the 4th and 5th the narrowest. X-rays of the carapaces of several individuals showed a continuous series of eight neural bones with the 1st touching or nearly touching the nuchal bone, and the reduced 8th neural widely separated from the suprapygal. There is no cervical scute, but there are 11 pairs of marginals. The marginals are not posteriorly serrated in adults, but may be slightly so in juveniles. *P. gabonensis* is easily separated from other species of *Pelusios* by its carapacial coloration and pattern. It is buff to gray yellow with a distinct black vertebral stripe which widens on the anterior marginals. The carapace has a tendency to darken with age by adding black radiations to the scutes until, in very old and large individuals, it may become almost totally black. The large plastron covers almost all of the carapacial opening, and is posteriorly notched. Its anterior movable lobe is long, and the abdominal scute short (less than half as long as the anterior lobe). The plastral formula is: hum > fem > an >< intergul > abd > pect > gul. Plastron and bridge are black with light-yellow seams. The head is moderately broad with a protruding snout and a slight notch, flanked by two toothlike cusps on the upper jaw. Two barbels are present on the chin. The head is buff colored with a broad, black Y-shaped stripe connecting the orbits and extending backward medially onto the neck. A second dark stripe may extend between the orbit and tympanum. Jaws and throat are tan in adults, but black in juveniles. Large, irregularly shaped scales occur on the anterior surfaces of the forelimbs. Limbs are blackish in juveniles, but gray to yellow in adults.

Males have slightly concave plastra, while those of females are flat. The male's tail is much longer and thicker than that of the female.

Distribution: *Pelusios gabonensis* lives in tropical West Africa where its range extends from Zaire northwestward to Liberia and Guinea.

Habitat: This terrapin is confined to tropical rain forests where it inhabits marshes, swamps, streams, and probably deeper rivers. The young prefer quiet waters but the adults are found in flowing waters.

Natural History: Cansdale (1955) reported they lay about a dozen eggs in a nest dug close to the water.

Ewert (1979) reported the average carapace length of eight hatchlings was 42.3 mm, and Schmidt (1919) found a juvenile of 42 mm with yolk sac and egg tooth (caruncle) still present, but no hinge on the plastron. The young have darker skin than adults, and also a distinct keel on the carapace.

Pelusios gabonensis is carnivorous, feeding on insects, worms, snails, and fish; it is attracted to traps baited with fish.

Pelusios nanus Laurent, 1956

African dwarf mud turtle

Recognition: *Pelusios nanus* is the smallest species of the genus. Its elongated oval carapace (to 12 cm) is flattened dorsally, and unserrated posteriorly (although a small notch may be present between the 12th pair of marginals). No cervical scute is present, and there is little, if any, evidence of a medial keel. Vertebral 1 is widely flared anteriorly and is broader than long in juveniles, but as long as or slightly longer than broad in adults. Vertebrals 2–4 are longer than broad in adults, but broader than long in juveniles; the 4th is the smallest vertebral. Vertebral 5 is flared posteriorly and is much broader than long. There are seven or eight neural bones; if only seven, it is the 8th which is lost. The 1st neural is anteriorly narrowed. Posterior marginals may be somewhat flared; lateral marginals are unkeeled. The carapace is brown, often with dark streaks. The plastron is large with a long anterior lobe (more than twice the length of the interabdominal seam), and a weakly developed hinge. Both the anterior and posterior plastral lobes are broad; the posterior lobe has an anal notch and is slightly constricted at the abdominal-femoral seam. The plastral formula is: fem > hum >< abd > an > intergul > pect > gul. Axillary and inguinal scutes are absent from the bridge; also the pectoral scute does not enter the bridge, which is formed entirely by the abdominal scute. The plastron is yellow with a black border; the bridge is black. The snout is short and blunt, and the upper jaw only slightly notched. There are two chin barbels. Forelimbs lack large transverse scales.

Males have concave plastra and long, thick tails; females have flat plastra and short tails.

Wermuth and Mertens (1961, 1977) designated *P. nanus* a subspecies of *P. adansonii;* however, additional evidence warrants that these be considered separate species. Laurent (1965) and Broadley (1981a) reexamined these species and found they differ in such features as size, coloration and pattern, shape of the carapace, development of the vertebral keel, width of the 3d vertebral versus measurements of other scutes, and the absence of either the 1st *(adansonii)* or 8th *(nanus)* neural from the carapace. Also, the two species occupy mutually exclusive, widely separated geographic ranges. Although closely related, *adansonii* and *nanus* should both be considered valid species.

Distribution: *Pelusios nanus* occurs in southern Africa, from northern Zambia westward along the southern border of the Congo Basin to Angola (Broadley, 1981a).

Habitat: Broadley (1981a) refers to it as an inhabitant of waterways in moist savannas.

Pelusios adansonii (Schweigger, 1812)
Adanson's mud turtle

Recognition: The carapace (to 18.5 cm) is elliptical, being broader posteriorly than anteriorly, and flattened across the vertebrals. A low medial keel is present on the first four vertebrals; all five vertebrals are broader than long in juveniles, but lengthen with age until they are of equal dimensions or slightly longer than broad. The 4th is the smallest. Seven or eight neurals may be present; if only seven, the 1st is missing. There is no cervical scute. The posterior marginal border is smooth and rounded. In color, the carapace is yellow brown to gray brown; some also have darker radiations or spots on the scutes. The plastron is smaller than the carapace and cannot completely cover the limbs when closed. Its anterior lobe is long and rounded, usually well over twice as long as the length of the abdominal scute. The posterior lobe is more tapered and contains a deep anal notch. Mesoplastra are only medially tapered posteriorly, forming a straight transverse hinge along the hyoplastra. The plastral formula is: fem > abd >< hum > intergul

Pelusios adansonii

> an > pect >< gul. The intergular is about twice as long as broad. Both the pectoral and abdominal scutes contribute to the bridge, and axillary and inguinal scutes are normally absent. Both plastron and bridge are yellow. The broad head has a short, only slightly projecting snout, and an upper jaw lacking either a hook or a notch. The frontal scale is very large. There is a pair of pronounced chin barbels. The head is gray brown dorsally with yellow vermiculations, and yellowish posteriorly. A yellow stripe may extend from the orbit to the tympanum. The jaws are yellow and other skin is yellow brown.

Males have long, thick tails; females have short tails.

Distribution: The range of *Pelusios adansonii* extends from the White Nile in the Sudan westward through Chad to Camaroon, Nigeria, Mali, Liberia, Senegal, and Gambia. It is also known from the Cape Verde Islands (the type locality).

Habitat: This is a river turtle of savanna regions outside of the rainforest (Loveridge, 1941).

Natural History: Werner (in Loveridge, 1941) dissected a female along the White Nile in March which contained seven eggs that measured from 29.5 × 18 to 33 × 19 mm. Ewert (1979) described the eggs as ellipsoidal with hard, expansible shells.

Pelusios adansonii is carnivorous and will take fish and a variety of meats in captivity.

Pelusios broadleyi Bour, 1986
Broadley's mud turtle

Recognition: This species has an elliptical carapace (to 15.5 cm) broadest behind the center and with a knobby keel on all five vertebrals; those on vertebrals 3 and 4 are highest, and the keel is most pronounced in juveniles. Juvenile vertebrals are broader than long; in adults, the 1st vertebral is flared anteriorly but long and narrow posteriorly, the 2d broader than long, the 3d longer than broad, the small 4th broader than long, and the 5th flared posteriorly. Six neurals are present, the 1st and 8th being absent. A small cervical scute may be present, and the posterior carapacial rim is smooth and rounded. Ground color of the carapace is grayish brown, and each scute contains many small, dark, radiating lines or dashes. The plastron is basically rigid as the pectero-abdominal hinge hardly moves. Its anterior lobe is long and rounded, usually well over twice as long as the length of the

Pelusios broadleyi

abdominal scute. The posterior lobe is tapered posteriorly at the femorals and anals, does not entirely cover the carapacial opening, and contains a deep anal notch. Mesoplastra are short. Medially, a large abdominal fontanelle persists in adults. The plastral formula is: fem > hum > abd >< intergul > an > pect > gul. The intergular is about 1.4 times longer than broad. Both pectorals and abdominals contribute to the bridge, axillary and inguinal scutes are normally absent, although a pseudoaxillary may be formed by subdivision of the pectoral scute near its lateral border. Plastron and bridge are brown to black; yellow pigment may be present medially at the abdominal fontanelle. Juveniles have more yellowish plastra with some dark-brown pigment. The broad head has a short, slightly projecting snout; its upper jaw is neither hooked nor notched. The frontal scale is large, and two chin barbels are present. Light vermiculations are present dorsally on the brown head. Ventrally, the chin and neck are gray to yellow; the yellow upper jaws contain dark spots or bars. Other skin is gray to yellowish brown. Large transverse scales lie on the anterior surfaces of the forelegs.

Males have concave plastra and long, thick tails; females have flat plastra and short tails.

Distribution and Habitat: *Pelusios broadleyi* is known only from the southeastern shores of Lake Rudolph (= Lake Turkana), Marsabit District, Kenya.

Natural History: This species has only recently been discovered, so little is known of its habits. Bour (1986) estimated that hatchlings have 25-mm carapaces. Captives eat fish and commercial dry trout chow.

Pelusios castaneus (Schweigger, 1812)

West African mud turtle

Recognition: The carapace (to 38 cm) is elongated, oval, broadest behind the center, and has a low vertebral keel (usually in the form of a protuberance on the 4th vertebral). In adults the 1st vertebral is flared anteriorly and the 5th is flared posteriorly; both are broader than long. The 2d through 4th vertebrals are as broad as long, or slightly longer than broad. Eight neurals are present in a continuous series; neural 1 is in contact with the nuchal bone, but the 8th neural is well separated from the suprapygal bones. The posterior marginals are not serrated. In color the carapace varies from yellowish brown to olive, dark brown, or black. The plastron is large, almost covering the entire carapacial opening, has a short anterior lobe less than twice as long as the interabdominal seam, and is deeply notched posteriorly. Its posterior lobe is only slightly constricted or not constricted at the level of the abdominal-femoral seam. The plastral formula is: abd >< fem > hum >< intergul > an >< pect > gul. The intergular varies from 1.3 to 1.5 times as long as broad. The bridge is broad and lacks an axillary scute. Usually the plastron is yellow, but it may contain some dark pigment medially or at the outer edges of the seams. The head is moderate in size with a slightly protruding snout and an upper jaw bearing two toothlike cusps. The seam between the frontal and temporal scales is long. The postocular scale normally touches the masseteric scale; the supralabial scale is usually absent, but, if present, is small. There are two chin barbels. The head is olive to brown with light vermiculations. The neck and limbs are yellow to gray, and each foreleg has several enlarged transverse scales on its anterior surface.

Males have longer tails and slightly concave plastra; the plastra of females are flat.

This species is closely related to *P. seychellensis* (Broadley, 1981a; Bour, 1983).

Pelusios castaneus

Distribution: *Pelusios castaneus* occurs in Guinea, Senegal, Mauritania, and Nigeria in West Africa, and in Congo and western Zaire.

Geographic Variation: Two subspecies are considered valid. *Pelusios castaneus castaneus* (Schweigger, 1812) is restricted to West Africa where it ranges from Guinea, Senegal, and Mauritania east to Nigeria. It has a chestnut-colored to black carapace to 24 cm in length, its plastron is mostly yellow, and the abdominal scute is longer than the femoral. *P. c. chapini* Laurent, 1965 occurs in the Congo River of northern Zaire. Its dark-brown to black carapace reaches 38 cm in length, its plastron has a medial dark blotch, and its femoral scute is longer than the abdominal. Bour (1983) elevated *chapini* to specific status, but we consider it only a large subspecies of *P. castaneus*. Laurent (1965) considered the form *Sternotherus derbianus* (Gray, 1844) to be a valid subspecies of *P. castaneus*; however, its distinctive characters are all within the variation of *P. c. castaneus*, thus we consider it synonomous with that race.

Habitat: *Pelusios castaneus* lives in a variety of habitats such as rivers, streams, marshes, swamps, lakes, and shallow ponds. Many of these water bodies are dry for much of the year and the turtles are forced to estivate buried either in the sand on shore or in the bottom mud.

Natural History: Much of the biology of this species has been confused with that of *Pelusios subniger*, with which it was combined for many years. Cansdale (1955) reported (as *P. subniger*) that in West Africa it nests in February and March laying 6–18 eggs (with soft, chalky surfaces) per clutch. He stated the eggs hatch in June or July. Villers (1958) reported a female (as *P. subniger*) laid 12 eggs, and that the eggs of a captive from Dakar measured 36 × 21 mm. He also stated that hatchlings are a little less than 30 mm long.

P. castaneus feeds mainly on large pulmonate snails and floating water lettuce (Broadley, 1981a).

Pelusios castanoides Hewitt, 1931

East African yellow-bellied mud turtle

Recognition: The yellowish, olive, or blackish carapace may have a marbled pattern of yellow and brown marks. It is oval, elongated (to 23 cm), and rather narrow in males (especially in the subspecies *P. c. intergularis*). A low vertebral keel may be present on the 4th and 5th vertebrals, and the posterior marginals may be slightly serrated. Vertebral 1 is anteriorly flared and is the largest; the 5th vertebral is posteriorly flared; the 4th is the smallest; and vertebrals 1–4 may be as long as or longer than broad. Five to eight neurals are present (the 1st, 7th and 8th are always reduced or absent) and well separated from both the nuchal and suprapygal bones (Broadley, 1983). The plastron is large, almost covering the entire carapacial opening, and its anterior lobe is much shorter than the posterior lobe, which is slightly constricted at the seam separating the abdominal and femoral scutes and has a deep anal notch. The plastral formula is variable: abd > fem > intergul > hum > pect >< gul >< an, and the intergular is 1.35–1.72 times as long as broad. The bridge is broad, but lacks an axillary scute. The plastron is yellow with some dark pigment along the seams. The brown to olive-black head is moderate in size with a slightly protruding snout. Two barbels occur on the chin. Other skin is yellow to brown. Several transverse rows of enlarged scales occur on the forelegs.

Males have narrower plastra and longer, thicker tails. The posterior lobe on the female plastron is broad.

Distribution: *Pelusios castanoides* is found in eastern Africa from Malawi and Mozambique southward to eastern South Africa. Populations also occur on the Seychelles Islands and Madagascar.

Geographic Variation: Two subspecies are currently recognized. *Pelusios castanoides castanoides* Hewitt, 1931 ranges from Malawi and Mozambique in East Africa south to Swaziland and eastern South Africa; it also occurs on Madagascar. It has a yellow, olive, or black carapace, and an intergular scute separated from the adjacent gular scutes by diagonal seams. *P. c. intergularis* Bour, 1983 occurs on the Seychelles Islands. It has a dark carapace with a marbled pattern of yellow and brown marks, and a pentagon-shaped intergular scute with longitudinally parallel (straight) seams separating it from the gulars. The Madagascar population named *P. castaneus kapika* by Bour (1978) is indistinguishable from *P. c. castanoides* (Broadley, 1981a; Bour, 1983).

Habitat: *Pelusios castanoides* inhabits marshes and swamps. During the dry season it estivates in the mud.

Natural History: Two captives from Malawi each laid 25 eggs (30–33 × 21.5–23.0 mm) at the end of September (Mitchell, in Broadley, 1981a).

Pelusios seychellensis (Siebenrock, 1906a)

Seychelles mud turtle

Recognition: The black carapace (to 16.5 cm) is elongated, oval, broadest at the center, and has a low vertebral keel. Vertebral 1 is flared anteriorly and the 5th is flared posteriorly; the first three and the 5th vertebrals are broader than long, the 4th may be as long as broad. Eight neurals are present; the 7th and 8th are shortened and isolated (Bour, 1983). The posterior marginals are not serrated. The plastron is large, almost covering the entire carapacial opening; its short anterior lobe is less than twice as long as the interabdominal seam. The posterior lobe is only slightly constricted at the level of the abdominal-femoral seams and has a deep posterior notch. The plastral formula is: abd > fem > intergul > hum > an > pect > gul. The pentagon-shaped intergular varies from 1.5 to 2.0 times as long as broad. The bridge is broad and lacks an axillary scute. The plastron either is totally black or may have a small medial area of yellow pigment along the hinge. A slightly protruding snout is present, and the upper jaw may lack toothlike cusps. The seam between the frontal and temporal scales is long. The postocular scale usually touches the masseteric scale; the supralabial scale is usually absent, but, if present, is small. The tympanum is oval in *P. seychellensis* but round to elliptical with the narrow end anterior in *P. castaneus* (Bour, 1983). Two chin barbels are present. The head is brown with yellow vermiculations. Both neck and limbs are yellowish brown; there are 8–10 enlarged transverse scales on the anterior surface of each foreleg as compared to 15–20 in *P. castaneus* (Bour, 1983).

Males have longer tails and slightly concave plastra; the plastra of females are flat.

This species is closely related to *P. castaneus* (Broadley, 1981a; Bour, 1983).

Distribution: *Pelusios seychellensis* is known only from the Seychelles Islands, Indian Ocean.

Habitat: This species inhabits slow-moving waters with soft bottoms.

Pelusios williamsi Laurent, 1965

Williams' mud turtle

Recognition: The oval, black to dark-brown carapace (to 25 cm) is elongated, widest behind the center, moderately depressed medially, and contains a low, blunt vertebral keel. The 1st and 5th vertebrals are flared, and vertebrals 1–4 are as long as wide or longer than wide. Seven or eight (usually eight) neurals are present; neural 8 is elongated and only narrowly separated from the suprapygal (Broadley, 1983). Anteriorly the neurals are well separated from the nuchal bone. The posterior marginals are not serrated. The large plastron covers most of the carapacial opening. Its anterior lobe is rounded and shorter than the posterior lobe. An anal notch is present, and the posterior lobe is slightly constricted at the level of the abdominal-femoral seam. The plastral formula is: abd > fem > hum > intergul > an > pect > gul. The intergular is broad (about 1.5 times broader than long). The bridge is wide but lacks an axillary scute. The plastron is either black with a yellow rim and midseam, or predominantly yellow. The head is broad with a slightly protruding snout, and the upper jaw may bear a pair of toothlike cusps. Two barbels occur on the chin. Head scalation is similar to that of *P. castaneus*. Head and limbs are brown; limb sockets yellow. Each foreleg has several enlarged transverse scales on its anterior surface.

Males have longer, thicker tails and slightly concave plastra.

Distribution: *Pelusios williamsi* is endemic to the upper Nile drainage about Lakes Victoria, Albert, and Edward.

Geographic Variation: Three subspecies are recognized. *Pelusios williamsi williamsi* Laurent, 1965 occurs in the upper Nile drainage about Lake Victoria. The plastron is mostly black with a yellow midseam, the posterior lobe as long as or shorter than the anterior lobe, and the intergular scute more than half as long as the length of the anterior lobe. *P. w. lutescens* Laurent, 1965 occurs in the Lake Edward–Semliki–Lake Albert drainage west of *P. w. williamsi*. Its plastron is yellow with some gray or brown spots, the posterior lobe as long as or longer than the anterior lobe, and the intergular scute less than half as long as the length of the anterior lobe. *P. w. laurenti* Bour, 1984 was described from Ukerewe Island, Lake Victoria. Its plastron is yellow with small dark spots along the sides of the gular scutes, the posterior lobe is longer than the anterior lobe, and the intergular is more than half as long as the anterior lobe. The 1st vertebral is very broad along the anterior seam but almost parallel sided posteriorly, thus presenting a T-shaped appearance.

Habitat: *Pelusios williamsi* inhabits lakes, rivers, and swamps.

Pelusios subniger (Lacépède, 1788)

East African black mud turtle

Recognition: This is a moderate-sized (to 20 cm) *Pelusios* with an elongated, oval, unkeeled, brown carapace that is unserrated posteriorly. In the adult, the vertebral scutes are all broader than long. Neural 1 touches the nuchal, but the 8th is separated from the suprapygal (occasionally it may be missing). The anterior plastral lobe is much broader than the posterior lobe; however, it is only slightly longer than the interabdominal seam (never more than twice as long). The posterior lobe is strongly constricted at the abdominal-femoral seam, and has an anal notch. The plastral formula is: abd > fem > intergul >< hum >< an > gul > pect. The intergular is about 1.5 times as long as broad. Axillary scutes are absent from the bridge. The plastron is yellow with dark seams or a dark border; the bridge is generally brown. A blunt, nonprotruding snout, an unnotched, nonbicuspid upper jaw, and two chin barbels occur on the large head. A short seam occurs between the frontal and temporal scales; a supralabial scale separating the postocular and masseteric scales is always present. The head is usually totally brown, but may contain some small black spots; the jaws are yellow. Neck, limbs, and tail are gray.

Males have longer, thicker tails and slightly concave plastra.

Distribution: *Pelusios subniger* is restricted to eastern Africa where it ranges from Burundi and Tanzania southward to Mozambique and westward to eastern Zaire, Zambia, and northern Botswana. It also occurs on the islands of Gloriosa, Mauritius, Madagascar, and the Seychelles (Broadley, 1981a).

Geographic Variation: Two subspecies are recognized. *Pelusios subniger subniger* (Lacépède, 1788) occurs in East Africa and on the islands of Gloriosa, Mauritius, and Madagascar. It has an entire (un-

divided) supraocular scale, little subdivision of the parietal scales, and an intergular scale that is not greatly enlarged. *P. s. parietalis* Bour, 1983 resides on the Seychelle Islands. Its supraocular and parietal scales are subdivided by numerous seams, and it has a very large intergular scale and correspondingly small gular scales.

Habitat: *Pelusios subniger* lives in swamps, marshes, lakes, rivers, and streams in the savannah over much of its range, but in southeastern Africa, Broadley (1981a) found it inhabiting pans and other temporary water bodies.

Natural History: Kaudern (1922) reported that a Madagascar female dug a flask-shaped nest cavity with her front legs and head (an obvious error), laid 12 eggs in an hour and then filled the nest. Broadley (1981a) reported a 17-cm captive female laid about 8 eggs during February and March. The eggs are elliptical (36 × 21 mm) with a thin, leathery shell. Eggs incubated by Ewert (1979) at about 30°C hatched in 58 days. Loveridge (1941) reported the carapace length of a recent hatchling as 30 mm; Ewert (1979) stated the average carapace length of eight hatchlings was 34.9 mm.

Foods reported by Loveridge (1941) indicate that *Pelusios subniger* is omnivorous: grass, crabs, worms, snails, insects, fish, and frogs.

Loveridge (1941) reported that these turtles are nocturnal, but they apparently bask and may wander about on land during the rainy season.

Pelusios bechuanicus FitzSimons, 1932

Okavango mud turtle

Recognition: This large species (to 33 cm) has a black, elongated, oval carapace which is broader behind the center and unserrated posteriorly. Juveniles have a knobby medial keel, but this disappears with age until it is entirely missing in larger adults. Vertebrals 1, 2, 4, and 5 are broader than long; the 3d is broader than long in juveniles, but as long as broad in adults. The eight neurals are separate from the suprapygal, and the 1st may not touch the nuchal bone. The strongly hinged, black plastron is large, closing all but a small part of the carapacial opening, and is notched posteriorly. Its anterior lobe is longer than the interabdominal seam, but not twice as long. The posterior lobe is constricted at the abdominal-femoral seam. The plastral formula is: abd > fem > hum > intergul >< an > pect > gul. The intergular is about 1.3 times as long as broad. The head is very broad with a blunt snout

Pelusios subniger

and an upper jaw which lacks cusps. There are two or three chin barbels, and the postocular scale is usually separated from the masseteric by a supralabial scale. A very short seam lies between the frontal and temporal scales. The head varies in color from black with yellow markings to tan with fine yellow vermiculations. Limbs, tail, and neck are yellow to gray.

Males have longer, thicker tails and slightly concave plastra.

Distribution: *Pelusios bechuanicus* ranges from the Fungwe and Lualaba river system of Zaire southward into western Angola, Namibia, Botswana, Zambia, and Zimbabwe.

Geographic Variation: Two subspecies have been described. *Pelusios bechuanicus bechuanicus* FitzSimons, 1932 has a black head with symmetrical yellow marks and three chin barbels. It occurs in the greater Okavango basin in Angola, Namibia, Botswana, Zimbabwe, and Zambia from the Cubango-Okavango River in the west to the Kafue Flats in the east. *P. b. upembae* Broadley, 1981a is distinguished from the nominate race by its light-brown head with fine yellow vermiculations and two chin barbels. It is known only from the Fungwe and Lualaba watersheds of Zaire. Bour (1983) considers these separate species.

Habitat: This turtle lives in deep, clear waters in rivers and swamps.

Natural History: Broadley (1981a) reported that a wild 23.9-cm female laid 21 eggs in moist soil on 16 October, and that three other females contained 28, 32, and 48 eggs, respectively. The elongated eggs (35–39 × 21–23 mm) have thin, leathery shells.

Pelusios rhodesianus Hewitt, 1927
Variable mud turtle

Recognition: Another species in the *subniger* group is *Pelusios rhodesianus*, which has an elongated, black, oval carapace (to 25.5 cm) that is broader posteriorly than anteriorly and weakly keeled at best. Its carapace is flattened across the vertebrals, and if a medial keel persists it usually is represented only by a slight knob on the 4th vertebral. Vertebrals are usually broader than long, although the 3d may be longer than broad. Neural 1 touches the nuchal and neural 8 often touches the suprapygal; neurals 7 and 8 may be reduced. Posterior marginals are smooth, not serrated. The plastron is large, only slightly smaller than the opening of the carapace, and bears a deep posteri-

or notch. Its anterior lobe is short, only about 1.5 times the length of the interabdominal seam. The posterior lobe is not indented at the abdominal-femoral seam. The plastral formula for the three specimens we examined was: abd > fem > hum > intergul > an > pect > gul. The intergular scute is about 1.5 times as long as broad. No axillary scutes occur on the broad bridge. Bridge and undersides of the marginals are black; the plastron is usually totally black, but Broadley (1981a) reported that a few Zimbabwean individuals have irregular yellow patches or uniformly yellow plastra. The head is small with a slightly projecting snout, a bicusped upper jaw, and two chin barbels. A long seam is present between the frontal and temporal scales, and a supralabial scale is usually absent or small with postocular and masseteric contact above. Broadley (1981a) reported that northern *P. rhodesianus* have brown heads with yellow vermiculations while heads of southern individuals are brown dorsally but yellow laterally. Skin of the neck and limbs is yellow, and the outer surfaces of the limbs are grayish brown.

Males have longer, thicker tails and slightly concave plastra.

Distribution: *Pelusios rhodesianus* occurs in central and southeastern Africa from northern Zaire and Uganda south to Angola, northern Botswana, Zimbabwe, and central Mozambique, with relect populations in Kwazulu and at Durban, Natal (Broadley, 1981a).

Habitat: This species occurs in rivers, swamps, and marshes. In Zimbabwe it seems to prefer quiet weed-choked backwaters behind dams (Broadley, 1981a).

Natural History: Broadley (1981a) reported that after the first rain of the season fell on 12 September 1958 at Lochinvar on the Kafue Flats many females were found walking about the grassland presumably to oviposit. He also reported that females contained eggs in September and October. Shelled oviducal eggs and those laid in captivity in April and May totaled 11 to 14, measured 33–37 × 20–23 mm, and had pliable, thin shells.

Pelusios niger (Duméril and Bibron, 1835)
West African black forest turtle

Recognition: The carapace (to 22.5 cm) is elongated and oval, flattened across the vertebrals, and unserrated posteriorly. A low medial keel is present, but there is no cervical scute. Vertebrals are broader than

Pelusios niger

long; the 1st flared anteriorly, the 5th posteriorly. Posterior marginals are slightly flared. There are eight neural bones. The carapace is black to reddish brown with light seams; some individuals have growth rings and light-colored radiations on the vertebrals and pleurals. The plastral anterior lobe is broad and short, only slightly longer or about the same length as the abdominal scute. The posterior lobe is also short, but narrow, contains a posterior notch, and cannot entirely close the shell. Also, there is no abdominal-femoral constriction. The plastral formula is: abd > fem > intergul > hum > an > gul > pect; as can be seen, the abdominal scute is very long. The pectoral scute does not contribute to the bridge, and axillary and inguinal scutes are absent. Both plastron and bridge are black with cream-colored to yellow seams. The head is narrow for the genus, with a protruding, pointed snout and a slightly hooked upper jaw. Two barbels are present on the chin. The head is yellow with numerous dark-brown or black vermiculations; the yellow jaws also are darkly vermiculated and the chin yellow. Neck, limb, and tail skin is yellow to gray. Three or four large transverse scales are present on the anterior surface of the forelimbs.

Males have longer, thicker tails than those of females.

This species has been confused with both *P. subniger* and *P. castaneus*. *P. niger* differs from the former by several characters including its unconstricted plastral posterior lobe. Laurent (1965) has shown that the length of the border of the intergular scute is much longer in *P. niger* than in *P. castaneus*, while the external border of the femoral scute is much longer in *P. castaneus* than in *P. niger*.

Distribution: *Pelusios niger* is confined to West Africa where it ranges from Ghana eastward to Cameroon and, according to Villiers (1958), possibly southward to Angola.

Habitat: *Pelusios niger* lives in permanent bodies of water with soft bottoms in the savannas.

Natural History: Although known since 1835, very little has been published on the life style of this species, and probably much has been confused with that of *Pelusios subniger*. Young individuals we have examined had more rounded carapaces with slightly serrated posterior borders. The medial keel is more pronounced than in adults, and the surfaces of the carapacial scutes are covered with small knoblike rugosities which often form radiating lines from the raised areola of the scute. None of those with plastra < 70 mm had the hinge developed. Also, several of these juveniles did not have as many dark vermiculations on the head as normally occur in adults. Cansdale (1955) reported this turtle (as *subniger*) digs a nest 10–12.5 cm deep in grasslands. A clutch comprises 6 to 18 eggs, but more normally 10 to 13; the eggs have pliable shells with chalky white surfaces. Cansdale also reported that they spend the dry season underground and that many are dug up while ploughing.

Like other *Pelusios*, *P. niger* is carnivorous, feeding on a wide variety of small aquatic animals.

Pelusios carinatus Laurent, 1956

African keeled mud turtle

Recognition: This little-known species from the Congo Basin has an elongated (to 23.2 cm), oval, black carapace which is broadest at the center and has a slightly serrated posterior rim. The vertebral keel is well developed; low on the 1st and 5th vertebrals, but higher on the 2d through 4th, especially as a raised protuberance at the rear of the 3d and 4th vertebrals. The keel is better developed in juveniles, but remains prominent in adults. Vertebrals are broader than long

Pelusios carinatus

in juveniles but may become longer than broad in adults; the 1st is flared anteriorly while the 5th is expanded posteriorly. An X-ray of a paratype (MCZ 57452) shows eight neurals with the 1st and 8th reduced; the narrow 1st neural touches the nuchal but the 8th neural is widely separated from the suprapygal. The plastral anterior lobe is less than twice as long as the interabdominal seam. Instead of being indented at the abdominal-femoral seam, as in *P. subniger*, the posterior lobe is convex and wider at this point but behind this tapers to the rear. It is posteriorly notched. The plastral formula is: abd > fem > intergul >< hum > an > pect >< gul. The intergular is about 1.5 times as long as broad. The bridge is moderate in length and lacks an axillary scute. The plastron is yellow with a black border usually confined to the anterior lobe. The head is moderate to large with a short protruding snout and an unnotched upper jaw. Two small chin barbels are present. A long seam lies between the frontal and temporal scales. The head is brown or black with yellow vermiculations in adults and marbled with yellow in juveniles. Other skin is grayish yellow. Each foreleg has several enlarged transverse scales on its anterior surface.

Males have long, thick tails and slightly concave plastra; the female's tail is flat, as is also her plastron.

Distribution: *Pelusios carinatus* is restricted to the Congo basin of Zaire.

Habitat: According to Broadley (1981a), *Pelusios carinatus* is apparently the Congo ecological equivalent of *P. sinuatus*, occupying rivers, lakes, lagoons, and backwaters.

Pelusios sinuatus (Smith, 1838)

East African serrated mud turtle

Recognition: *Pelusios sinuatus* is a large (to 46.5 cm) *Pelusios* with an elongated oval carapace that is strongly serrated posteriorly in juveniles, less so in adults. Broadest behind the bridge, the carapace is flattened or concave across the vertebrals in adults. All vertebrals are strongly keeled in juveniles, less so in adults. In adults the keel is reduced to raised protuberances at the rear of the first four vertebrals. Vertebrals are broader than long in juveniles, usually longer than broad or of equal dimensions in adults. Four to seven neurals are present, with six or seven the usual number; the 8th is always absent, the 7th is reduced or absent, the 5th may be absent, and the 1st is reduced and does not touch the nuchal. Growth annuli and radiations are often present on the carapacial scutes, giving the shell a roughened appearance in ju-

Pelusios sinuatus

veniles and young adults. The carapace is black, but the seams may be yellow in adults; juveniles have brown to olive carapaces. The plastron is large, only slightly smaller than the opening of the carapace, and bears a deep posterior notch. Its anterior lobe is short, less than twice as long as the interabdominal seam, and the posterior lobe is not indented at the abdominal-femoral seam. The plastron formula is: abd > fem > intergul >< an > hum > gul >< pect. The intergular scute is usually about twice as long as broad. A small axillary scute is present on the broad bridge. The bridge and the undersides of the marginals are black; the plastron is yellow with a black border. A yellow streak extends backward on each anal scute. The head is broad, but not overly long, with a protruding pointed snout and notched, often bicuspid, upper jaw. A large frontal scute covers much of the dorsal surface of the head. Two barbels are present on the chin. Dorsally, the head is yellowish gray to olive, sometimes with small darker markings. Chin, throat, and underside of the neck are yellow, and the limbs, tail, and dorsal surface of the neck grayish.

Males have longer, thicker tails and slightly concave plastra.

Distribution: *Pelusios sinuatus* lives solely in East Africa where it ranges from Somalia southward to Zululand, Transvaal, and Natal, and westward to Lake Tanganyika and Victoria Falls.

Habitat: *Pelusios sinuatus* occurs only in permanent bodies of water, such as rivers and lakes, which are found from the coastal plain to elevations of over 1500 m in the upland savanna (Loveridge, 1941).

Natural History: Loveridge (1941) reported native fishermen told him that *Pelusios sinuatus* nests in July, but he questioned this since very small individuals (51 mm) were collected in March and July, indicating a very long developmental period. Ewert (1979) described the eggs as elongated with a parchmentlike shell and a long vitelline sac. Loveridge (1941) described the plastron in the very young as brick red edged with black, the scute seams being broadly edged with white.

In the wild, *P. sinuatus* feeds on worms, insects, snails, fish, and frogs.

Podocnemis Wagler, 1830

This pleurodiran genus comprises six neotropical riverine turtles and is closely related to the genera *Erymnochelys* and *Peltocephalus*. The oval carapace is flattened or slightly domed and with or without a middorsal keel in adults. There are six or seven neural bones well separated from the suprapygal bone, but in contact with the nuchal. The large plastron is rigid, without a hinge, and has a pair of small, lateral mesoplastral bones lying between the hyo- and hypo-

plastra. Its axillary and inguinal buttresses are large and well developed. No cervical scute is present. The skull is elongated and in some cases broad, with a well-developed dermal roof which is slightly emarginated posteriorly. The postorbital bone is very small and well separated from the quadratojugal and squamosal bones. The quadratojugal lies between the quadrate and jugal bones and prevents their contact. The quadratojugal touches the parietal bone. The parietal also has anterolateral contact with the jugal. A small vomer may be present. One, two, or three ridges occur on the triturating surface of the maxilla, and the upper jaw is either medially notched, rounded, or squared off. There is a single, large scale between the orbits, and an interorbital groove may be present. One to three chin barbels are present. The toes are webbed.

The karyotype is 2n = 28 (Ayres et al., 1969; Huang and Clark, 1969; Rhodin et al., 1978; Bull and Legler, 1980).

Species are compared in Table 3.

Podocnemis erythrocephala (Spix, 1824)
Red-headed Amazon River turtle

Recognition: A small *Podocnemis* species, the carapace is only 32 cm long. The adult carapace is oval, somewhat domed, broadest behind the center, and has a smooth posterior rim. There is only a weak cervical indentation. A medial keel is best developed on the 2d or 3d vertebrals. Vertebrals are broader than long in juveniles, but in adults the first two may be as long as broad or even longer than broad. The 4th and 5th are the smallest, and the 5th is posteriorly expanded. The carapace is brown to dark chestnut brown; there is a lighter border around the rim. The plastron is large, but does not completely cover the carapacial opening. The anterior lobe is longer but slightly narrower than the posterior lobe, which is

Key to the species of *Podocnemis*:

1a. Carapace flattened — 2
 b. Carapace not flattened, but somewhat domed — 4
2a. Upper jaw notched; suboculars large; interfemoral seam longer than interpectoral seam: ***P. vogli***
 b. Upper jaw lacking a notch, square or rounded; suboculars, if present, not enlarged; interpectoral seam longer than interfemoral seam — 3
3a. 2d vertebral longer than broad, upper jaw squared, interparietal long, suboculars absent: ***P. expansa***
 b. 2d vertebral broader than long, upper jaw rounded, interparietal short and heart shaped, suboculars present: ***P. lewyana***
4a. Interparietal completely separates parietals, no red or yellow spots on head, a raised tubercle on each pectoral scute: ***P. sextuberculata***
 b. Interparietal not completely separating parietals, red or yellow spots on head (at least in males), no raised tubercles on pectorals — 5
5a. Carapace broadest at center; no interorbital groove present; head spots yellow, diffused or tan; usually only one chin barbel: ***P. unifilis***
 b. Carapace broadest behind center; an interorbital groove present, head spots, if present, usually red, reddish orange or diffuse yellow beige; two chin barbels: ***P. erythrocephala***

Podocnemis erythrocephala

Table 3. Characteristics separating adult turtles of the genus *Podocnemis*. See text for further explanations of characters

	Species					
Character	*erythrocephala*	*expansa*	*lewyana*	*sextuberculata*	*unifilis*	*vogli*
Carapace:						
max. length	32 cm	107 cm	44.6 cm	31.7 cm	68 cm	36 cm
shape	domed	flat	flat	domed	domed	flat
widest point	behind center	behind center	center	behind center	center	center
cervical indentation	slight	slight	absent	if present, slight	deep	slight
keel	present	usually absent	usually absent	present	present	usually absent
2nd vertebral	W >< L	L > W	W > L	W > L	L > W, L = W	W > L
Plastron:	abd > pect > fem	abd > pect > fem	abd >< pect > fem	abd > pect > fem	pect > abd > fem	abd > fem > pect
Skull:						
upper jaw	notched	square	rounded	notched	notched	notched
maxillary ridges	2	2–3	2	1	2	3
premaxillae	short	short	short	long	short	short
vomer	present	absent	present	absent	present	present
Head:						
interorbital groove	present	present	present	present	absent	present
interparietal separating parietals	short, not separating parietals	long, may separate parietals	short, not separating parietals	long, separating parietals	long, not separating parietals	long, not separating parietals
suboculars	present	absent	present, large	present, small	present, large	present
tympanum width	= orbital W	> orbital W	= orbital W	= orbital W	> orbital W	= orbital W
chin barbels	2	2	2	1–2	1–2	2
pattern (if present)						
color	males = red-orange females = brown	yellow	yellow	none	yellow	yellow
transverse postorbital bar	present	absent	absent	absent	absent	absent
Hind-foot scales:	3, rarely 2	2–3	3	3	3	3

notched posteriorly. The bridge is broad, as wide or slightly narrower than the posterior lobe. The plastral formula is: abd > pect > fem > intergul > an > gul >< hum. The intergular is long and narrow, and completely separates the gular scutes. The plastron is brownish gray along the outside but yellow, orange, pink, or red along the midseam. The bridge is brown to gray. The head is elongated with a protruding snout and a distinctly notched upper jaw. This notch continues to the nostril as a groove, and there is also an interorbital groove. Two long parallel ridges occur on the triturating surface of the maxilla. Premaxillae are short, do not extend posteriorly as far as the choanal rim, and join with the small vomer to separate the maxillae. The incisive foramen is completely within the premaxilla. The tympanum is as broad as the orbit. The interparietal scute is short and very broad, at times even heart shaped; the parietals usually meet behind it. Suboculars are present, and there are two chin barbels. The head is dark brown to dark gray in ground color. Juveniles < 12 cm contain large red or reddish-orange (rarely yellow) spots on the snout and tympanum, and a similarly colored transverse band stretches across the head from tympanum to tympanum. Adult males retain this coloration, but the brightest colors fade to dull brown in adult females. Jaws are dark and the neck and limbs gray to brown. There are usually three, rarely two, large scales on the posterior margin of the hind foot.

Males have larger, thicker tails and deeper anal notches; females have shorter tails and shallow, V-shaped anal notches.

Distribution: Mittermeier and Wilson (1974) reported that *Podocnemis erythrocephala* occurs in the Río Negro and Río Casiquiare drainages of Colombia, Venezuela, and Brazil; the Casiquiare forms a link between the Ríos Negro and Orinoco, and the single verified Colombian specimen is a juvenile from the mouth of the Río Inírida of the Orinoco drainage (Lamar, 1986).

Habitat: *Podocnemis erythrocephala* seems to prefer blackwater streams and rivers, but is also known from some whitewaters.

Natural History: Nesting occurs from August to November with the peak in September and October (Mittermeier and Wilson, 1974). The eggs are usually laid in sandy, brushy areas, and probably more than one clutch is laid each season. A typical clutch includes 5 to 14 eggs (although Freiberg, 1981, reported the clutch size to be 30–40). The eggs are ellipsoidal (43.0 × 27.0 mm) with either rigid calcareous or slightly flexible shells. Hatchlings have approximately 40-mm carapaces.

Mittermeier and Wilson (1974) found this species to be primarily herbivorous, feeding on aquatic plants and fallen fruits, but it is also taken on lines baited with fish, and captives will accept fish. Rhodin, Medem, and Mittermeier (1981) reported that juvenile *Podocnemis erythrocephala* sometimes practice an inertial feeding mechanism, termed neustophagia, for swallowing fine particulate matter from the water's surface (Belkin and Gans, 1968). This is a coordinated mechanism in which the turtle opens its jaw at the surface, keeping the cutting edge of the mandible 1–2 mm below and parallel to the surface, while rapidly expanding the hyoid and pharynx to the maximum, creating a low pressure area in the mouth and throat which sucks in only the surface water and its contents. Contraction of the jaws forces the water out while the particulate matter is filtered and retained.

Podocnemis expansa (Schweigger, 1812)
Giant South American river turtle

Recognition: This is the largest living *Podocnemis*. Its oval (to 107 cm) adult carapace is flattened, broadest behind the center, and has a smooth posterior rim. At best there is only a weak cervical indentation. A medial keel is usually absent, but if present, it appears as a raised area on the 2d (and very rarely the 3d) vertebral. Keeling is more prominent in juveniles. Vertebrals are broader than long in juveniles, but in adults the 2d is longer than broad. The 4th is the smallest, and the 5th is posteriorly expanded. Posteriorly, the marginals are flared over the hind limbs. The surface of the carapacial scutes usually lacks ridges or raised annuli. The carapace is olive to dark gray or brown and may have some dark spots and a light border in younger individuals. The plastron is large, but does not completely cover the carapacial opening. Its anterior lobe is broader than the posterior lobe and

Podocnemis expansa

rounded anteriorly. The posterior lobe tapers toward the rear and contains a posterior notch. The broad bridge is wider than the plastral posterior lobe. The plastral formula usually is: abd > pect > fem > intergul > an > gul > hum. The interhumeral seam may be very short. The intergular is long and narrow; it completely separates the gulars and almost separates the humerals. Plastron, bridge, and undersides of the marginals are yellow. The head is broad with a protruding snout and an upper jaw which is neither notched nor rounded, but rather is squared-off (Williams, 1954a). An interorbital groove is present. Two or three ridges occur on the triturating surface of the maxilla. Premaxillae are short, do not separate the maxillae or reach the choanal rim, and have foramina incisiva at the posterior margin. There is no vomer bone. The tympanum is wider than the orbit. The large interparietal scute tapers posteriorly and may or may not separate the parietals. Subocular scales are absent. Usually two chin barbels are present. The head is gray brown with yellow markings. Two yellow spots occur on the interparietal scutes, and one on each side of the head; these spots fade with age. Jaws are tan; the chin yellow. The neck is gray dorsally and yellow ventrally. Small rounded scales or tubercles occur on the dorsal surface of the neck. Two or three enlarged scales occur on the posterior margin of the hind foot; limbs are gray.

Males have longer, thicker tails, deeper anal notches, and more obtusely rounded heads than do females. Males also retain the juvenile head markings.

Distribution: *Podocnemis expansa* occurs in the Caribbean drainages of Guyana and Venezuela and in the upper Amazon tributaries in Boliva, Peru, Colombia, Venezuela, and Brazil. It is also occasionally found on Trinidad, especially after floods of the adjacent mainland Orinoco River.

Habitat: *Podocnemis expansa* lives in the larger rivers and their tributaries (both blackwater and whitewater) and in adjacent lagoons and forest ponds.

Natural History: Nesting occurs at night, but may extend past sunrise during the dry season. Vanzolini (1977) reported nesting occurs in Brazil in September and October on the Rio Purus, October on the Rios Trombetas and Tapajos, and in December on the Rio Negro. Pritchard (1979) stated the nesting season is February to April in Venezuela. The nests are dug in low sand beaches and bars which are vulnerable to flooding. The nesting process involves several distinct phases (Roze, 1964; Alho and Padua, 1982). Suitable nesting sites are apparently rare, and large congregations of *Podocnemis expansa* gather there at the beginning of the dry season. Copulation occurs in the water,

and the females then begin a period of basking for as much as six or more hours a day, presumably hastening egg development. After several weeks of this, they enter phase two where they retreat into the water after sunset but only for a few hours. Then they emerge to lay eggs in groups; the first nights may only be used for exploring the beach with actual nesting occurring a few nights later. A body pit 80–100 cm deep, similar to that of sea turtles, is dug first and then the actual nest is excavated in the bottom of this pit. Nesting beaches may become so crowded that several females may utilize the same nest cavity. The flask-shaped nest may be 75–80 cm deep (Alho and Pudua, 1982). From 63 to 136 eggs are laid in each nest and possibly more than one clutch is produced each season. Contrary to the eggs of other *Podocnemis* species, those of *P. expansa* are spherical (32–54 mm) and leathery shelled. Under natural conditions hatching may occur in about 50 days. Hatchlings have carapace lengths of 40–45 mm and are more brightly marked than adults.

In the wild, *P. expansa* is predominantly herbivorous, eating fruits, flowers, roots, and soft vegetation of aquatic plants and also those of the flooded riverine forests during the wet season. They seem to fast during the dry season. In captivity some will take beef and fish. Rhodin, Medem, and Mittermeier (1981) reported that juvenile *P. expansa* sometimes skim fine particulate matter from the water's surface by neustophagia.

This species is now endangered since it has been greatly overexploited for its meat, oil, and eggs.

Podocnemis lewyana A. Duméril, 1852

Magdalena River turtle

Recognition: This medium-sized *Podocnemis* reaches a carapace length of 44.6 cm. Its oval adult carapace is flattened, usually keelless, lacks a cervical indentation, and has a smooth posterior rim; however, the

Podocnemis lewyana

juvenile carapace may bear a faint keel and is posteriorly serrated. In both juveniles and adults all vertebrals are broader than long; the 3d is usually the broadest, the 1st and 5th are smaller than the other three, and the 5th is posteriorly flared. The carapace is gray to olive brown or pinkish brown and may bear some dark spots. The plastron is much smaller than the carapacial opening. Its anterior lobe is shorter than the posterior lobe and is rounded in front. The posterior lobe is about as broad as the anterior lobe and contains an anal notch. The bridge is broad. The plastral formula usually is: ab >< pect > fem > intergul > an >< gul > hum. The intergular is broader than the gulars are long (Williams, 1954a). Both bridge and plastron are olive gray. A protruding snout and an unnotched, slightly rounded upper jaw are present on the narrow head. Two longitudinal ridges occur on the triturating surface of the maxilla; the premaxillae do not separate the maxillae and do not extend posteriorly to the choanal margin. The incisive foramen lies well within the premaxilla. A small vomer bone may be present. The tympanum is as broad as the orbit. The interparietal scale is broad and heart shaped; the parietals touch behind it. Subocular scales are present, as also are two chin barbels. The head is gray to olive with a yellowish band extending backward from the orbit dorsal to the tympanum and yellowish to horn-colored jaws. Neck and limbs are gray to olive; three large scales occur on the posterior margin of the hind foot.

Males have longer, thicker tails and more obtusely rounded heads than do the females.

Distribution: *Podocnemis lewyana* occurs in the Río Magdalena and Río Sinu drainages in Colombia.

Habitat: *Podocnemis lewyana* is a riverine species which also enters lagoons, swamps, and flood-plain marshes.

Natural History: Clutches may contain 15 to 30 eggs. The eggs are ellipsoidal, and one measured 40 × 34 mm (Vanzolini, 1977).

The diet is probably similar to that of *Podocnemis unifilis*.

The present status of this species requires attention as populations appear to be decimated severely throughout its range, especially in the Río Magdalena.

Podocnemis sextuberculata Cornalia, 1849

Six-tubercled Amazon River turtle

Recognition: This is one of the smaller species of *Podocnemis*, reaching a carapace length of 31.7 cm. Its domed carapace is elliptical and broader behind the center. The posterior rim is serrated in juveniles, but only slightly so or smooth in adults; a cervical indendation may be present. A blunt medial keel is present on vertebrals 2 and 3. All vertebrals are broader than long with the 1st and 5th the smallest, and the 5th posteriorly expanded. Surfaces of the carapacial scutes are usually smooth and show few if any growth annuli. The carapace is gray to olive brown. The plastron is large but does not completely cover the carapacial opening. Its anterior lobe is rounded in front and broader than the posterior lobe. The anal scutes are decidedly tapered and much narrower than the femorals. A shallow posterior notch is present. A unique character (which gives this turtle both its scientific and common names) is the presence of six pairs of prominent swelled tubercles on the plastron of juveniles. These occur at the base of the bridge on the pectoral and abdominal scutes and at the outer posterior point of the femorals. These swellings disappear with age, although those on the pectoral scutes may persist into adulthood. The plastral formula usually is: abd > pect > fem > intergul > an > gul > hum. The long intergular separates the gulars. The bridge is not as broad as the width of the plastral posterior lobe. Both plastron and bridge are yellow to gray or brown. The broad head has a protruding snout and notched upper jaw. There is only a single weak ridge on the triturating surface of the maxilla. The premaxillae separate the maxillae and extend to the choanal rim. The incisive foramina lie completely within the premaxillae. There is no vomer. The interparietal scale is elongated and widely separates the parietals, which, although also elongated, do not touch behind the interparietal. Large subocular scales are usually

Podocnemis sextuberculata

present. A deep groove lies between the orbits, and the tympanum is about as broad as the orbit. One or two chin barbels are present. The head is olive to reddish brown with cream-colored jaws. The neck is dark gray to olive dorsally but lighter colored ventrally; limbs are gray to olive. Three large scales occur on the posterior margin of the hind foot.

Males have longer, thicker tails than do females.

Distribution: *Podocnemis sextuberculata* occupies the Amazon drainages of Brazil, Peru, and Colombia.

Natural History: Vanzolini (1977) reported that in Brazil the nesting season is June and July on the Rio Purus and August and September on the Rio Trombetas, and Pritchard (1979) stated that, in the upper Amazon, nesting occurs in October on the Río Caquetá and in November and December on the Río Putumayo. The nests are dug in low beaches and bars of sand which are particularly vulnerable to flooding; probably two clutches are laid each season (Vanzolini, 1977). Clutches include about 8 to 19 ellipsoidal eggs (34–44 × 24–30 mm). Hatchlings have carapace lengths of 45–47 mm, and their vertebral keel and six plastral tubercles are well developed.

Foods in the wild include aquatic plants and fish.

Podocnemis unifilis Troschel, in Schomburgk, 1848

Yellow-spotted Amazon River turtle

Plate 1

Recognition: The oval, adult carapace (to 68 cm) is slightly domed, broadest at the center, and has a smooth posterior rim and a cervical indentation. A medial keel is usually present, pronounced in juveniles but reduced to a low mound on the 2d and 3d vertebrals in adults. Vertebrals are broader than long in juveniles, but the 2d and 3d may lengthen to become as long or slightly longer than broad in adults. Vertebral 5 is the smallest and it is posteriorly expanded. The posterior marginals are flared over the hind limbs and tail. Surfaces of the scutes are rough with raised annuli. The carapace of juveniles is brown to greenish gray with a narrow yellow border, that of adults is olive to dark gray or brown. The plastron does not cover the entire carapacial opening. Its large anterior lobe is slightly broader than the posterior lobe, and is rounded in front. The posterior lobe is slightly tapered toward the rear, and there is an anal notch. The bridge is broad but not as wide as the plastral posterior lobe. The plastral formula usually is: pect > abd > fem > intergul > an > gul > hum. The

intergular scute is long and narrow; it completely separates the gulars and partially separates the humerals. The yellow plastron and bridge may develop dark blotches with age. The elongated head has a protruding snout and a distinctly notched upper jaw. No interorbital groove is present, and the interorbital breadth is less than the height of the orbit. Two ridges occur on the triturating surface of the upper jaw. Premaxillae are short and do not separate the maxillae or extend backward to the choanal rim. The foramina incisiva occur completely within the premaxillae. A small vomer may be present. The tympanum is broader than the orbit. The interparietal scale is elongated, but does not completely separate the parietals. Small suboculars are usually present. One or two chin barbels are present (Medem, 1964)—one in Orinoco populations, two in Amazonian populations. The head is gray to olive or brown with yellow spots—one on top of the snout, one on each side of the snout extending to the upper jaw rim, another on each side of the head extending from the lower posterior edge of the orbit to the corner of the mouth, and one on each tympanum. Jaws are dark brown or black, but the chin has a transverse yellow bar and a yellow spot on each side below the corner of the mouth. Limbs are gray to olive brown. Usually there are three large scales on the posterior margin of the hind foot.

Males have longer, thicker tails than do females and retain more of the juvenile head pattern. Females are larger (to 68 cm) than males (to 35 cm) (Medem, 1964).

Distribution: This turtle lives in the Caribbean drainages of the Guianas, Venezuela, and Colombia, and the upper tributaries of the Amazon River in Colombia, Ecuador, Peru, northern Bolivia, southern Venezuela, and Brazil.

Habitat: *Podocnemis unifilis* lives in lakes, ponds, flood-plain pools, swamps, lagoons, and oxbows along major rivers.

Natural History: In Brazil, the nesting season is June and July on the Rio Purus, September and October on the Rio Trombetas, and December on the Rio Negro (Vanzolini, 1977). In Colombia, the season varies from July to December in the several rivers (Medem, 1964). The nests are dug by solitary females in a variety of soil types, sometimes at great distances from water. Probably at least two clutches of 15 to 25 eggs are laid each season. The eggs are ellipsoidal (41–53 × 25–37 mm) with hard calcareous shells. Hatchlings have carapaces about 45 mm long.

Podocnemis unifilis is primarily a vegetarian, but some captives will eat fish. They sometimes use an inertial feeding mechanism, termed neustophagia, for extraction of fine particulate matter from the water surface (see the account of *Podocnemis erythrocephala* for a description of this process).

Podocnemis vogli Müller, 1935
Savannah side-necked turtle

Recognition: This is one of the smaller species of *Podocnemis*, with a recorded carapace length of 36 cm (Pritchard, 1979). The adult carapace is oval, very flattened, not broadened posteriorly, and has a smooth marginal rim with a slight cervical indentation. Adults usually lack a medial keel although a low one is found in juveniles. Vertebrals are broader than long and the 5th is posteriorly expanded. In adults, the carapace is olive to brown, but that of juveniles is brown with dark seams and a narrow yellow border. The plastron is large but does not completely cover the carapacial opening. Its anterior lobe is rounded in front and broader than the posterior lobe, which tapers toward the rear and has a posterior notch. The broad bridge is wider than the posterior lobe. The plastral formula normally is: abd > fem > pect > intergul >< an > gul > hum. The intergular is long and completely separates the gular scutes. Plastron and bridge are yellowish with some faded dark blotches. The broad head has a protruding snout and medially notched upper jaw. Three denticulate ridges occur on the triturating surface of the maxilla. Premaxillae are short, do not extend to the choanal rim, but join with the vomer to separate the maxillae. Foramina incisiva are located entirely within the premaxillae, but are almost hidden by extensions of the triturating ridges. The tympanum is as broad as the orbit. The interparietal scale is elongated, but the parietals still meet behind it. Large subocular scales and two chin barbels are present. The head is gray to brown; large yellow spots on the interorbital groove and tympanum are pronounced in juveniles. The jaws

Podocnemis vogli

are yellow, and the neck and limbs gray. Three large scales occur on the posterior margin of the hind foot.

Males have longer, thicker tails than do females and retain more of the juvenile head pattern.

Distribution: *Podocnemis vogli* lives in the Orinoco drainage in the llanos of Venezuela and adjacent Colombia. It is rumored also to exist in the upper Río Yari, Caquetá, Colombia, a fact which, if true, places *P. vogli* in the Amazon drainage as well (William W. Lamar, pers. comm.).

Habitat: *Podocnemis vogli* inhabits small streams, rivers, swamps, and isolated ponds, often in heavily settled regions.

Natural History: The nesting season lasts from October to February and the nests may be dug considerable distances from water. One clutch comprises 5 to 20 eggs, and at least two clutches are produced each season (Vanzolini, 1977; Pritchard, 1979). The eggs have brittle calcareous shells and are elliptical (37–48 × 21–29 mm) (Pritchard and Trebbau, 1984). Hatchlings emerge from the nest in late April and May, and one we measured had a carapace length of 43 mm.

Podocnemis vogli is omnivorous, feeding on aquatic plants as well as small aquatic invertebrates, insects, amphibians and fish. Rhodin, Medem, and Mittermeier (1981) reported that *P. vogli* practices neustophagia.

P. vogli is predominantly diurnal and basks on sunny days.

Erymnochelys Baur, 1888

This is one of the two monotypic genera of podocnemine turtles, and is the only member of the group now found outside the neotropics. Its sole living species is *Erymnochelys madagascariensis*.

Erymnochelys madagascariensis (Grandidier, 1867)
Madagascar big-headed side-necked turtle

Recognition: *Erymnochelys madagascariensis* has an oval carapace (to 43.5 cm) which is flattened dorsally, lacks a vertebral keel, and is unserrated posteriorly in adults. Hatchlings and juveniles have a medial vertebral keel. Vertebrals are broader than long; the

Erymnochelys madagascariensis

shortest is the 5th which is flared posteriorly. No cervical scute is present, and the posterior marginals may be somewhat flared or upturned over the tail and legs. Rugose radiations are usually present on the surface of the carapacial scales, although the scales of very old turtles may be worn smooth. The carapace is olive to grayish brown. The plastron is long and narrow, not completely covering the carapacial opening. Its anterior lobe is rounded anteriorly and is broader than the posterior lobe, which tapers toward the rear and has a shallow anal notch. The intergular scale is small and does not completely separate the gulars. The plastral formula is: abd > pect > fem > an > intergul > hum > gul. The broad bridge is approximately equal to the width of the plastral posterior lobe. Plastron and bridge are yellow to brown. The large, broad head has a protruding snout, a slightly hooked upper jaw, and a hooked lower jaw. Only a single weak ridge is present on the triturating surface of the maxilla. In the skull, the jugal and quadrate are in contact, but the quadratojugal extends to the parietal separating the postorbital and squamosal bones. Premaxillae extend backward to separate the maxillae and usually reach the choanal rim; the incisive foramen lies completely within the premaxilla. Also, the fossa housing the geniculate ganglion is larger than in other podocnemines. No groove occurs between the eyes, and the interparietal scale is large but does not separate the parietals. Usually only one chin barbel is present, but some individuals may have two. The head is brown to reddish brown dorsally, yellow laterally, and has yellow jaws. The neck is olive, gray, or brown dorsally, but yellow ventrally. Limbs are olive or gray with webbed toes. Three large scales occur on the posterior border of the hind foot.

Rhodin et al. (1978) and Bull and Legler (1980) reported the diploid condition to be 28 biarmed chromosomes (3 large to medium-sized metacentrics or submetacentrics, 2 large to medium-sized subtelocentrics, and 9 small to very small metacentrics and submetacentrics).

Males have longer, thicker tails than do females.

Distribution: *Erymnochelys* is restricted to the island of Madagascar.

Habitat: *Erymnochelys* lives in slow-flowing rivers, streams, swamps, lagoons, and marshes.

Comment: Although often included in the genus *Podocnemis*, this species differs from those turtles, and also *Peltocephalus*, in having an additional central bone in its foot and an iliac-carapacial connection contacting the suprapygal. Frair, Mittermeier, and Rhodin (1978) showed that *E. madagascariensis* has a distinct blood chemistry.

Peltocephalus Duméril and Bibron, 1835

This is the second monotypic podocnemine genus; the sole species is *Peltocephalus dumeriliana* (Schweigger, 1812). There has been some recent debate as to whether the trivial name *tracaxa* Spix, 1824 was the correct one for *Peltocephalus*. This was based on the idea that the missing type of *Emys dumeriliana* Schweigger, 1812 was either a specimen of *Podocnemis unifilis* or *P. expansa*. However, Schweigger's description of the head clearly shows this to be the separate big-headed species, so *dumeriliana* is the correct name (see Pritchard and Trebbau, 1984, for further discussion).

Peltocephalus dumeriliana (Schweigger, 1812)

Big-headed Amazon River turtle

Recognition: The oval carapace (to 68 cm) is domed rather than flattened and has a medial keel and smooth posterior rim. The keel is most pronounced in juveniles, but becomes lower with age, and may be

Peltocephalus dumeriliana

absent from very large individuals. Vertebrals are usually broader than long in adults, and the 5th is posteriorly flared. A cervical scute is normally present. The flared posterior marginals may be slightly raised over the tail. Carapacial scutes of young adults contain growth annuli and rugose striae, but scutes of older turtles are often worn smooth. The carapace is gray to olive, brown, or nearly black. The plastron is large, but does not cover the entire carapacial opening. Its anterior lobe is rounded in front and broader than the posterior lobe, which is not strongly tapered to the rear but is notched posteriorly. The broad bridge is at least equal to the width of the posterior lobe. The intergular scute is longer than the gulars, which it separates. The usual plastral formula is: fem> abd > intergul > an > pect >< hum > gul. Plastron and bridge are yellow to brown. The large head is decidedly triangular in shape when viewed from above, and has a protruding snout and a strongly hooked upper jaw. Only a single weak ridge develops on the triturating surface of the maxilla. In the skull, the quadrate usually touches the jugal, but the quadratojugal extends to the parietal and separates the postorbital and squamosal bones. Premaxillae extend backward to the choanal rim, usually separating the maxillae. The incisive foramen lies completely within the premaxilla. No vomer is present. The forehead is not grooved between the eyes, but is more or less convex. The interparietal is large and expanded posteriorly, widely separating the parietals. The tympanum is as large as or larger than the breadth of the orbit. Only a single chin barbel is present. The head is usually gray to olive, but the tympanic area may be lighter in color, and in old adults the head may become noticeably white. Jaws are tan, neck and limbs gray to olive. Three large scales occur on the posterior border of the hind foot; all toes are webbed.

The karyotype is 2n = 26: 8 large to medium-sized metacentric and submetacentric chromosomes, 18 small to very small metacentric and submetacentric chromosomes (Ayres et al., 1969; Rhodin et al., 1978).

Males have longer, thicker tails than do females.

Distribution: *Peltocephalus dumeriliana* occurs in the Amazon watershed from Peru, Colombia, and western Venezuela eastward to the mouth of the great river. It also lives in the Orinoco drainage of Colombia, where William W. Lamar has taken it from the Ríos Vichada, Tuparro, Tomo, Elivita, and Muco.

Habitat: This is a riverine species living predominantly in blackwater streams, but also in whitewaters, and adjacent flood-plain lagoons, oxbows, and swamps.

Natural History: William W. Lamar has informed us that in the Orinoco drainage of Colombia, *Peltocephalus dumeriliana* nests in the dry season beginning in mid-December. The flask-shaped nests are 15–19 cm deep (Medem, 1983b), and are dug in flood-plain forests. Clutch size is 7 to 25. The large ellipsoidal eggs measure 61–53 × 38–34 mm, and incubation lasts about 100 days (Pritchard, 1979; Medem, 1938b). Hatchlings have dark-brown to gray or black carapaces (47–53 mm) and several yellow spots on the sides of the brown to black head (Medem, 1983b).

Pritchard (1979) reported that dissected specimens contained mostly the fruits of various palms, but Medem (1983b) reported that hatchlings will feed on fish and vegetable matter, and William W. Lamar has taken adults on long lines baited with fish and other types of meat.

Chelidae

Key to the genera of Chelidae:

1a. Four claws on forefeet   2
  b. Five claws on forefeet   3

2a. Intergular scute extends to anterior plastral rim, completely separating gular scutes: ***Hydromedusa***
  b. Intergular scute does not extend to anterior plastral rim, as gular scutes meet anterior to it: ***Chelodina***

3a. Neural bones usually present   4
  b. Neural bones usually absent   5

4a. Intergular scute completely separating gular scutes; usually six or fewer neural bones: ***Phrynops***
  b. Gular scutes meet posterior to intergular; seven neural bones: ***Chelus***

5a. Intergular scute separates humeral scutes: ***Pseudemydura***
  b. Intergular scute does not separate humeral scutes   6

6a. 1st vertebral scute broader or about as broad as 2d vertebral scute   7
  b. 1st vertebral scute much narrower than 2d vertebral scute   8

7a. Medial vertebral groove present; hind side of thigh may contain enlarged pointed tubercles: ***Platemys***
  b. Medial vertebral groove absent; hind side of thigh without enlarged tubercles: ***Rheodytes***

8a. Temporal region of head covered with smooth skin; cervical scute present: ***Emydura***
  b. Temporal region of head covered with scales; cervical scute usually absent: ***Elseya***

This is the more advanced of the two pleurodiran families. These semiaquatic side-necked turtles are now found in South America and the Australia–New Guinea region. Their fossil record extends back only to the Miocene and is restricted to sites in South America and the Australian region. The oldest fossil genera are *Chelus* from South America and *Emydura* (including *Elseya*, Gaffney, 1979a) from Australian deposits. Both of these genera are extant.

The skull is flattened and either broad or long. It is but little emarginated posteriorly, and the squamosal still touches the parietal (except in *Chelodina*) but is

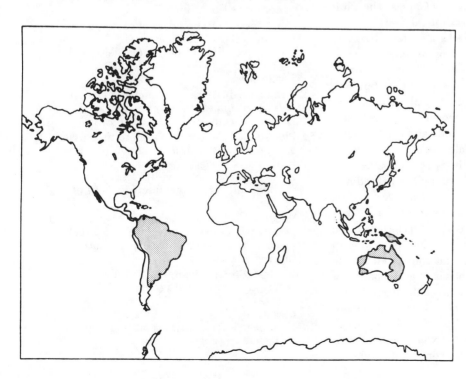

Distribution of Chelidae

cut away ventrally with the loss of the quadratojugal. The small prefrontals are usually medially divided by the frontal bones. Nasal bones are present (except in *Chelus*). In the roof of the mouth, the palatines are separated by the vomer, and the premaxillae are usually not fused. In the lower jaw, the unfused dentaries are not as well developed as in the Pelomedusidae, but a splenial bone is present. The carapace varies from arched with a medial keel (sometimes also a pair of lateral keels) to depressed with a medial groove. Neural bones are absent in several genera, but may total eight in others; where the neurals are missing the costal bones meet dorsally. A cervical scute is usually present (absent in *Elseya*). On the hingeless plastron there is a single intergular scute which may separate or lie behind the gulars, which may meet anterior to it. No mesoplastral bones are present. The 5th and 8th cervical vertebrae are biconvex, all others are procoelous. The hind toes are heavily webbed. The family includes 9 genera with 36 species.

Gaffney (1977) has proposed the subfamily Pseudemydurinae for the species *Pseudemydura umbrina* on the basis of skull characters he believes are derived rather than primitive. He places all the other genera in the subfamily Chelinae.

Chelodina Fitzinger, 1826

Chelodina includes the semiaquatic, long-necked chelid turtles of Australia and New Guinea. The adult carapace is flattened, oval to elliptical, usually has a smooth marginal rim, and lacks a vertebral keel. A vertebral groove is often present. A small cervical scute is present and the 1st vertebral scute is broader than the 2d. Usually neural bones are absent, but in *C. oblonga* up to eight may occur. The plastron is rigid and either large and broad, almost completely covering the carapacial opening, or long and narrow. Axillary buttresses are strong while those of the inguinal are weak and extend only to the 5th costal bones. The intergular scute is large, but usually lies behind and does not separate the gular scutes. However, it does separate the humerals and anterior portion of the pectorals. The cervical vertebrae are longer than those beneath the carapace. A temporal arch is lacking in the skull, so the parietal does not contact either the quadrate or supraoccipital bones. Nasal bones are separated by the anterior process of the frontal; prefrontals are only narrowly exposed along the dorsal margin of the external nares. Frontal bones are fused. Beneath the cranium there is extensive quadrate-basisphenoid contact, and the vomer and palatine meet. The jaws lack ridges on the triturating surface, and the upper jaw is not hooked. The lower jaw lacks an external horny sheath. The toes are webbed, and there are four claws on each foot.

The karyotype is 2n = 54: 22 macrochromosomes, 32 microchromosomes (Bull and Legler, 1980).

Burbidge et al. (1974) divided the genus into three groups as follows: (1) neck comparatively short and thin; skull small, not flattened (height/width of the skull at rear of maxillae < 1:2.5); carapacial neurals usually absent (Rhodin and Mittermeir, 1977, showed *C. novaeguineae* may have up to four neurals on occasion); carapace length of adult female < 25 cm—*C. longicollis, C. novaeguineae, C. steindachneri.* (2) neck very long and thick; skull large and flattened (height/width > 1:2.5); carapacial neurals absent; carapace length of adult female > 25 cm—*C. expansa, C. parkeri, C. rugosa, C. siebenrocki.* (3) neck very long

Key to the species of *Chelodina*:

1a. Plastron broad, almost covering all of carapacial opening; forelobe broader than hind lobe; intergular scute at least twice as long as interpectoral seam 2
  b. Plastron narrow; forelobe scarcely broader than hind lobe, allowing much of carapacial opening to remain uncovered; intergular scute less than 1.5 times longer than interpectoral seam 4
2a. Forelobe of plastron very broad, extending laterally to or beyond marginal scutes of carapace: *C. longicollis*
  b. Forelobe of plastron not extending laterally to carapacial marginals 3
3a. Carapace round, broadest at center, and very flat: *C. steindachneri*
  b. Carapace oval to elliptical, broadest behind center, and not very flattened: *C. novaeguineae*
4a. Head heavily marked with cream, yellow, or pale-green markings: *C. parkeri*
  b. Head uniformly dark with no spotting 5
5a. Plastron long and narrow, more than twice as long as width immediately in front of bridge: *C. oblonga*
  b. Plastron moderate in width, less than twice as long as width immediately in front of bridge 6
6a. Plastral hind lobe broad across femorals, not greatly posteriorly tapered: *C. siebenrocki*
  b. Plastral hind lobe distinctly posteriorly tapered at femorals 7
7a. Plastral forelobe beginning to taper anteriorly immediately in front of bridge: *C. rugosa*
  b. Plastral forelobe not tapering anteriorly until near level of seam separating humeral and pectoral scutes: *C. expansa*

and thick; skull large and flattened (height/width > 1:2.5); five to eight neurals present in carapace; carapace length of adult female > 25 cm—*C. oblonga* (Goode, 1967, includes this species in group 2, but, in addition to having no neural bones, the shells of those species are deeper and more heavily built than that of *C. oblonga*).

Chelodina longicollis (Shaw, 1802)

Common snake-necked turtle

Recognition: The oval, dark-brown to black carapace (to 27.5 cm) is broadest behind the center, has a smooth posterior rim, depressed vertebrals, and a pronounced medial groove on the 2d to 4th vertebrals. Adult vertebrals are broader than long; the 1st is largest and flared anteriorly, the 4th is smallest, and the 5th is flared posteriorly. Neurals are not usually present. Marginals lying immediately over the tail are raised, and lateral marginals are often upturned. The plastron is large, almost covering the entire carapacial opening, and has a deep posterior notch. Its forelobe is as broad as or broader than the hind lobe. A slight lateral indentation may occur at the level of the abdominal-femoral seam, and the anal scutes taper toward the midline. The plastral formula usually is: intergul > abd > an > fem >< pect > hum > gul. The intergular is more than twice as long as the interpectoral seam. Plastron, bridge, and undersides of the marginals are cream to yellow with dark-brown or black pigment bordering the seams. The head becomes broader with age. Its snout is slightly upturned and protruding, and the upper jaw is unnotched. The thin neck is long (60% of the carapace length) and covered with pointed tubercles. Skin of the head and neck is brown to dark gray dorsally and laterally, but yellow ventrally. Jaws are cream to yellow. Exposed skin of the limbs is gray to brown, the undersides are

cream colored. Each foreleg has four or five large transverse scales on the anterior surface.

Bull and Legler (1980) found the diploid karyotype to be 54.

Males have longer, thicker tails and concave plastra.

Distribution: This species is found in eastern Australia from northern Queensland southward to southern South Australia.

Geographic Variation: No subspecies are currently recognized, although the carapacial shape does vary between populations in different drainage systems. The rough-scuted *Chelodina sulcifera* Gray, 1855 is now known to be only a growth form of *C. longicollis* (Goode, 1967).

Habitat: *Chelodina longicollis* lives in slow-flowing rivers, streams, swamps, and lagoons. It may hibernate in the southern portions of its range.

Natural History: Mating occurs in September and October. The male pursues the female from the rear with his snout near her vent. He then touches his chin to her carapace and, by swimming faster, follows her vertebral groove with his chin until he reaches the anterior rim of her shell, where he grabs the rim of her carapace with his toes and mounts, bending his tail beneath hers (Murphy and Lamoreaux, 1978). Most nesting occurs in November and December, but some may oviposit in January. The nests are dug in grassy or sandy areas at night during or after rains. Twelve nests measured by Vestjens (1969) were 82–127 mm in depth ($\bar{x} = 113$) and flask shaped. A single clutch of 6 to 24, brittle-shelled, oval (20–42 × 12–29 mm) eggs are laid each year. Natural incubation periods range from 130 to 168 days and the young emerge from January to late April. Hatchlings have carapace lengths of 25–30 mm. They are black to dark gray with an orange spot at the rim of each marginal and on each plastral scute. A yellow to orange horseshoe-shaped mark occurs on the chin, and a broad yellow or orange longitudinal stripe runs from each corner of the mouth backward onto the ventral surface of the neck.

Chelodina longicollis is carnivorous and eats fish, amphibians, crustaceans, insects, worms, and mollusks.

Wild *C. longicollis* often emit strong-smelling volatile fluids from their musk glands when first caught, but long-term captives seldom do this. Woolley (1957) reported that this species has the ability to match the color of its background by expanding or contracting the melanophores of its skin.

Chelodina longicollis

Chelodina novaeguineae Boulenger, 1888b

New Guinea snake-necked turtle

Recognition: The oval carapace (to 30 cm) is broadest behind the center, and has a smooth posterior rim, depressed vertebrals, and a pronounced medial groove on the 2d to 4th vertebrals. Vertebrals are broader than long in adults; the 1st is the largest and is flared anteriorly, the 4th is the smallest, and the 5th is flared posteriorly. Although neural bones are usually absent, Rhodin and Mittermeier (1977) found 5 of 20 New Guinea specimens had one to four scattered neurals. Surfaces of the carapacial scutes, at least in young individuals, are covered with rugose radiations. Lateral marginals are not upturned as in *C. longicollis*, and those over the tail are at best only slightly raised. The carapace is chestnut to dark brown. The plastron is large, covering most of the carapacial opening, and is deeply notched posteriorly. Its forelobe is broad, but does not reach the marginals; the hind lobe tapers toward the rear. The plastral formula is: intergul > an > fem > hum > pect > gul, with the intergular at least twice as long as the interpectoral seam. The bridge is moderate in size. Plastron, bridge, and undersides of the marginals vary from cream to yellow; their seams may be narrowly bordered in black. The head is broad and flat with a slightly protruding snout and an unnotched upper jaw. The jaws are wide and the maxillae compose more of the palate than in other *Chelodina*. The neck is comparatively short and thin, only about 55–60% of the carapace length, and its dorsal surface is covered with large, rounded tubercles. Head and neck are olive brown to brown above and cream to yellow below. Limbs are olive to brown. Each foreleg has a series of five enlarged transverse scales on its anterior surface.

Males have longer tails than do females.

Distribution: In Australia, *Chelodina novaeguineae* occurs in northeastern Queensland on the Cape York Peninsula from the northern tip to the vicinity of Bowen, and in rivers of the Gulf country in western Queensland and Northern Territory. It is also known from southeastern New Guinea and the island of Rotti near Timor.

Habitat: *Chelodina novaeguineae* apparently prefers slow-moving streams and swamps. Cann (1978) reported it has been taken in shallow lagoons which are dry for most of the year. Apparently these turtles estivate beneath the ground during the dry period.

Natural History: Engberg (1978), Cann (1978), and Feldman (1979) have reported their observations on

Chelodina novaeguineae

courtship and mating in *Chelodina novaeguineae*. The male swims around the female and approaches from the rear. He rubs his chin over her carapace, progressing forward from her posterior marginals. By the time his chin touches the base of her neck, his forefeet are able to reach her carapace. He climbs onto her back and places his hind feet over hers and continues to caress her neck with his chin. She then bobs her head in response. Intromission soon follows. Feldman's (1979) male placed his feet at the base of his tail as if to aid in extruding his penis. Cann (1978) reported that once intromission occurred, the male relaxed the grip of his forefeet and drifted slowly upward to an almost vertical position (apparently somewhat similar to that of male *Terrapene carolina*). Cann (1978) collected females in West New Guinea that had clutches of 17 to 21 eggs in September. He found several emerging from eggs after a 9-week incubation period. Goode (1967) reported that Australian *C. novaeguineae* lay clutches of 9 to 12 eggs. Ewert (1979) gave the average measurements of the eggs as 29 × 20 mm.

Natural foods include snails, prawns, insects, amphibians, and probably small fish. Various aquatic plants are also consumed.

This species emits a rather pungent defensive odor from its scent glands when first captured, but soon ceases this objectionable habit in captivity.

Chelodina steindachneri Siebenrock, 1914

Steindachner's side-necked turtle

Plate 1

Recognition: This side-necked turtle has a flattened, almost round carapace (to 21 cm) with a medial groove on the 2d to 4th vertebrals and an unserrated rim. All vertebrals are broader than long. The 1st is slightly flared anteriorly and is the largest; the 5th vertebral is flared posteriorly. There are no neural bones in the carapace. Lateral marginals are not upturned; those over the tail not raised. Carapacial color varies from light to dark brown. The plastron is narrow; the forelobe is broader than the hind lobe. A wide posterior notch is present on the hind lobe. The plastral formula is: intergul > an > fem > hum > pect >< abd > gul. Plastron, bridge, and undersides of the marginals are yellow with black seam borders. The head is small with a protruding snout, an unnotched upper jaw, and laterally placed eyes. The neck is comparatively short and thin and is covered with small granular scales which have a reticulated, wrinkled appearance. Dorsally the head and neck are gray to olive, beneath they are cream colored. Each forelimb has three enlarged transverse scales on the anterior surface. Limbs are gray to olive on the outer surfaces but yellowish beneath.

Bull and Legler (1980) found the karyotype to be 2n = 54.

Males have long thick tails and slightly concave plastra; the females have shorter tails and flat plastra.

Distribution: *Chelodina steindachneri* is restricted to Western Australia where it occurs in coastal drainages from the De Gray River system in the north to the Murchison watershed in the south.

Habitat: Over much of its range, *Chelodina steindachneri* lives in streams that are completely dry for long periods between rains. Apparently the turtles estivate in the dried-up river beds during these times, and are adapted to prevent desiccation.

Natural History: This is a poorly known species. It is apparently carnivorous, as captives feed on chopped meats, fish, and prawns, but Pritchard (1979) reported they also feed on carrion and plants.

Pritchard (1979) also reported that nesting occurs in the spring, September to October, and that the young hatch in February.

Chelodina expansa Gray, 1856

Giant snake-necked turtle

Plate 2

Recognition: This is the largest of Australia's chelid turtles, reaching a carapace length of 48 cm. Its carapace is oval, slightly broader behind the center, has a smooth posterior rim, and lacks both a medial keel or groove. The carapace becomes deeper with age. Its posterior marginals may be slightly flared over the tail, but the lateral marginals are not upturned as in *C. longicollis*. Vertebrals are variable in shape and dimension. The 1st is the largest and is anteriorly flared, the 2d and 3d are elongated and may be longer than broad in adults, the 4th is the smallest and is broader than long, and the 5th is broader than long and posteriorly flared. No neural bones occur in the olive to brown carapace. The plastron is quite narrow (about twice as long as wide), and does not nearly cover the carapacial opening. Its forelobe is rounded in front and slightly broader than the hind lobe, which tapers toward the rear and contains a deep anal notch. The bridge is also narrow. The plastral formula is: intergul > pect >< abd > fem > an > hum > gul. The intergular never divides the gular scutes and is about 1.5 times as long as the length of the interpectoral seam. Both plastron and bridge are gray to cream or yellow. The broad head is large and flattened with a slightly upturned, protruding snout and an unnotched upper jaw. Two to four small chin barbels are present. The neck is very long (over 65% of the carapace length) and thick; it is covered with wrinkled skin but lacks tubercles. Skin of the head and neck is dark gray to olive dorsally and laterally, but yellow ventrally. The jaws and chin are cream to yellow. Exposed skin of the limbs is gray to olive or brown, that beneath cream to yellow. The forelegs have seven or eight large transverse scales on the anterior surface.

Males have long, thick tails and concave plastra.

Distribution: *Chelodina expansa* occurs in Queensland from Rockhampton southwest through the coastal and interior areas to western New South Wales and the Murray River drainage, and on Fraser Island off Queensland (Cann, 1978).

Geographic Variation: Unstudied, but Cogger (1975) stated there are distinctive populations in central and eastern Queensland.

Habitat: *Chelodina expansa* lives in permanent bodies of fresh water such as waterholes and rivers. Legler (1978) found them in medium to large, very turbid rivers.

Natural History: Courtship has not been described; however, Legler (1978) reported frequent head bobbing in this species, and some of this may be associated with courtship. Nesting usually occurs in the fall from March to May, but Goode and Russell (1968) also reported that eggs are sometimes deposited in September. The nests are dug in sandy soil after heavy rains or during periods of high humidity, and probably three or four clutches are deposited each year. The brittle-shelled oval eggs average 37 × 25 mm and a clutch may contain 5 to 25 eggs ($\bar{x}$ = 15.4) (Ewert, 1979; Goode and Russell, 1968). The eggs undergo incubation for an extremely long period, sometimes over a year before hatching. Some young may hatch, but remain within the nest over winter; Cann (1978) opened one nest after 664 days which contained live hatchlings. Hatchlings have 35–40-mm carapaces that are olive to dark gray with a yellow border. Their skin is also dark, but the jaws, chin, and throat are yellow.

Chelodina expansa is strictly carnivorous, consuming insects, shrimp, small fishes, and frogs. Legler (1978) observed that feeding is by a gape and suck method and always involves a forceful strike.

When handled, this turtle seldom bites, but may scratch with its feet and flail its neck and limbs. It does not omit an offensive odor, as do some other species of *Chelodina*.

Chelodina parkeri Rhodin and Mittermeier, 1976

Parker's side-necked turtle

Recognition: This New Guinea species is closely related to *Chelodina siebenrocki*, but can be distinguished by its unique head pattern of light vermiculations. Its carapace is oval, flattened, and elongated (to 26.7 cm) with a smooth posterior border. Laterally it is nearly parallel. No medial keel or depression are present in adults. Vertebrals 1–4 are broader than long; the 1st is flared anteriorly and is the largest, and the 2d to 5th decrease in size until the 5th is the smallest. The 5th is flared posteriorly, but is longer than broad. Neural bones are absent. The marginals are not flared. The carapace is uniformly brown to dark brown, or there may be some yellow reticulations along the anterior and posterior borders. The plastron is elongated and narrower than the carapacial open-

ing. Its forelobe is anteriorly rounded and usually broader than the hind lobe, which tapers to the rear and has a shallow posterior notch. The plastral formula of three individuals examined was intergul > pect > abd > an > fem > hum > gul. The intergular is broad and not as elongated as in *C. siebenrocki*, being only about 1.5 times as long as broad. Both plastron and narrow bridge are cream to yellow. The broad head is elongated and flattened with a slightly protruding snout and an unnotched upper jaw. Its dorsal surface is covered with many small irregularly shaped scales. Usually only two small chin barbels are present, but the number is variable. The head is gray to dark brown with an extensive pattern of fine white, cream, yellow, or green vermiculated lines or dots. There is also a large, bright blotch on the mid-posterior surface of the tympanum. The neck is long and thick (about 75% of the carapace length), with small pointed tubercles on the dorsal surface. The forelimbs have several large transverse scales on the anterior surface. The dorsal surfaces of the neck and limbs are gray, the undersides are white to pink.

Males have longer, thicker tails and flatter shells than do females.

Distribution: Parker's sideneck is endemic to New Guinea where it is restricted to the area surrounding Lakes Murray and Balimo and the Fly and Aramia rivers. Verified collection localities are restricted to the Western District's savannah regions north of the Fly River (Rhodin and Mittermeier, 1976).

Habitat: *Chelodina parkeri* is restricted to large inland grass swamps in savannah areas; it inhabits the open edges of lakes and rivers where a dense growth of grass and aquatic vegetation occurs in shallow water. These areas are subject to periodic inundation and partial or complete drying during very dry seasons (Rhodin and Mittermeier, 1976).

Natural History: Nothing is known of its behavior in the wild. In captivity it feeds on fish (Rhodin and Mittermeier, 1976).

Chelodina parkeri

Chelodina rugosa Ogilby, 1890

Northern Australian snake-necked turtle

Plate 2

Recognition: This is a large *Chelodina* reaching a carapace length of at least 35 cm; Cogger (1975) reported the maximum length to be 40 cm. The carapace is oval, somewhat elongated, and usually is broadest behind the middle. Its surface is covered with small rugosities. The posterior carapacial rim is smooth to slightly serrated, and there is no medial vertebral keel or depressed groove in adults. All vertebrals are broader than long in adults, although the 2d is almost as long as broad. The 1st is the largest and is flared anteriorly, the 4th is the shortest, and the 5th is flared posteriorly. No neurals are present. The carapace is brown to black and the seams may be darkly outlined. Some individuals have lighter brown or yellowish pigment on the dorsal surface of the marginals. The plastron is moderate, but does not cover the entire carapacial opening. Its forelobe is broad and tapers to a rounded front; the hind lobe is broad across the femorals but tapers to the rear and is deeply notched posteriorly. The plastral formula is: intergul > pect > abd > fem > an > hum > gul. The intergular is almost twice as broad as long, and may extend anteriorly to separate the gulars. Surfaces of the plastron and narrow bridge are covered with rugosities, and the undersides of the marginals, plastron, and bridge are yellow with dark seams. The large head is broad and flat, the snout does not protrude, and the upper jaw is neither notched nor cusped. Chin barbels are present but variable in number. The neck is very long and thick, over 75% as long as the carapace. The dorsal surface of the head is covered with irregularly shaped scales and that of the neck with small blunt tubercles. Head and neck are olive, brown, or gray; the jaws may be cream colored. Limbs and tail are gray. A series of large transverse scales occur on the anterior surface of the forelimbs.

Males have much longer, thicker tails than do females.

Distribution: *Chelodina rugosa* occurs only in northern Australia, where it is known from the Cape York Peninsula westward to the Kimberley District of Western Australia (Cogger, 1975).

Geographic Variation: The species is little known and poorly described. In the past it has been confused with both *Chelodina oblonga* and *C. siebenrocki*, and in fact it may be synonomous with *C. siebenrocki*; additional study of this relationship is badly needed.

Habitat: Cogger (1975) reported the habitat as swamps, billabongs, waterholes, and slow-flowing rivers.

Natural History: Cann (1978) reported that nesting occurs in March and that a 29-cm female deposited a clutch of 14 elongated (35 × 25 mm) eggs. When dissected she contained a second clutch that would have been deposited a few weeks later. In captivity, *Chelodina rugosa* feeds on aquatic insects, crustaceans, and fish.

Chelodina siebenrocki Werner, 1901
Siebenrock's side-necked turtle

Recognition: This little-known *Chelodina* from New Guinea has been confused with *C. rugosa* of northern Australia and *C. parkeri* of New Guinea. It has an oblong dark-brown or black carapace (to 30 cm) which is broader behind the center and flattened dorsally. No vertebral keel is present, and females may have a slight medial depression. All vertebrals are broader than long; the 1st is flared anteriorly and is the largest, the 4th is the shortest, and the 5th is flared posteriorly. Normally no neural bones are present. The posterior marginals are neither flared nor serrated. The plastron is short, narrow, and elongated, leaving much of the carapacial opening uncovered. Its forelobe rounds to a point and the hind lobe is deeply notched posteriorly. The plastral formula is: intergul > pect > fem >< an > abd > hum > gul. The intergular scute is long and narrow (about twice as long as broad). The bridge is narrow. The plastron is yellow to light brown, and some individuals may have darkened seams. The broad head is large and flattened with a slightly protruding snout and an unnotched upper jaw. Numerous small irregularly shaped scales cover the dorsal surface. A variable number of chin barbels is present, often four or more. The neck is long and thick, reaching about 75% of the length of the carapace, and is covered with blunt tubercles. Head and neck are gray to brown dorsally, cream colored ventrally. The forelimbs contain enlarged transverse scales on their anterior surfaces, and the toes are webbed. The limbs and tail are gray.

Males have longer, thicker tails than do females, and are flatter shelled than are the more domed females.

Distribution: *Chelodina siebenrocki* occurs on the southern coast of New Guinea west of the Fly River, and on certain islands in the Torres Strait (Rhodin and Mittermeier, 1976).

Habitat: *Chelodina siebenrocki* inhabits the tidal areas of small streams, swamps, marshes, and offshore islands (Rhodin and Mittermeier, 1976).

Natural History: All we know of the biology of this turtle has been reported by Rhodin and Mittermeier (1976). Mating takes place in water, with the male mounting the female from the rear. The female digs a nest at the end of the wet season in May and lays 4 to 17 elongated eggs (35–36 × 28–29 mm). Hatching occurs at the start of the next wet season in November or December. Hatchlings have carapace lengths of about 36 mm. Their carapaces are brown with small black spots; each marginal tends to have one spot but the number varies on the vertebrals and pleurals.

Chelodina siebenrocki is apparently carnivorous, using its long neck to strike for food. Despite this, it is unaggressive and does not bite when handled.

It does not bask, but instead spends most of the time on or in deep mud.

Chelodina oblonga Gray, 1841

Narrow-breasted snake-necked turtle

Recognition: This species has a narrow, oval carapace (to 40 cm), broadest behind the center, with a smooth posterior rim. The vertebrals are depressed with a pronounced medial groove on the 2d to 4th; a low medial keel may also occur. Vertebrals 1, 4, and 5 are broader than long; 2 and 3 may be longer than broad. Vertebral 1 is largest and anteriorly flared, 4 is the smallest, and 5 is posteriorly flared. Five to eight neural bones are present. Surfaces of the carapacial scutes of juveniles and young adults may be rugose with radiating striations, but very old, large individuals are often smooth. Lateral marginals may be slightly upturned. The carapace ranges in color from light brown to black, and some dark flecking may be present. The long, narrow plastron is smaller than the carapacial opening. Its forelobe is rounded anteriorly; the hind lobe tapers to the rear and bears a large anal notch. The forelobe is broader than the hind lobe. The plastral formula is: intergul > fem > pect > abd > an > hum > gul. The intergular is about 1.2–1.5 times as long as the length of the interpectoral seam. Both plastron and narrow bridge are cream to yellow. The head is large and flat with a protruding snout and an unnotched upper jaw. Several chin barbels may be present. Head and jaws are olive to gray with numerous dark mottlings. The olive to gray neck is thick, with blunt rounded tubercles, and so long (over 75%

of the carapace length) that it cannot be tucked entirely under the carapace. Limbs are olive to gray and the forelegs have seven large transverse scales on their anterior surfaces.

Males have slightly longer, thicker tails than do females, and the male plastron is concave.

Distribution: *Chelodina oblonga* is confined to the southwestern corner of Western Australia (Cann, 1978).

Habitat: Swamps and streams with permanent water seem to be the primary habitat. Under natural conditions it apparently does not estivate.

Natural History: Most nesting occurs from September to early November, and a second nesting period may occur in December and January (Clay, 1981). Nests examined by Clay (1981) were 20–105 m from water; distances traveled were longer during September–November ($\bar{x}$ = 86.56 m), shorter during December–January ($\bar{x}$ = 25.38 m). Nest sites were usually open and free from thick vegetation, and once the maximum daily air temperature remained above 17.5°C the females came ashore to nest. Clay also found a strong relationship between nesting and the approach of a rain-bearing atmospheric depression. Clutches ranged from 3 to 12 eggs with a mean of 8.2 eggs during September–November and a mean of 4.0 during December–January. The white eggs are elongated (36.7–30.0 × 23.4–18.0 mm). The natural incubation period ranges from 183 to 222 days, depending on weather conditions. Hatchlings are approximately 31 mm in carapace length (Ewert, 1979).

Cann (1978) reported that his captives ate fish and tadpoles which they stalked. The neck and head were retracted toward the carapace and held there until the prey moved within range, then the head was struck forward with amazing speed, seizing the victim.

Chelodina oblonga

Elseya Gray, 1867

This is the first of three short-necked chelid genera from Australia and New Guinea. The adult carapace is oval, depressed to somewhat domed, broadest behind the center, and with either a smooth or a serrated posterior rim. Juveniles have a vertebral keel which is lost in adults, and large adults may develop a medial groove. Neural bones are usually absent, but Rhodin and Mittermeier (1977) found four in an *E. latisternum* and three in an unidentified *Elseya*. A cervical scute is present in some species but absent in others. The plastron lacks a hinge and is usually long and somewhat narrow. Its hind lobe is notched posteriorly. Axillary buttresses are strong; inguinal buttresses are smaller and weaker, barely reaching the 5th costal bone. The intergular scute is narrow and rectangular; it completely separates the gulars, but not the humerals or pectorals. Cervical vertebrae are shorter than those beneath the carapace. The temporal arch is moderate, formed by the squamosal and expanded parietal bones; however, the parietals do not touch either the supraoccipital or quadrate bones. The nasal bones are not completely separated by the anterior process of the frontal bone. The frontals are not fused, and the prefrontals are not exposed along the dorsal margin of the external nares. Beneath the cranium there is no quadrate-basisphenoid contact, but the vomer touches the palatine. The triturating surface of the maxilla bears a medial ridge in only one species, *E. dentata*. The upper jaw is either unnotched medially or bears a slight medial indentation. The snout is long and usually protrudes, and there are usually a few prominent, large, conical tubercles behind the orbits. Dorsally, the head is covered with a single, well-developed, horny plate. The toes are webbed, and there are five foreclaws and four hind claws.

The diploid karyotype is 50: 22 biarmed macrochromosomes, 28 microchromosomes (Bull and Legler, 1980).

Three species have been described, but both Cann (1978) and Bull and Legler (1980) refer to several other newly discovered unnamed species. McDowell (1983) has recently placed *Elseya* in the synonymy of *Emydura* Bonaparte, 1836, but this needs further confirmation.

Key to the species of *Elseya*:

1a. Sides of horny plate on head turned downward toward tympanum; posterior rim of carapace strongly serrated in adults: *E. latisternum*
 b. Sides of horny plate on head not bent downward toward tympanum; posterior rim of carapace smooth or only slightly serrated in adults 2
2a. Triturating surface of maxilla with a medial ridge; posterior rim of carapace smooth; vertebral keel usually absent; Australia: *E. dentata*
 b. Triturating surface of maxilla without a medial ridge; posterior rim of carapace slightly serrate; vertebral keel usually present; New Guinea: *E. novaeguineae*

Elseya dentata (Gray, 1863b)
Northern Australian snapping turtle

Recognition: The oval (to 40 cm), flattened adult carapace usually is broadest behind the center, lacks a medial keel, and has a smooth posterior rim. Hatchlings and juveniles have a medial keel and strongly serrated posterior marginals on their round carapaces. Neural bones are absent, as is also a cervical scute. Vertebral scutes are usually broader than long, but the 2d through 4th may be longer than broad in large females; the 1st and 5th are the smallest; the 1st is somewhat flared anteriorly, and the 5th is expanded posteriorly. Carapacial scutes are very rugose in texture. The carapace is olive gray to dark brown or black. Both sexes may become melanistic with age (Goode, 1967). The plastron is long and narrow, leaving much of the carapacial opening uncovered. Its forelobe is broader at its base than is the hind lobe, tapers toward the front, and is rounded or pointed anteriorly. The hind lobe tapers toward the rear and is posteriorly notched. The bridge is broad. The intergular scute is long and narrow (more than twice as long as broad), and it completely separates the gulars. The plastral formula is: pect >< fem > abd >< an > intergul > gul > hum. Plastron and bridge change from cream or yellow to gray brown or black with age. The head is large with a projecting snout and an unnotched to slightly notched upper jaw. The triturating surface of the maxilla bears a medial ridge. Dorsally, the head is covered with a large horny plate instead of smooth skin, and there are two chin barbels. The dorsal surface of the neck is covered with large blunt tubercles. Head, neck, and limbs are gray to olive or dark brown, and on the side of the head a broad, light stripe extends from below the orbit to the neck. The jaws are yellow to horn colored.

Males have longer, thicker tails than do females.

Elseya dentata

Distribution: *Elseya dentata* occurs in the rivers of northern Australia from the Kimberley district of Western Australia eastward through the Northern Territory to the Burnett River in southeastern Queensland.

Geographic Variation: Unknown. McDowell (1983) has placed *novaeguineae* in the synonymy of *dentata* (in the genus *Emydura*), but we feel that additional evidence must be presented before this designation is accepted.

Habitat: *Elseya dentata* inhabits large rivers and associated lagoons and oxbow lakes. Legler and Cann (1980) found them in riffles, deep pools, stretches with slow current, and areas with dead wood. Apparently it estivates underground during dry periods (Cann, 1978).

Natural History: Nesting occurs in October or November, with hatching about six months later. The brittle-shelled eggs are elongated (48–50 × 27–28 mm; Legler and Cann, 1980), and three to five constitute a normal clutch.

   Elseya dentata is omnivorous, eating aquatic plants, fruits which drop into the water (especially figs), aquatic insects, crustaceans, amphibians, and fish.

   This turtle may be very aggressive when first captured, and large individuals can deliver severe bites; however, after a while in captivity they usually become tame.

Elseya novaeguineae (Meyer, 1874)

New Guinea snapping turtle

Recognition: The round to oval, medially keeled, deep adult carapace (to 30 cm) is broadest behind the center, and has at least a slightly serrated posterior rim. The medial keel and posterior serrations are well de-

Elseya novaeguineae

veloped in juveniles, and, although they become less pronounced with age, are never entirely lost. The serrations are never as prominent in adults as they are in *Elseya latisternum*. No neurals are present. The cervical scute is well developed. All vertebrals are broader than long; the 5th is the smallest and is expanded posteriorly; the 2d is the largest. Lateral and posterior marginals are outwardly expanded. The adult carapace is uniformly brown to black. The plastron is long and narrow, allowing much of the carapacial opening to remain uncovered, and is posteriorly notched. Its forelobe tapers gradually toward the front, is rounded anteriorly, and broader at the bridge than is the hind lobe, which tapers toward the rear. The bridge is well developed. The intergular is very narrow, almost three times as long as broad, and completely separates the gulars. The plastral formula is: pect >< fem > an > abd > intergul > gul > hum. Plastron and bridge are cream to yellow. The head is small and narrow with a projecting snout and an unnotched upper jaw. There is no medial ridge on the maxillary triturating surface. Dorsally, the head is covered with a large horny plate instead of smooth skin; there are two small chin barbels. Flattened tubercles are present in front of the tympanum; those on the neck are small and pointed. The skin of the head, neck, and limbs is gray.

   Males have much longer tails than do females.

Distribution: *Elseya novaeguineae* is restricted to New Guinea, where it occurs over much of the island.

Habitat: Rivers and swamps are preferred, especially those along the coast.

Natural History: The large eggs are ellipsoidal (55 × 33 mm) and brittle shelled. Hatchlings have carapaces ranging from 45 to 48 mm that are brown with a small black blotch on each carapacial scute. An ecological study of this species would be rewarding.

Elseya latisternum Gray, 1867

Serrated snapping turtle

Recognition: The broad, depressed, oval adult carapace (to 28 cm) is broadest behind the center, has a vertebral groove, and is distinctly serrated along the posterior rim. Juveniles have keeled carapaces. Neural bones are absent. A cervical scute may be present or absent. Vertebrals are broader than long; the 5th is posteriorly flared. All carapacial scutes are slightly rugose. The carapace varies from olive or chestnut brown with dark mottlings to dark brown or black.

Elseya latisternum

The plastron is long and narrow, exposing much of the carapacial opening. Its forelobe is broader than the hind lobe and is rounded anteriorly. The hind lobe tapers toward the rear and has a posterior notch. The intergular is usually 1.5 to more than 2.0 times longer than broad, and it completely separates the gulars. The plastral formula is: an > pect > fem > intergul > abd > gul > hum. The plastron is cream to yellow with brown pigment on the lateral edges of the scutes. The head is large with a projecting snout and an unnotched to slightly notched upper jaw. No medial ridge is present on the triturating surface of the upper jaw. The top of the head is covered with a large horny plate instead of smooth skin; unlike other *Elseya* species, the sides of this plate turn downward toward the tympanum. Two chin barbels are present, and there are long pointed tubercles on the back of the neck. The head varies from chestnut brown to olive gray or dark brown; the legs and neck are similarly colored. The jaws are yellow to tan.

Males have much longer tails than do females.

Distribution: *Elseya latisternum* ranges in northeastern Australia from the Cape York Peninsula southward to northern New South Wales.

Geographic Variation: Although no subspecies have been described, variation does exist. Goode (1967) reported that *Elseya latisternum* from different locations may be unlike in coloration: those from Cape York have chestnut-brown carapaces, but individuals from the Flinders River are mottled dark and pale sepia. Legler and Cann (1980) found that the condition of the cervical scute varies geographically: it is usually absent in northern populations (Cape York, 93% absence) but usually present and well developed in populations south of 29°S (98% presence).

Habitat: This sideneck lives in rivers, streams, lagoons, and marshes. Legler and Cann (1980) did not find it in large rivers such as the Fitzroy and Dawson, at 23°S, but it was common in smaller tributary streams.

Natural History: Courtship involves the male approaching the female with a series of dorsoventral head bobs, touching her cloacal area with his snout, then moving to the front of the female and attempting to align their gular barbels. This is followed by extended stroking of the female's barbels and snout with his forefeet and claws (Murphy and Lamoreaux, 1978). Nesting occurs in Queensland in September and October (Cann, 1978). Probably several clutches of 9 to 17 eggs are laid each year (Cann, 1978). Eggs have brittle shells and are elongated (21–25.7 × 33–40.8 mm; Ewert, 1979; Legler and Cann, 1980).

Legler and Cann (1980) reported *Elseya latisternum* as chiefly carnivorous, with insectivorous tendencies but no mollusk-eating tendencies. They also reported that vegetation was not present in the stomachs they examined; however, Cann (1978) reported that *E. latisternum* in one small waterhole ate weeds, and Worrell (1964) also reported plants are eaten. Animal foods taken include aquatic insects, crayfish, and frogs.

Large *E. latisternum* can bite viciously when provoked and care should be taken when handling wild individuals. Also, they have a tendency to emit a foul-smelling odor from their musk glands.

Emydura Bonaparte, 1836

The short-necked sidenecks of the genus *Emydura* are semiaquatic and restricted to Australia and New Guinea. The adult carapace is oval to elliptical, somewhat domed, usually has a smooth posterior rim, and lacks a vertebral keel. A slight medial groove may be present on the carapace of larger adults, and that of hatchlings and juveniles is keeled and posteriorly serrated. Usually no neural bones are present, but a well-developed cervical scute is present. The plastron is hingeless, usually long and somewhat narrow, and notched posteriorly. Much of the carapacial opening is left uncovered. Both axillary and inguinal buttresses are strong; the inguinal buttress extends to the 5th or between the 5th and 6th costal bones. The intergular scute is longer than broad and is dilated posteriorly; it completely separates the gular scutes but not the humerals or pectorals. Cervical vertebrae are not longer than those beneath the carapace. The temporal arch is moderate, formed by the squamosal and parietal bones; however, the parietal does not touch either the supraoccipital or the quadrate. Nasal bones are not completely separated by the anterior frontal process and the frontals are not fused. The prefrontals are not exposed along the dorsal margin of the external nares. Beneath the cranium there is no quadrate-basisphenoid contact, but the vomer and palatine meet. The jaws lack ridges on the triturating surfaces and the upper jaw is neither hooked nor notched. The snout is short and flattened (only slightly projecting), and there are many small flattened tubercles behind the orbits. Smooth skin covers the dorsal surface of the head. The toes are webbed, and there are five foreclaws and four hind claws.

The diploid karyotype is 50: 22 macrochromosomes, 28 microchromosomes (Killebrew, 1976; Bull and Legler, 1980).

Five species are currently recognized, although both Cann (1978) and Bull and Legler (1980) refer to several other undescribed species, and the status of at least one other form, *E. victoriae* (Gray, 1842), remains uncertain.

Wells and Wellington (1984) have resurrected the name *Chelymys* Gray 1844 for this genus; however, *Emydura* Bonaparte, 1836 has priority, and should also be recognized owing to its common usage for over a century (Stimson, 1986).

Key to the species of *Emydura*:

1a. A postocular light stripe runs backward
   from orbit to neck 2
 b. No postocular light stripe present 4
2a. Lower jaw stripe pink to red, some red
   pigment on neck and plastron 3
 b. Lower jaw stripe gray or greenish yellow, no
   red pigment on neck and plastron: *E. kreffti*
3a. Hard palate extensively developed; length of
   mandibular symphysis about 1.5 times the
   horizontal diameter of tympanum;
   postocular stripe red or pink: *E. australis*
 b. Hard palate not extensively developed;
   length of mandibular symphysis about equal
   to the horizontal diameter of tympanum;
   postocular stripe yellow: *E. subglobosa*
4a. Light stripe continuing posteriorly from
   corner of mouth barely touches lower rim of
   tympanum; female carapace may be longer
   than 26 cm: *E. macquarrii*
 b. Light stripe continuing posteriorly from
   corner of mouth passes through lower third
   of tympanum; female carapace length barely
   reaches 25 cm: *E. signata*

Emydura australis (Gray, 1841)
Australian big-headed side-necked turtle
Plate 2

Recognition: The oval carapace (to 30 cm) is broadest at the center, has a smooth or only slightly serrated posterior rim, and a medial keel which is best developed in the male, but becomes lower with age. Neural bones are absent, but a well-developed, cervical scute is present. Vertebrals 1 and 5 are broader than long with the 5th expanded posteriorly; vertebrals 2–4 may be as long as or longer than broad in adults. Surfaces of the carapacial scutes are often rugose with various longitudinal striations. The carapace is uniformly light to dark brown. The long plastron is narrow and much of the carapacial opening is uncovered. Its forelobe is rounded anteriorly and is broader than the posterior lobe which gradually tapers to the rear and is posteriorly notched. The bridge is narrow. The intergular scute separates the gulars, but is less than twice as long as broad. The plastral formula is: pect > fem > abd > an > intergul > gul > hum. Both plastron and bridge are cream to yellow. The head is broad with a slightly projecting snout and an unnotched upper jaw. The hard palate is very broad and extends backward beyond the maxillae covering over half of the roof of the mouth (the palate is not as extensively developed in other species of *Emydura*). Also, the diameter of the mandibular symphysis is very broad, much more so than the greatest diameter of the orbit. Only rudimentary chin barbels are present, at best. The dorsal surface of the head is covered with smooth skin, and the neck tubercles are poorly developed. Head, neck, and limbs are gray to olive or brown. On each side of the head are two light stripes: one extends from the orbit to the neck and the other from the corner of the mouth to the neck. In juveniles these stripes are pinkish, but they become more yellow with age.

Males have longer, thicker tails than do females.

Distribution: *Emydura australis* occurs in the drainages of the Kimberley region of Western Australia and the Northern Territory of Australia.

Geographic Variation: Cann (1978) remarked that variation in coloration, scute texture, and head size is considerable throughout the range. He also pointed out that in the coastal waterways of the Northern Territory there is a yellow-faced turtle with a less developed hard palate which has been confused with *Emydura australis* and may be the species described as *E. victoriae* by Gray (1842). This relationship needs confirmation. McDowell (1983) has synonymized *E.*

kreffti and *E. subglobosa* with *E. australis*. Although they are undoubtedly closely related, we still feel that they probably represent different species.

Habitat: *Emydura australis* lives in shallow rivers and streams.

Natural History: Worrell (in Pritchard, 1979) reported that copulation occurs while in a plastron-to-plastron position, very unusual for a turtle. Hatchlings have pinkish-red facial stripes and some red on the neck, limbs, and plastron. Their carapaces are peaked with a prominent medial keel and somewhat serrated posteriorly. The broad adult head develops with age.

The well-developed hard palate is apparently an adaptation for crushing molluscan shells; freshwater snails and bivalves form an important part of the diet. *Emydura australis* also feeds on insects, amphibians, and some fruits.

Emydura kreffti (Gray, 1871)
Krefft's river turtle

Recognition: The oval, deeply domed adult carapace (to 25.6 cm) is broadest at the center, has a smooth posterior rim, but lacks a medial keel. A cervical scute is usually present, but neural bones are lacking. Vertebrals 1, 4, and 5 are broader than long, and the 5th is expanded posteriorly; vertebrals 2 and 3 may be as long as or longer than broad. Carapacial scutes are translucent and often rugose with raised longitudinal striations. The carapace ranges from olive brown to dark brown, with or without darker flecking. The plastron is long and narrow and the bridge is also narrow, leaving much of the carapacial opening uncovered. Its

Emydura kreffti

forelobe is rounded anteriorly and is broader than the hind lobe, which tapers to the rear and has an anal notch. The intergular is longer than broad and completely separates the gulars. The plastral formula is: pect > fem > abd > an > intergul > gul > hum. Both plastron and bridge are cream to yellow. The head is moderate with a slightly upturned, projecting snout and an unnotched upper jaw. The mandibular symphysis is broader than the greatest diameter of the orbit. Chin barbels are absent or only poorly developed. Smooth skin covers the dorsal surface of the head; the neck has numerous low rounded scales. Head, neck, and limbs are dark gray. A yellow to greenish-gray stripe extends backward from the orbit to the tympanum, and a second stripe of the same color passes backward along the lower jaw to the neck. Both jaws and the chin are light gray.

Females are larger and have higher domed shells and shorter tails than do the males.

Distribution: *Emydura kreffti* occurs in Queensland generally east of the Great Dividing Range from the Cape York Peninsula southward to northern New South Wales.

Geographic Variation: Carapacial coloration is quite variable.

Habitat: *Emydura kreffti* lives in rivers, streams, swamps, and lagoons. Legler and Cann (1980) commented that in rivers it occurs in riffles, but is most likely to be found associated with dead wood or undercut banks and never in the fastest water.

Natural History: Nesting occurs from October to December. Cann (1978) reported his captives always chose a nesting site among ferns and shrubs. The elongated eggs have brittle shells; 82 measured by Legler and Cann (1980) averaged 36.45 × 21.13 mm. Normal clutches may vary from 7 to about 18 eggs; Legler and Cann (1980) reported 82 eggs from five clutches, an average of 16 eggs per clutch. Hatchlings vary from 24 to 29 mm in carapace length, and have high-peaked, medially keeled carapaces with serrated posterior rims. Natural incubation may take 80 or more days.

Emydura kreffti is omnivorous, feeding on aquatic plants, fruits which drop into the water, aquatic bivalves and snails, small crustaceans, insects (both aquatic and terrestrial), and frogs. Small individuals are carnivorous, larger turtles more omnivorous.

Emydura subglobosa (Krefft, 1876)

Red-bellied short-necked turtle

Plate 2

Recognition: The brown oblong carapace (to 25.5 cm) is broader posteriorly than anteriorly. It is somewhat domed but lacks a vertebral keel in adults; small juveniles are keeled. The cervical scute is well developed, and the vertebrals are broader than long. Posteriorly, the marginal rim is smooth and somewhat flared; undersides of the marginals are red. The narrow plastron is notched posteriorly and lacks a hinge. It is yellow with a broad lateral border of red pigment. The intergular scute is longer than broad, longer than the length of the interhumeral seam. The plastral formula is: pect > fem > abd > an >< intergul < gul > hum. The bridge is composed of parts of the pectoral and abdominal scutes with no axillary and only small inguinal scutes present. It is yellow with some reddish markings. On the olive head, a yellow stripe runs from the tip of the snout through the orbit to and above the tympanum, and another yellow stripe may be present along the upper jaw. A broken red stripe runs along the lower jaw to the neck, often extending to the plastron. Two yellow barbels are present on the chin, and the light-colored maxillary and mandibular sheaths are prominent. The neck is dark gray dorsally, but ventrally lighter gray with red streaks. Limbs and tail are gray anteriorly and white posteriorly with red streaks. A series of enlarged narrow horizontal scales occurs on the anterior surface and lateral margin of the forelegs and on the outer margins of the hind legs. The red pigment is more prominent in juveniles, but fades to a pinkish salmon with age.

Males have long, thick tails and narrow posterior plastral notches, while females have short tails and wide posterior plastral notches.

Distribution: *Emydura subglobosa* ranges mainly in southern New Guinea, but it has also been found in the Jardine River at the northern tip of the Cape York Peninsula, Queensland, Australia.

Geographic Variation: Unknown. *Emydura albertisii* Boulenger, 1888b is considered a synonym of *E. subglobosa.*

Habitat: Basically a riverine dweller, *Emydura subglobosa* also occurs in lakes and lagoons.

Natural History: Very little is known of the biology of *Emydura subglobosa.* Cann (1978) reported that large numbers occur in some New Guinea rivers, and that the females lay an average clutch of 10 eggs in September. A 21-cm female kept by Ernst laid 5 eggs between 17 and 21 November 1980. These had white, brittle shells and were elongated. Their measurements ranged from 31 × 19 mm to 44 × 20 mm (all but one was over 40 mm in length).

In the wild they are probably carnivores, feeding primarily on mollusks, crustaceans, and aquatic insects, but in captivity they thrive on a fish diet. Captives kept by us were shy and remained hidden most of the time; they seldom basked.

Emydura macquarrii (Gray, 1831b)

Murray River turtle

Recognition: The broad, oblong, olive to brown carapace (to 31 cm) is broadest behind the center, and may have either a smooth or a slightly serrated posterior rim. The posterior marginals are flared and those along the sides may be upturned. A cervical scute is present. Juveniles and males may have a medial keel, which becomes lower with age; a medial groove may develop in larger individuals, especially females. No neurals are present. Vertebrals 1, 4, and 5 are broader than long; the 1st may be expanded anteriorly and the 5th is flared posteriorly. Vertebrals 2 and 3 may be as long as or longer than broad. The surface of the carapacial scutes is roughened with numerous longitudinal striations. The plastron is long and narrow, leaving much of the carapacial opening uncovered. Its forelobe is rounded in front and only slightly broader than the hind lobe, which gradually tapers toward the rear and has an anal notch. The intergular scute is longer than broad and totally separates the gulars. The plastral formula is: fem > pect > abd > intergul >< an > gul > hum. Plastron, bridge, and undersides of the marginals are cream to yellow. The moderate-sized head has a slightly pro-

Emydura macquarrii

jecting snout and an unnotched upper jaw. The diameter of the mandibular symphysis is broader than the greatest diameter of the orbit. Two chin barbels are present. Dorsally, the head is covered with smooth skin, and the small neck tubercles are blunt and rounded. Head, neck, and limbs are grayish to olive brown. A yellow to cream-colored stripe runs backward from the corner of the mouth to the neck, and a yellow spot occurs on each side of the chin.

Females are larger than males, and have higher domed carapaces and shorter tails.

Distribution: *Emydura macquarrii* lives in southeastern Australia from south-central Queensland southward to northern Victoria and southeastern South Australia.

Habitat: This turtle inhabits large rivers, lagoons, and waterholes. It hibernates during the winter.

Natural History: During courtship the male approaches the female with a series of head bobs, touches her cloacal vent with his snout, swims to the front and faces her, and then attempts to align his chin barbels with hers; finally he strokes her barbels and snout with his foreclaws (Murphy and Lamoreaux, 1978). Copulation soon follows. The eggs are laid from late October through December in nests dug in the stream banks, usually after a rainstorm. The eggs are elongated (33 × 23 mm) with brittle shells; a single clutch may contain 6 to 24 eggs, but usually about 15. Natural incubation takes about 75 days (66–85). Hatchlings have carapaces about 30 mm in length with a high medial keel and a serrated posterior rim.

Foods taken from the stomach of this turtle include fish, various aquatic worms, crustaceans, insects, bivalves, snails, algae, and parts of aquatic plants, indicating it is omnivorous (Chessman, 1986). *Emydura macquarrii* is fond of basking on river banks.

Emydura signata Ahl, 1932

Brisbane short-necked turtle

Recognition: *Emydura signata* is similar to *E. macquarrii,* of which it is sometimes considered a subspecies. The oblong, olive to brown carapace (to 25 cm) is broadest behind the center and is usually posteriorly serrate in smaller individuals but smoother in larger turtles. Posterior marginals are flared, and lateral marginals may be slightly upturned. A cervical scute is present. Juveniles and males may have a medial keel which becomes lower with age; adult females

Emydura signata

may develop a medial groove. No neurals are present. In adults, the 2d vertebral may be longer than broad; other vertebrals are broader than long. The 1st vertebral is expanded anteriorly while the 5th is flared posteriorly. Longitudinal striations occur on each carapacial scute, giving it a rugose appearance. The long, narrow plastron does not cover the carapacial opening. Its forelobe is rounded anteriorly and slightly broader than the hind lobe, which gradually tapers toward the rear and contains an anal notch. The intergular scute is longer than broad, and totally separates the gulars. The plastral formula of the two specimens examined was: fem > pect > intergul >< abd > an > hum > gul. Plastron, bridge, and undersides of the marginals are yellow. The head is of moderate size with a slightly projecting snout and an unnotched upper jaw. The mandibular symphysis is broader than the greatest diameter of the orbit. Two chin barbels are present. Dorsally, the head is covered with smooth skin, and the small, blunt neck tubercles are rounded. All skin is gray to olive brown. A yellow to cream-colored stripe runs backward from the corner of the mouth to the neck, passing through the lower third of the tympanum. A yellow spot occurs on each side of the chin.

Males are shorter than females, have more flattened carapaces and longer, thicker tails.

Distribution: *Emydura signata* is found in coastal drainages from the vicinity of Brisbane, Queensland, southward to northern New South Wales.

Habitat: This turtle lives in rivers and large streams.

Natural History: Nesting occurs from late September to early October (Cann, 1978).

Rheodytes Legler and Cann, 1980

Rheodytes is most closely related to the short-necked genera and *Emydura*. Its only species is *R. leukops*.

Rheodytes leukops Legler and Cann, 1980
Fitzroy turtle

Recognition: This is a medium-sized (26.2 cm) short-necked chelid turtle with a white ring around its iris and huge cloacal bursae. The adult carapace is elliptical and has a smooth marginal border; that of juveniles (to 9.5 cm) is nearly round and extremely serrated at the edges. Juveniles bear a distinct keel, but adults are only slightly peaked or middorsally flattened. Neurals are absent, and the rib tips of costals 2–4 articulate with the centers of peripherals 4–6. Carapacial scutes are thin and translucent allowing the underlying intercostal sutures to be seen; their surfaces contain a series of sharp parallel ridges. A cervical scute is usually present. The interpleural seams touch the posterior parts of marginals 6 and 8. The adult carapace is medium to dark brown with some olive coloration present in pale individuals, and a few black spots may be present. Females tend to become slightly paler with age. Hatchlings are tan to pale brown with dark flecks, and a series of black dots occurs along their dorsal and lateral keels. The yellowish to brown plastron is narrow and tapered to a blunt point anteriorly and notched posteriorly. It is slightly concave in both sexes and lacks a hinge. The plastral formula is: fem > an > abd > pect > hum > intergul > gul. The bridge is broad and may be slightly darker than the plastron. The head is narrow and high with a short snout. The orbits are small in diameter, and there is no splenial bone. The triturating surface of the upper jaw is narrow and bears no ridge. The head is uniformly brown or olive dorsally, but yellow to orange ventrally; the male head becomes more brightly colored with age. The iris is ivory white in adults, but silver in juveniles. The neck is covered with large tubercles and rounded warts, and the chin bears one pair of barbels. Other skin is olive gray. The toes are heavily webbed.

Bull and Legler (1980) found the diploid number of chromosomes to be 50, with all macrochromosomes biarmed.

Males have longer, thicker tails with the vent beyond the edge of the carapace. The female's tail is short with the vent beneath the carapace.

Rheodytes leukops

Rheodytes leukops, hatchling

Rheodytes leukops breathing through its cloaca

Distribution: This turtle is known only from the Fitzroy River and its tributaries, Queensland, Australia.

Habitat: *Rheodytes leucops* is a riffle dweller, preferring waters with some current. The bottom may be of sand or gravel, and if submerged logs are present, the turtles often hide about them. Apparently, the Fitzroy turtle seldom comes to the surface to either bask or breathe. It seems to extract enough oxygen from the water by a form of cloacal breathing. The cloacal vent is held continuously open (see the photograph) while the huge cloacal bursae pulsate, drawing water in and then squirting it out the vent after the exchange of gases takes place. This is an excellent adaptation for living in the oxygen-rich waters of riffles.

Natural History: Observations by Legler and Cann (1980) indicate nesting occurs in September and October. Up to five clutches and 46–59 eggs may be laid each season. The small eggs are elongated, ranging from 23 to 33 mm in length and 19 to 24 mm in width. Two nests found by Legler and Cann were on a flat sand and gravel bar approximately 9 m from the water, but less than a meter above the water level. One was a slanting cavity 17 cm deep. The incubation period at 30°C ranged from 44 to 50 days. Small juveniles are quite secretive and spend most of their time hiding; adults are quite placid when caught (Cann, 1978).

Stomach contents examined by Legler and Cann (1980) consisted of insects and a few fragments of freshwater sponges. Food is apparently found by probing the substrate with the snout, which involves both olfactory and visual cues. A female observed by Cann (1978) used her beak to scrape aquatic life from rocks in her aquarium.

Pseudemydura Siebenrock, 1901

The single species of this genus, *Pseudemydura umbrina*, is one of the earth's most endangered turtles; its wild population contains only about 200 individuals.

Pseudemydura umbrina Siebenrock, 1901
Western swamp turtle

Recognition: This small sideneck (to 14 cm) has a flat, rectangular carapace with smooth posterior margi-

nals and a medial depression on the 2d to 4th vertebrals. All vertebrals are broader than long in adults. The 1st is the longest, and the 5th is the shortest and usually flared posteriorly. No neurals are present. The posterior marginals are elevated over the tail, and the surface of the scutes is wrinkled and leathery. The carapace varies from light brown to black. The plastron is very large, covering almost all of the carapacial opening. It has a posterior notch. The intergular scute is very large, separating both the gulars and humerals and also partially separating the pectoral scutes. The plastral formula is: intergul > an > pect > abd > fem > gul > hum. The bridge is broad, about one-third the length of the plastron. Plastron, bridge, and undersides of the marginals are yellow with dark seams. The head is broad and flat with a short, slightly projecting snout and an unnotched upper jaw. Gaffney (1977) lists the following skull characters as unique to *Pseudemydura:* contact between the quadrate and parietal bones, lateral expansion of the supraoccipital and parietal bones, ventrolateral expansion of the postorbital bone, anterior extension of the squamosal, separation of the coronoid and splenial bones by the prearticular, and separation or near separation of the premaxillae into anterior and posterior parts by the medial approximation of the maxilla. The head is covered with rough tuberculate skin and there are two small chin barbels. Dorsally, the neck is covered with numerous large conical tubercles. The forelimbs are covered with large scales, but lack transverse scales on the anterior surface. All toes are webbed. The head, neck, and limbs are brown, and the jaw surfaces are yellowish.

Bull and Legler (1980) found the karotype of *Pseudemydura* to be distinct from those of other chelid turtles with a diploid number of 50 (22 macrochromosomes, 28 microchromosomes) in that macrochromosomes 6 and 10 are acrocentric and pair 10 has a secondary constriction adjacent to the centromere.

Males have longer tails than do females.

Distribution: *Pseudemydura umbrina* is restricted to southwestern Australia where it ranges southward from Bullsbrook about 25 km to the marshy areas in the suburbs of Perth. Two reserves, Ellen Brook (53 ha) and Twin Swamps (142 ha) have been established within this area to protect the species.

Habitat: This rare species lives in swamps and marshes which fill during the winter rains, but dry out completely during the hot arid summers. At that time the turtles dig terrestrial burrows in sandy soil under plant debris and estivate (Spence et al., 1979).

Natural History: Cann (1978) reported that *P. umbrina* nests in October and November and the young hatch in about six months. Three to five elongated (36–44 × 19–22 mm), brittle-shelled eggs constitute a clutch.

Pseudemydura umbrina is carnivorous, eating aquatic insects, crustaceans, and tadpoles.

Hydromedusa Wagler, 1830

Hydromedusa is the genus of semiaquatic South American turtles most closely resembling the long-necked Australian chelids; its neck is longer than its carapacial vertebral column. The carapace is flattened and may bear conical protuberances on the vertebrals and pleurals throughout life. There are six to nine neural bones, and in one species, *H. maximiliani*, the nuchal bone lies behind the anterior peripherals and does not reach the carapacial rim. The 7th or the 7th and 8th pairs of costal bones meet at the midline. The large cervical scute lies behind the anterior marginals and does not extend to the carapacial rim. The plastron is extensive, lacks a hinge, and has a large intergular scute separating the gulars. Its axillary buttresses are strong, but the inguinal buttresses are weak. The skull is long and flattened with the prefrontal bones meeting at the midline so that the separate frontal bones are exposed anterior and posterior to the point of contact. A rather slender temporal arch is formed by the squamosal and supraoccipital bones. The dorsal part of the parietal is absent, and its lateral edges are reduced and tapered posteriorly. The parietal contacts the supraoccipital, but there is no vomer-palatine contact. There is a large bony opening formed by the reduction of the palatines. Only limited contact occurs between the basisphenoid and quadrate bones. The jaws lack triturating ridges. A valvelike flap occurs at the corner of the mouth in *H. maximiliani*. No chin

barbels are present. The toes are fully webbed and each foot bears four claws.

Bull and Legler (1980) reported the diploid chromosomes total 58 (22 macrochromosomes, 36 microchromosomes).

Hydromedusa maximiliani (Mikan, 1820)
Maximilian's snake-necked turtle

Recognition: The oval, dark-brown carapace (to 21 cm) has nearly parallel sides and is dorsoventrally flattened. A low medial keel is present in the form of posterior protuberances on vertebrals 1–4 and as an anterior protuberance on the 5th; the keel may disappear with age. The cervical scute lies behind the first two anterior marginals. Similarly, the nuchal bone lies behind the anterior peripheral bones and does not reach the carapacial border. Vertebrals are broader than long throughout life; the 1st is the largest and is anteriorly flared, the 4th is the shortest, and the 5th is posteriorly expanded. Seven to nine neurals lie beneath the vertebral scutes, and the 8th pair of costal bones may touch at the midline. Peripheral bones total 22–24, and the lateral marginals are slightly upturned. Plastron, narrow bridge, and undersides of the marginals are cream to yellow with some dark seams or brown areas. The plastral forelobe is longer than the hind lobe and is rounded in front; the hind lobe has a shallow posterior notch. The plastral formula is quite variable: intergul > an >< fem > pect >< hum >< gul > abd. The head is moderate in size with a short, slightly protruding snout and an upper jaw lacking a medial notch or hook. Its dorsal surface is covered with smooth skin, but the skin behind the eyes is divided into scales. A prominent valvelike flap of skin is present at each corner of the mouth. The head is brown to olive gray dorsally and cream colored laterally and ventrally; there is a sharp break between these colors just below the level of the tympanum. The tympanum in some contains a dark central spot surrounded by white pigment. The jaws are cream to yellow. The neck is longer than the vertebral column, colored like the head, and has numerous spiny tubercles on its sides. Limbs are olive gray on the outside, but cream to yellow beneath. Forelimbs have large transverse scales on the anterior surface, while similar scales are found at the heel and up the posterior surface of the hind leg. The toes are webbed and there are four claws on each foot. The tail is olive dorsally and ventrally, but yellow laterally.

Males have concave plastra with deeper anal notches, and longer, thicker tails with the vent beyond the carapacial rim.

Key to the species of *Hydromedusa*:

1a. Valvelike flap present at corner of mouth; no black-bordered light stripe on side of face; midseam separating abdominal scutes shortest along plastral midline; conical protuberances absent from pleural scutes: **H. maximiliani**
  b. Valvelike flap not present at corner of mouth; black-bordered light stripe begins on upper jaw and extends backward to neck; midseam separating humeral and pectoral scutes shortest along plastral midline; conical protuberances present on pleural scutes: **H. tectifera**

Distribution: *Hydromedusa maximiliani* is restricted to the states of Espírito Santo, Rio de Janeiro, and São Paulo, Brazil.

Habitat: This turtle lives in slow-moving water bodies.

Natural History: Very little is known of the life style of this turtle. It is carnivorous, eating aquatic insects, amphibians, and possibly fish.

Hydromedusa tectifera Cope, 1870
South American snake-necked turtle

Recognition: The oval, dark-brown carapace (to 30 cm) has nearly parallel sides and is relatively flattened. A medial keel is present in the form of posterior conical protuberances on vertebrals 1–4 and as an anterior conical protuberance on the 5th; these are well developed in juveniles but become lower with age and may disappear entirely in very old turtles. Similar protuberances are found on the pleural scutes, and, although they also become lower with age, evidence of these can be found in even old individuals. The cervical scute lies behind the first two anterior marginals, but the nuchal bone beneath reaches the rim of the carapace by lying partially between the first two anterior peripheral bones. Vertebral scutes are all broader than long in juveniles, but in adults the 1st vertebral is as long as or slightly longer than broad, and the 4th vertebral may be as long as broad. Vertebral 1 is the largest and is anteriorly flared, the 4th is the smallest, and the 5th is posteriorly expanded. Since there are only six neurals, the 7th and 8th pairs of costal bones meet at the midline. There are 22 peripheral bones, and the lateral

marginals are slightly upturned. The undersides of the marginals are brown and the narrow bridge is either brown or yellow with large brown blotches. The plastron varies from immaculate yellow to yellow with brown blotches. Its forelobe is longer than the hind lobe and more or less rounded anteriorly; the hind lobe has a deeper posterior notch than in *H. maximiliani*. The plastral formula is quite variable: intergul > an >< fem > gul >< abd > pect >< hum (for other shell comparisons with *H. maximiliani*, see Wood and Moody, 1976). The head is moderate in size with a short, slightly protruding snout and an upper jaw that is neither notched nor hooked. Its dorsal surface is covered with numerous irregularly shaped scales. No prominent valvelike flaps of skin occur at the corners of the mouth, as in *H. maximiliani*. The head is olive to gray with a broad, black-bordered white or cream-colored stripe extending from the upper jaw backward along the neck. Chin and undersides of the neck are yellow with dark spotting or vermiculations; the jaws are yellow or tan. The neck is longer than the vertebral column and its lateral surfaces contain numerous spiny tubercles. Limbs are olive or gray on the outer surfaces but cream to yellow beneath. The forelimbs have large transverse scales on the anterior surface; similar scales are found at the heel and up the posterior surface of the hind leg. There are four claws on each foot. The tail is olive gray.

Males have concave plastra with deep anal notches and longer, thicker tails with the vent beyond the carapace. Females grow larger.

Distribution: *Hydromedusa tectifera* ranges from the states of São Paulo and Rio Grande do Sul in southeastern Brazil westward and southward through eastern Paraguay and the Chaco and Formosa of northeastern Argentina to Uruguay.

Habitat: *Hydromedusa tectifera* is found in slow-moving, soft-bottomed ponds, marshes, lakes, streams, and rivers with some aquatic vegetation. In coastal regions, brackish waters may be entered. In the colder parts of its range, it may hibernate underwater, buried in the soft bottom.

Natural History: Courtship and nesting have not been described. The eggs are elongated (34 × 20 mm), white, and brittle shelled (Freiberg, 1981). Hatchlings are about 30 mm long and have more rugose carapaces than adults. *Hydromedusa tectifera* is carnivorous, eating snails, aquatic insects, fish, and amphibians; it seems to prefer snails. It has a timid disposition.

Hydromedusa tectifera

Phrynops Wagler, 1830

The genus *Phrynops* contains 11 variable South American freshwater species, the taxonomy of which is still debated. The carapace varies from flat topped, with or without a medial keel, to peaked with a prominent medial keel. A pair of lateral keels may also occur along the pleural scutes. Neural bones vary from none (in some *P. gibbus* and female *P. zuliae*) to three to six, and the nuchal bone extends to the anterior carapacial rim. Where there are no neural bones to separate them, costal bones meet at the midline. The cervical scute is rather small, and when present reaches the carapacial rim. The plastron is large, lacks a hinge, and has the intergular separating the gulars but not entirely separating the humerals. Axillary buttresses are very strong; inguinal buttresses are also strong and ankylosed to the 5th costal bones. The skull is broad and somewhat flattened. Its prefrontal bones do not meet at the midline, and the temporal arch is moderate and formed by the squamosal and parietal bones. The narrow dorsal portion of the parietal covers little of the adductor fossa, and while always present, may be greatly reduced; the parietal does not touch the supraoccipital. The dorsal horizontal portion of the supraoccipital is not expanded. On the roof of the mouth there is vomer-palatine contact, as the palatines are not reduced. There is no contact between the quadrate and basisphenoid. No triturating ridges are present. Chin barbels are present. The toes are fully webbed and each forefoot has five claws.

Some variations in karyotype occur; these are reported in the species accounts.

Phrynops has been divided into three subgenera (Zangerl and Medem, 1958). (1) *Mesoclemmys* Gray, 1873c includes *P. gibbus* and *P. vanderhaegei*, small species with only rudimentary or no neural bones and having moderately sized heads, small chin barbels, and relatively short intergular scutes (shorter than or about the same length as the combined lengths of the interhumeral and interpectoral seams). (2) *Batrachemys* Stejneger, 1909 includes the species *P. nasutus*, *P. dahli*, *P. tuberculatus*, and *P. zuliae*. These are larger species usually with three or four neural bones in males, and a very large broad head, long chin barbels, and an intergular scute longer than, or almost as long as, the combined lengths of the interhumeral and interpectoral seams. (3) *Phrynops* Wagler, 1830 includes the remaining five species, *P. geoffroanus*, *P. hilarii*, *P. hogei*, *P. rufipes* and *P. williamsi*, which generally fall between the other two subgenera in carapace length and head size, have up to six neural bones, very long chin barbels, and an intergular scute as long or almost

as long as the combined lengths of the interhumeral and interpectoral seams. Serious doubts as to the validity of this subgeneric arrangement have been raised by Bour (1973) and Gaffney (1977), and further study into the relationships within this genus is needed.

Key to the species of *Phrynops*:

1a. Head breadth 20% or less of carapace length — 2
 b. Head breadth greater than 20% of carapace length — 3
2a. Jaws with dark bars; keel present on vertebrals 3–5; dark pigment occurs on all plastral scutes: **P. gibbus**
 b. Jaws without dark bars; medial groove usually present on carapace; dark pigment on plastron usually restricted to area from pectorals to femorals: **P. vanderhaegei**
3a. Chin with a dark horseshoe-shaped mark; intergular much shorter than combined lengths of interhumeral and interpectoral seams: **P. williamsi**
 b. Chin lacking a dark horseshoe-shaped mark; intergular almost as long or longer than combined lengths of interhumeral and interpectoral seams — 4
4a. Head with bright-red pigment; carapace medially peaked: **P. rufipes**
 b. Head lacking bright-red pigment; carapace not medially peaked — 5
5a. Plastron with scattered small dark spots: **P. hilarii**
 b. Plastron without scattered small dark spots — 6
6a. Plastron with an extensive red and black pattern: **P. geoffroanus**
 b. Plastron without an extensive red and black pattern — 7
7a. Head relatively narrow: **P. hogei**
 b. Head very large and broad — 8
8a. Cervical scute broader than long: **P. tuberculatus**
 b. Cervical scute small or longer than broad — 9
9a. Posterior lobe of plastron broad (breadth greater than 45% of carapace length): **P. nasutus**
 b. Posterior lobe of plastron narrow (breadth less than 42% of carapace length) — 10
10a. Black stripe extends from snout through eye to rear of head; breadth of posterior plastral lobe less than 35% of carapace length: **P. zuliae**
 b. No black stripe on side of head; breadth of posterior plastral lobe greater than 36% of carapace length: **P. dahli**

Phrynops gibbus (Schweigger, 1812)

Gibba turtle

Recognition: The ellipsoidal, slightly bowed carapace (to 23 cm) has a medial keel, is somewhat serrated, has a shallow supracaudal notch, and is usually broadest at the level of the 8th marginals and highest on the 3d vertebral. Its surface is smooth or slightly roughened. Vertebrals are broader than long and the 3d to 5th may bear a small posterior projection on a low keel. Neural bones vary from none to five, but, if present, they are rudimentary and never contact the nuchal (Pritchard and Trebbau, 1984). The carapace is chestnut brown to dark gray or black. The plastron is well developed and slightly upturned anteriorly. Its intergular scute completely separates the gulars, but not the humerals, and is shorter than its distance from the abdominals. The posterior plastral lobe curves inward and the anals are deeply notched posteriorly. The plastral formula is variable, but usually is: fem > intergul > abd > hum > an > pect > gul. The plastron is red brown to yellow with a brown blotch on each scute, and a narrow yellow border may occur anteriorly and posteriorly. Bridge and undersides of the marginals are brown to yellow. Together, the head and neck are considerably shorter than the carapace. Dorsally and laterally, the head is covered with numerous convex scales; those between the orbit and tympanum are smaller than those on the top and sides of the head. The snout protrudes and the upper jaw is not notched or serrated. Head and neck are red brown to dark gray dorsolaterally and grayish to pale yellow ventrally; the jaw may contain dark spots and the two small chin barbels are yellow. The upper jaw is often yellow to white with black bars. The toes are heavily webbed and there is a fringe of large scales on the outer border of the forelimbs and hind limbs. Limbs and tail are gray black, limb sockets yellow.

Barros et al. (1976) reported the karyotype of *Phrynops gibbus* to be 2n = 60, but listed 4 metacentric, 2

submetacentric, and 16 acrocentric macrochromosomes, and 40 microchromosomes, which adds up to 62; however, Killebrew (1976) found the diploid number to be only 50: 26 macrochromosomes (16 metacentric, 6 submetacentric, and 4 acrocentric) and 24 microchromosomes.

Males have slightly longer, thicker tails with the vent nearer the tip and a deep posterior plastral notch; females have a wide, shallow posterior plastral notch.

Distribution: *Phrynops gibbus* ranges from central and northeastern Peru, eastern Ecuador, and southeastern Colombia northward through eastern Colombia to the Río Negro in southwestern Venezuela. It also occurs east of the Sierra Nevada de Mérida in northeastern Venezuela and on Trinidad, eastward through Guyana, Surinam, and French Guiana to northeastern Brazil; recently, specimens have been taken in Paraguay.

Geographic Variation: No variation has been described, but *Phrynops vanderhaegei* may prove to be a subspecies. It differs in having no dark bars on its jaws, the dark plastral pattern ending on the pectorals and femorals, and a slightly narrower parietal width on the skull.

Habitat: This turtle lives in shallow marshes, pools and ponds, streams, and blackwater rivers usually located under the tree canopy of primary forests.

Natural History: The nesting period extends from July to November. Usually two to four elongated (41–50 × 25–32 mm), white, hard-shelled eggs are laid in a cavity about 10 cm deep. Some females, however, lay their eggs on the surface of the ground in leaf litter or among roots. Incubation may last up to 200 days. The 45–48-mm hatchlings range in carapacial coloration from totally black to cream with black flecks.

Phrynops gibbus is an omnivore, eating plant materials, aquatic insect larvae, and tadpoles. In captivity these turtles accept a wide variety of meats. They are shy and basically nocturnal, although some bask in the daytime. When first captured they emit a foul-smelling musk and often bite.

Phrynops vanderhaegei Bour, 1973

Vanderhaege's turtle

Recognition: *Phrynops vanderhaegei* is closely related to *P. gibbus* and may eventually be shown to be merely a subspecies of *gibbus*. It was first described by Bour

Phrynops gibbus

Phrynops vanderhaegei

(1973) as a subspecies of the broad-headed *P. tuberculatus*, but this designation is certainly in error. *P. vanderhaegei* may be a transitional form between the subgenera *Mesoclemmys* and *Batrachemys*.

The ellipsoidal carapace (to 27 cm) is similar to that of *P. gibbus*, but has a low medial groove, is somewhat serrated with a shallow subcaudal notch, and usually broadest at the 8th marginals and highest on the 3d vertebral. Some rough striations may occur on the scutes. Vertebrals are broader than long. The carapace is brown to gray or black. The plastron covers much of the carapacial opening, is slightly up-turned anteriorly, and is posteriorly notched. The intergular scute completely separates the gulars, but not the humerals, and is slightly shorter than, or about the same length as, its distance from the abdominals. The plastral formula is variable, but the femoral, abdominal and intergular scutes are usually longest. The plastron is yellow with a brown to black pattern which usually extends between the pectorals and femorals. Head and neck are considerably shorter than the carapace. The snout slightly protrudes and the upper jaw is neither notched nor serrated. Head and neck are gray, throat and chin yellow, and the yellowish upper jaws are seldom marked with dark pigment. Some orange vermiculations may occur on the head, and the lower jaw may be red. Other skin is gray to olive, and there is a fringe of large scales on the outer border of the forelimbs.

Males have slightly longer, thicker tails with the vent nearer the tip, and a deep plastral notch. Females have a shallow plastral notch.

Distribution and Habitat: *Phrynops vanderhaegei* is known only from rivers and swamps in Paraguay.

Remarks: Nothing is known of the natural history of this turtle. Its relationship to *Phrynops gibbus* needs further clarification.

Phrynops nasutus (Schweigger, 1814)
Toad-headed side-necked turtle

Recognition: The oval to elliptical (usually broadest behind the center), flattened, brown to olive-gray carapace (to 32.3 cm) has a smooth posterior rim. There is usually no medial groove, but a low vertebral keel may be present on younger individuals. The flared 1st vertebral is the largest and broader than long; the 5th, which is also flared and broader than long, is second largest; vertebrals 2–4 are broader than long in juveniles, but become longer with growth until they may be slightly longer than broad; the 4th is smallest. The cervical scute varies from small to long and narrow. The plastron is well developed and has an anal notch. Its forelobe is rounded anteriorly and is broader than the hind lobe. The bridge is relatively broad. The plastral formula is usually: intergul > fem >< abd >< an >< pect >< hum > gul; the intergular completely separates the gular scutes. Plastral coloration is variable, consisting of a yellow ground color with or without brown pigmented areas. The head is large and very broad with a slightly projecting snout and an upper jaw with only a slight notch. There are two chin barbels. The upper surface of the head is covered with small to large, irregularly shaped scales. The head is distinctly bicolored (brown or gray dorsally, yellow ventrally) or brown to olive gray all over. Dark stripes may be present if the head is not bicolored. The neck is darker dorsally and lighter ventrally, and may bear some small blunt tubercules. Limbs are gray on the front and sides but cream colored beneath.

Killebrew (1976) reported the karyotype as 2n = 50 (26 macrochromosomes, 24 microchromosomes), but both Gorman (1973) and Bull and Legler (1980) found the diploid number to be 58 (22 macrochromosomes, 36 microchromosomes).

Phrynops nasutus nasutus

Males have longer, thicker tails and may have concave plastra.

Distribution: *Phrynops nasutus* ranges from southeastern Colombia, eastern Peru, and northern Bolivia eastward through the Amazon basin, and it also lives in the Caribbean drainages of the Guianas.

Geographic Variation: Two subspecies are presently assigned to *Phrynops nasutus* (Pritchard and Trebbau, 1984); however, the variation within each is poorly known, and further study is needed. Each may prove to be a separate species. *Phrynops nasutus nasutus* (Schweigger, 1814) ranges from the Caribbean drainages of the Guianas southward through most of Amazonian Brazil and Bolivia. The hind lobe of its plastron is broad across the femorals and not strongly tapering toward the rear, tubercles are present on its neck, and its head is distinctly bicolored. *P. n. wermuthi* Mertens, 1969b is found in southern Colombia, northern Ecuador, and northern Peru. It has a broad plastral hind lobe not strongly tapering toward the rear, no tubercles on its neck, and a unicolored grayish head.

Habitat: *Phrynops nasutus* has been taken from a variety of slow-moving aquatic habitats: small forest streams, lakes, marshes, creeks, and muddy ditches (Pritchard and Trebbau, 1984).

Natural History: Medem (1960) reported that six to eight spherical eggs are laid in shallow nests. Hatchlings are 58–60 mm in carapace length. *Phrynops nasutus* is presumably carnivorous, as are other *Phrynops*, but no wild foods have been reported.

Phrynops dahli Zangerl and Medem, 1958

Dahl's toad-headed turtle

Recognition: The olive to brown carapace is oval to elliptical (to 21.5 cm) and widest behind the middle with a slightly serrated posterior rim. It is somewhat flattened dorsally, and the lateral marginals are upturned. A poorly developed vertebral keel may be present in juveniles and some adults. The flared 1st vertebral is the largest and broader than long, as is also the flared 5th. The 2d and 3d vertebrals are also usually broader than long, but the 4th, the smallest of the series, may be slightly longer or as long as broad. The cervical scute is normally longer than broad. The well-developed plastron is notched posteriorly. Its forelobe is longer and broader than the hind lobe; the narrowness of the posterior lobe is particularly notice-

Phrynops dahli

able in males, and its breadth is only 36–38% of the plastral length. The bridge is relatively broad. We have seen few specimens of *Phrynops dahli*, but in those measured the plastral formula was: intergul > fem > abd > pect > an > hum >< gul. The intergular completely separates the gulars. Plastron, bridge, and undersides of the marginals are cream to yellow with gray pigment outlining the seams. The head is large and broad with a slightly projecting snout and slightly notched upper jaw. There are two chin barbels, and the dorsal surface of the head is covered with small to large, irregularly shaped scales. Dorsally, the head is gray to olive brown, but the upper jaw, tympanum, and sides are cream to yellow. Lower jaw, chin, and barbels are yellow. The neck is gray dorsally but lighter ventrally. No horny tubercles are present on the neck. Limbs and tail are gray to olive brown on the outside but lighter beneath.

Males have longer, thicker tails, narrower posterior plastral lobes, and narrower heads. The head of the female is swollen behind the eyes.

The karyotype is 2n = 58: 22 macrochromosomes, 36 microchromosomes (Bull and Legler, 1980). It differs from that of *P. nasutus* in having the 8th chromosome pair biarmed.

Distribution: This turtle is only known from Bolivar State, Colombia.

Habitat: Medem (1966) reported the habitat was originally ponds and small brooks within forests, but with the clearing of the woodlands the entire area around the type locality has been transformed into pastures. There, turtles seem to prefer shallow, quiet water bodies, where they fill a bottom-dwelling niche. Estivation occurs during dry periods.

Natural History: Medem (1966) reported the mating season in Colombia as June and July with nesting occurring mainly in September and October, but it may extend through the year. Apparently several clutches of one to six eggs are laid by each female. The white eggs are ellipsoidal (29–35 × 23–28 mm) with brittle shells, and hatchlings have carapaces about 28–30 mm long. *Phrynops dahli* is predominantly carnivorous, feeding on snails, aquatic insects, other aquatic invertebrates, fish, and amphibians; carrion is also eaten.

Phrynops tuberculatus (Luederwaldt, 1926)
Tuberculate toad-headed turtle

Recognition: The oval, slightly domed carapace (to 25 cm) is medially keeled, broadest behind the center, and has a smooth to slightly serrated posterior rim. The cervical scute is broad. Vertebrals are broader than long; the 1st is expanded anteriorly and the 5th is laterally flared posteriorly. The vertebral keel is discontinuous and a shallow longitudinal groove parallels it on each side; in old individuals these grooves may join posteriorly to form a single medial groove. Posterior marginals may be slightly raised above the tail. The carapace is light to dark brown or black; the surface of each scute may be roughened with raised striations. The plastron is well developed and covers much of the carapacial opening. Both lobes are of about equal breadth; the forelobe is rounded anteriorly, while the anal scutes of the hind lobe are tapered toward the rear and posteriorly notched. The intergular scute is well developed and usually longer than its distance from the abdominals; it completely separates the humerals. The plastral formula is quite variable: intergul >< fem >< an >< abd > gul > pect > hum. On the bridge, the inguinal scute is larger than the axillary, if one is present. The plastron varies in color from uniformly yellow to yellow with dark seam borders to almost totally dark brown. The head is broad with a projecting snout, an unnotched upper jaw, and two chin barbels. Behind the orbits, the sides of the head bear numerous irregularly shaped scales; the skin between the orbits is unbroken. The neck may or may not contain conical tubercles. In color, the head varies from uniformly gray to gray with numerous pink or yellow speckles. Jaws are light brown and there may be light bars on the upper jaw. Neck and limbs are gray to brown, and the neck may be lightly speckled.

Males have longer tails and concave plastra.

Distribution: *Phrynops tuberculatus* occurs in eastern Brazil, mainly in the Rio San Francisco drainage, and in the Ríos Paraná and Paraguay in Paraguay.

Geographic Variation: Unknown; *Phrynops vanderhaegei* was originally described as a subspecies but is now considered a full species.

Habitat: Presumably, *Phrynops tuberculatus* is an inhabitant of small rivers and streams.

Natural History: Nothing is known of its life in the wild, but Bour (1973) reported that captives will eat earthworms, insects, and fish.

Phrynops zuliae Pritchard and Trebbau, 1984
Zulia toad-headed turtle

Recognition: We have not seen this turtle, so the description is adapted from the original.

The oblong, somewhat flattened carapace (to 28 cm) is broadest behind the center in males, but parallel sided in females. There is no medial keel, and females may have a medial longitudinal groove. The flared 1st vertebral is the largest and broader than long; the 5th, which is also flared and broader than long, is second largest; vertebrals 2–4 may be as long as or slightly longer than broad (the 3d is the shortest and the 4th the narrowest). The cervical scute is long and narrow. The carapace is an unpatterned dark gray to black. The yellow plastron is well developed, has an anal notch, and is rounded anteriorly. Its forelobe is broader than the hind lobe, and the bridge is relatively broad. Pritchard and Trebbau (1984) did not give the plastral formula, but that of the female illustrated (Plate 33G) is fem > intergul > abd > pect > hum > an = gul; that of the male (Plate 33H) is fem > intergul > hum > abd > gul > an > pect. The intergular completely separates the gular scutes. The head is large and very wide with a slightly projecting snout and a notched upper jaw. Two chin barbels are pres-

Phrynops tuberculatus

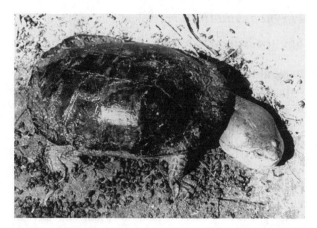

Phrynops zuliae

ent. Dorsally, the head is covered with smooth skin divided into small irregularly shaped scales. It is distinctly bicolored (gray dorsally, cream to white ventrally) with a narrow black stripe extending from the snout through the orbit to over the tympanum. The neck is darker dorsally and lighter ventrally and has soft, wrinkled skin. Limbs are gray.

Males are smaller (to 21 cm), with narrower heads and long, thick tails.

Distribution: *Phrynops zuliae* is confined to an area in the southwestern part of the Maracaibo basin in Venezuela (Pritchard and Trebbau, 1984).

Habitat: Swamps, sometimes in forested areas, are inhabited.

Natural History: Pritchard and Trebbau (1984) reported a female laid seven ($\bar{x}$ = 35.3 × 30.2 mm) brittle-shelled eggs in September. They also reported a newly obtained individual excreted parts of a giant water bug, and that captives readily ate fish.

Phrynops hogei Mertens, 1967
Hoge's side-necked turtle

Recognition: This rare species is known from only a few specimens collected in southeastern Brazil (Mertens, 1967; Rhodin, Mittermeier, and Rocha e Silva, 1982). The domed carapace is elongated (to 34.7 cm) and oval with unserrated posterior marginals. It lacks any keel, medial groove, or flaring of the posterior marginals. In the adult, vertebral 1 is large and much broader than long; vertebral 5 is expanded posteriorly and broader than long; and vertebrals 2–4 are longer

than or at least as long as broad. Carapaces vary in color from light to dark brown. The plastron is well developed with a posterior notch. Its forelobe is broader than the hind lobe, and the bridge is broad. The plastral formula of the holotype is: abd > fem > intergul > hum > gul = an > pect. The plastron is uniformly yellow or has irregular, indistinct gray-brown blotches. The head is very narrow with a slightly projecting snout and an unnotched upper jaw. There are two rather long chin barbels. The head is distinctly bicolored: uniformly dark dorsally and yellow ventrally. Dorsal coloration is gray to dark brown, and females have dark-red pigment along the sides. Jaws are yellow with a small gray blotch on the posterior part of the mandible. The iris is brown with a narrow, irregular yellow border around the pupil. The neck is gray to brown dorsally, yellow ventrally. Limbs and tail are dark on their outer and upper surfaces, but pinkish orange to cream beneath.

Males have larger, thicker tails and concave plastra. The plastron of the female is flat, and her tail is shorter.

Distribution: *Phrynops hogei* seems to be restricted to a small area in the Rio Paraíba drainage of the states of Rio de Janeiro and Minas Gerais, and in the nearby Rio Itapemirim in southern Espírito Santo, Brazil (Rhodin, Mittermeier, and Rocha e Silva, 1982).

Habitat: The waterways occupied by *Phrynops hogei* are in low-lying areas under 500-m elevation.

Comment: *Phrynops hogei* is rare and is listed in the *Red Data Book.*

Phrynops rufipes (Spix, 1824)
Red Amazon side-necked turtle

Plate 3

Recognition: The oval, high-peaked brown carapace (to 25.6 cm) is medially keeled and has a slightly serrated posterior rim. Its cervical scute is long and very narrow. Vertebrals are broader than long; the 1st is slightly flared anteriorly, and the 5th is distinctly expanded posteriorly. Posterior and lateral marginals are outwardly flared. The plastron is large, covering much of the carapacial opening. Its forelobe is broader than the hind lobe, and on the anterior rim the long, broad, intergular scute projects in front of the gulars. The posterior rim bears a distinct notch. The intergular scute completely separates the smaller gulars, but is shorter than its distance from the abdominal scutes.

The plastral formula of the three specimens we examined was: intergul >< fem > an > hum >< pect > abd > gul. Plastron, bridge, and undersides of the marginals are cream to yellow. The head is broad and flattened with a projecting snout and an unnotched upper jaw. Behind the orbits, the sides of the head bear numerous irregularly shaped scales, but the skin on top is smooth and unbroken. Two small chin barbels are present. The head is bright red with a black stripe extending backward from each orbit to the neck and a wide medial black stripe beginning on top of the snout and running backward between the orbits to the nape. This medial stripe may be broken on the snout or between the orbits; also the two lateral stripes may pass through the orbit to join the medial stripe in front of the eye. Outer surfaces of the limbs and neck are red, but there may be extensive dark-brown or black pigment along the outer edge of the forelimbs and on the heel of the hind foot. However, all red pigment may fade with age.

Males have longer and thicker tails and concave plastra; the plastra of females are flat. Females grow larger (Lamar and Medem, 1982).

Distribution: This turtle is restricted to northwestern Brazil and adjacent southeastern Colombia.

Habitat: *Phrynops rufipes* lives in flowing creeks in rainforests. It is known from both blackwater and whitewater streams and spends most of its time on the bottom foraging or hiding in the substrate of leaf debris. It seldom basks.

Natural History: Pritchard (1979) reported it nests during July and August, but Lamar and Medem (1982) stated that oviposition occurs from early June until August and from December to February at two Colombian sites. Three to 12 small, spherical (41–42 × 37–38 mm), brittle-shelled eggs are laid at one time. Hatching may occur in September; the young are light brown with a well-developed vertebral keel and a strongly serrated carapace.

Phrynops rufipes is omnivorous, eating a variety of aquatic animals (crabs and shrimp, amphibians, and fish) and palm fruits and seeds that drop into the water.

P. rufipes does poorly in captivity as it is prone to develop bacterial or fungal infections of the skin and quickly dies.

Serological tests conducted by Frair (1982a) have shown *P. rufipes* to be most closely related to *P. geoffroanus* and *P. hilarii*, which supports its inclusion in the subgenus *Phrynops* Wagler, 1830.

Phrynops hilarii (Duméril and Bibron, 1835)
Spotted-bellied side-necked turtle

Recognition: This large species (to 40 cm) has an oval, flattened, carapace with parallel sides, is broadest near the center, and has a smooth posterior rim. There is a slight medial groove on the depressed dorsal surface, and a weak keel may be present on younger individuals. Vertebral 1–3 and 5 are broader than long in adults, but the 4th is as long as or longer than broad (all are broader than long in juveniles); the 1st is expanded anteriorly and largest, the 5th expanded posteriorly. The cervical scute is usually long and narrow. The carapace is dark brown, olive, or gray, with a yellow border; the scutes are often rugose. The plastron is well developed and has an anal notch. Its forelobe is much broader than the hind lobe, and the bridge is relatively broad. The plastral formula is quite variable: intergul >< fem >< abd >< an > gul > hum > pect; the intergular completely separates the gulars. In life, the undersides of the marginals, bridge, and plastron are yellow with numerous irregularly shaped black spots. The head is large and broad with a projecting snout, two bicolored chin barbels, and an upper jaw lacking both a notch and a hook. Dorsally, the head is covered with irregularly shaped scales, but the frontal scute is prominent. The head is gray to olive above with a pronounced black stripe on each side which begins at the nostril and runs backward through the orbit and over the tympanum to the neck. Below this stripe the head and neck are white to cream. Another black stripe is usually present on each side of the throat, beginning behind the barbel and running backward to sometimes join the upper black stripe. The jaws are yellow, and the neck may contain some blunt tubercles. Limbs are gray to olive on the front and sides but cream colored beneath; soles and palms of the feet may be black.

Males have longer, thicker tails and concave plastra.

Phrynops hilarii

Distribution: *Phrynops hilarii* ranges in Brazil from Buenos Aires southward and westward into Uruguay and Argentina.

Habitat: This turtle is found in oxbows, swamps, lakes, and ponds with soft bottoms and abundant aquatic vegetation.

Natural History: Freiberg (1981) reported that the spherical eggs are about 31 mm in diameter and are laid in nests dug in beaches. Hatchlings we have examined had keeled, oval carapaces of 33–35 mm with very rugose scutes; coloration and patterns were like those of the adult but brighter. Freiberg (1981) reported that wild *Phrynops hilarii* feed on fish and mollusks, but captives will eat meat.

Phrynops geoffroanus geoffroanus

Phrynops geoffroanus (Schweigger, 1814)

Geoffroy's side-necked turtle

Recognition: The oval, flattened carapace (to 35 cm) is broadest behind the center and has a smooth posterior rim (there may be a slight supracaudal notch). A slight medial groove is present on the depressed dorsal surface in adults, and a weak medial keel may be present on younger individuals. The flared 1st vertebral scute is largest and broader than long; the 2d through 4th are broader than long in juveniles, but lengthen with age until they may be longer than broad in adults; the 5th is also broader than long and flared. The cervical scute is usually long and narrow. The carapace is brown to black with gray mottlings and has a yellow border; its scutes are often rugose with raised striations. The plastron is well developed and has an anal notch. Its forelobe is only slightly broader than the hind lobe, and the bridge is relatively broad. The variable plastral formula is: fem >< intergul > an >< abd > pect >< gul >< hum; the intergular completely separates the gulars. In older adults, the undersides of the marginals, bridge, and plastron may be uniformly yellow to light brown; juveniles and young adults have an extensive red and black plastral pattern. The large, broad head has a projecting snout, an upper jaw lacking a notch or hook, and two yellow chin barbels. Dorsally, the head is covered with small, irregularly shaped scales. It is gray to olive dorsally, often with black vermiculations; a broad black stripe runs backward on each side from the nostril through or over the orbit and tympanum to the side of the neck, and a second black stripe runs along the upper jaw to the side of the neck; between these stripes is a yellow or cream-colored band. On the yellow chin and underside of the throat is a series of black streaks or

Phrynops geoffroanus tuberosus

stripes. The jaws are yellow. The neck may contain some blunt tubercles. Limbs are gray to olive on the outside, but have cream-colored areas beneath; soles and palms of the feet may be black.

Males have longer, thicker tails and shallowly concave plastra.

Distribution: *Phrynops geoffroanus* ranges from southwestern Venezuela, southeastern Colombia, eastern Ecuador, and eastern Peru, southward and eastward through southwestern Brazil and northern Bolivia to Paraguay and northeastern Argentina, then northward through eastern Brazil. It also occurs in eastern Venezuela, and probably in adjacent Guyana. Its presence in the central Amazon basin is unknown.

Geographic Variation: Two subspecies are recognized. *Phrynops geoffroanus geoffroanus* (Schweigger, 1814) occurs in central and southern Brazil, Paraguay, northeastern Argentina, and Uruguay. It has only a medial vertebral keel, and its intergular scute is shorter than its distance from the abdominal scute. *P. g. tuberosus* (Peters, 1870) occurs in the Guianas and eastern Brazil southward to Bahia. A population of this subspecies also occurs in Colombia, but Pritchard (1979) feels it may represent a different subspecies. *P. g. tuberosus* has a medial vertebral keel flanked on each side by a lateral keel formed by a series of raised knobs along the dorsal surface of each pleural scute, and its intergular is usually longer than its distance from the abdominal. Lamar and Medem (1982) have noted that those Colombian turtles referred to this race differ considerably from the nominate form. Their status needs clarification. The southernmost populations of *P. geoffroanus* have recently been designated as a separate species, *P. williamsi* (Rhodin and Mittermeier, 1983).

Habitat: This sideneck frequents rivers, lakes, and lagoons with slow current, soft bottoms, and abundant aquatic vegetation.

Natural History: In Colombia, the nests are dug from December to February, and in Venezuela in March and April (Pritchard and Trebbau, 1984). The eggs are almost spherical (34 × 32 m) and are laid in clutches of 10 to 20. Hatchlings have carapace lengths of about 40–45 mm. Their carapaces are rugose with a slight vertebral keel and a slightly serrated posterior rim; the undersides of the marginals, bridge, and plastron are pinkish red with a pattern of irregular black marks.

The food consists mostly of fishes, aquatic insects, and other small aquatic invertebrates; these turtles do well in captivity on fish and chopped beef. In the wild they are often observed basking (Prichard and Trebbau, 1984).

Phrynops williamsi Rhodin and Mittermeier, 1983

Williams' side-necked turtle

Recognition: This newly described species is most closely related to *Phrynops geoffroanus* and *P. hilarii*. Its oval, moderately domed carapace (to 33 cm) is broadest at the center, and the entire marginal rim is weakly serrated. A small supracaudal notch is also present. Vertebral 1 is largest; all vertebrals are broader than long. No keel or medial groove occurs along vertebrals 1–4 in adults, but a slight raised area is found posteriorly on the 5th. The cervical scute is about twice as long as broad. The carapace is brown with radiating, narrow black stripes on all scutes, and a narrow yellow or orange border. The yellow plastron is well developed and has an anal notch. Its fore-

lobe is slightly broader than the hind lobe, and the bridge is relatively wide. The plastral formula is: fem > intergul >< an > abd >< pect >< hum > gul; the intergular completely separates the gulars. Unlike that of other members of the subgenus *Phrynops*, the short, broad intergular is not nearly as long as the combined lengths of the interhumeral and interpectoral seams. The moderately narrow head has a projecting snout, an upper jaw lacking a notch or hook, and two yellow chin barbels. Skin on top of the snout, interorbital region, and middle third of the dorsal surface of the head is incompletely divided into scales. The rest of the dorsal surface of the head is covered with small irregularly shaped scales. The head is black dorsally with several white stripes, and the ventral surface of the chin and neck is reddish yellow or yellow. Three subparallel broad black stripes occur on the head: (1) a dorsal stripe extending from the nostrils through the orbit and upper part of the tympanum onto the neck; (2) a medial stripe passing posteriorly from the angle of the jaws along the lower edge of the tympanum and onto the ventrolateral surface of the neck; and (3) a ventral stripe forming a posteriorly directed horseshoe-shaped mark on the lower surface of the chin. The jaws are dark. The skull differs from other members of the subgenus *Phrynops* in having a wide parietal roof, a short, thick temporal arch, no exoccipital contact above the foramen magnum, widely divergent trochlear processes, broad anterior maxillary triturating surfaces with short ridges, and a shovel-shaped mandible. The short neck may contain a few well-defined blunt tubercles. The black to brown dorsal surface of the limbs has an indistinct vermicular pattern; the undersides are reddish yellow or yellow, and the toe webbing may be reddish.

Rhodin and Mittermeier (1983) reported that sexual dimorphism was not apparent in the type series.

Distribution: *Phrynops williamsi* occurs only in Santa Catarina and Rio Grande do Sul in southeastern Brazil, and in adjacent Uruguay.

Habitat: This turtle apparently prefers rocky forest streams.

Natural History: Rhodin and Mittermeier (1983) reported that a 25-cm female contained nine white, oval, brittle-shelled oviducal eggs (32.9–34.2 × 26.7–27.6 mm), and thought that nesting probably occurs in either November or December. Nothing is known of the feeding habits of *Phrynops williamsi*, but its jaw modifications may be adaptations for bottom feeding and for crushing mollusks.

Platemys Wagler, 1830

Platemys includes five relatively small South American freshwater turtles and is the morphologically closest genus to the Australian short-necked forms. The flat, oval to elliptical carapace bears a shallow to rather deep medial groove; if deep, the groove is bordered by two prominent longitudinal ridges. The cervical scute is small, and the nuchal bone extends to the anterior carapacial rim. Usually no neural bones are present and the costals meet at the midline. The hingeless plastron is large, and the intergular scute entirely separates the gulars but not the humeral scutes. Both the axillary and inguinal buttresses are strong; the inguinal buttresses are ankylosed to the 5th costals. The extensively roofed skull is moderately wide, parallel sided, and somewhat flattened. Its prefrontal bones do not meet at the midline, and the temporal arch is moderate and formed by the squamosal and parietal bones. The dorsal part of the parietal covers the central area of the adductor fossa but not the lateral parts; the parietal does not touch the supraoccipital. The dorsal horizontal part of the supraoccipital is slightly expanded. In the palate there is vomer-palatine contact; the latter are not reduced. There is no quadrate-basisphenoid contact. Triturating ridges are absent. Chin barbels are present. The toes are fully webbed and each forefoot has five claws.

Major biochemical, karyological, and morphological differences exist between *Platemys platycephala* and the other four species (Derr et al., 1987), and this genus probably represents two divergent genera. Iverson

(1986) assigned the species *macrocephala, pallidipectoris, radiolata,* and *spixii* to the genus *Acanthochelys* Gray, 1873c.

Platemys platycephala (Schneider, 1792)

Twist-necked turtle

Plate 3

Recognition: This is a medium-sized (to 18 cm), side-necked turtle with a flattened, elliptical carapace bearing a pronounced medial groove between the posterior part of the 1st vertebral and the anterior part of the 5th. In juveniles, all vertebrals are broader than long, but in adults the 3d is usually longer than broad; the 4th and 5th are smallest. Posterior marginals are flared (slightly serrated in juveniles); those lateral are upturned. The surface of the scutes of turtles < 90 mm is usually roughened with numerous rounded rugosities. The carapace is highest just anterior to the 2d intervertebral seam and broadest at the level of the 7th or 8th marginals. It is yellow with dark-brown or black pigment covering varying amounts of the surface (see the section on geographic variation for a more concise description of the two distinct patterns). The plastron is dark brown or black with a yellow border; the bridge is yellow with a dark transverse bar. The forelobe is slightly upturned and longer and slightly broader than the hind lobe, which contains a wide posterior notch. The intergular scute is about half as long as the length of the plastral forelobe. The plastral formula is usually: intergul > abd > fem > an > gul > hum >< pect. The head is orange to yellow brown dorsally, but dark brown to black laterally and ventrally; the light dorsal pigment extends downward on the sides to the midpoint of the orbit and tympanum. The unnotched jaws are dark brown. Dorsally, the head is covered with smooth, undivided

Key to the species of *Platemys*:

1a. Dorsal surface of head covered with smooth skin; carapace with a deep medial groove: **P. platycephala**
 b. Dorsal surface of head covered with scales; carapace with a shallow medial groove   2
2a. Carapace yellowish brown to olive; thigh has a series of large tubercles with at least one larger than the rest: **P. pallidipectoris**
 b. Carapace dark gray to black or blackish brown; thigh has moderate tubercles, none greatly enlarged   3
3a. Dorsal surface of neck has numerous long pointed tubercles: **P. spixii**
 b. Dorsal surface of neck has numerous low rounded tubercles   4
4a. Carapace length usually greater than 5.0 times the tympanic head width: **P. radiolata**
 b. Carapace length usually less than 4.9 times the tympanic head width: **P. macrocephala**

Platemys platycephala melanonota

skin; one to three rows of large scales occur laterally. The snout is short and only slightly protruding. Two small brown chin barbels are present, and the iris is brown. The neck is colored similarly to the head, and its dorsolateral surface contains numerous blunt tubercles. Anterior surfaces of the black limbs are covered with large scales, and small blunt tubercles are present on the thighs. The black tail is short.

The normal 2n karyotype is probably 64 (Barros et al., 1976); however, Kiester and Childress (in Gorman, 1973) reported it to be 68, and Bull and Legler (1980) discovered that two specimens they examined had an unusual triploid karyotype of 96.

Males are slightly larger, have concave plastra and longer tails with the vent extending beyond the carapacial rim.

Distribution: *Platemys platycephala* is restricted to northern South America where it occurs in the Caribbean drainages of Venezula and the Guianas and the Amazon drainages from northeastern Bolivia, eastern Ecuador and Peru and southeastern Colombia eastward to the vicinity of Belem, Brazil.

Geographic Variation: Two subspecies are recognized. *Platemys platycephala platycephala* (Schneider, 1792) (Plate 3) occurs in the Caribbean drainages of Venezuela and the Guianas and the Amazon drainages as far upstream as the Ríos Purus, Juruá, Yavari, and Putumayo. The dark pigment on the yellow carapace is restricted to the border of the seam separating the vertebrals and pleurals and to an incomplete band extending on each side of the medial groove downward through the 2d and 3d pleurals to the lateral carapacial rim (the medial groove remains yellow); the dark bar crosses less than 80% of the bridge. *Platemys p. melanonota* Ernst, 1984 is restricted to the upper Amazon drainages of the Ríos Santiago and Cenepa in Peru and the Ríos Napo and Curaray in Ecuador. It intergrades with *P. p. platycephala* in the Río Mamoré drainage, Boliva, and the Ríos Madre de Dios, Purus, and Ucayali of Peru. It is a dark subspecies with the yellow pigment on the brown carapace restricted to the vertebral groove, in some on the extreme anterior part of the 1st pleurals, and on the posterior part of the 4th pleurals; the dark bar crosses more than 90% (usually 100%) of the bridge.

Habitat: *Platemys platycephala*, a poor swimmer, is an inhabitant of shallow rainforest streams, pools, and marshes; it frequently wanders about the forest floor, but does not enter large rivers.

Natural History: Courtship and mating take place primarily during the rainy season (late March to early December) in or out of water (Medem, 1983a). During courtship, the male pursues the female and mounts from behind. He bends his head over hers, touches his barbels to the top of her head, and swings his head back and forth from side to side, occasionally expelling a stream of water from his nostrils over her face (Harding, 1983). Oviposition occurs early in the dry season (primarily from August to February). The female does not dig a nest cavity but instead makes a shallow groove or lays her egg directly on the ground. The egg is always laid under rotten leaves and may then be partially covered by sand or dirt (Medem, 1983a). Usually only one oblong to elliptically tapered, brittle-shelled egg (51–61 × 26–29 mm) is laid at one time (Medem, 1983a). Hatchlings are 43–57 mm in carapace length and have numerous small rounded rugosities on each carapacial scute; coloration is like that of adults.

Food preferences of captives include worms, snails, slugs, insects, amphibians, fish, and some vegetation.

Platemys pallidipectoris Freiberg, 1945
Chaco side-necked turtle

Recognition: This species is a medium-sized (17.5 cm) Argentine sideneck with several enlarged pointed tubercles on its upper thighs. Its elliptical adult carapace appears flattened because of a shallow dorsal groove running between the posterior part of the 1st vertebral and the anterior part of the 5th vertebral. Vertebrals 1–3 are broader than long, the 4th is narrower than long in adults (broader than long in juveniles), and the 5th is equal in length and breadth or only slightly longer than broad. Marginals are not serrated, and the anterior- and posteriormost are broadest while those lateral are narrow and upturned. Posterior marginals are depressed over the thighs. The carapace is highest just behind the center and broadest at the level of the 8th marginals. It varies from yellow brown to gray brown or olive, often with dark seam borders. Plastron and bridge are yellow with a broad, dark seam-following pattern. In some, the dark pigment may become so extensive that the yellow is restricted to the outer edges of the scutes. Mertens (1954b) speculated that this dark pigmentation fades with age. The intergular scute is approximately half as long as the length of the plastral forelobe. The plastral formula is: intergul > fem > abd >< an > gul > hum > pect. The plastral forelobe is broader than the hind lobe, which contains a wide posterior notch. The head has a wide, yellowish medial stripe bordered on each side by a gray-brown lateral stripe. Its tympanum is yellow, and its dorsal surface is covered with large scales. The iris is white. The neck is gray

Platemys pallidipectoris

brown dorsally, blending into grayish yellow ventrally. Its dorsal and lateral surfaces contains large conical tubercles. The yellow legs are covered with large scales and the toes are webbed. On the inside of each thigh, near the tail, is a series of enlarged conical tubercles, at least one of which is larger than the rest.

Males have concave plastra, longer, thicker tails, and better developed thigh spines than do females.

Distribution: *Platemys pallidipectoris* is known from the central Chaco region of Argentina, and probably also occurs in southern Paraguay and Bolivia.

Habitat: This species usually frequents rivers, ponds, lagoons, and other slow-moving, shallow water bodies, but may venture onto land for short periods.

Natural History: *Platemys pallidipectoris* is said to lay its eggs on mud and sand banks and to be very wary (Freiberg, 1945). Captives have fed on a variety of meats and fishes.

Platemys radiolata (Mikan, 1820)
Brazilian radiated swamp turtle

Recognition: *Platemys radiolata* is a medium-sized (to 20 cm) South American sideneck with radiating striations on the carapace. Its flattened elliptical carapace has a very shallow dorsal groove between the 2d and 4th vertebrals. In adults, the 1st and 5th vertebrals are much broader than long; the 1st is broadest of all, the 2d and 3d are usually only slightly broader than long, and the 4th may be longer than broad. Anterior- and posteriormost marginals are broadest; those lateral are narrowest and slightly upturned. Posterior marginals are slightly flared and serrated; there is a slight

posterior notch. Both vertebrals and pleurals contain numerous striations radiating anteriorly from the posterior margin of each scute. The carapace is highest at the level of the seam separating vertebrals 1 and 2, and broadest at the level of marginals 6–8. The carapace is uniformly dark olive, gray, or black. Plastron and bridge are yellow with either a large dark blotch or dark mottlings on each scute, or with dark seam borders. The slightly upturned forelobe is broader than the hind lobe, which contains a posterior notch. The length of the intergular scute is slightly less than half (48–49%) that of the plastral forelobe. The plastral formula is: intergul > fem >< abd > hum > gul > an > pect. The head is olive to grayish brown dorsally and yellow laterally and ventrally. The unnotched jaws are yellow to tan and may contain some dark mottling. Dorsally, the head is covered with numerous irregularly shaped scales. The snout is short and only slightly projecting. Two small yellow barbels are present on the chin, and the iris is white. The neck is olive to brown dorsally and yellow ventrally. Its dorsal surface contains numerous very short, rounded tubercles. Tubercles on the sides are fewer and smaller. Anterior surfaces of the limbs are covered with large scales. The thighs have a few scattered, small, pointed tubercles. Outer limb surfaces are olive or brown and the inner surfaces yellow. The olive to brown tail is short.

Males have concave plastra and longer, thicker tails with the vent beyond the carapacial margin; females have flat plastra and short tails with the vent under the carapace.

Distribution: *Platemys radiolata* occurs entirely within Brazil, ranging from the states of Bahia, Minas Gerais, and Mato Grosso southward to the vicinity of the city of São Paulo.

Geographic Variation: None has been reported. However, some authors (Pritchard, 1979) consider *Platemys spixii* a subspecies of *P. radiolata*. There are fundamental differences between these forms (see recognition sections of each), and it is best to consider them separate species.

Habitat: *Platemys radiolata* lives in slow-moving waters with soft bottoms and abundant aquatic vegetation.

Natural History: This is a rather shy turtle in captivity, often hiding much of the time. It seems to be exclusively carnivorous, accepting a variety of animal foods such as fish, amphibians, aquatic insects, snails, and worms. Sometimes it climbs onto the land to bask or roam about.

Ewert (1979) reported that two hatchlings had a

mean carapace length of 41.9 mm, but we have seen a 30.5-mm hatchling. Hatchlings have grayish-brown carapaces with a wedge-shaped yellow mark on each marginal. The plastron is yellow with a large, dark central blotch which extends outward along the seams. The yellow bridge contains two dark spots. The underside of the throat and neck is yellow to cream with large dark blotches. Due to their relatively large heads, *Platemys radiolata* hatchlings may be confused with *Phrynops* hatchlings, but on each hind leg there is a pretibial flap of large scales that is never found in *Phrynops*.

Platemys macrocephala Rhodin, Mittermeier, and McMorris, 1984

Large-headed Pantanal swamp turtle

Recognition: This is the largest of the South American *Platemys*, growing to 23.5 cm in carapace length. It has a broad, oval to moderately elongated, deep carapace with a shallow dorsal groove extending along the 2d to 4th vertebrals. The 1st and 5th vertebral scutes are very broad, the 2d through 4th may be slightly longer than broad, and the 5th is laterally expanded. Marginals are unserrated; 1, 2, and 8–10 are slightly expanded but not flared, and 3–7 are often slightly upturned. The carapace is highest just behind the center and broadest at the level of the anterior part of the 8th marginals. The carapace is dark to blackish brown, but may be light brown in some. Juveniles often have lighter brown radiations on their carapacial scutes. The broad plastron and bridge are yellow with some dark pigment extending along the seams (sometimes covering most of a scute, but usually not the areola); this pigment fades with age. The forelobe is broader than the hind lobe, which contains a deep posterior notch. The intergular scute is approximately half as long as the length of the forelobe. The

plastral formula is: intergul > fem > abd > hum > an > gul > pect. The head is extremely broad; the carapace length averages only 4.4 times the tympanic head width, and older females may have massive heads. It is dark grayish brown above, yellow or cream below; the area of demarcation is indistinct. The tympanum and posterior part of the lower jaw are yellow with a few gray blotches and orange spots. Jaws are grayish yellow; the iris is brown tan. Dorsally, the head is covered with large distinct scales. There are two chin barbels. The neck is grayish brown dorsally, yellow ventrally, and has a few scattered blunt, conical tubercles on the dorsal surface. Limbs are gray on the outside, yellow beneath, and covered with large scales. Large conical tubercles are present on the inside of each thigh.

The karyotype is 2n = 48 (Rhodin, Mittermeier, and McMorris, 1984).

Females are larger and more domed; males have slightly concave plastra and longer, thicker tails.

Distribution: *Platemys macrocephala* is known only from the upper Río Mamoré drainage of central Bolivia and the Pantanal region and other swamplands of the upper Rio Paraguay drainage in southwestern Mato Grosso, Brazil, but may also occur in other parts of Bolivia and in northeastern Paraguay.

Geographic Variation: Unknown. This species is most closely related to *Platemys radiolata*.

Habitat: *Platemys macrocephala* inhabits marshes, swamps, and slow-flowing streams.

Natural History: Practically nothing is known of this newly described turtle. The holotype contained large, nearly round, hard-shelled eggs when collected in April (Rhodin, Mittermeier, and McMorris, 1984). Snails form a large part of their diet.

Platemys macrocephala

Platemys spixii Duméril and Bibron, 1835

Black spiny-necked swamp turtle

Recognition: *Platemys spixii* is a medium-sized (to 17 cm) side-necked turtle with several elongated pointed tubercles on its neck. Its flattened elliptical carapace has a shallow dorsal groove between the posterior part of the 1st vertebral and the anterior part of the 5th vertebral. In adults, the 1st and 5th vertebrals are much broader than long, with the 1st broadest of all; the 2d to 4th vertebrals are about as broad as long or longer than broad. Anterior- and posteriormost marginals are broadest, those lateral narrowest. Posterior

Platemys spixii

marginals are unserrated, or only slightly serrated. Carapacial scutes often contain concentric and radiating striations. The carapace is highest just anterior of the 2d intervertebral seam, and broadest at the level of the 7th or 8th marginals. The adult carapace is dark gray to black. Occasionally some yellow occurs at the base of the pleurals. In adults, the plastron and bridge are usually uniformly dark gray or black, but some individuals have small yellow spots along the plastral midseam. The forelobe is broader than the hind lobe and slightly upturned. The hind lobe contains a wide posterior notch. The intergular scute is over half (55–60%) as long as the length of the plastral forelobe. The plastral formula is: intergul > fem > abd > an > gul > hum > pect. The head is olive to gray and the unnotched jaws are yellow to tan. Dorsally, the head is covered with numerous, variably shaped scales which become arranged into three or four lateral rows above the tympanum. The snout is short and only slightly projecting. Two small gray barbels are present on the chin. The iris is white. On the dorsal surface of the neck are numerous long pointed tubercles, which become fewer and shorter on the sides. The toes are webbed, and the anterior surfaces of the limbs are covered with large scales. Several rows of spiny tubercles occur on the thighs. The tail is relatively short, and all soft parts of the body are colored olive to gray.

Males have concave plastra and longer, thicker tails with the vent beyond the carapacial rim; females have flat plastra and shorter tails with the vent beneath the carapace. Females have the anal scutes upturned.

Distribution: *Platemys spixii* ranges from the vicinity of São Paulo southward through Rio Grande do Sul, Brazil, to the provinces of Rocha and Tacuarembó in Uruguay, and westward to the territories of Formosa and Chaco in Argentina.

Geographic Variation: No subspecies have been described. Pritchard (1979) considers *Platemys spixii* to

be a subspecies of *P. radiolata*, but there are several differences, and they should be considered separate species.

Habitat: This turtle lives in slow-moving waters with soft bottoms and abundant aquatic vegetation. Freiberg (1967) collected *Platemys spixii* on land in vegetation that had been flooded after intense rains.

Natural History: Very little life history is known. A clutch of seven spherical eggs, 25 mm in diameter, was shown by Freiberg (1981). Juveniles do not have the medial vertebral groove found in adults, but instead may actually have a low medial keel. Also, the scutes of both their carapace and plastron contain numerous radial and concentric striations. In the wild, *Platemys spixii* feeds on snails and larval amphibians, but accepts fish and meat in captivity. Captives are often extremely shy, keeping their head, neck, and limbs tucked under the shell for long periods. When in this position they resemble small pincushions owing to the elongated spiny tubercles on their neck and thighs, and possibly these evolved as an adaptation to discourage predators.

Chelus Duméril, 1806

The sole species in this genus, *Chelus fimbriatus*, the matamata, is one of the strangest creatures on earth; its large flattened head with long narrow snout and cutaneous fringes and the three knobby keels on its algae-covered carapace effectively conceal it as it lurks in or prowls along the bottom of some blackwater South American stream.

Chelus fimbriatus (Schneider, 1783)

Matamata

Plate 3

Recognition: The brown or black carapace is rather oblong (to 44.9 cm), with a straight (rather than rounded) anterior rim, almost parallel sides, and a heavily serrated, rounded posterior rim. Lateral marginals are somewhat serrated and downturned. A small cervical scute is present. Vertebrals 1–4 are broader than long in juveniles, and the 5th is longer than broad; however, with age the 2d to 4th lengthen to become about as long as broad or slightly longer than broad. Vertebral 1 is anteriorly expanded; the

Chelus fimbriatus, juvenile

5th is expanded in the rear. Beneath the vertebrals are a series of seven (sometimes eight) neurals, and the 8th pair of costals meet medially behind the neurals. Three knobby longitudinal keels are present; one extending along each row of pleural scutes and a medial keel along the vertebrals. The highest knobs on these are those most posterior. All carapacial scutes are rugose with growth annuli, providing good places of attachment for algal filaments. The hingeless, narrow plastron is somewhat truncated anteriorly and deeply notched posteriorly. Its bridge is relatively narrow, giving the plastron a crosslike appearance; however, both the axillary and inguinal buttresses are well developed. The intergular scute does not always separate the gulars, which often meet behind it. If the intergular is elongated and separates the shortened gulars, the plastral formula is: fem > pect > abd > hum > an >< gul > intergul, or intergul > hum > an > gul. The axillary and inguinal scutes are indistinct. Plastron and bridge are cream to yellow or brown. The triangular-shaped head is large and extremely flattened with numerous tubercles and cutaneous flaps, a long, tubular snout, and an upper jaw lacking either a medial notch or hook. Nasal bones are absent, so prefrontals form much of the rims of the external nares. The prefrontals do not meet at the midline. The temporal arch is moderate and formed by the squamosal and parietal bones; there is no supraoccipital or quadrate-parietal contact. Medial parts of the jugal and postorbital are entirely on the surface of the skull, and face more laterally than posteriorly. The tympanic cavity is extended laterally. The reduced jaws are weak and ridgeless, but there is a slight ridge on the palatine. The anterior pterygoid process prevents the palatine from reaching the vomer. Two conical chin barbels are present, and two other filamentous barbels occur at the angles of the jaws. Head, neck, limbs, and tail of adults are grayish brown. The neck is longer than the vertebral column,

and there is a lateral fringe of small cutaneous flaps on each side. Each forefoot has five claws and all toes are webbed.

The karyotype is 2n = 50 (22 macrochromosomes, 28 microchromosomes). Pairs 1–3 are biarmed and there are 3 acrocentric and 5 biarmed chromosomes among pairs 4–11 (Barros et al., 1976; Bull and Legler, 1980).

Males have concave plastra and longer, thicker tails than do females.

Distribution: *Chelus fimbriatus* occurs in most major drainages in northern Bolivia, eastern Peru, Ecuador, eastern Colombia, Venezuela, the Guianas, and northern and central Brazil. It has also been found on Trinidad, apparently washed there by mainland floods.

Geographic Variation: There is color variation between populations, and Schmidt (1966) thought that various morphological characters varied enough over the range to designate separate subspecies, but has not published adequate descriptions nor given names to these forms. A thorough study of geographic variation is needed. Pritchard and Trebbau (1984) further discuss geographic variation.

Habitat: *Chelus fimbriatus* prefers slow-moving, blackwater streams, oxbows, muddy lakes, stagnant pools, marshes, and swamps where it remains at shallow depths which enable it to push its snout just above the surface for a breath of air. It is able to hold its breath for long periods, allowing it to remain motionless on the bottom, mimicking an algal-covered rock. When it does move, *Chelus fimbriatus* creeps slowly about the bottom rather than actively swimming. Juveniles can swim poorly, but possibly adults are too ponderous to swim. It apparently seldom, if ever, basks.

Natural History: Carpenter and Ferguson (1977) reported the male courtship act consists of a rapid forward lunging of the head toward the female with a gaping and opening and closing of the large mouth, a moving of the lateral cutaneous flaps on the head, and a hyperextension of the legs from the shell. Nesting occurs from October through December in the Upper Amazon, and 12 to 28 eggs may be deposited in one clutch. Eggs are almost spherical, about 35 mm (34–37.5) in diameter, with brittle shells. Hatchlings are more colorful than adults; their carapace is light brown with a black line along the medial keel, an orange to pink blotch at the outer edge of each marginal scute, and a dark-brown to black spot on each pleural at the posterior seam. The surface of each pleural and vertebral is quite wrinkled. The plastron, bridge, and undersides of the marginals are pink with dark seams. The head is tan with three narrow, dark, dorsal stripes extending onto the neck, the medial being the shortest. Three wide pink to red stripes extend from the neck along the venter almost to the chin. The barbels at the corner of the mouth are brown, while those on the chin are pink. The limbs are tan on the outside and pink with small black flecks on the inner surface. The limb sockets are tan with some dark flecking, and a large dark blotch may occur between the neck and foreleg. The tail is tan dorsally and dark brown ventrally.

Chelus fimbriatus is carnivorous, feeding on aquatic invertebrates and fish. Its bizarre structure and the ease with which algae attach to the shell allow it to occupy the role of a wait-and-ambush predator. The loose flaps of skin on its head and neck sway like tufts of aquatic vegetation, attracting small animals within reach; then the head and neck are thrust forward accompanied by a rapid distension of the throat and a gaping of the huge mouth. The prey is sucked into the mouth by the low pressure created. Excess water is then expelled, but the prey swallowed. The jaws are weak and the mandibles are not attached medially, so the turtle has difficulty chewing prey. The hyoid apparatus and well-developed neck musculature play an active role in this gape-and-suck method of feeding. However, this is not the only method employed to gather food. Holmstrom (1978) observed adult *C. fimbriatus* slowly herd fish into shallow water, trap and eat them. This may be an important alternate feeding strategy, especially for juveniles.

Kinosternidae

Kinosternidae is a New World family of small to medium-sized semiaquatic mud or musk turtles ranging from Canada to South America. These turtles apparently evolved in the Americas, as their fossils have been found only there. Their center of diversification is Mexico, but the oldest known fossil kinosternid is *Xenochelys formosa* from Chadronian Oligocene deposits in South Dakota. This species is closely related to the Recent genus *Staurotypus*, but also shares some characters with the Dermatemydidae. Of the modern genera, the oldest fossil is a *Kinosternon flavescens arizonense* from Pliocene (Blancan) deposits in Cochise County, Arizona (Gilmore, 1922), and *K. baurii* and *K. subrubrum* are known from Pleistocene (Wisconsin) remains. Today there are 3 genera and 22 species recognized.

The elongated skull of this group has the temporal region only moderately emarginated. Frontal bones are reduced and do not enter the orbits. There is maxilla-quadratojugal contact, and the triturating surface lacks a ridge. The squamosal is separated from the postorbital, and the quadrate bone does not enclose the stapes. A secondary palate is present. The carapace may be flattened or slightly domed with one to three longitudinal keels. The nuchal bone has riblike lateral processes which extend below the marginals. There are 10 pairs of peripheral bones and 23 marginal scutes (including the cervical). Neural bones range from five to seven. The 10th dorsal vertebra lacks ribs, and there is only one biconvex cervical vertebra; caudal vertebrae are procoelous. The plastron is variable in shape: small and cross shaped in some, large and well developed in others. One or two hinges may be present on the plastron, and the number of plastral scutes varies from 7 or 8 to 10 or 11. An entoplastron is not present in all species. Limbs are developed for bottom crawling, but always have some toe webbing; the phalanges have condyles. Musk glands, associated with the bridge, exude malodorous secretions when the turtle is disturbed. The family is divided into two subfamilies.

Subfamily Staurotypinae (staurotypines; 2 genera, 3 species). The genera, *Claudius* and *Staurotypus*, range

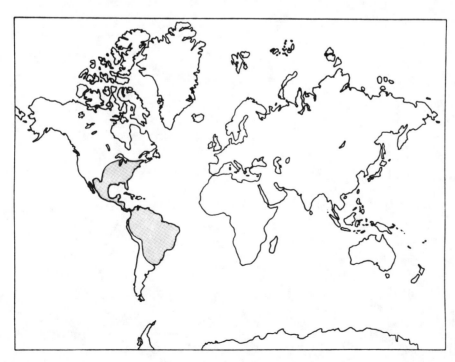

Distribution of Kinosternidae

from central Mexico into northern Central America. These are rather large musk turtles with reduced, cross-shaped plastra, having only seven or eight scutes. An entoplastron is present. The carapace is three keeled and contains seven neural bones. On the pelvic girdle, the pubic and ischiadic symphyses are widely separated. Bickham and Carr (1983) would elevate this subfamily to family status, Staurotypidae, on the basis of chromosomal differences from the Kinosterninae.

Key to the genera of Staurotypinae:

1a. Large axillary and inguinal scutes are present on bridge, which is connected to carapace by an osseous suture; plastron has a movable hinge: **Staurotypus**
 b. Axillary and inguinal scutes are usually absent, but if present small and not well developed; bridge connected to carapace by ligaments; plastron lacks a movable hinge: **Claudius**

Subfamily Kinosterninae (mud and musk turtles; 1 genus, 19 species). The single genus, *Kinosternon*, ranges from southern Canada to South America. These are small to medium-sized turtles, usually with well-developed, large plastra having 10 or 11 scutes, and no entoplastron. The carapace may lack a keel, or have one or three; five to seven neural bones are present. The pubic and ischiadic symphyses are in contact.

Electrophoretic data gathered by Seidel, Iverson, and Adkins (1986) indicate that Kinosterninae is monophyletic and divergent from Staurotypinae.

Morphological studies by Lamb (1983) and those on the karyotype by Sites et al. (1979) and the biochemistry by Frair (1972) and Seidel, Iverson, and Adkins (1986) indicate that *Kinosternon baurii* and *K. subrubrum* of the United States are more closely related to the sympatric species of *Sternotherus* (sensu Zug, 1986) than to the other more tropical species of *Kinosternon*. It appears that the genus *Kinosternon* is a paraphyletic group, and that *K. baurii* and *K. subrubrum* share a more immediate ancestor with the four species of *Sternotherus*. Separation of *Sternotherus* from *Kinosternon* on the basis of shell hinges is dubious, in view of the studies of Bramble, Hutchison, and Legler (1984) showing that *K. herrerai* exhibits no more structural evidence of posterior plastral lobe movement than does *Sternotherus*. Consequently Seidel, Iverson, and Adkins (1986) have placed *Sternotherus* in the synonymy of *Kinosternon*.

Claudius Cope 1865

Claudius is the only monotypic genus of the kinosternid turtles. Its sole species is *C. angustatus*.

Claudius angustatus Cope, 1865
Narrow-bridged musk turtle

Plate 3

Recognition: The oval carapace (to 16.5 cm) usually bears three longitudinal keels which become obscured with age. Posterior marginals are unserrated and unflared; the 10th, and sometimes also the 11th, is elevated above the preceding marginals. Vertebrals 1–4 are broader than long, but the 5th is longer than broad or about equal in these dimensions. The 1st vertebral is widely flared anteriorly and touches the first two pairs of marginals. There are eight neurals. The posterior pair of costals meet at the dorsal midline, but are separated from the posteriormost peripheral by a large suprapygal. Carapacial scutes may be roughened due to growth annuli and radiations. The carapace is dark brown or yellowish brown with dark seams. Juveniles and young adults may also have a pattern of dark radiations. The hingeless plastron is very small and cruciform. The bridge is extremely narrow (only about 5% of the plastron length; Smith and Smith, 1979), and the rest of the plastron is connected to the carapace by a ligament. Axillary and inguinal scutes may be absent. Gular and humeral scutes are absent; the plastral formula is: pect > fem > an > abd. Both the anterior and posterior lobes are triangular in shape. In the bony plastron, the hyoplastra are fused with the hypoplastra, and an entoplastron is present. Iverson and Berry (1980) pointed out that growth of the abdominal scute is directed anteriorly instead of posteriorly as in *Kinosternon*. Both plastron and bridge are yellow; the juvenile plastron has a dark median blotch extending outward along the seams. The head is large with a slightly projecting snout and a sharply hooked upper jaw. In addition, a pair of cusps are present on the upper jaw just below the anterior margin of the orbit. The lower jaw also has a very long medial hook. The round to oval rostral shield is relatively small but does extend posteriorly to the orbit. Only one pair of barbels is present on the chin. In the skull, the squamosal does not touch the jugal, the frontal bone does not contribute to the orbital rim, and the pterygoid barely touches the maxilla. Small processes from the prootic and pterygoid bones divide the foramen of the trigeminal nerve into two distinct holes. Only a slender bar of bone separates the orbit from the emarginated temporal fossa. The head is yellowish brown to gray with dark mottlings, and the yellowish jaws contain dark streaks. The neck has several rows of tubercles and is gray with dark mottlings. The limbs are gray brown with webbed toes. Three large transverse scales occur on the anterior surface of the forelegs.

Moon (1974) reported the diploid chromosome number as 54.

Males are larger (to 16.5 cm) than females (to 15 cm) and have long, thick tails tipped with a horny spine and patches of roughened scales (vinculae) on their thighs and crura.

Distribution: *Claudius* ranges, at low elevations, from central Veracruz and northern Oaxaca southward through northern Guatemala to Belize, but excluding the Yucatán Peninsula.

Habitat: *Claudius angustatus* lives in shallow water bodies with soft bottoms, such as marshes, ponds, and small streams. Pritchard (1979) thought that its rather thin shell and exposed soft parts caused *Claudius* to select shallow waters to avoid predation from crocodiles. At times it wanders on land, and has been found estivating during the dry season in underground burrows (Duellman, 1963; Pritchard, 1979).

Natural History: Nesting seems to occur at the onset of the dry season in November (Pritchard, 1979), and several clutches may be deposited. A typical clutch is two to eight, hard-shelled, elongated (32 × 18 mm; Ewert, 1979) eggs. Incubation is rather long, 115 to 150 days. Hatchlings are about 35 mm in carapace length (Ewert, 1979) with three well-developed keels. Their carapaces are dark brown or black, and their plastra are yellow.

Claudius, like other kinosternids, is carnivorous, eating worms, snails, aquatic insects, amphibians, and fish. These turtles have a nasty temperament and bite viciously when first caught. After being in captivity for some time, they often mellow and allow themselves to be handled without snapping.

Staurotypus Wagler, 1830

Included in the genus *Staurotypus* are the large (to 37.9 cm) aquatic musk turtles of Mexico and northern Central America. Their oval carapaces are strongly three keeled and lack serrated marginals. There are 23 marginals, including the cervical scute. Seven neurals and ten pairs of peripheral bones are also present. The plastron is slightly hinged, and small and cruciform with a narrow, triangular posterior lobe. The anterior lobe is rounded. A large entoplastron is present. Only seven or eight scutes cover the plastron, and large axillary and inguinal scutes occur on the bridge. The head is large, with a projecting beak and only slightly hooked upper jaw. In the skull, the temporal region is moderately emarginated. The maxilla and quadratojugal touch, but the squamosal is separated from the postorbital. The quadrate does not enclose the stapes. There is a secondary palate, but no ridge is present on the triturating surface of the jaws.

Staurotypus is unique among kinosternids in that males of its two species, *S. salvinii* and *S. triporcatus*, possess sex chromosomes (Moon, 1974; Bull, Moon, and Legler, 1974; Sites, Bickham, and Haiduk, 1979). Males are heteromorphic (heterogametic, XY) while females are homomorphic (homogametic, XX). Moon (1974) and Bull, Moon, and Legler (1974) found that males have a single subtelocentric chromosome (X), bearing a secondary constriction in the long arm, which is the fourth largest and is homologous to one (Y) of three smaller telocentric chromosomes (Sites, Bickham, and Haiduk, 1979, reported the Y chromosome is acrocentric). Karyotypes of the females of both species possess two subtelocentric and two telocentric chromosomes. It is not known if the male's heteromorphic chromosomes actually influence sex determination, but they are definitely sex related since they do not occur in females. Sites, Bickham, and Haiduk (1979) conducted various staining and band-pattern determination studies on these chromosomes in *S. salvinii*, and concluded that, unlike previously described systems in most other vertebrates in which the Y or W is derived and the homomorphic sex represents the primitive condition, the opposite is true for *S. salvinii*. The X chromosome is derived, so the homomorphic female is more derived than the

Key to the species of *Staurotypus*:

1a. Dorsolateral carapacial keels extend the length of carapace from anterior to posterior marginals; anterior plastral lobe shorter than posterior lobe; head and jaws with bold pattern of dark reticulations: **S. triporcatus**

b. Dorsolateral carapacial keels do not extend from anterior to posterior marginals, but end on 1st and 4th pleurals; anterior plastral lobe longer than posterior lobe; head unicolored or with few dark reticulations, jaws unmarked yellow: **S. salvinii** heteromorphic male. The male is actually intermediate between the female and the ancestral condition of other turtles. The 2n chromosome number is 54: *S. salvinii* has 24 macrochromosomes and 30 microchromosomes, *S. triporcatus* has 26 macrochromosomes and 28 microchromosomes (Moon, 1974; Killebrew, 1975b; Sites, Bickham, and Haiduk, 1979).

Staurotypus salvinii Gray, 1864b

Chiapas giant musk turtle

Recognition: The elongated (to 25 cm), oval, adult carapace has three longitudinal keels which may become lower with age. The medial keel extends from the posterior portion of the 1st vertebral to the seam between the 11th pair of marginals. The dorsolateral keels extend along the dorsal portions of all four pleurals. Vertebrals 1–4 are longer than broad and the 5th is broader than long. Posterior marginals are somewhat flared and the 10th and 11th are elevated over the preceding nine, with the 11th highest of all; the marginal rim is smooth. The carapace of *S. salvinii* appears to be wider and more flattened than its congener, *S. triporcatus*. It is dark brown to olive gray, with or without faded mottled spots. Most of the distinguishing features are on the plastron, which is small and cross shaped and contains a hinge that allows movements of the anterior lobe. The anterior lobe is longer than the posterior lobe, and its sides converge anteriorly throughout; the posterior lobe is pointed and unnotched. Gular and humeral scutes are absent; the plastral formula is: pect > an >< fem > abd. The abdominal and femoral scutes are much wider than long. The bridge is narrow, only 10–20% of the plastral length. Large axillary and inguinal scutes are present. Both bridge and plastron are yellow to gray. The head is large, broad across the temples, has

a projecting snout and a weak hook on the upper jaw. It is gray with fine orange or yellow markings and plain yellow jaws. The head becomes uniformly dark with age. Two barbels are on the chin. Limbs and tail are grayish brown, two rows of conical tubercles occur on the tail, and the toes are webbed.

Males have long, thick tails and roughened patches of scales (vinculae) on the thighs and crura. The female tail is short and she lacks the roughened scales on the hind legs.

Distribution: *Staurotypus salvinii* occurs in the lowland Pacific drainages of Oaxaca and Chiapas, Mexico, southward into Guatemala and El Salvador.

Habitat: Slow-flowing waterways with soft bottoms and abundant aquatic vegetation are preferred by this species.

Natural History: Schmidt (1970) observed courtship in captivity; it occurred in water during January. While copulating, the male maintained his position on the female's carapace with his hind legs, the roughened patches of scales aiding his grip. The female actively bit at the male's jaws. Sachsse and Schmidt (1976) reported that in addition to visual cues, olfactory releasers play a major role in sexual stimulation. Several clutches of 6 to 10 eggs are laid each season, probably during the fall or early winter months. The brittle-shelled eggs are elongated (40 × 20 mm), and incubation periods reported by Sachsse and Schmidt (1976) ranged from 80 to 210 days. Hatchlings have carapace lengths of 25–31 mm; their plastra are variably patterned with dark mottled spots.

Like other kinosternids, *S. salvinii* is carnivorous. In the wild it probably feeds on aquatic insects and other invertebrates, amphibians, and small fish.

This species is well known for its vile temper and sharp jaws!

Staurotypus salvinii

Staurotypus triporcatus (Wiegmann, 1828)

Mexican giant musk turtle

Plate 4

Recognition: The elongated (to 37.9 cm), oval adult carapace has three strongly developed longitudinal keels throughout life. Its medial keel extends along all five vertebrals, and the dorsolateral keels are also more extensive than in *S. salvinii*, running along the entire length of the pleurals. The five vertebrals are usually longer than broad in adults. The posterior marginals are unserrated and somewhat flared, and the 9th through 11th are elevated over the preceding eight, with the 11th highest of all. In appearance, the carapace of *S. triporcatus* seems to be longer and more elevated than that of *S. salvinii*. The carapace is brown with yellow seams, dark radiations and spots. The plastron is small and cross shaped with a movable hinge between the pectoral and abdominal scutes. Its anterior plastral lobe is shorter than the posterior lobe (the opposite of *S. salvinii*), and its sides are almost parallel. The posterior lobe is narrow and pointed to the rear, without an anal notch. Gular and humeral scutes are absent; the plastral formula is: fem > pect > an >< abd. Abdominal scutes are about as broad as long; femorals are narrow. The bridge is broad, extending over 25% of the plastral length. Large axillary and inguinal scutes are present. Both bridge and plastron are yellow, sometimes with dark seams. The head is large, broad across the temples, and has a projecting snout and an upper jaw that is not hooked or only slightly so. The head is yellowish to olive with numerous well-marked dark reticulations that extend onto the jaws. There are two chin barbels. Limbs and tail are grayish brown, and two rows of conical tubercles occur on the tail. The toes are webbed.

Males have long, thick tails and roughened patches of scales (vinculae) on their thighs and crura. The female tail is short and she lacks the roughened scales on the hind leg.

Distribution: *Staurotypus triporcatus* ranges, at elevations of usually less than 300 m, in Gulf and Caribbean drainages from central Veracruz southward through the base of the Yucatán Peninsula, Belize, and northeastern Guatemala to western Honduras.

Habitat: This turtle lives in slow-moving waterways such as lakes, weedy marshes, and the lagoons of large rivers.

Natural History: Holman (1964) reported an unsuccessful mating attempt. The male was mounted on the posterior part of the female's carapace and was trying to grasp her lateral keels with his forefeet. His hind legs and tail were thrust under her carapace. A normal clutch consists of three to six ellipsoidal, brittle-shelled eggs measuring about 35–44 × 21–26 mm (Ewert, 1979). Nesting probably occurs in September. Hatchlings have carapace lengths of 30–35 mm.

Staurotypus triporcatus is a voracious predator, capturing and eating many types of small invertebrates (aquatic insects, worms, clams, snails, crustaceans), fishes, and amphibians. Pritchard (1979) reported it also feeds on the smaller mud turtles that occur with it, such as *Kinosternon acutum* and *K. leucostomum*. Holman (1964) observed that during feeding a captive would not pursue live fish; when taking food it moved its head slowly and involved much hyoid action both in drawing in and swallowing.

This species, like its congener *S. salvinii*, has an evil temper and can inflict a severe bite on a careless handler.

Kinosternon Spix, 1824

The mud turtles are small to medium-sized (to 27 cm), aquatic New World turtles with elongated to oval carapaces having either three longitudinal keels, a single middorsal keel, or no keel at all. The carapace may be high peaked or flattened dorsally, and has a smooth to slightly serrated posterior rim. A cervical scute is present along with 11 marginal scutes on each side. There are only five to seven neural bones, allowing some of the posterior costals to meet at the midline. Ten peripheral bones occur on each side of the carapace. The plastron varies from relatively large and broad to rather narrow. It may contain either two transverse movable hinges bordering the abdominal scute which allow the plastron to close upon the carapace, or an inconspicuous single hinge located between the pectoral and adominal scutes. The hind lobe is usually longer than the forelobe, and it may or may not contain a shallow posterior notch. There is no entoplastron. Plastral scutes may be completely or almost completely cornified at the seams, or much skin may be evident between the scutes. There is only a single gular scute (absent in *K. carinatum*). The bridge normally contains an inguinal scute, but the axillary may be absent. The head is moderate to large, with a projecting snout and, usually, a hooked upper jaw. The temporal region is moderately emarginated. The maxilla contacts the quadratojugal, and its triturating surface lacks a ridge. No contact occurs between the squamosal and postorbital, and the quadrate does not enclose the stapes. A secondary palate is present. All toes are webbed.

The karyotype consists of 56 chromosomes (24–26 macrochromosomes; 30–32 microchromosomes (Barros et al., 1972; Stock, 1972; Moon, 1974; Killebrew, 1975b; Sites et al., 1979; Bickham and Carr, 1983).

Kinosternon ranges from New England and southern Ontario west to Nebraska, southward through the Great Plains, Mexico, and Central America to Ecuador, Peru, Bolivia, Brazil, and northern Argentina in South America. Nineteen species and 25 subspecies are recognized, but other variation occurs and probably several additional taxa will be described.

The following key to the species of *Kinosternon* is in part courtesy of John B. Iverson.

Key to the species of *Kinosternon*:

1a. Some skin showing along seams separating plastral scutes; pectoral scutes squared; gular scute (if present) is the shortest scute along plastral midline ... 2

b. No skin showing along plastral seams; pectoral scutes usually triangular (much narrower along plastral midseam than at edge of plastron); gular scute never the shortest scute along plastral midline ... 5

2a. Two light stripes on side of head; barbels on chin and throat; nonoverlapping carapacial scutes: *K. odoratum*

b. Light stripes absent from head; barbels on chin only; overlapping carapacial scutes ... 3

3a. Gular scute absent; very prominent vertebral keel: *K. carinatum*

b. Gular scute present; vertebral keel not strongly developed (but two dorsolateral keels may be present on young individuals) ... 4

4a. Carapace wide and flattened, its sides slope at an angle greater than 100°; mean angle/height ratio 8:1 or greater: *K. depressum*

b. Carapace not greatly flattened, its sides slope at an angle less than 100°; mean angle/height ratio about 5:1 in those with a vertebral keel: *K. minor*

5a. 9th marginal much higher than 8th: *K. flavescens*

b. 9th marginal about the same height as 8th ... 6

6a. Carapace with three longitudinal light stripes; abdominal scute long: *K. baurii*

b. Carapace lacking three longitudinal light stripes; abdominal not necessarily long ... 7

7a. Posterior plastral lobe immovable (akinetic): *K. herrerai*

b. Posterior plastral lobe hinged and movable ... 8

8a. Nasal scale furcate ... 9

b. Nasal scale not furcate ... 10

9a. Plastron reduced in size (much smaller than carapacial opening); carapace with one or three keels, but with medial keel evident at least posteriorly; 1st vertebral scute broad: *K. hirtipes*

b. Plastron not so reduced in size, at least anteriorly; carapace usually smooth (always posteriorly lacking a distinct medial keel); 1st vertebral scute narrow: *K. subrubrum* (in part)

10a. 1st vertebral scute narrow; carapace without obvious keels: *K. subrubrum* (in part)

b. 1st vertebral scute broad; carapace with some evidence of keeling ... 11

11a. Anterior pair of chin barbels very long, subequal to orbit diameter: *K. sonoriense*

b. Anterior pair of chin barbels not long, never approaching orbit diameter in length ... 12

12a. Bridge very narrow, its length less than 21% of carapace length: *K. angustipons*

b. Bridge not so narrow, its length more than 21% of carapace length ... 13

13a. Plastron with distinct posterior notch ... 14

b. Plastron without a distinct posterior notch ... 16

14a. Males with roughened scale patches (clasping organs or vinculae) on their thighs and crus; carapace lacking a keel; 11th marginal usually as high as 10th: *K. dunni*

b. Males lacking roughened scale patches on their thighs and crus; carapace typically with three keels, but obscure in some older individuals; 11th marginal rarely as high as 10th ... 15

15a. Width of plastral forelobe at anterior hinge more than 67% of greatest carapace width; maximal width of plastral hind lobe greater than 59.5% of greatest carapace width in males and greater than 62% in females; interfemoral seam length less than 46% of bridge length, and less than 12% of maximum plastron length: *K. integrum*

b. Width of plastral forelobe at anterior hinge less than 67% of greatest carapace width; maximal width of plastral hind lobe less than 59.5% of greatest carapace width in males and less than 62% in females; interfemoral seam length greater than 38% of bridge length, and more than 9% of maximum plastron length: *K. oaxacae*

16a. Gular scute broader on dorsal surface of plastron than on ventral surface; males with clasping organs (vinculae) present; usually with a single, broad, light postorbital stripe on head: *K. leucostomum*

b. Gular scute no broader on dorsal surface of plastron than on ventral surface; males lack clasping organs; no single, broad, light postorbital stripe on head ... 17

17a. Abdominal scute very long, more than 33% of total plastron length; 1st vertebral in contact with 2d marginal: *K. acutum*

b. Abdominal scute not extremely long, less than 33% of total plastron length; 1st vertebral may contact seam between 1st and 2d marginals, but seldom touches much of marginal 2 ... 18

18a. Carapace with three keels ... 19

b. Carapace lacking keels: *K. alamosae*

19a. Anterior margin of posterior plastral lobe straight across: *K. scorpioides*

b. Anterior margin of posterior plastral lobe not straight across, but instead angled posteriorly to midline: *K. creaseri*

Kinosternon carinatum (Gray, 1855)

Razor-backed musk turtle

Recognition: The deep, steeply sloping, oval carapace (to 16 cm) has a prominent medial keel and is slightly serrated posteriorly; each vertebral scute usually overlaps the one behind it. This high-peaked shell is triangular when viewed from the front; its sides always slope at an angle less than 100°. No dorsolateral keels are present on adults. The 1st vertebral scute is long and narrow but flared anteriorly; however, it never touches the 2d marginal scutes. The 2d vertebral varies from longer than broad, to as broad as long, or broader than long; the last three vertebrals are usually broader than long, and the last is laterally expanded. Carapacial scutes are light brown to orange, with dark spots or radiating streaks and dark posterior borders on each; this pattern usually fades with age. The immaculate yellow plastron lacks a gular scute and has only 10 instead of the normal 11 scutes found in *Kinosternon*. An indistinct hinge is located between the pectoral and abdominal scutes, and there is a slight anal notch. The plastral formula is: an > abd > pect > hum >< fem. The head is moderate with a protruding tubular snout and a slightly hooked upper jaw. The rostral shield is deeply furcated posteriorly, and there is usually only one pair of chin barbels. All skin is gray to light brown or pink, with small dark spots; the jaws are tan with dark streaks.

Males have thick, long, spine-tipped tails with the vent located posterior to the carapacial rim, and roughened patches of scales (vinculae) on their thighs and crura.

Distribution: *Kinosternon carinatum* ranges from southeastern Oklahoma, central Arkansas, and Mississippi south to the Gulf of Mexico; the range is almost completely within the Gulf Coastal Plain.

Kinosternon carinatum

Habitat: This species lives in rivers, slow streams, and swamps. Little current, soft bottoms, abundant aquatic vegetation, and some basking sites are the preferred conditions.

Natural History: Courtship and mating occur in the spring, and the acts are identical to those of *Kinosternon odoratum* (which see for details). Presumably, nesting occurs from April through June, as females with oviducal eggs have been found during this period. At least two clutches of two to four white, elongated, brittle-shelled eggs are laid each year. Hatchlings emerge in August and September with carapace lengths of 23–31 mm and often three carapacial keels.

K. carinatum is omnivorous, feeding on insects, crustaceans, snails, clams, amphibians, and aquatic plants.

Seidel and Lucchino (1981) electrophoretically examined 11 protein systems (17 loci) and discriminantly analyzed 15 morphological characters in a study of speciation among *K. carinatum*, *K. minor* and *K. depressum*, and found *K. carinatum* to be strongly divergent from the other two species.

Kinosternon depressum (Tinkle and Webb, 1955)

Flattened musk turtle

Recognition: The oval carapace (to 11.5 cm) is wide and very flattened. Its sides always slope at an angle greater than 100°, and in juveniles the mean angle/height ratio is about 9.5:1. There is a blunt middorsal keel, and the posterior marginals are serrated. Each vertebral overlaps the one behind (vertebrals are juxtaposed). The 1st is very long, and never touches the 2d marginals; the other four vertebrals are broader than long, and the 5th is posteriorly expanded. The carapace is yellowish brown to dark brown with small, dark-brown or black spots or streaks and dark seams. The immaculate pink to yellowish-brown plastron contains a single gular scute, and has an indistinct hinge between the pectoral and abdominal scutes. There is a shallow posterior notch. The plastral formula is: an > abd > pect > fem >< hum > gul. The head is moderate with a projecting tubular snout and a slightly hooked (sometimes notched) upper jaw. The rostral shield is furcated posteriorly, and there are two pairs of chin barbels. Fine black mottling is present on the olive head, and dark bars on the jaws. A yellow stripe from the nostril to the orbit may occur on some individuals. Other skin is olive with fine black mottling.

Males have thick, long, spine-tipped tails with the vent posterior to the carapacial rim, and roughened

Kinosternon depressum

patches of scales on their thighs and crura.

Kinosternon depressum was originally described by Tinkle and Webb (1955) as a full species; however, Wermuth and Mertens (1961) and Ernst and Barbour (1972) designated it a subspecies of the closely related *K. minor*. Specimens intermediate between these species exist, but are probably interspecific hybrids rather than intergrades. Recent morphological studies and studies of the electrophoretic properties of proteins of these two turtles have shown them to be closely related but distinct species (Iverson, 1977b; Seidel and Lucchino, 1981; Seidel, Reynolds, and Lucchino, 1981).

Distribution: *Kinosternon depressum* is restricted to the Black Warrior River system in west-central Alabama.

Habitat: Clear, rock-bottomed to sandy, permanent streams above the fall line seem to be the preferred habitat.

Natural History: Little has been published on its biology. Estridge (in Mount, 1975) reported a female laid an oblong (32.0 × 16.4 mm) brittle-shelled egg in June which hatched in October. The hatchling had the following carapacial measurements: length, 25 mm; width, 20 mm; height 7.8 mm.

The food consists of mostly clams, snails, insects, crayfish, arachnids, and isopods. Older specimens often develop enlarged heads, presumably correlated with a mollusk diet.

Kinosternon minor (Agassiz, 1857)

Loggerhead musk turtle

Recognition: The oval carapace (to 13.5 cm) is generally deep, serrated posteriorly, and has a vertebral keel and two dorsolateral keels (which may disappear with age). Its sides always slope at an angle less than 100°, and the mean angle/height ratio is about 5:1. The vertebral scutes overlap the one behind. The 1st vertebral is long, and never touches the 2d marginals; the other four vertebrals are slightly broader than long, and the 5th is posteriorly expanded. The carapace is dark brown to orange with dark-bordered seams and often a pattern of scattered dark spots or radiating dark streaks which fade with age. The immaculate pink to yellowish plastron has a single gular scute, and contains an indistinct hinge between the pectoral and abdominal scutes. There is only a shallow posterior notch. The plastral formula is: an > abd > pect > fem > hum > gul. The head is moderate to large (especially in older turtles) with a protruding snout and a slightly hooked to nonhooked upper jaw. The rostral shield is furcated posteriorly, and there are two chin barbels. The head is grayish to pink with

Kinosternon minor minor

Kinosternon minor peltifer

dark dots; jaws are tan with dark streaks. Other skin is pinkish, gray, or orange with dark streaks.

Males have thick, long, spine-tipped tails with the vent behind the carapacial rim, and roughened patches of scales on their thighs and crura.

Morphological and electrophoretic studies by Tinkle (1958), Zug (1966), Iverson (1977b), Seidel and Lucchino (1981), and Seidel, Reynolds, and Lucchino (1981) have shown *Kinosternon minor* to be distinct, but more closely related to *K. depressum* and *K. odoratum* than to *K. carinatum*.

Distribution: *Kinosternon minor* ranges from southwestern Virginia, eastern Tennessee, and central Georgia south to central Florida and west to the Pearl River system in south-central Mississippi.

Geographic Variation: Two subspecies are recognized. *Kinosternon minor minor* (Agassiz, 1857) ranges from central Georgia and southeastern Alabama to central Florida. It usually lacks stripes on the head and neck, and the juvenile has three carapacial keels (these may disappear with age). *K. m. peltifer* (Smith and Glass, 1947) ranges from eastern Tennessee and Alabama to the Pearl River in south-central Mississippi. It has only a single, middorsal carapacial keel in juveniles (which disappears with age), and distinct wide stripes on the neck.

Habitat: *Kinosternon minor* inhabits rivers, creeks, spring runs, oxbows, and swamps. It occurs most commonly around snags, fallen trees, or rocky outcrops.

Natural History: Mating may occur throughout most of the year, as Cox and Marion (1978) and Iverson (1978a) found oviducal eggs in females from September to July, but observations of copulation have only been in March and April (Cox, Nowak, and Marion, 1980). Mating takes place under water in shaded, partially concealed areas. The male follows the female with his head and neck fully extended, and after smelling the posterior edge of her bridge (inguinal glands?) attempts to mount. No biting, caressing, or titillating occurs (Cox, Nowak, and Marion, 1980). Nesting occurs from October through July, often at the base of a tree or beside a log. Probably two to three clutches of one to five eggs are laid each year. The eggs are elongated ($\bar{x}$ = 28.5 × 17.2 mm, Iverson, 1978a) with brittle white shells. Hatchlings are about 25 mm in carapace length.

Kinosternon minor is primarly a mollusk eater but it also feeds on filamentous algae, vascular aquatic plants, aquatic insects, millipedes, spiders, crayfish, and fishes. In *K. m. minor*, increasing size brings a shift from an insectivorous to a molluscivorous (snail, clam) diet. This subspecies develops heavy lower-jaw musculature and an expanded crushing surface on both jaws—apparently adaptations to eating mollusks.

Kinosternon odoratum (Latreille, in Sonnini and Latreille, 1802)

Common musk turtle

Recognition: The deep carapace (to 13.6 cm) is highly arched, elongated and narrow. Younger individuals may have a prominent middorsal keel, but this is lost in adults; two low dorsolateral keels are also present on hatchlings and small juveniles. The posterior carapacial rim is unserrated, and the vertebral scutes do not overlap. The 1st vertebral is long and never touches the 2d marginals. The other four vertebral scutes are usually broader than long, and the 5th is expanded posteriorly. In adults, the carapace is plain gray-brown to black, but juveniles have a pattern of scattered spots or radiating dark streaks. The unmarked plastron ranges in color from yellow to brown. There is only a single gular scute present, and an indistinct hinge lies between the pectoral and abdominal scutes. Posteriorly, there is a shallow anal notch. The plastral formula is: abd > an > pect >< fem >< hum > gul. The head is moderately elongated with a protruding snout and a nonhooked upper jaw. A posteriorly furcate rostral shield is present, and there are one or two pairs of chin barbels. Barbels also occur on the throat. Skin is gray to black, and the sides of the head and neck usually have a pair of conspicuous yellow or white stripes, which begin on the snout and extend backward, passing above and below the eye; these stripes may be faded or broken, and absent in some Florida specimens.

Kinosternon odoratum

Males have longer, thicker, spine-tipped tails with the vent behind the carapacial rim, and roughened patches of scales on their thighs and crura.

Kinosternon odoratum seems more closely related to *K. carinatum* than to *K. depressum* or *K. minor* (Seidel, Reynolds, and Lucchino, 1981).

Distribution: *Kinosternon odoratum* ranges from New England and southern Ontario south to Florida and west to Wisconsin and central Texas. There are also scattered records from south-central Kansas, western Texas, and Chihuahua, Mexico.

Geographic Variation: No subspecies have been described; however, Seidel, Reynolds, and Lucchino (1981) reported there is much heterozygosity, as shown by their protein electrophoretic studies, but little morphological variation occurs between populations.

Habitat: This musk turtle occurs in almost any waterway with a slow current and soft bottom: rivers, streams, lakes, ponds, sloughs, canals, swamps, bayous, and oxbows. It may, however, occasionally occur in almost any sort of stream; for example, we have taken them in a gravel-bottomed, fast-flowing stream in northwestern Arkansas. Tinkle (1959) showed that the fall line usually limits the distribution of this species; it is found above it only in the rivers draining into the Gulf of Mexico.

Natural History: Courtship and mating occur sporadically throughout the year, with peaks in the spring and fall. Most mating occurs in April and May, before the nesting period. A second period of mating occurs in September and October but may extend into December. There is evidence that sperm from the late matings may be retained through the winter in viable condition in the oviducts. Mating occurs under water, in shallows, at night or in the early morning. Mahmoud (1967) studied the sexual activities of *Kinosternon odoratum*, *K. carinatum*, *K. subrubrum*, and *K. flavescens*; he has given a most detailed composite account of their courtship and mating. There were three phases: tactile, mounting and intromission, and biting and rubbing. During the tactile phase a male with head extended approached another turtle from behind and felt or smelled its tail, apparently to determine sex. Courtship usually proceeded no further if the approached turtle was a male; if a female, the male, with head still extending forward, moved to her side and nudged the region of her bridge with his nose. This movement apparently was directed at the musk glands there. If the female was not receptive she moved away. The male responded either by giving chase or by going elsewhere. If chasing occurred, the male, with head fully extended, persistently attempted to nudge or bite the female about the head as he followed. The chase was either by walking or swimming and was sometimes followed by mounting, a few seconds later. If initially receptive, the female remained immobile while the male, with head fully extended, gently nudged her just behind the eye and a few seconds later assumed the mounting position. This tactile phase varied in length from a few seconds to three minutes. The mounting phase usually followed the tactile phase. Males approached females either from behind or from the side. The male positioned himself with his plastron directly over the female's carapace by grasping the margins of her carapace with the toes and claws of all four feet. By flexing one knee the male held the female's tail between the two scaly patches on the opposing posterior surfaces of the upper and lower leg throughout coitus. The male's tail was looped so that the terminal nail was in touch with the skin at one side of the female's cloaca; this brought the vents together, and insertion of the penis followed. The male's head extended forward to gently touch the top of the female's head and neck. These actions occurred simultaneously. When the coital position was attained the rubbing-and-biting phase began. The time between mounting and penial insertion was 5–10 sec.

Nesting season varies with latitude: in the south, egg laying lasts from February through July, in the north, from May through August. Nesting takes place from early morning into the night, and nests have been dug as far as 45 m from the water. Some females lay their eggs on the open ground; others dig well-formed nests as deep as 10 cm. Most nests, however, are shallow and are formed by scraping away debris such as decaying vegetable matter, leaf mold, or rotting wood. Many eggs are laid under stumps and fallen logs and in the walls of muskrat houses. Females are noted for sharing nesting sites; often several oviposit at the same place. Eggs are elliptical (24–32 × 14–17 mm), with a thick, white, brittle shell that appears slightly glazed when dry. One to nine eggs (usually two to five) constitute a clutch, and four clutches may be laid each year. There is a correlation between clutch size and the body size of the female: the largest females lay the most eggs per clutch. Hatchlings emerge in August and September in Florida and in September and October in the north after an incubation period of about 75 or 80 days. The hatchling carapace (20–22 mm) is rough and has a prominent vertebral keel and two smaller lateral keels. It is black, with a light spot on each marginal. The dark, light-mottled plastron lacks hinges and is rough in texture. The skin is black and the two light stripes on the head are prominent.

Kinosternon odoratum is omnivorous. Those under 5

cm in carapace length feed predominantly on small aquatic insects, algae, and carrion, whereas those over 5 cm feed on any kind of food. They are bottom feeders, often walking about, with head extended, in search of food. They probe soft mud, sand, and decaying vegetation with their heads, apparently looking for food, and prefer slow-moving prey. They are known to eat earthworms, leeches, clams, snails, crabs, crayfish, aquatic insects, fish eggs, minnows, tadpoles, algae, and parts of higher plants.

Kinosternon subrubrum subrubrum

Kinosternon subrubrum (Lacépède, 1788)

Common mud turtle

Recognition: This is a small (to 12.5 cm) turtle with an oval, smooth, unkeeled (in adults) and often depressed carapace. Its sides are straight, the posterior marginals unserrated, and the carapace drops abruptly behind. Hatchlings have three longitudinal keels. The 1st vertebral scute is long and widely separated from the 2d marginal scutes. Vertebrals 2–5 are usually broader than long, and the 5th may be expanded. The 10th marginal scute is elevated above the rest. The patternless carapace varies from yellowish brown to olive or black. The immaculate yellow to brown plastron has a single gular scute and two well-marked transverse hinges (which may ossify in old individuals). Its forelobe is longer than the hind lobe in two of the three subspecies; there is a posterior notch. The gular scute is short, and much less than half the forelobe length. The plastral formula is: an > abd > hum > fem >< gul > pect. Axillary and inguinal scutes meet on the bridge. The head is medium sized with a slightly protruding snout and a hooked upper jaw. Its rostral scute may or may not be posteriorly furcated. The head is usually brown with some yellow mottling; sometimes there are two light lines on each side of the head and neck. Skin is brown to olive or grayish and may exhibit some markings. The tail is spine tipped.

Kinosternon subrubrum hippocrepis

Males have longer, thicker tails, and a patch of roughened scales on the posterior surface of each thigh and crus.

Kinosternon subrubrum steindachneri

Distribution: *Kinosternon subrubrum* ranges from Connecticut and Long Island south to the Gulf Coast, west to east-central Texas, and north to Missouri, southern Illinois, and southern Indiana. Isolated colonies exist in northwestern Indiana and in west-central Missouri.

Geographic Variation: Three subspecies are recognized. *Kinosternon subrubrum subrubrum* (Lacépède,

1788) ranges from southwestern Connecticut and Long Island to the Gulf Coast and northwest through Kentucky to southern Indiana and Illinois. This subspecies has a wide bridge, a spotted or mottled head, and the anterior lobe of the plastron shorter than the posterior lobe. *K. s. hippocrepis* Gray, 1855 ranges in the Mississippi Valley from Louisiana and eastern Texas northward to Missouri and western Kentucky. It has a wide bridge, two distinct light lines on each side of the head, and the anterior lobe of the plastron shorter than the posterior lobe. *K. s. steindachneri* (Siebenrock, 1906c) is restricted to peninsular Florida. It has a narrow bridge, a plain or mottled head, and the anterior lobe of the plastron often longer than the posterior lobe. See Ernst et al. (1974) for a more detailed analysis of variation.

Habitat: This mud turtle prefers slow-moving bodies of shallow water with soft bottoms and abundant aquatic vegetation. Frequently it inhabits the lodges of muskrats. We have found it in ditches, sloughs, wet meadows, ponds, marshes, bayous, lagoons, and cypress swamps. *Kinosternon subrubrum* shows a marked tolerance for brackish water and is often abundant in salt marshes.

Natural History: Mating occurs from mid-March through May; copulations are earliest in the south. Courtship and mating are as in *Kinosternon odoratum* (which see). Mating usually takes place under water but sometimes occurs on land. Most nesting occurs during June but has been observed from February through September in various parts of the range. The nesting site usually is open ground not far from water. Sandy, loamy soils are preferred, but piles of vegetable debris also are used. In some localities mud turtles often nest in muskrat tunnels. Eggs have been found on the surface of the ground and under piles of boards, and, in the south, alligator nests are often used. The completed nest usually is a semicircular cavity 75–125 mm deep and entering the ground at about a 30° angle. Clutches vary from one to nine eggs; normally two to five, and most commonly three to five eggs are laid. At least three clutches are laid annually. The eggs are elliptical, pinkish white or bluish white, and brittle shelled, 22–29 mm long and 13–18 mm wide. Hatching occurs after about 100 days. The hatchling carapace (20–27 mm) is shaped like that of the adult but has a vertebral keel and two weak dorsolateral keels, is rough, and is not depressed anteriorly or sharply turned down posteriorly. It is dark brown or black, with light spots along the marginals. The plastron is irregularly mottled with orange or red, and the hinges are poorly developed. Hatchling skin is brown or black, and in the subspecies *K. s. hippocrepis* there

may be two faint yellow lines on each side of the head and neck.

Kinosternon subrubrum is an omnivorous feeder. Mahmoud (1968) reported the following percentages of frequency and volume, respectively, of food items: Insecta 98.3, 30.4; Crustacea 15.0, 1.4; Mollusca 93.1, 31.8; Amphibia 30.0, 2.2; carrion 68.6, 11.9; and aquatic vegetation 89.6, 22.3. Mud turtles under 5 cm in carapace length feed predominantly on small aquatic insects, algae, and carrion; those above 5 cm feed on almost any kind of food. Captives feed readily on canned and fresh fish, canned beef, hamburger, dog food, snails, insects, tomatoes, and watermelon. *K. subrubrum* occasionally feeds at the surface but is predominantly a bottom feeder, frequently walking along the bottom, probing into soft mud, sand, and decaying vegetation.

Kinosternon subrubrum is quite terrestrial. It leaves the water in early summer, forages on land for a short period, and then burrows to estivate during the hot weather. Many remain underground until the next spring.

Kinosternon baurii (Garman, 1891)

Striped mud turtle

Recognition: This small (to 12 cm) mud turtle has a broad, smooth carapace which is keelless and unserrated in adults, and usually widest and highest behind the middle. A middorsal keel is present in juveniles. The 1st vertebral is elongated but widened anteriorly so that it extends to the seam separating the 1st and 2d marginals on each side. Vertebrals 2–5 are usually broader than long. The vertebrals may be depressed, forming a broad, shallow middorsal groove. Marginal 10 is more elevated than the rest. The carapace varies from tan to black; some have rather transparent scutes; three variable light-yellow or cream-colored longitudinal stripes are usually present. The double-

Kinosternon baurii

hinged plastron is broad and only slightly notched posteriorly. Its hind lobe is larger than the forelobe, and the abdominal scute is almost as long as the forelobe. The plastral formula is: an > abd > hum > gul > fem > pect. Axillary and inguinal scutes meet on the bridge. The plastron is olive to yellow and either plain or with dark seam borders. The small, conical head has a slightly protruding snout and a nonhooked upper jaw. Its rostral scale is, at best, only slightly posteriorly furcated. Two light stripes extend posteriorly from the orbit, one above and one below the tympanum. All skin is tan to black, and the neck and head may show dark mottling. A horny spine occurs at the tip of the tail.

Males have longer, thicker tails and a patch of rough scales on the inner surface of each thigh and crus.

Distribution: *Kinosternon baurii* is restricted to the extreme southeastern United States from near Aiken, South Carolina, along the Atlantic Coastal Plain of Georgia southward through peninsular Florida and the Keys.

Geographic Variation: No subspecies are currently recognized. Formerly, the mainland population and that from the Florida Keys were considered separate subspecies, but Iverson (1978b) showed that variation occurs throughout the range and that no populations are distinct.

Habitat: The striped mud turtle is most often found in quiet water with a soft bottom—swamps, sloughs, and ponds. It also frequents wet meadows, and enters brackish water.

Natural History: Mating apparently occurs throughout the year, as Iverson (1977d) found oviducal eggs or fresh corpora lutea in every month. Its courtship has not been adequately described. Nesting occurs from April to June. Nests are dug in sand or in piles of decaying vegetation. A clutch varies from one to five eggs, and possibly as many as three clutches may be laid each season. Eggs are elliptical, whitish, and brittle shelled (22.8–31.8 × 13.6–19.3 mm, Iverson, 1977d). Incubation takes about 100–130 days. Hatchlings are black with yellow spots on each marginal. Their rough carapace (15–25 mm) has three weak keels, one beneath each stripe, and the yellow plastron is marked with a dark central blotch and dark-bordered seams. Plastral hinges are not functional until about the third month after hatching. The yellow head stripes are pronounced.

Kinosternon baurii is omnivorous. Natural foods include seeds of the cabbage palm (*Sabal palmetto*), juniper leaves, various algae, snails, insects, and dead fish. It is easily caught on a hook and line baited with liver, grasshoppers, worms, or dough. They also have some scavenging tendencies, and while on land may forage in cow dung, perhaps seeking insects.

Kinosternum acutum Gray, 1831b

Tabasco mud turtle

Recognition: This medium-sized (to 12 cm) mud turtle usually has a single medial keel present on the carapace (a faint pair of dorsolateral keels may be present in juveniles). Its vertebral scutes are broader than long, and the 1st is in contact with the first two marginals. The 4th pleural scute normally touches the 11th marginal, and marginals 10 and 11 are higher than those preceding them. The carapace is brown to black with dark seams. The double-hinged plastron is not notched posteriorly. When closed, the plastron almost completely covers the shell opening. Its posterior plastral lobe is slightly constricted at the hinge. The plastral formula is: an > abd > gul > hum > fem pect (the length of the gular exceeds 50% of the plastral forelobe length). On the wide bridge the large inguinal scute usually touches (but may be slightly separated from) the much smaller axillary scute. Plastron and bridge are yellow to light brown with dark seams. The head is not greatly enlarged, and the rostral shield covers most of the dorsal surface. Head and limbs are gray to yellow or reddish. Yellow to red marks are always present on the head from the temporal region posteriorly to the neck and on the limbs. The chin is cream colored with dark mottling. Brown or black mottling may also occur on the head, neck, and limbs. No raised patches of horny scales (vinculae) are on the thigh and crus, and both sexes have a spine at the tip of the tail.

Males have long, thick tails and are shorter (10.5 cm) than the females (12 cm), which have short tails.

Kinosternon acutum

Distribution: *Kinosternon acutum* occurs in the Caribbean lowlands of Mexico from central Veracruz southeastward to northern Guatemala and Belize. It is apparently absent from the Yucatán Peninsula (Smith and Smith, 1979).

Habitat: It inhabits lakes, streams, and temporary pools in low-altitude, humid forests not exceeding 300 m elevation.

Natural History: Cope (1865) reported that nesting occurs in March and April and that "few" eggs are laid.

Kinosternon creaseri Hartweg, 1934
Creaser's mud turtle

Recognition: *Kinosternon creaseri* is poorly known. Its oval, dark-brown carapace is of medium size (to 12.1 cm) and highest behind the center. The posterior carapacial profile is nearly vertical. A weak median keel is present in adults, and a pair of low dorsolateral keels is also present on juveniles. The 1st vertebral is about as broad as or slightly broader than long, vertebrals 2–4 are broader than long, and the 5th is small and almost as long as broad. Marginals 10–11 are elevated above the preceding nine. The double-hinged plastron is long, wide, and practically capable of closing completely. It is unnotched posteriorly. The anterior plastral lobe is longer than the abdominal scute, but slightly shorter than the posterior lobe, and the gular scute is more than half the length of the anterior plastral lobe. The plastral formula is: an > abd > gul > hum > fem >< pect. Axillary and inguinal scutes barely touch across the bridge. Plastron and bridge are yellowish brown with dark seams. The head is large with a slightly projecting snout and a strongly hooked upper jaw. Both head and neck are dark brown to black above with very fine light speckles; the sides and ventral surfaces are lighter. The jaws contain dark streaking, and the limbs are grayish to brown. Vinculae are absent.

Males have long, thick tails with a terminal horny spine. Females have short tails.

Distribution: *Kinosternon creaseri* is restricted to the northern and central portions of the Yucatán Peninsula in the states of Yucatán, Quintana Roo, and Campeche, Mexico.

Geographic Variation: *Kinosternon creaseri* is a monotypic species; however, Duellman (1965) collected two specimens from a solution cave which had pale tan carapaces, pale cream-colored plastra, and pale grayish-brown limbs and heads. Perhaps this was an adaptation to their dimly lighted habitat.

Habitat: This turtle lives in the drier part of the Yucatán Peninsula, a region where no surface streams exist, and is abundant in temporary pools in limestone areas.

Kinosternon leucostomum (Duméril, Bibron, and Duméril, 1851)
White-lipped mud turtle

Recognition: The dark-brown or black, oval carapace (to 17.4 cm) is flattened to concave across the vertebrals and drops off abruptly to the rear. Juveniles and young adults have a single vertebral keel, which becomes lower with age until it may totally disappear. The 1st, 3d, and 4th vertebrals are broader than long, the 1st broadest of all; the 2d and 5th may be longer than broad in adults. The 1st vertebral touches the first two marginals, and the 4th pleural scute usually touches the 11th marginal. The cervical scute is

Kinosternon leucostomum leucostomum

Kinosternon leucostomum postinguinale

very narrow. Lateral marginals are downturned, but those posterior are somewhat flared. Marginals 10 and 11 are higher than the previous nine; usually these two are the same height, but occasionally the 10th may be slightly higher than the 11th. Carapacial seams become deeply grooved with age. A single hinge lies between the pectoral and abdominal scutes which allows the large plastron to completely close the shell; it is not notched posteriorly, or only slightly so. The plastral formula is: an > abd > hum > gul > fem > pect; the gular scute is usually less than 55% of the length of the plastral forelobe, and the interabdominal seam length is less than 27% of the plastral length. The axillary scute is short, the inguinal much longer; these usually touch on the bridge. Plastron and bridge are yellow with darker seams. The head is moderate in size with a protruding snout and a hooked upper jaw. It is brown with cream-colored jaws (some dark streaking may occur on the jaws). A broad yellowish stripe extends backward from each orbit to the neck. This stripe may be bronze to light brown with small gold flecks in some individuals. Two large barbels followed by a pair of smaller barbels occur on the chin. The limbs are grayish brown with lighter gray sockets, and each foreleg has several large transverse scales just above the wrist and a rough fringe of small scales along its outer border. On the heel of the hind foot is a series of horizontal scales; vinculae occur on the thighs and crura of males. A horny spine adorns the tip of the tail.

Males grow larger (17.4 cm) than females (15.8 cm) and have long, thick tails. The female tail is very short.

Distribution: *Kinosternon leucostomum* ranges from central Veracruz, Mexico, southward in the Atlantic drainages to Nicaragua, and then southward in both Atlantic and Pacific drainages to Colombia, Ecuador, and northern Peru.

Geographic Variation: Two subspecies are currently recognized by Berry (1978). *Kinosternon leucostomum leucostomum* (Duméril, Bibron, and Duméril, 1851) occurs in the Atlantic drainages of southern Mexico at elevations of less than 300 m, from central Veracruz southward across the base of the Yucatán Peninsula to Belize, Guatemala, and northern Nicaragua. Its carapace is high and the plastron is large; the mean width of the plastron at the anterior hinge is 73% of the maximum carapace width in both sexes, and 69% in males and 70% in females at the midfemoral width. Gular scute length is about 14–15% of the carapace length. The inguinal scute is long and usually touches the axillary. Vinculae are poorly developed, and the light postorbital stripes are obscure or disappear in adults. *K. l. postinguinale* (Cope, 1887) occurs in the

Atlantic and Pacific drainages from the Río San Juan in Nicaragua southward to Colombia, Ecuador, and northern Peru. This subspecies has a relatively flattened carapace and a narrower plastron; the mean width of the plastron at the anterior hinge is only 69% of the maximum carapace length in males and 71% in females, and at the midfemoral width, only 66% in males and 68% in females. The gular scute is short, only about 12% of the carapace length. The inguinal is set well back on the bridge and is separated from the axillary. Vinculae are well developed in males, and the postorbital stripes are usually well marked.

Habitat: This species occurs in most quiet waters with soft bottoms and abundant vegetation: marshes, swamps, streams, lagoons in rivers, and ponds. Over the greater part of the range these waterways are within low-altitude forests. *Kinosternon leucostomum* is not confined to water and often wanders extensively on land; Medem (1962) reported it enters brackish water.

Natural History: Courtship and mating have not been described. Nesting occurs throughout the year and multiple clutches are laid (Medem, 1962; Moll and Legler, 1971). Eggs are either deposited in a shallow nest or laid on the surface of the ground and then covered with leaf litter. A typical clutch includes one to three eggs, with single-egg clutches most common. Shells are brittle and eggs are elongated (37 × 20 mm). Eggs naturally incubated by Moll and Legler (1971) hatched in 126–148 days. A typical hatchling has a carapace length of about 33 mm.

Kinosternon leucostomum is nocturnal and apparently omnivorous. Medem (1962) found them to be highly carnivorous in Colombia, feeding on worms, aquatic mollusks, insects, and carrion; however, Moll and Legler (1971) reported they also eat aquatic plants. Captives readily eat fish, shrimp, and dry trout chow. Moll and Legler believed they probably feed only in water.

Kinosternon scorpioides (Linnaeus, 1766)
Scorpion mud turtle

Recognition: The elongated, oval carapace (to 27 cm) is high domed, has three well-developed longitudinal keels (these may become lower with age until they are barely noticeable in large individuals), and is widest behind the middle. The 1st vertebral is flared anteriorly and is broader than long; vertebrals 2–5 are usually longer than broad. The 10th and 11th marginals are elevated above those preceding; the 10th is

Kinosternon scorpioides scorpioides

Kinosternon scorpioides cruentatum

Distribution: *Kinosternon scorpioides* ranges at low elevations from southern Tamaulipas, Mexico, southward to northern Argentina, Bolivia, and northern Peru.

Geographic Variation: *Kinosternon scorpioides* is quite variable, and many different names have been applied to populations throughout its range. Currently, however, six subspecies are considered valid. *Kinosternon scorpioides scorpioides* (Linnaeus, 1766) ranges from southern Panama over most of northern South America, being found in Ecuador and northern Peru in the west, eastward to the Guianas and Brazil. It is large (to 27 cm), has three well-developed carapacial keels, a narrow cervical scute, and a moderate-sized head. In this subspecies the anal notch is prominent (lacking or only shallow in the other subspecies), and the plastral hind lobe is narrow. The small (to 13 cm) *K. s. carajasensis* Cunha, 1970 is known only from the Sierra dos Carajas (Sierra Norte) plateau in the southern portion of the state of Pará, Brazil. It has a high-peaked carapace which lacks, or has only poorly developed, dorsolateral keels and an extremely small cervical scute, and its head is large and broad. *K. s. seriei* Freiberg, 1936 occurs in northern Argentina and adjacent Bolivia. This subspecies has a moderate-sized head, a carapace with three longitudinal keels (the median is best developed), and a cervical scute which is wider posteriorly than anteriorly. Also, its interanal seam is at least three times longer than the interfemoral seam, and there is much cartilage along the abdominal-femoral seam. We follow Berry (1978) in considering *K. s. pachyurum* Müller and Hellmich, 1936 a synonym of *K. s. seriei*. *K. s. albogulare* (Duméril and Bocourt, 1870) ranges from Panama northward to Honduras. It is tricarinate, but somewhat flattened dorsally, usually lacks an anal notch on the plastron, and has a plain yellow lower jaw and chin in the female. *K. s. cruentatum* (Duméril, Bibron, and Duméril, 1851) ranges from northeastern Nicaragua and Honduras northward to Veracruz and Tamaulipas, Mexico. This may be the most strikingly marked mud turtle. The sides of its head usually have bright red or orange spots, giving it the common name red-cheeked mud turtle. Also, the carapace may be yellow or slightly orange with dark seam borders, and the plastron orange. The three carapacial keels become lower with age, and its plastron can completely close. The last subspecies, *K. s. abaxillare* Baur (in Stejneger, 1925b), occurs on the central plateau of Chiapas, Mexico, at elevations over 800 m. It is best identified by the absence of an axillary from the bridge causing the axillary-abdominal seam to be incomplete; all other *K. scorpioides* normally have an axillary scute. Also, it is a flatter turtle than its nearest relative, *K. s. cruentatum*.

highest of all. Posterior marginals may be slightly flared. The carapace is light to dark brown or black; lighter individuals may have darkened seam borders. The plastron is well developed with a single movable hinge between the pectoral and abdominal scutes, and no posterior anal notch, or only a slight one. It is not always large enough to completely cover the shell opening; the hind lobe is longer than the forelobe. The plastral formula is: abd > an > gul > hum > fem > pect. Axillary and inguinal scutes usually touch. Plastron and bridge are brown. The head is moderate to large in size with a slightly projecting snout and a hooked upper jaw. On the chin are two large anterior barbels followed by two or three smaller pairs. The head is grayish brown, darker dorsally, lighter laterally with small irregular dark spots; the jaws may be plain yellow or have some dark streaks. Neck, limbs, and tail are gray brown. Several enlarged transverse scales are on the anterior surface of the foreleg and on the heel of the hind foot. No vinculae are present.

The male has a slightly concave plastron and a long, thick tail with a terminal spine. The female plastron is flat, and her tail short.

Habitat: This species lives in streams, rivers, lakes, and ponds. If its waterway dries up, it will bury itself in the mud bottom until the next rain.

Natural History: During courtship the male pursues the female with his head extended and normally bites her rump and legs. If she remains in position, he will often circle her, occasionally biting at her limbs or head. Finally he mounts from the rear, hooking his foreclaws under the rim of her shell for support, and intertwining her tail with his, inserts his penis. These observations on captives essentially agree with courtship behavior reported by Sexton (1960) and Fretey (1976). Nesting occurs from March to May in Mexico (Alvarez del Toro, 1960) and Venezuela (Sexton, 1960). Clutches vary from 6 to 16 eggs (Alvarez del Toro, 1960; Freiberg, 1981), but this latter number seems high and was perhaps based on a count of follicles rather than shelled eggs. Eggs are elongated (40 × 18 mm) with white, brittle shells. Alvarez del Toro (1960) reported the incubation period is about 3 months. Hatchlings have 30-mm carapaces.

According to Vanzolini et al. (1980) this turtle is omnivorous, eating fish, snails, adult amphibians, insects, algae, and other plants. In captivity, it readily adapts to a fish diet.

Kinosternon oaxacae Berry and Iverson, 1980a

Oaxaca mud turtle

Recognition: *Kinosternon oaxacae* is a large (to 17.5 cm) mud turtle with a depressed, strongly tricarinate carapace. The carapace is brownish to black, but may be darkly mottled or have dark seams. Its scutes overlap slightly. Vertebrals are variable in width and length; the 1st or 3d through 5th are broadest, and either the 1st or 4th and 5th are shortest. The 1st touches the 2d marginals. The cervical is small, and

Kinosternon oaxacae

marginal 10 is elevated higher than the other marginals. The small plastron is notched posteriorly and double hinged, with both the anterior and posterior lobes freely movable. When closed, it does not completely cover the hind limbs (width of hind lobe is only 57–63% of carapace width). The anterior hinge is straight, but the posterior curved, and the plastral hind lobe is slightly constricted at the hinge. The plastral formula is either abd > an > gul > hum > fem > pect (33%) or abd > an > gul > fem > hum > pect (67%). Plastron and bridge are yellow to brown with dark seams; some brown stains may be present. Axillary and inguinal scutes touch. The head is small, darkly mottled dorsally, reticulated laterally, and yellowish with a few spots ventrally. A V- or bell-shaped rostral shield is present, and the upper jaw is hooked. The jaws may be darkly streaked. Three or four pairs of barbels are on the chin. Limbs are brown to gray dorsally, cream colored ventrally. Vinculae are absent. The tail is totally brown or gray and terminates in a horny spine.

Males have concave plastra and long, thick tails. The female plastron is flat and the tail short.

Distribution: *Kinosternon oaxacae* occurs only in the Río Colotepec and Río Tonameca basins on the Pacific coast of Oaxaca, Mexico.

Habitat: The Oaxaca mud turtle lives in semipermanent water bodies on the coastal plain and permanent flowing waters in the uplands. It is often washed onto the coastal plain by seasonal upland flooding.

Kinosternon integrum LeConte, 1854

Mexican mud turtle

Recognition: *Kinosternon integrum* has an oval carapace (to 20.2 cm) which gradually slopes posteriorly and is depressed over the vertebrals. It may be tricarinate in juveniles, but lack keels or has only a faintly marked medial keel on the posterior vertebrals in adults. The cervical is very narrow. The 1st vertebral is broader than long (wide anteriorly, but narrower posteriorly), touching the first two pairs of marginals; vertebrals 2–4 are longer than broad; and the 5th is also broader than long, but broader posteriorly than anteriorly. Lateral marginals are downturned, but those posterior are flared; the 10th is highest, the 11th next highest. The carapace is gray to yellow brown or dark brown; lighter individuals may have some dark spotting. The double-hinged plastron is large, and may completely, or almost completely, close the shell. It is usually notched posteriorly. The

Kinosternon integrum

Kinosternon alamosae

Geographic Variation: No subspecies are currently recognized, but it is strongly possible that some populations will prove to be subspecifically distinct.

Habitat: *Kinosternon integrum* prefers low-gradient streams with deep pools, at altitudes to 3000 m, but may also be found in roadside ponds and ditches. Hardy and McDiarmid (1969) often found them crossing roads at night.

Natural History: Courtship and nesting have not been described. The elongated (30 × 16 mm) eggs have white, brittle shells. Hardy and McDiarmid (1969) found hatchlings in Sinaloa in late July, August, and September; and Ewert (1979) reported the average carapace length of 14 hatchlings was 27 mm.

plastral formula is: abd > an > gul > hum > fem > pect; the interanal seam is greater than 67% of the length of the posterior plastral lobe. On the wide bridge, the large inguinal usually touches (often barely) the smaller axillary. Plastron and bridge are yellow with dark seams. The head is large with a protruding snout and hooked jaws. Its rostral shield is pointed anteriorly and wide posteriorly. There are two large barbels on the chin followed posteriorly by two to four pairs of smaller barbels. The head is dark brown dorsally and lighter gray or yellow brown laterally and ventrally. Some dark mottling may occur on the sides of the head, and the yellow jaws may have narrow dark streaks or speckles. The neck is dark brown or gray dorsally, but yellow to pinkish with dark speckles on the sides and bottom. Limbs are gray brown with cream-colored sockets. There are no vinculae. Body skin is relatively smooth, lacking the raised tubercles found in other species.

Males are larger (to 20.2 cm) than females (to 18.8 cm) and have long, thick tails terminating in a horny spine. Females have short, spineless tails.

Kinosternon integrum is closely related to *K. scorpioides*, but Berry (1978) found no evidence of intergradation or introgression between the two in Mexico. They are not sympatric, but come within 40 km of each other in the Río Pánuco and Río Papaloapan drainages. *K. integrum* is also often confused with *K. hirtipes*, but males of the latter species have the roughened scaly vinculae on their thighs and crura.

Distribution: The Mexican mud turtle ranges from southern Sonora and extreme southwestern Chihuahua southward along the Pacific drainages to central Oaxaca, on the central plateau from central Durango and southern Nuevo León southward to central Oaxaca, and in the east it occurs in southwestern Tamaulipas.

Kinosternon alamosae Berry and Legler, 1980

Alamos mud turtle

Recognition: This medium-sized (to 13.5 cm) mud turtle has a narrow, oval, rounded or flat-topped, keelless carapace. Its scutes are imbricate, and the 1st vertebral does not usually touch the 2d marginals. The 10th and 11th marginals are higher than the rest, with the 10th highest. The posterior margin of the carapace is straight and often vertical, but not flared or recurved. The carapace is tan or brown to olive with dark seams; often the sutures of the underlying bones can be seen through the translucent scutes. The plastron is double hinged, and the movable anterior and posterior lobes are extensive, closing or nearly closing the shell completely so the head, limbs, and tail are hidden. If present, the anal notch is very small. The plastral scute formula is: abd > an > gul > hum > fem > pect. The plastron is yellowish with brown seams and growth annuli. The bridge is long, 26–33% of the carapace length, and the axillary and inguinal are widely separated. The inguinal touches the 6th marginal, but not the 5th. On the broad head are a short snout and a slightly hooked upper jaw. The rostral shield is neither concave posteriorly nor V-shaped. Two yellow barbels occur on the chin. Dark spotting occurs dorsally and mottling laterally on the gray heads. A pale stripe extends from the orbit to the corner of the mouth. The jaws are cream to gray and are faintly streaked with brown in males. The neck is gray dorsally and yellow ventrally. No vinculae occur on the thighs or crura, but the tail ends in a spine.

Males are slightly larger but narrower than the females, and have concave plastra and long, thick tails. The females have flat to slightly concave plastra and short tails.

Kinosternon hirtipes murrayi

Distribution: *Kinosternon alamosae* occurs on the Pacific coastal lowlands of Mexico from the vicinity of Hermosillo, Sonora, southward into Sinaloa to at least Guasave. It is found at elevations ranging from sea level to 1000 m in the Sierra de Alamos.

Habitat: The Alamos mud turtle lives in aquatic habitats which are seasonally dry. During these dry periods it apparently estivates under the dried mud. It is most active during the wet months, July through September.

Natural History: Berry and Legler (1980) examined the reproductive tracts of several females and concluded that a clutch comprises about five eggs and that some females may lay more than one clutch per season. They hypothesized that follicular development begins when the females become active early in the wet season and that nesting is probably completed by the end of the wet season in October or November. The eggs are relatively small; two oviducal eggs measured 25.7 × 16.5 and 27.6 × 16.4 mm.

Kinosternon hirtipes (Wagler, 1830)

Mexican rough-footed mud turtle

Recognition: *Kinosternon hirtipes* has an elongated (to 18.2 cm), somewhat elevated, usually keeled carapace. Three longitudinal keels may be found on the young, but the lateral keels often disappear with age, and the medial keel may become restricted to the posterior portion of the carapace. The 1st vertebral is flared anteriorly, touching the first two marginals on each side, the 2d through 4th are longer than broad, and the 5th is flared posteriorly. Marginals are narrow except for the 10th, which is elevated, and those posterior to the bridge, which are somewhat flared. The

carapace varies from olive or light brown to dark brown or nearly black; lighter individuals may have dark seams. The hinged plastron is short and narrow, not completely closing the carapacial opening. There is a posterior anal notch. The plastral formula is: abd > an > hum >< gul > fem >< pect. On the short bridge (23% of carapace length in males, 27% in females), the axillary and inguinal scutes are nearly always in broad contact. The plastron is yellow or tan to brown with dark seams (dark staining may cover the plastron of some individuals). The head is of moderate size with a projecting snout and a hooked upper jaw. Its rostral scale may be V-shaped and there are usually three pairs of relatively short chin barbels. Head and neck are tan to black, with light reticulations if black and dark reticulations or spotting if tan. Jaws are tan to gray and may be finely streaked with dark brown or black. Limbs and tail are gray, olive, or brownish.

Males are longer (to 18.2 cm) than females (to 15.7 cm), with vinculae, long spine-tipped tails, concave plastra, and the posterior marginals more flared.

Distribution: *Kinosternon hirtipes* ranges southward from the Big Bend region of Texas and adjacent Chihuahua on the Mexican Plateau, at elevations to 2600 m, to Mexico City.

Geographic Variation: Six subspecies are recognized; the following descriptions are taken largely from Iverson (1981a). *Kinosternon hirtipes hirtipes* (Wagler, 1830) occurs only in the drainages of the Valley of Mexico. Adults have a triangular, rhomboidal, or bell-shaped rostral scale; a mottled head pattern with a light streak extending posteriorly from the corner of the mouth; one or typically two pairs of mental chin barbels with the anterior pair the largest; medium carapace length (both sexes reach 14 cm); a short bridge (17.6% of the carapace length in males, 21.7% in females); a relatively short interfemoral seam length (6.9% of carapace length in males, 7.1% in females); and a relatively long interanal seam length (20.6% of carapace length in males, 25.8% in females). *Kinosternon h. murrayi* Glass and Hartweg, 1951 ranges from the Big Bend region of Texas and Chihuahua south in Mexico to northern Jalisco, northern Michoacán, and the eastern portion of the state of México. The rostral scale is V-shaped (posteriorly bifurcate); the head pattern is quite variable (mottled to reticulated); there are two pairs of mental chin barbels with the anterior pair largest; the carapace is rather large (males to 18.2 cm, females to 15.7 cm); the bridge is long (20% of carapace length in males, 23.7% in females); and a long gular length (14.7% of carapace length in males, 15.8% in females). *K. h. chapalaense* Iverson, 1981a is confined to the Chapala and Zapotlan (and possibly the Duero) basins of Jalisco and Michoacán in Mexico. It has a reduced crescent-shaped rostral scale; reduced dark pigment on the head and neck (confined to isolated spots or reticulations dorsally, but sometimes as two dark lateral postorbital stripes); the neck and chin nearly immaculate and the jaws with few dark streaks; one to three pairs of chin barbels present (anterior pair largest); medium carapace length (males 15.2 cm, females 14.9 cm); a long bridge (20.3% of carapace length? in males, 25.3% in females); and a long interanal seam (19.1% of carapace length in males, 25.2% in females). *K. h. tarascense* Iverson, 1981a is known only from the Logo de Patzcuaro basin of Michoacán, Mexico. It has a finely mottled to spotted head; a V-shaped rostral scale; two pairs of chin barbels; small to medium body size (males to 13.6 cm, females to 13.2 cm); a short bridge (18.0% of carapace length in males, 21.4% in females); a short gular length (10.6% of carapace length in males, 12.6% in females), and a long interpectoral seam (10.1% of carapace length in males, 8.5% in females). *K. h. magdalense* Iverson, 1981a is endemic in the Magdalena Valley of Michoacán, Mexico. It has a finely mottled to spotted head pattern with little or no jaw streaking; a large V-shaped rostral scale; two pairs of chin barbels; a small carapace (males 9.4 cm, females 9.1 cm); a small plastron; a short bridge (18.5% of carapace length in males, 19.7% in females); short gular scutes (9.9% of carapace length in males, 11.0% in females); and a long interpectoral seam (8.7% of carapace length in males, 11.0% in females). The last subspecies, *K. h. megacephalum* Iverson, 1981a, is known only from two localities in southwestern Coahuila, and may now be extinct since natural water bodies apparently no longer exist in the area. Iverson (1981a) described it as having an enlarged head with broad triturating jaw surfaces and enlarged musculature. Other characters

are a V-shaped rostral scale; a head pattern like that of *K. h. murrayi;* three or four pairs of chin barbels (the anterior pair largest) and another small pair at the level of the anterior edge of the tympanum; small body size (males to 9.9 cm, females to 11.7 cm); a very small plastron; a short bridge (17.3% of carapace length in males, 23.9% in females); short gular scutes (11.0% of carapace length in males, 12.8% in females), and a short interanal seam (15.9% of carapace length in males, 20.9% in females).

Habitat: The rough-footed mud turtle inhabits bodies of water in arid mesquite grasslands. It usually is found in lakes, ponds, streams, or rivers flowing into lakes, but also enters temporary pools, stock ponds, and marshy areas.

Natural History: Nesting occurs from May to August in Mexico, and from two to four clutches of four to seven eggs may be laid each season. The eggs are elliptical (28.0–31.8 × 16.2–18.3 mm) and brittle shelled (Ernst and Barbour, 1972; Iverson, 1981a). Hatchlings are from 25 to 30 mm long with a brown carapace and a red or orange, dark-centered plastron.

Kinosternon hirtipes is carnivorous, feeding on a variety of insects, worms, and other small invertebrates, amphibians, and small fish. It is chiefly nocturnal but sometimes active by day, and seems more aquatic than some other species of *Kinosternon;* Seidel and Reynolds (1980) found it consistently demonstrated greater evaporative water loss compared to the more terrestrial *K. flavescens.*

Kinosternon herrerai Stejneger, 1925b

Herrera's mud turtle

Recognition: This poorly known species has a relatively long (to 17 cm), slightly domed carapace which is wider behind the middle. The 1st vertebral scute is elongated and extremely narrow; it never touches the 2d marginals. Vertebrals 2–5 are broader than long or at least as broad as long. Adults have a single low medial keel; juveniles may have two lateral keels as well. Marginals 10 and 11 are elevated above those preceding. Only four neurals are present, separating only costals 2–5. The carapace is olive to brown with darkened seams. The plastron is narrow (smaller than the carapacial opening) and notched posteriorly. Its anterior lobe is longer than the posterior lobe in males, but the opposite in females. The posterior hinge is akinetic (has little ability to move). The rigid abdominal scute is shorter than either anterior or posterior lobe, being about 20–30% of the maximum plastral length. The gular scute is usually less than half the length of the anterior lobe, and the interpectoral seam less than 10% of the maximum plastral length. The plastral formula is: an > abd > hum > gul > fem > pect. Bridge length is approximately 18–20% that of the plastron; there is axillary and inguinal contact. Both plastron and bridge are yellow to light brown, unmarked or with darker seams. The large head has a slightly projecting snout, a strongly hooked upper jaw, and its rostral scale is bifurcated posteriorly. Skin of the head is grayish brown and spotted dorsally and laterally; the cream-colored jaws are darkly streaked. Two chin barbels are present. The limbs are grayish brown and spotted with darker brown. The tail ends in a horny spine.

Males have long, thick tails and a patch of raised horny scales on their thighs and crura. Females have short tails.

Distribution: *Kinosternon herrerai* occurs in Gulf drainages in Tamaulipas, Veracruz, San Luis Potosí, Hidalgo, and Puebla, Mexico.

Habitat: *Kinosternon herrerai* lives in free-flowing to seasonally intermittent waterways at elevations below 800 m. It occasionally travels about on land, but Poglayen (1965) reported that a captive seemed disturbed by too much light, seldom left the water, and did not bask. His captive ate meats and fish.

Kinosternon sonoriense LeConte, 1854

Sonora mud turtle

Recognition: This medium-sized (to 17.5 cm), dark mud turtle has an elongated, tricarinate carapace with an unserrated posterior margin. Some individuals have three well-developed keels, but these may be lower in others, or nonexistent in very old turtles. In

Kinosternon sonoriense sonoriense

adults, the first four vertebrals are longer than broad, but the 5th (which is posteriorly flared) is usually broader than long. The 1st is anteriorly flared and usually touches the 2d marginals. Posterior marginals are flared, and the 10th is higher than the others. The carapace is olive brown to dark brown with dark seams. The hinged plastron is large and capable of almost closing completely; at best, it is only slightly notched posteriorly. The plastral formula is: abd > an > gul >< hum >< fem > pect (but see under geographic variation). Axillary-inguinal contact occurs on the bridge. The plastron is yellow to brown (sometimes mottled) with dark seams; the bridge usually brown. The head is moderate in size with a slightly projecting snout and a hooked upper jaw. There are three to four pairs of relatively long chin or neck barbels, and the rostral scale is not bifurated posteriorly. Gray, mottled skin occurs on head, neck, and limbs. The jaws are cream colored and may be dark flecked.

Males have concave plastra; long, thick, spine-tipped tails; vinculae; and are shorter (to 15.5 cm) than females (to 17.5 cm). Females have short tails and flat plastra.

Distribution: *Kinosternon sonoriense* ranges from the lower Colorado and Gila rivers in Arizona and New Mexico southward to the Río Yaqui basin of the Continental Divide, and eastward through the Río Casas Grandes basin of northwestern Chihuahua (Iverson, 1981a).

Geographic Variation: According to Iverson (1981a), two subspecies exist. *Kinosternon sonoriense sonoriense* LeConte, 1854 is found in New Mexico, Arizona, Sonora, and western Chihuahua. It has a long interanal seam (19.5% of carapace length in both sexes), a 1st vertebral scute of medium width (24.4% of carapace length in males, 25.5% in females), and a relatively wide gular scute (20.0% of carapace length in males, 19.4% in females). *K. s. longifemorale* Iverson, 1981a is known only from the Río Sonoyta basin in Arizona and Sonora, Mexico. It has a short interanal seam (14.4% of carapace length in males, 18.5% in females), a long interfemoral seam (12.8% of carapace length in males, females 13.5%), a wide 1st vertebral scute (28.9% of carapace length in males, 28.8% in females), and a narrow gular scute (17.7% of carapace length in males, 17.8% in females).

Habitat: This turtle occurs at elevations to 2042 m in streams, creeks, ditches, ponds, springs, and waterholes, usually in woodlands. Degenhardt and Christiansen (1974) reported that in New Mexico it migrates considerable distances from one body of water to another and may be present in farm ponds as far as 8 km from permanent water.

Natural History: Iverson (1981a) observed mating in the field on 4 May, but copulation probably also occurs earlier in the spring. Nesting takes place from May to September; two clutches may be laid. Up to nine eggs are laid per clutch (Hulse, 1982), and the elongated eggs (33 × 19 mm; Ewert, 1979) have brittle shells. Hatchlings (22–28 mm) have a low, broad, medial keel and two elongated lateral keels. The upper posterior edge of each marginal often has a black smudge, and the 10th marginal is not noticeably elevated. Their plastra are cream colored with a dark central blotch extending outward along the seams. Hatchlings may also have a pair of yellow lateral head stripes.

Kinosternon sonoriense is omnivorous, eating algae and higher aquatic plants as well as insects, snails, crustaceans, worms, fish, and amphibians. They are predominantly nocturnal.

Kinosternon flavescens (Agassiz, 1857)

Yellow mud turtle

Recognition: Yellow to brown in color, *Kinosternon flavescens* is a moderate-sized (to 16.5 cm) mud turtle with a broad, smooth, unserrated, medially depressed carapace lacking a medial keel. Its 1st vertebral scute is elongated, expanding anteriorly to touch the 2d marginals. Vertebrals 2–5 are usually broader than long, and the 5th is somewhat expanded. Both marginals 9 and 10 are elevated. Dark pigment occurs along the seams. The plastron is large with two well-developed hinges; it is yellowish to brown, with dark pigment along the seams. The forelobe is as long as or slightly longer than the hind lobe; there is a shallow posterior notch. The gular is much shorter than half the forelobe length. The plastral formula is: an > abd > hum > gul >< fem > pect. Seldom do the axillary and inguinal scutes touch on the bridge. The head is

Kinosternon flavescens flavescens

moderate in size with a slightly protruding snout and a hooked upper jaw in males (only weakly hooked at best in females). Its rostral scale is strongly furcate posteriorly. The head is yellow to grayish; the jaws are whitish to yellow and may bear small dark spots. There are two pairs of barbels. Other skin is yellow to gray, and the tail is spine tipped.

Males have long, thick tails and vinculae.

Distribution: The yellow mud turtle occurs in northwestern Illinois and adjacent Iowa and Missouri, and from Nebraska south through Texas, New Mexico, and southern Arizona to Sonora, Durango, Tamaulipas, and Veracruz, Mexico. It has been found at elevations as great as 6500 m.

Geographic Variation: Four subspecies are recognized. *Kinosternon flavescens flavescens* (Agassiz, 1857) ranges from southern Nebraska southward to Texas, New Mexico, and southeastern Arizona, and to immediately south of the Río Pánuco basin in Veracruz, Mexico. It has a short gular scute (to 59% of forelobe length); a long interhumeral seam (8–25% of carapace length); a wide plastral hind lobe (36–49% of carapace length in males, 38–53% in females); and a short bridge (16–25% of carapace length). *K. f. arizonense* Gilmore, 1922 is found in southern Arizona and Sonora, Mexico. It has a long gular (54–70%); a short interhumeral seam (8–15%); a narrow plastral hind lobe (35–42%); and a long bridge (21–28%). This subspecies was formerly known by the name *stejnegeri*, but Iverson (1979a) has shown it identical to fossils of *arizonense* described by Gilmore. *K. f. spooneri* Smith, 1951 is known only from isolated populations in Illinois, eastern Iowa, and northeastern Missouri. It is dark colored and has a long gular (40–55%); a short interhumeral seam (11–19%); a wide plastral hind lobe (36–45%); and a short bridge (17–22%). There is some controversy over its validity. Houseal et al. (1982) found it morphologically indistinguishable from *flavescens*, but Dodd (1982) questioned their findings. We have always been able to distinguish these two subspecies, and continue to do so here. *K. f. durangoense* Iverson, 1979a is known from the Mexican states of Chihuahua, Coahuila, and Durango. It has a much shorter plastral forelobe (31–33% of carapace length; usually over 33% in the other subspecies); a long gular (54–69%); a short interhumeral seam (6–12%); a narrow plastral hind lobe (39–40% in males, 42–47% in females); and a long bridge (20–24%). The characters used are basically those of Iverson (1979a).

Habitat: *Kinosternon flavescens* inhabits almost any quiet water within its range: swamps, sloughs, sinkholes, rivers, creeks, ponds, lakes, reservoirs, cisterns, and cattle tanks in semiarid grasslands or open woodlands. A mud or sand bottom is preferred, and aquatic vegetation is often present.

Natural History: The courtship pattern is identical to that of *Kinosternon odoratum* (which see for details). *K. flavescens* may stay in coitus for 10 min to 3 hr. Courtship usually takes place in water varying in depth from one inch to several feet, but a mating pair has been seen on land (Mahmoud, 1967). Nesting occurs from May to August but is usually completed by 1 July. Clutches vary from one to six eggs, with two to four being most common; apparently only one clutch is laid each season (Christiansen and Dunham, 1972). Eggs are elliptical, hard, and white, 24.0–28.5 mm in length and 13.5–16.5 mm in width. The hatchling carapace is 21–30 mm long, and the width nearly equals the length. It is slightly keeled, and marginals 9 and 10 are the same height as or slightly lower than marginal 8. Cahn (1937) reported that elevation of the marginals first appears at a carapace length of about 67 mm.

Kinosternon flavescens is omnivorous. Mahmoud (1968) found the following percentages of frequency and volume, respectively, of food items in *K. flavescens*: Insecta 94.7, 27.8; Crustacea 99.2, 27.7; Mollusca 93.7, 23.5; Amphibia 91.2, 9.2; carrion 13.2, 3.2; and aquatic vegetation 37.2, 8.5. Turtles under 5 cm in carapace length fed predominantly on small aquatic insects, algae, and carrion, whereas those over 5 cm ate a greater variety of items. Yellow mud turtles are essentially bottom feeders but sometimes feed at the surface or on land. Their acute sense of smell and taste under water aids them in locating food.

With the onset of cold weather these turtles burrow into natural depressions, such as old stumpholes, and beneath shrubs, brushpiles, logs, or leaf litter. Some dig burrows in loose sandy soil; others hibernate in muskrat dens or in the mud at the bottom of pools. During the hottest parts of summer they seek out sheltered spots to estivate.

Kinosternon angustipons Legler, 1965

Narrow-bridged mud turtle

Recognition: The low, oval, uniformly brown carapace (to 12 cm) is depressed and unkeeled in adults, but may be weakly tricarinate in juveniles. The posterior margin of the carapace is slightly notched, and the scutes overlap slightly in older individuals. Vertebrals 1–3 are about as broad as long; the 4th and 5th are much broader than long. The 1st or 3d vertebral is longest and broadest; the 5th is the shortest and narrowest. The 1st vertebral touches the 2d marginals;

Kinosternon angustipons

only the 10th marginal is elevated. The plastron is narrow, double hinged, and notched posteriorly. Its anterior and posterior lobes are, respectively, less than 39% and 34% of the maximum carapace length. In older individuals the plastral seams contain much soft pale tissue. The plastral formula is: abd > an > hum > fem > gul > pect. Width of the very narrow bridge is less than 21% of the maximum carapace length. The axillary and inguinal scutes are in contact on the bridge. Plastron and bridge are yellow, sometimes with dark seams. The head is only slightly broadened, the upper jaw is neither hooked nor notched, and the blunt snout does not protrude. Usually three to six cream-colored barbels are present in a longitudinal row on each side of the throat. Dorsally, the head is dark brown, changing gradually to tan and then cream on the sides of the snout. Lateral and ventral skin is pale cream. The iris is brown with gold flecks and the jaws are unmarked or only lightly marked with a few pale streaks. The limbs are gray to gray brown. Vinculae are present on the thighs and crura of males, and the tail of both sexes lacks a terminal horny spine.

Males are slightly shorter (to 11.5 cm) than females (to 12 cm) and have a bosslike enlargement of the snout in the prefrontal region and a long, thick prehensile tail. The female tail is short.

Distribution: *Kinosternon angustipons* occurs on the Caribbean lowlands from the Río San Juan on the Nicaraguan–Costa Rican border southward to Bocas del Toro Province, Panama.

Habitat: This is a relatively rare denizen of shallow swamps with slow currents, soft bottoms, and warm temperatures.

Natural History: Legler (1966) hypothesized (based on dissected females) that as many as four eggs may be laid, possibly singly, so several clutches may be laid each year. The only known egg was white, elliptical (40 × 22 mm), and had a brittle, semiglazed and irregularly lumpy (but not granular) shell. Stomachs examined by Legler (1966) contained plant remains and an orthopteran insect. Adults were attracted to traps baited with canned sardines and bananas, and captives ate ground beef.

This species seems most closely related to *Kinosternon dunni*.

Kinosternon dunni Schmidt, 1947

Dunn's mud turtle

Recognition: The elongated, oval, uniformly dark-brown carapace (to 17.5 cm) is slightly depressed, and unkeeled in adults, or slightly tricarinate in juveniles. Carapacial scutes are slightly overlapping, and the carapace is constricted at the bridge. Vertebrals 1 and 5 are broader than long, but 2–4 are longer than broad; the 1st touches the 2d marginals. The marginals form a raised rim to the shell and are sharply set off from the pleural scutes. Marginal 10, and occasionally also the 11th, is elevated above the level of those preceding it. The narrow plastron is double hinged and notched posteriorly. Plastral forelobe width is less than 40% and the hind lobe width is less than 35% of the maximum carapace length. The plastral formula is: abd > an > hum > gul > fem > pect. Bridge length is 21–26% of the maximum carapace length. Axillary and inguinal contact occurs. Plastron and bridge are yellow with dark seams. The head is broad, with a projecting snout, and the upper jaw is moderately hooked. An extensive rostral scale covers most of the dorsal skull, but is not posteriorly bifurcated. There are four chin barbels. The head is dark brown dorsally but lighter laterally (with some mottling) and ventrally. Neck, limbs, and tail are gray brown. Vinculae are present on the thighs and crura of the male, and the tail of both sexes ends in a horny spine.

Kinosternon dunni

In addition to the clasping organs on their hind legs, males grow larger and have long, thick tails. The shorter female (to 15 cm) has a small tail.

Distribution: *Kinosternon dunni* is known only from waterways in the Department del Chocó, Colombia.

Habitat: This species lives in small streams, where it burrows beneath the leaf litter covering the bottom.

Natural History: *Kinosternon dunni* apparently reproduces throughout the year, laying several clutches of about two eggs each (Medem, 1961). Eggs are ellipsoidal (45 × 25 mm) with brittle shells. Medem (1961) found mollusks remains in the digestive tract.

K. dunni is most closely related to the rare *K. angustipons*.

Dermatemydidae

Fossil dermatemydids have been found in Cretaceous to Miocene deposits from North America, Europe, and eastern Asia, but today this family is only known from one living species, *Dermatemys mawii*.

Dermatemys mawii Gray, 1847

Central American river turtle

Plate 4

Recognition: *Dermatemys mawii* is large (to 65 cm), almost totally aquatic, with a row of inframarginals on the bridge. The oval, uniformly olive-gray, flat-tened adult carapace is only slightly domed, not marginally serrated, and lacks a well-defined vertebral keel. Its nuchal bone contains well-developed lateral extensions (costiform processes) in hatchlings, but these, with age, are apparently covered by costals and peripherals. Neurals 6 and 7 are reduced, and consequently the posterior costals meet along the midline. The shell is thick, and in old adults the bones become so tightly fused that the sutures are obliterated. Vertebral scutes are longer than broad in adults, the opposite in juveniles. Thinness of the scutes allows the shell surface to be easily injured by abrasion, and suture lines may disappear with age. Twelve pairs of marginals are present, and the 10th dorsal rib does not contact the costal bones. The cream-colored bridge is broad and covered by three to six (usually four or five) inframarginal scutes; the inguinal is the largest, the axillary the smallest in this series. The solid, hingeless, cream-colored plastron is well developed, but the buttresses are not greatly developed. Posteriorly there is a wide plastral notch. Either 11 or 12 scutes are present on the plastron, as the gular may be single or divided. Occasionally, small extra scutes may lie along the mid- or lateral seams. The plastral formula is usually: abd > pect > fem > hum > an > gul. For such a large turtle, the head is relatively small. It contains a slightly upturned, somewhat tubular snout, and is olive gray laterally and lemon yellow to reddish brown dorsally. Some dark vermiculations may occur laterally, and the lower jaw is white. The skull has the temporal region cut away posteriorly so that the squamosal is separated from the parietal and broad postorbital bones. Frontal bones enter the orbits. The jugal articulates with both the pterygoid and palatine bones, but the quadratoju-

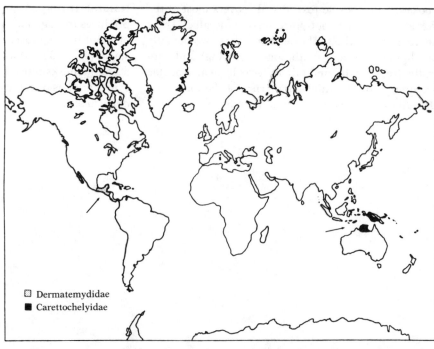

□ Dermatemydidae
■ Carettochelyidae

Distribution of Dermatemydidae and Carettochelyidae

gal does not touch the maxilla, and the quadrate does not surround the stapes. A concavity separates the premaxilla from the rest of the triturating surface, which contains a strong medial ridge; there is a moderately developed secondary palate. Several skull features are absent: the stapedio-temporale foramen, the stapedial artery and the structures housing it, and the postlagenar hiatus. The unpatterned limbs are dark gray, and the toes are webbed; a conspicuous fringe of large scales occurs on the outer edge of each foot. On the femur, the trochanteric fossa is widely open, a primitive condition. In the pelvic girdle, the pubes and ischia contact those from the opposite side ventrally, but the pubic and ischiadic symphyses are widely separated.

Dermatemys mawii has a diploid number of 56: 7 pairs of metacentric to submetacentric macrochromosomes, 5 pairs of telocentric or subtelocentric macrochromosomes, and 16 pairs of microchromosomes (Carr, Bickham, and Dean, 1981).

Males have long tails which extend beyond the carapacial rim; females have very small tails that barely reach the rim. Also, the dorsal surface of the head is yellowish to reddish brown in males, olive gray in females.

Dermatemys mawii is most closely related to kinosternid, carettochelyd, and trionychid turtles, being included with them in the superfamily Trionychoidea (Gaffney, 1975b).

Distribution: *Dermatemys* ranges from central Veracruz eastward through Tabasco, northern Chiapas, southern Campeche, and southern Quintana Roo in Mexico (it is absent from most of the Yucatán Peninsula) to northern Guatemala and Belize. There is also a single record from Tetela, Oaxaca (Smith and Smith, 1979) and it possibly occurs in northwestern Honduras.

Habitat: As its common name implies, *Dermatemys mawii* lives primarily in large rivers, lagoons, and lakes. It has also been taken at sites ranging from deep, clean, permanent holes in these water bodies to muddy backwaters, oxbows, and temporary seasonal pools. Apparently, as long as an abundance of aquatic food plants exist, it can live in almost any freshwater body within its range. Brackish water can be tolerated, and, evidenced by the occurrence of barnacles attached to the shell of some specimens, it may spend some time in the tidal areas near the mouths of large rivers.

This species is almost entirely aquatic, and it spends long periods of time under water. Apparently it takes sufficient dissolved oxygen from the water through the nasopharyngeal lining, as submerged individuals continually suck in water through the mouth and then expel it from the nostrils.

This turtle is most active at night and spends the daylight hours under water or floating (perhaps basking) at the surface.

Dermatemys is so adapted to a buoyant liquid medium that its limbs cannot well support its weight out of water, and it has difficulty walking any distance on land or holding its head off the ground.

Natural History: Since this turtle is so aquatic, it is no surprise that much of its life history is unknown. The courtship and mating acts have not been described, but may involve aggressive behavior since members of the opposite sex often fight when kept together in captivity. Alvarez del Toro et al. (1979) reported that *Dermatemys mawii* nests twice a year, in April and December in Chiapas, Mexico, and Lee (1969) found mature oviducal eggs in Guatemalan females slaughtered in February and March. Smith and Smith (1979) reported the nesting season to be continuous from April to September. Fall nestings are aided by flooding, which allows the female to select secluded areas away from the normal river channel that she could not reach by walking. However, Smith and Smith (1979) stated that nests are excavated at the very margin of inhabited waters, or but a few meters from it. Eggs may be deposited in a shallow excavation or merely covered with decaying vegetation. Six to 20, white, hard-shelled, elongated (57–70 × 30–34 mm) eggs are laid in each clutch. Juveniles have a brownish-olive carapace which bears a well-developed vertebral keel and a highly serrated posterior rim. Their olive-colored heads have a yellow stripe extending posteriorly from each orbit; in some this stripe may also run anteriorly from the orbit to the nostril.

Wild adults are entirely herbivorous, although captives sometimes eat fish; they consume both submerged and emergent plants, predominantly *Paspalum peniculatum*, fallen leaves, and fruits (Alvarez del Toro et al., 1979). Captive juveniles more readily accept animal foods, leading to speculation that they may be naturally more carnivorous than adults.

These turtles are highly prized for their meat and can be found in most markets close to large river bodies. They are hunted to the point of overexploitation. Alvarez del Toro et al. (1979) commented:

Dermatemys is capable of converting otherwise unused aquatic vegetation into high quality protein. Carefully harvested on a sustained yield basis, it could continue to be a valuable protein source; it has considerable economic potential for southern Mexico and other countries where it occurs. To let it become extinct is senseless.

Moll (1986) has determined that although the status of *Dermatemys* is relatively good in Belize, it is not currently protected by law there, and is vulnerable to exploitation by an ever-increasing human population.

Carettochelyidae

This formerly widespread family is today represented by only one living species, *Carettochelys insculpta*. Carettochelyids are known from the Eocene of North America and Europe, and may have been present in Asia and South America as well, but the family is now restricted to southern New Guinea and northern Australia.

Carettochelys insculpta Ramsey, 1887

Pig-nosed turtle

Plate 4

Recognition: *Carettochelys insculpta* is an aberrant freshwater species resembling the marine turtles in having paddlelike forelimbs. Its gray to olive carapace (to at least 55 cm) lacks scutes but has a rugose granulated surface. In this respect it resembles the softshells of the family Trionychidae, to which it is related. The adult carapace is smooth bordered and usually lacks a vertebral keel; that of juveniles has a serrated border and a pronounced knobby medial keel. The underlying bony carapace is fairly well developed with six to seven slender neurals partly separated along the midline by costals, and 10 complete peripherals. Vertebral centra are opisthocoelous except the 8th which is biconvex. There may be a series of white blotches along the lower sides of the carapace. The white bridge and plastron are well developed, but the entoplastron and epiplastra are only loosely attached to the hyoplastra. No scutes are present on the bridge or plastron, and the underlying bones and sutures may show through the skin covering. Emargination occurs posteriorly in the temporal region and the parietals no longer touch the squamosals; the skull is unique in having a well-developed concavity on the posterior surface of the articularis process. The quadratojugal extends forward to the maxilla, and the premaxillae are fused. The small jugal is separated from the parietal by a large postorbital, and only its medial process articulates with the palatine. Also, the vomer is so small it does not separate the palatines, allowing the choanae to be confluent. In the region behind the premaxillae, a large opening, the intermaxillary foramen, prevents the vomer from touching the premaxillae. Also the vomer does not touch the maxillae. The palatine touches the prefrontal and has a dorsal process that is incorporated into the lateral wall of the braincase. Palatines contact the basisphenoid which separates the broad pterygoids. The occipital condyle is composed only of exoccipitals. No ridge is present on the crushing surface of the white jaws. Dorsally, the head is granulated and gray to olive in color; it is whitish ventrally. A white blotch occurs behind each orbit. The thick snout protrudes like that of trionychids, but it is shorter, wrinkled, and has the nares somewhat laterally placed, giving it a piglike appearance. Each unpatterned, paddlelike forelimb contains two claws and is dark anteriorly but lighter posteriorly. The hind feet are heavily webbed and colored like the forelimbs. A series of large horizontal scutes lie on the dark dorsal surface of the tail.

The karyotype is 2n = 68 (Bickham and Carr, 1983).

Males have longer, thicker tails with the vent near the tip.

Distribution: *Carettochelys insculpta* inhabits the Fly, Strickland, Morehead, Lorentz, and Stekwa rivers and Lake Jamur in Papua New Guinea, and the Daly, Victoria, and Alligator drainages in the Northern Territory of Australia. It has only recently been discovered in Australia, but Cann (1978) thought it probably occurs in many rivers along the northern coast.

Geographic Variation: No subspecies have been described, but those from Australia should be critically compared to the New Guinea populations.

Habitat: Rivers, streams, lakes, and lagoons with soft bottoms and slow currents seem to be preferred, but *Carettochelys insculpta* has also been taken in the estuary of the Fly River, and may be able to withstand brackish conditions.

Natural History: The nesting season extends from September through November. In the evening, females crawl out onto mud or sand banks and deposit 15 to 30 round (39 × 40 mm), smooth, white, brittle-shelled eggs in shallow nests. The hatchling has a medially keeled carapace (57 mm) with a distinctly serrated margin. Growth, at least in captives, is slow and the dorsal keel may still be noticeable on individuals over 20 cm in carapace length.

Carettochelys feeds on the fruits of figs and pandanus which drop into the water, and the leaves, stems, roots, and seeds of aquatic plants. Animal foods include viviparid snails, insects, and small fish.

Due to the paddlelike structure of its forelimbs, *Carettochelys* does not swim like other freshwater turtles, but instead rows itself through the water by simultaneous movements of its forelimbs, as do marine turtles.

Cann (1978) remarked that this species often aggressively bites when first captured and that its powerful jaws can inflict severe wounds; with time, however, captives become calmer.

Trionychidae

The semiaquatic softshell turtles are found today in Africa, Asia, the Indo-Australian archipelago, and North America. Their fossil record indicates they were previously more widespread, once occurring in Europe and South America. The oldest known fossil trionychid is *Sinaspideretes wimani*, which dates possibly from the Late Jurassic. Although numerous fossil species are known, only 22, belonging to 14 genera, live today (we follow the generic designations proposed in Meylan, 1987).

Softshells are a highly derived group with rounded, flattened carapaces lacking horny scutes and covered by a leathery skin. Their neck is long and retractile, and limbs are paddlelike, with three claws on each. The snout is usually a long proboscis. Unfortunately, since the shell lacks cutaneous scutes, much emphasis has been placed on its bones. The arrangement of the typical carapacial and plastral bones is shown in the accompanying illustrations. Bony elements of the shell are secondarily reduced; pygals are absent, as are also peripherals, except in the genus *Lissemys*, and the distal ends of the ribs project freely. There are normally 7 or 8 neurals and 7 to 10 pairs of costals, with the last pair often meeting at the midline. A central lacunae and lateral fontanelles occur on the plastron. Several callosities (superficial ossifications closely connected to the underlying plastral bones) occur on the adult plastron. Cutaneous femoral valves (flaps), which cover the hind limbs when they are withdrawn, are present on the plastron of several Old World genera. Neck vertebrae are slender, and none is biconvex. The 4th digit has four to six phalanges; the 5th has two to four. The temporal region of the skull is widely open. The premaxillae are fused, and neither the parietal nor the postorbital touches the squamosal. Reduction of the vomer allows palatine contact. It is usually separated from the maxillae and prefrontals (in contact in *Chitra indica* and *Cyclanorbis elegans*), but touches the premaxillae and separates the internal nares. The jugal touches the parietal. A large nasopalatine foramen is present; the palatine fenestra is small. The maxilla touches the pterygoid but not the quadratojugal, and the basisphenoid is separate from the pterygoid but touches the palatines. The quadrate encloses the stapes, and an epipterygoid is present. The jaw is very deep at the level of the coro-

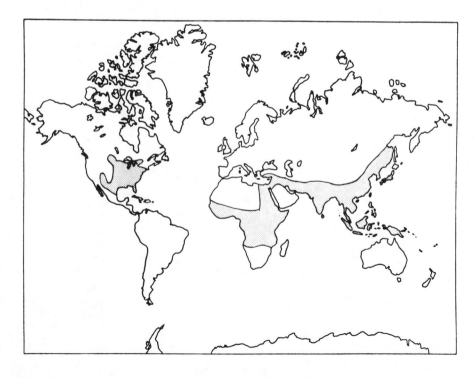

Distribution of Trionychidae

noid process, and the mandible reaches laterally nearly to the posterior end of the jaw. The family is divided into two subfamilies.

Subfamily Cyclanorbinae (Old World flap-plastroned softshells; 3 genera, 5 species). These turtles have hyo- and hypoplastra fused, the xiphiplastra surrounding the middle prong of the posteromedial process of the hypoplastra, seven or more callosities, and cutaneous femoral flaps on the plastron; the nuchal bone of the carapace has a conspicuous ventral ridge on each side which extends under the 1st costal bone; and the posterior border of the pterygoids has an ascending process which forms a suture with the opisthotic and greatly restricts the postotic fenestra. Members of this subfamily are restricted to Africa, India, and Burma.

Key to the genera of Cyclanorbinae:

1a. Peripheral bones present; India and Burma:
    Lissemys
 b. Peripheral bones absent; Africa   2
2a. Neurals form a continuous series;
    postorbital arch wider than greatest
    diameter of orbit; preplastra short and
    straight: ***Cycloderma***
 b. Neural series interrupted and not
    continuous; postorbital arch narrower than
    greatest diameter of orbit; preplastra long
    and angular: ***Cyclanorbis***

Subfamily Trionychinae (typical softshells; 11 genera, 17 species). These softshells have the hyo- and hypoplastra separated and distinct, the xiphiplastra surrounding the lateral prong of the posteromedial process of the hypoplastra, four or five callosities, and lack cutaneous femoral flaps on the plastron. The nuchal bone of the carapace lacks a conspicuous ventral ridge and lies over the 1st costal bone; the posterior border of the pterygoids lacks an ascending process that touches the opisthotic. Members of this subfamily occur in Southeast Asia, Indonesia, New Guinea, the Middle East, Africa, and North America.

In the following key, characters are taken from Meylan (1984, 1987).

Key to the genera of Trionychinae:

1a. Snout (proboscis) as long as greatest
    diameter of orbit   2
 b. Snout shorter than greatest diameter of
    orbit: ***Pelochelys***
2a. Postorbital arch much broader than
    greatest diameter of orbit; triturating
    surfaces narrow and sharp: ***Chitra***
 b. Postorbital arch narrower than greatest
    diameter of orbit; triturating surfaces
    broad and flat   3
3a. Anterior preplastral extensions long   4
 b. Anterior preplastral extensions short or
    medial in length   7
4a. All eight pairs of costals prevented from
    touching medially by neural bones:
    Dogania
 b. 7th or 8th pair of costals touch medially,
    not separated by neurals   5
5a. Seven plastral callosities: ***Pelodiscus***
 b. Four or five plastral callosities   6
6a. Four plastral callosities; nuchal less than
    three times as broad as long: ***Palea***
 b. Five plastral callosities; nuchal at least
    three times as broad as long: ***Amyda***
7a. Anterior preplastral extensions of medial
    length: ***Aspideretes***
 b. Anterior preplastral extensions short   8
8a. Nuchal less than three times as broad as
    long: ***Nilssonia***
 b. Nuchal three or more times broader than
    long   9
9a. 8th pair of costals complete, not reduced:
    Trionyx
 b. 8th pair of costals reduced or absent   10
10a. Only two callosities on plastron: ***Rafetus***
  b. Four or more callosities on plastron:
    Apalone

Dogania Gray, 1844

Only one species, *Dogania subplana*, is assigned to this genus.

Dogania subplana (Geoffroy, 1809)

Malayan softshell turtle

Plate 6

Recognition: The oval, very flat carapace (to 35 cm) is black to olive or dark brown with a black medial stripe and two or three pairs of black-centered, slightly yellow-bordered ocelli. This pattern fades with age. There are several longitudinal rows of small tubercles on the carapace, but only small blunt tubercles, if any, are present on the anterior carapacial rim above the neck. A single neural occurs between the 1st pair of costals. The 8th pair of costals is well developed but, like the preceding seven pairs, separated by a neural bone. Costals and neurals are roughened with granulated rugosities and small pits. The plastron is whitish to cream or gray, and bears four weakly developed (hyo-hypoplastral, xiphiplastral) callosities. Often only the xiphiplastral callosities are distinguishable. Preplastral arms are long and slender and nearly meet anterior to the epiplastra. The epiplastra have long rami which lie at obtuse or right angles to the plastral midline. The skull is large with a bony snout at least as long as the diameter of the orbit and a mandibular symphysis narrower than the diameter of the orbit; it lacks a median ridge. Triturating surfaces of the maxilla also lack longitudinal ridges. The head is brown to olive or black with some yellow and black spots. A black medial stripe may pass along the top of the snout and between the orbits, another may pass on each side from the snout through the orbit, and a small black stripe may pass diagonally backward from the orbit. In the young, a reddish blotch occurs behind the eye and over the tympanum on each side of the head; this fades in adults. The chin has black vermiculations, and the neck and outer surface of the limbs are olive to blackish with some small yellow spots.

Males have long, thick tails; the female tail is short.

Distribution: *Dogania subplana* ranges from southern Burma and Thailand southward through Malaya to Java, Sumatra, Borneo, and the Philippines.

Habitat: Smith (1931) reported that *Dogania subplana* prefers clean hill streams rather than slow-flowing mud-bottomed rivers. It often hides beneath rocks and large stones, its flattened shell allowing it to do so.

Palea Meylan, 1987

Palea steindachneri is the only species in this Asian genus.

Palea steindachneri (Siebenrock, 1906b)
Wattle-necked softshell turtle

Recognition: The oval (longer than broad) carapace (to 42.6 cm) has numerous longitudinal rows of small, raised tubercles in younger individuals; the carapacial surface becomes more nearly smooth with age. There is a well-defined marginal ridge along the anterior rim of the carapace, and on it are enlarged, blunt tubercles. No preneural bone is present and only a single neural bone separates the most anterior of the eight pairs of costals; the 8th pair meets at the midline. Carapacial bones are pitted. Adult carapaces are brown, olive brown, or gray brown, and patternless. The yellow to cream or grayish plastron usually lacks dark markings. It has callosities on the hyo- and hypoplastra, xiphiplastra, and epiplastra. The long preplastra are in contact. The epiplastra lie at right angles to the plastral midline and have pointed branches. There is usually a suture between the hyo- and hypoplastra. The skull is moderate in size with a bony snout which is slightly longer than the greatest diameter of the orbit. No symphysial ridge occurs on the mandible, and the width of the symphysis is less than the greatest diameter of the orbit. Head and limbs are olive to brown. There are black preorbital, suborbital, and postorbital streaks, and shorter black streaks and dots on top of the head. A pale yellow stripe begins behind the eye and runs backward on the

Palea steindachneri, hatchling

side of the neck, becoming narrower as it proceeds toward the body. A yellowish spot also occurs at the corner of the jaws. Head and neck markings are lost with age. At the base of the neck is a large clump of coarse tubercles (wattles) which is diagnostic for the species. A lateral ridge projects from each side of the nasal septum.

Males are smaller than females and have longer, thicker tails with the vent closer to the tip.

Distribution: *Palea steindachneri* ranges in China from Kwangtung, Kwangsi, and Hainan Island southwestward into Vietnam. It has been introduced and is now established on the islands of Kauai and Oahu, Hawaii (Webb, 1980b; McKeown and Webb, 1982).

Habitat: In Hawaii this softshell occurs in marshes and drainage canals. It is found to an altitude of 1500 m in Tongking, China (Pope, 1935).

Natural History: The sparse life history data known for *Palea steindachneri* were reported by McKeown and Webb (1982). Nesting in Hawaii probably occurs in June, and hatching in late August or September. Clutch size varies from 3 to 28, and the spherical (22 mm in diameter) eggs have brittle shells. Hatchlings have rounded carapaces (54–58 mm) which are orangish brown with scattered black spots. The longitudinal rows of tubercles on the carapace are prominent, as also are the dark head stripes and yellow neck stripe.

Both juveniles and adults are primarily carnivorous. Captives at the Honolulu Zoo have eaten fish, raw beef, horse meat, chicken parts, mice, crickets, crawfish, mollusks, amphibians, and some plant materials. The basking habit is poorly developed in adults, although juveniles may venture onto land.

Pelodiscus Gray, 1844

Pelodiscus sinensis is the only species in this genus.

Pelodiscus sinensis (Wiegmann, 1835)

Chinese softshell turtle

Recognition: There is a marginal ridge on the oval, slightly longer than wide, carapace (to 25 cm). Juveniles exhibit longitudinal rows of blunt tubercles, but

Pelodiscus sinensis

the adult carapace is smooth, except for some enlarged, blunt knobs on the anterior rim above the neck. No preneural bone is present and only a single neural bone separates the anterior of the eight pairs of costals. The 8th costals meet at the midline. Carapacial bones are finely pitted. The carapace is olive to gray and in juveniles is patterned with round, light-bordered black spots. The white to yellow plastron is immaculate in adults but has large black blotches in juveniles. Callosities, totaling seven, occur on the hyo- and hypoplastra, the xiphiplastra, and sometimes the epiplastra. The preplastra are separated, and the epiplastra form obtuse angles to the midline. There is usually a suture between hyo- and hypoplastra. The skull is of moderate size with a bony snout which is longer than the greatest diameter of the orbit. The mandible lacks a symphysial ridge, and the width of the symphysis is greater than the greatest diameter of the orbit. Head and limbs are olive to yellowish white, and the head and neck may have fine black lines. The throat is either light with vermiculations or dark with yellow spots. There often are fine black lines radiating from the eyes. The tubular snout has a lateral ridge projecting from each side of the nasal septum; the lips are fleshy and the jaws are sharp.

Males differ from females in being shallower and having long, thick tails, with the vent near the tip.

Distribution: This species inhabits central and southern China, Vietnam, and the islands of Hainan and Taiwan, and has been introduced into the Hawaiian Islands, one of the Bonin Islands, Timor, and Japan.

Geographic Variation: No subspecies are currently recognized as valid; however variation in patterns does exist, and the Chusan Island population has been named *tuberculatus* (Cantor, 1842), and those on Japan, *japonicus* (Temminck and Schlegel, 1835). These are usually placed in the synonomy of *sinensis*, but, as Stejneger (1925a) noted, a lack of specimens has prevented a thorough study of variation in *sinensis*.

Habitat: In China, *Pelodiscus sinensis* is found in rivers, lakes, ponds, canals, and creeks with slow currents (Pope, 1935). On Kauai, Hawaii, it occurs in marshes and drainage ditches. The basking habit is not well developed.

Natural History: In Japan, nesting begins in late May and continues to mid-August (Mitsukuri, 1905). Licent (in Pope, 1935) discovered quantities of eggs on 14 June in southeastern Kansu, China. The nest is a squarish hole with the corners rounded out; it generally is about 7.5–10 cm across the entrance. The white, spherical eggs average about 20 mm in diameter, but may be as large as 24 mm. Clutch size is 15 to at least 28 eggs, and two to four clutches are laid each year. Incubation takes about 60 days (40–80). Hatchlings average 27 mm in carapace length and are about 25 mm wide. The carapace is olive, and it may have a pattern of small, dark-bordered ocelli. Its marginal fold is prominent, as are the longitudinal rows of spiny tubercles. The plastron is white to yellow, with large dark blotches. Limbs and head are olive above and lighter below. The head has dark flecks, and dark lines radiate from the eyes. The throat is mottled, and the lips may have small, dark bars. There is a pair of dark blotches in front of the tail and a black band on the posterior side of each thigh.

  Pelodiscus sinensis is predominantly carnivorous. Heud (in Pope, 1935) found the remains of fish, crustaceans, mollusks, insects, and seeds of marsh plants in stomachs he examined. Mitsukuri (1905) reported that juveniles feed on fish and that adults eat fish and bivalves. Captives eat canned and fresh fish, canned dog food, raw beef, mice, frogs, and chicken.

Comment: The genera *Pelodiscus*, *Palea*, and *Dogania* are closely related (Meylan, 1987).

Trionyx Geoffroy, 1809

This monotypic genus contains only the species *Trionyx triunguis*.

Trionyx triunguis (Forskål, 1775)
African softshell turtle

Recognition: This is the largest species of *Trionyx*; its oval carapace reaches a length of 95 cm (Loveridge and Williams, 1957). The carapace is olive to dark brown, sometimes without markings, but usually, at least in juveniles, with some light-centered, dark-bordered spots. It is often bordered with yellow. In juveniles, several longitudinal rows of tubercles lie on the carapace, but in large adults the dorsal surface is smooth. The anterior rim of the carapace is thickened over the neck. A single neural bone separates the 1st pair of costals. The 8th pair of costals are complete (not reduced), and these and sometimes the 7th pair are at least partially in contact at the midline. Surfaces of all carapacial bones are coarsely pitted. The plastron is white to cream colored, usually unpatterned, but with a few faded anterior vermiculations in some. The hyo-hypoplastral and xiphiplastral callosities are well developed and pitted in adults. Preplastra are separated and the epiplastra are bent at right angles to the plastral midline. The skull is small, considering the size attained by adults, with its bony snout longer than the diameter of the orbit. The intermaxillary foramen is small, only about a third as long as the primary palate. No longitudinal ridge occurs on the maxilla, and a symphysial ridge is also absent from the mandible. The width of the mandibular symphysis is equal to or longer than the diameter

Trionyx triunguis

of the orbit. Head and limbs are olive and heavily marked with small yellow or whitish spots and vermiculations. Both chin and throat contain a network of large white spots. The undersides of the limbs are yellow.

Males have long, thick tails with the vent near the tip.

Distribution: African softshells range over most of the continent except the waterways of southern and northwestern Africa. They are known from both the White (below Murchison Falls) and entire Blue Nile drainages, Lakes Rudolf and Albert, the tributaries of the Congo River, and most drainages in West Africa. They have not been found in Lake Victoria. *Trionyx triunguis* also occurs along the coasts of Israel, Lebanon, Syria, and Turkey.

Habitat: The usual habitat is a slow-moving freshwater body, such as a large river, stream, pond, or lake, but *Trionyx triunguis* also enters brackish waters where its range meets the sea. A large specimen was caught in the ocean 3 or 4 km from the mouth of the Gaboon River (Loveridge and Williams, 1957), and it has been taken in waters of 33–42 ppm salinity in Israel and Turkey, where it is a nuisance to fishermen.

Natural History: Courtship and mating have not been described. Nesting occurs from late March to July, depending on latitude, and nest cavities are dug in sand and earthen banks and on islands. Those living in brackish coastal waters may lay on the adjacent sea beaches. The white brittle-shelled eggs are spherical, about 32 mm in diameter, and a single female may contain 25 to more than 100 eggs. Hatchlings have carapaces about 30 mm long.

Trionyx triunguis is omnivorous, feeding on such animals as aquatic insects, crustaceans, mollusks, fish, and amphibians, as well as palm nuts and dates. Both live and dead animals are accepted. Cansdale (1955) reported it will lie in ambush and then suddenly grab its prey when it comes within reach.

Cansdale (1955) also reported that it sometimes leaves the water to bask, but is shy and difficult to observe. When first caught it is a vicious biter. Much of its life is spent under water, and it is capable of aquatic breathing. Girgis (1961) showed that a quiet, submerged *T. triunguis* used only 6.5 ml oxygen/kg/hr; approximately 30% (2.04 ml oxygen/kg/hr) was attributed to pharyngeal breathing, the other 70% entered through the skin. However, this probably allows only a very low level of metabolism.

Rafetus Gray, 1864a

This genus contains two little-known Asiatic species with only two plastral callosites. *Rafetus* and the North American genus *Apalone* form a natural group (Meylan, 1987).

Key to the species of *Rafetus*:

1a. Epiplastra form right angles: ***R. swinhoei***
 b. Epiplastra form acute angles: ***R. euphraticus***

Rafetus swinhoei (Gray, 1873d)
Swinhoe's softshell turtle

Recognition: The oblong carapace (to 33 cm) is olive green with numerous yellow spots and many small yellow dots between them (sometimes encircling the larger spots, or forming narrow stripes). These marks are particularly prominent along the anterior parts of the sides. A single neural separates the 1st pair of costals. The 8th pair of costals is reduced and does not meet at the midline. Carapacial bones are coarsely pitted. The plastron is grey with only two poorly developed callosities on the hyo- and hypoplastra. The preplastra are separated and the epiplastra form right angles to the plastral midline. The skull is moderate in size with a short bony snout. The mandibular symphysis is narrower than the greatest diameter of the orbit and lacks a median ridge. The jugal touches the squamosal, and the basisphenoid contacts the palatines. The head, neck, and dorsal surface of the limbs are black to olive, but yellow ventrally. Numerous large yellow spots occur on the head, neck and chin.

Males have long, thick tails with the vent near the tip.

Distribution: This species occurs in southern China, and probably also Vietnam.

Comment: *Rafetus swinhoei* has only recently been resurrected by Meylan and Webb (1988). Nothing has been published on its habitat requirements or natural history.

Rafetus euphraticus (Daudin, 1802)

Euphrates softshell turtle

Recognition: The round to oval carapace (to 40 cm) is olive green, flecked with yellow, cream, or white in juveniles but without markings or with a few dark blotches in adults. Numerous longitudinal rows of small tubercles occur on the juvenile carapace, and some may persist into adulthood. A series of enlarged blunt tubercles lies above the neck on the anterior rim of the carapace. A single neural bone separates the 1st pair of costals. The 8th pair of costals is reduced in size and does not meet at the midline. Carapacial bones are coarsely pitted. The plastron is white to cream colored and has only two poorly developed callosities (on the hyo- and hypoplastra). Its pre-plastra do not touch and the epiplastra form acute angles to the plastral midline. The skull is moderate in size with a short, bony snout. The triturating surface on the maxilla is ridgeless but rugose; the mandibular symphysis is narrower than the diameter of the orbit and lacks a median ridge. The basisphenoid is medially constricted and not in contact with the palatines. The head, dorsal surface of the neck, and exposed parts of the limbs are green. Juveniles have numerous light spots on the head. The chin and undersides of the neck and limbs are whitish.

Males have long, thick tails with the vent near the tip.

Distribution: *Rafetus euphraticus* is known from the Tigris and Euphrates drainages in southern Turkey, Syria, Iraq, and Iran. It has also been reported from northeastern Israel.

Habitat: It is found in permanent rivers and streams with mud or sand bottoms.

Natural History: *Rafetus euphraticus* is fond of basking but dashes into the water at the slightest disturbance. Like other softshells it is carnivorous, eating a variety of small aquatic animals such as insect larvae, mollusks, crustaceans, amphibians, and fish.

Apalone Rafinesque, 1832

North American softshell turtles having pronounced sexual dimorphism, four or more plastral callosities, short preplastral extensions, the nuchal at least three times broader than long, the 8th pair of costals reduced or lost, and eight or nine neurals. The karyotype

is 2n = 66: 12 macrochromosomes, 54 microchromosomes (Atkin et al., 1965; Stock, 1972; Gorman, 1973).

Key to the species of *Apalone*:

1a. Marginal ridge present, at least anteriorly on carapace: **A. ferox**
 b. Marginal ridge absent   2
2a. Nostrils round with no ridge projecting from septum; anterior rim of carapace without tubercles: **A. mutica**
 b. Nostrils crescentic, with a ridge projecting from medial septum; anterior edge of carapace with tubercles: **A. spinifera**

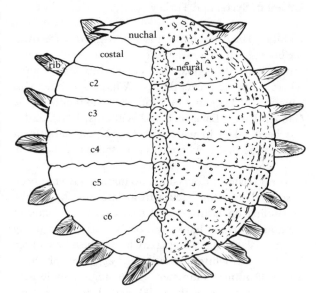

Carapacial bones of *Apalone*

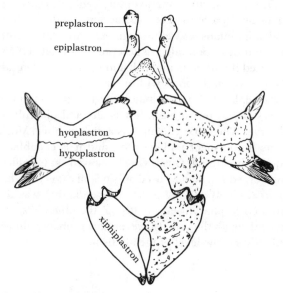

Plastral bones of *Apalone*

Apalone ferox (Schneider, 1783)

Florida softshell turtle

Recognition: The oval carapace (to 60 cm) is gray to brown and may have darker blotches, especially in younger individuals. A series of enlarged, blunt tubercles lies along the anterior rim of the carapace and on the well-defined marginal ridge. Often there are longitudinal rows of indentations and small raised tubercles on the dorsal surface. No preneural bone is present and only a single neural bone separates the anterior pair of costals. Seven or eight neurals and seven or eight pairs of costals are present; the last pair of costals meet at the midline. Carapacial bones are strongly pitted. The gray to white plastron has hyo-hypoplastral and xiphiplastral callosities, but callosities are usually lacking or poorly developed on the preplastra and epiplastra. The preplastra are separated, and the epiplastra lie at right angles to the midline. Often there is no suture between the hyoplastra and hypoplastra. The moderately sized skull has a bony snout slightly longer than the diameter of the orbit. The mandibular symphysis is shorter than the diameter of the orbit. No ridge occurs on the maxillary triturating surface, and this surface may be broadened in older males. Head and limbs are gray to brown and sometimes bear light mottlings or reticulations. Often a red or yellow stripe extends from the posterior corner of the eye to the base of the lower jaw. Each nostril contains a lateral ridge that projects from the nasal septum.

Adult females are 20–60 cm in carapace length, males 15–29 cm. Males have long, thick tails with the vent near the tip. Older males may develop an expanded crushing surface in the upper jaw.

Distribution: *Apalone ferox* ranges from southwestern South Carolina and southern Georgia southward through peninsular Florida and westward through the Florida Panhandle and southern Alabama to Mobile Bay.

Apalone ferox

Habitat: This softshell occurs in all freshwater habitats, but in the northern parts of the range it seems to be more at home in still waters than in the southern part of the range. It prefers deep water with sand or mud bottoms, or bubbling mud-sand springs where there is foliage overhead. It sometimes occurs in brackish water near the mouths of streams, and the tides occasionally carry it out to sea. The Florida softshell spends much time buried in the soft bottom in either shallow or deep water, with only its head protruding. When buried in the mud it appears to consider itself perfectly concealed and protected. It can burrow into and tunnel through mud with amazing speed.

Natural History: Nesting occurs from mid-March to July in Florida and from June through July in the northern parts of the range. A typical nest has a total depth of about 12.5 cm. The spherical, white eggs have thin, brittle shells; they are 24–32 mm in diameter. Clutches consist of 4 to 23 eggs, most commonly 17 to 22, and possibly more than one clutch is laid each season (Iverson, 1977d). Hatching occurs in August and September, after about 60–70 days of incubation. Hatchlings are less rounded than are those of other North American species of *Apalone*. The carapace is yellowish olive with dusky spots and a narrow yellow or orange border. The spots are large, and the narrow light lines separating them give a reticulated appearance. The plastron is dark gray, and the olive skin is mottled with lighter pigment. A Y-shaped figure extends from the anterior edge of each orbit to the middle of the snout. Carr (1952) gave the following measurements of a very young *A. ferox:* carapace length 38 mm, carapace width 33.5 mm, and depth 12.5 mm.

Apalone ferox is thought to be omnivorous but not overly fond of plant material. The bulk of the diet apparently consists of invertebrates, and the turtle is known to do some scavenging. Its natural foods include crayfish, snails, mussels, frogs, fish, and waterfowl. Captives readily feed on canned and fresh fish, canned dog food, raw beef, and chicken. The expanded crushing surface of the jaws in some large individuals may be an adaptation for crushing mollusks.

Apalone spinifera (Le Sueur, 1827)

Spiny softshell turtle

Plate 6

Recognition: The round, leathery carapace (to 49.8 cm) has a rough, sandpaperlike surface. Conical projections or spines are present along the anterior edge of the shell. No preneural bone is present and only a

Apalone spinifera spinifera

Apalone spinifera pallida

single neural separates the anterior pair of costals. Seven or eight neurals and seven or eight pairs of costals present; if present, the 8th pair of costals is reduced, but the 7th pair meets at the midline. Carapacial bones are strongly pitted. The carapace is olive to tan, with a pattern of black ocelli or dark blotches and a dark marginal line. The plastron is immaculate white or yellow. It has well-developed callosities on the hyoplastra, hypoplastra, and xiphiplastra; poorly developed callosities occur on the preplastra and epiplastra less frequently. The epiplastral angle is about 90°, and there is usually a suture between the hyo- and hypoplastra. The moderate-sized skull has a bony snout about as long as the greatest diameter of the orbit. The mandibular symphysis is shorter than the diameter of the orbit. There is no maxillary ridge and the maxillae touch above the premaxillae (not in contact in *Apalone mutica*). Head and limbs are olive to gray, with a pattern of dark spots and streaks. Two separate, dark-bordered, light stripes are found on each side of the head: one extending backward from the eye, the other backward from the angle of the jaw. The tubular snout has large nostrils, each of which

contains a septal ridge; the lips are yellowish with dark spotting, and the jaws are sharp.

Adult males (12.7–21.6 cm) have long, thick tails with the vent near the tip, and retain the juvenile pattern of ocelli, spots, and lines. Adult females (16.5–49.8 cm.) develop a mottled or blotched pattern.

Distribution: This species ranges from northwestern Vermont, southern Ontario, and Quebec to North Dakota and Montana and south to the Gulf Coast States and New Mexico. It also occurs in the Gila and lower Colorado rivers in California, Nevada, Arizona, and New Mexico and in the northern parts of Tamaulipas, Nuevo Léon, Coahuila, and eastern Chihuahua, Mexico. Individuals have also been taken from the lower Otay Reservoir and San Diego River in California (Stebbins, 1985) and from the Red River of the North along the Minnesota–North Dakota line. *Apalone spinifera* has also become established in Salem County, New Jersey.

Geographic Variation: Seven subspecies are recognized. *Apalone spinifera spinifera* (Le Sueur, 1827) ranges east of the Mississippi River, from Vermont and extreme southeastern Canada west to Wisconsin and south to North Carolina, western Virginia, and Tennessee. It can be distinguished from the other subspecies by the presence of large black ocelli in combination with only one dark marginal line. *A. s. hartwegi* (Conant and Goin, 1948) ranges west of the Mississippi River, from Minnesota to Montana and south to northern Louisiana, Oklahoma, and northeastern New Mexico. It has uniform small dots and ocelli on the carapace and only one dark marginal line. *A. s. asper* (Agassiz, 1857) (Plate 6) occurs from southern North Carolina to southeastern Louisiana; its range includes the Florida Panhandle but not peninsular Florida. This race has more than one black line paralleling the rear margin of the carapace, and there is often a fusion of the postlabial and postocular stripes on each side of the head. *A. s. pallida* (Webb, 1962) occurs west of the Mississippi in the upper Red River drainage and in rivers that drain into the Gulf of Mexico east of the Brazos River in Texas. It is pale and has white tubercles on the posterior half of the carapace; the tubercles gradually decrease in size anteriorly and are indistinct or absent on the anterior third of the carapace, and they are not surrounded by black ocelli. *A. s. guadalupensis* (Webb, 1962) occurs in the Nueces and Guadalupe–San Antonio drainage systems of south-central Texas. It is dark and has white tubercles surrounded by narrow black ocelli on the anterior third of the carapace; some tubercles are as large as 3 mm in diameter. Small black dots are sometimes interspersed among the white tubercles. *A. s. emoryi* (Agassiz, 1857) occurs in the Rio Grande drain-

Apalone spinifera ater

age in Texas and New Mexico and the Colorado River drainages in Arizona, New Mexico, southwestern Chihuahua, central Coahuila, northern Nuevo Léon, and Tamaulipas. It has a pale rim on the carapace that is four or five times wider posteriorly than laterally. There is a dark, slightly curved line connecting the anterior margins of the orbits, and the postocular stripes usually are interrupted, leaving a pale blotch behind each eye. *A. s. ater* (Webb and Legler, 1960) is restricted to permanent ponds in the Cuatro Ciénegas basin, Coahuila, Mexico. It is dark gray or brownish black with a gritty to smooth carapacial surface but no tubercles on the anterior rim. The rear carapacial margin is rugose and the edge ragged. Many black flecks occur on the plastron. Originally described as a separate isolated species, *ater* is now considered only a subspecies of *spinifera* by Smith and Smith (1979). Drainage canals constructed in the Cuatro Ciénegas basin have allowed adjacent populations of *A. s. emoryi* to invade the range of *ater*, and Smith and Smith feel this has resulted in interbreeding, and that the turtles of the basin are losing their *ater*-like characters. However, a 1983 field trip by Kenneth Darnell, George Grall, and Anthony Wisnieski yielded several "pure" *ater*; thus a more thorough examination of the relationship between *ater* and *emoryi* is needed.

Habitat: *Apalone spinifera* inhabits a great variety of aquatic habitats, including marshy creeks, large swift-flowing rivers, bayous, oxbows, lakes, and impoundments. A soft bottom with some aquatic vegetation is essential, and sandbars and mud flats are preferred. Fallen trees with spreading underwater limbs are frequented.

The spiny softshell is highly aquatic, spending most of its time in the water, either foraging, floating at the surface, or buried in the soft bottom with only the head and neck protruding. It buries itself by flipping silt over its back until it is entirely concealed. Often it lies buried under water shallow enough to allow the

nostrils to reach the surface when the neck is extended, but it also burrows in deep water.

Natural History: Mating takes place in April or May. Legler (1955) observed a captive male *Apalone spinifera* swim behind or above a female *A. mutica* and nip at the anterior part of her carapace. During these movements the posterior edge of her carapace was turned up slightly, whereas his was turned down. They frequently surfaced to breathe, and she occasionally followed him. When they settled to the bottom he crawled onto her carapace from the rear but did not clasp it with his feet. June and July are the usual months of nesting, but the nesting season may begin in May and extend into August. The nests are flask shaped and extend to a depth of 10–25 cm. The white, brittle eggs are spherical or nearly so and average about 28 mm (24–32) in diameter. Clutch size is 4 to 32 eggs, and probably more than one clutch is laid each year. Hatching occurs from late August to October. The young resemble adult males in shape and color. The pale, olive to tan, rounded carapace has a well-marked pattern of small, dark ocelli or spots and a yellowish border set off by a black line. The granulation of its surface is pronounced, but the spines along the anterior edge are small and poorly developed. Hatchlings are 30–40 mm in carapace length. Bull and Vogt (1979) showed that sex determination in *A. spinifera* is independent of the incubation temperature; clutches they incubated at 25°C and 30.5°C both yielded approximately 1:1 sex ratios.

A. spinifera is predominantly carnivorous; however, it consumes some plant material, possibly by accident. It eats crayfish, aquatic insects, mollusks, earthworms, fishes, tadpoles, and frogs.

Apalone mutica (Le Sueur, 1827)

Smooth softshell turtle

Recognition: The round, smooth carapace (to 34.5 cm) is olive to orange brown, with a pattern of darker dots, dashes, or blotches; often there is a lighter marginal band bordered on the inside by a darker area. The leathery carapace lacks spines or raised knobs, and there is no marginal ridge. There is no preneural bone, and only a single neural separates the anterior pair of costal bones. Seven or eight neurals and seven or eight pairs of costals are present; the posterior costals may or may not meet at the midline. Carapacial bones are strongly pitted. The white or gray, patternless plastron has callosities on the hypoplastra and xiphiplastra, which seem larger than those in *A. ferox* or *A. spinifera*. Adults also frequently have callosities on the preplastra and epiplastron; that on the epi-

Apalone mutica mutica

Apalone mutica calvata

plastron sometimes covers the entire surface. The epi-plastron is not divided, and its median angle is obtuse and somewhat greater than 90° (Webb, 1962). The hyo- and hypoplastra may be fused, or may be connected by a suture. The moderately sized skull has a bony snout which is about as long as the greatest diameter of the orbit. The mandibular symphysis is shorter than the greatest diameter of the orbit. No medial ridge occurs on the maxilla. Head, neck, and limbs are olive to orangish above and gray to white below; there may be a few scattered dark spots on the limbs, but they are generally without a distinct pattern. A black-bordered light stripe extends through the eye and onto the neck. The tubular snout usually terminates somewhat obliquely, and the round nostrils are slightly inferior; no septal ridge is present.

Adult males (11.0–17.5 cm) have long, thick tails with the vent near the tail tip. Females have longer hind claws but males have longer foreclaws. Adult females (17.0–34.5 cm) may have a mottled or blotched carapace.

Distribution: *Apalone mutica* ranges from Ohio, Minnesota, and North Dakota south to the western Florida Panhandle, southeastern Texas, and eastern New Mexico. It is now considered extirpated in western Pennsylvania.

Geographic Variation: Two subspecies are recognized. *Apalone mutica mutica* (Le Sueur, 1827) ranges in the central United States from Ohio to southern Minnesota and South Dakota, south to Tennessee, Louisiana, and Oklahoma and west into Texas and New Mexico. It is distinguished by a juvenile pattern of dusky dots and short lines, ill-defined pale stripes on the snout, and pale postocular stripes with black borders less than half their width, except in some individuals in the Colorado River drainage of Texas. *A. m. calvata* (Webb, 1959) occurs along the Gulf coast from the Escambia River system of Alabama and the Florida Panhandle west to Louisiana and Mississippi, including the Pearl River drainage. It has a juvenile pattern of large circular (often ocellate) carapacial spots, no stripes on the dorsal surface of the snout, a pattern of fine markings on the dorsal surface of the limbs, and pale postocular stripes having thick black borders approximately half their width on adult males.

Habitat: The smooth softshell occurs in large rivers and streams having moderate to fast currents. *A. m. calvata* has been taken only from such habitats, but *A. m. mutica* is also known from lakes, impoundments, and shallow bogs. Waterways with sandy bottoms and a few rocks or aquatic plants are preferred.

Natural History: Plummer (1977a) saw courtship behavior in April, May, June, and August. Males approached other basking turtles with neck fully extended and placed their snouts under the edge of the other turtle's carapace and probed around to the sides and back. If the other turtle was a female, she often would spin around, charge and bite at the male. Receptive females remained passive while the male investigated and mounted. Mountings were only successful in deeper water. The nesting period lasts from late May through July. Nests are dug on sandbars, banks, and islands in full sunlight, and are 15–22 cm deep. Clutches consist of 3–33 eggs (11–22 are usual), and two to three clutches may be laid each season. Eggs are spherical (20–23 mm in diameter) with thick, brittle, white shells. Incubation lasts 65–77 days. The dull-olive carapace (34–45 mm) of the hatchling is marked with numerous short, black lines and bordered with a pale margin, which broadens posteriorly. Like adults, the young are nearly circular in outline.

Smooth softshells are decidedly carnivorous; some plant material is eaten, however, perhaps accidentally. They obtain food by prowling about on the bottom or on submerged debris, actively pursuing and capturing prey, or by ambushing it by lying concealed on the bottom. *Apalone mutica* feeds on fish, frogs, tadpoles, mudpuppies *(Necturus)*, crayfish, aquatic insects, snails, bivalves, and worms. Plant material reportedly taken includes algae, fruits, and "hard nuts" (Carr, 1952). Captives feed readily on canned or fresh fish and various meats.

Aspideretes Hay, 1904b

This genus comprises four closely related Asiatic species which are unique in having two pairs of neurals between the 1st pair of costals. Five callosities are present on the plastron, and the anterior preplastral arms are intermediate in length. In the skull, the jugal touches the squamosal in some individuals, and the basisphenoid is medially constricted. It is most closely related to the genus *Nilssonia* (Meylan, 1987).

Key to the species of *Aspideretes*:

1a. Head totally black: **A. nigricans**
 b. Head olive with or without black markings 2
2a. Head with or without narrow black stripes, but no yellow mottling 3
 b. Head mottled with yellow, but lacking black stripes: **A. hurum**
3a. Triturating surface of mandible flat at symphysis and not raised along inner margin; mandibular symphysis equal to or longer than diameter of orbit; a dense patch of tubercles at midanterior rim of carapace: **A. leithii**
 b. Triturating surface of mandible projecting into a ridge at symphysis and raised at inner margin; mandibular symphysis shorter than diameter of orbit; no large tubercles on anterior rim of carapace: **A. gangeticus**

Aspideretes leithii (Gray, 1872)

Leith's softshell turtle

Recognition: The oval to rounded carapace (to 49 cm) is olive with yellow vermiculations in adults. In juveniles four to six dark-centered, light-bordered ocelli are present on the gray carapace, but these fade with age and the carapace darkens to olive. Several longitudinal rows of tubercles are on the juvenile carapace,

but the surface becomes smoother with age until only a dense patch of large tubercles remains at the midanterior edge of the carapace. Another patch of tubercles usually occurs along the midline posterior to the bony portion of the carapace. A preneural and one or two neural bones separate the 1st pair of costal bones; the 8th pair of costals is well developed and meets at the carapacial midline. In the cream-colored plastron, the hypoplastral bones lie close at the midline, but do not touch. The preplastra are also only narrowly separated, and the epiplastra form obtuse angles to the plastral midline. Plastral callosities are large. The skull is moderate in size and the narrow pointed bony snout is longer than the diameter of the orbit. Triturating surfaces of the maxillae are flat with a well-marked median groove between them. The symphysis of the lower jaw is short and flattened, its length equal to or longer than the diameter of the orbit. In juveniles, the head is greenish with a black longitudinal streak between the eyes extending backward to the neck; two or three pairs of black lines extend outward from this streak toward the sides of the head, and another black streak extends backward from the eye. These black lines fade with age and may be absent in adults. A yellow spot may occur at the corner of the mouth. The outer surface of the limbs is green, the underside cream.

Males have long, thick tails with the vent near the tip; females have short tails.

Distribution: *Aspideretes leithii* occurs in the Bhavani, Godaveri, and Moyer rivers of peninsular India (Moll and Vijayi, 1986).

Habitat: This turtle probably lives in reservoirs and shallow, mud-bottomed stretches of streams and rivers, and is sometimes maintained in tanks within cities and villages.

Natural History: Annandale (1915) reported the small eggs had a diameter of only 31 mm, and found shelled eggs ready for deposition in June and in January.

Satyamurti (1962) reported this species is carnivorous, feeding on fish, crabs, and freshwater mollusks.

Aspideretes hurum (Gray, 1831b)

Indian peacock softshell turtle

Plate 5

Recognition: The rounded juvenile carapace (to 60 cm) is olive with usually four, but up to six dark-centered, yellow-bordered ocelli, and numerous yellowish spots forming a border about the rim of the shell. In adults the carapace is more oval and becomes darker green with black reticulations; the ocelli and yellow spots fade with age. Several longitudinal rows of tubercles occur on the juvenile carapace and some of these persist in adults. Also, a series of enlarged blunt tubercles occur above the neck on the anterior marginal rim. A preneural and a neural lie between the 1st pair of costals, the other seven pairs of costals are well developed, with the last pair touching at the midline. Carapacial bones have pitted surfaces. The plastron is tan to gray, and has rather large callosities on the hyo- and hypoplastra and xiphiplastra and in older individuals on the epiplastra. The preplastra nearly touch. The epiplastra are straight or form an obtuse angle to the midline. The skull is moderate in size with a strongly downturned, sharply pointed bony snout which is much longer than the diameter of the orbit. Triturating surfaces of the maxillae are ridgeless, but there is a median groove. The surface of the mandible bears a well-developed longitudinal ridge at the symphysis; its diameter at the symphysis is longer than the diameter of the orbit. Skin of the head, neck, and limbs is olive to green. The head is marked with black reticulations and yellow spots in the young; the largest of the spots cross the snout and occur on the side of the head at the tympanum; these spots fade with age.

Males have long, thick tails with the vent situated near the tip.

Distribution: *Aspideretes hurum* is found in the Brahmaputra and Ganges rivers of Nepal, India, and Bangladesh.

Geographic Variation: Variation does occur and possibly valid subspecies exist; a thorough study is needed.

Habitat: *Aspideretes hurum* is found in rivers and streams with mud or sand bottoms.

Natural History: Courtship involves much biting by the male, and Flower (1899) reported males emit sounds. Chaudhuri (1912) described the eggs as round. Both wild and captive *Aspideretes hurum* are fond of burying themselves under mud or sand.

Annandale (1912) found that *A. hurum* kept in tanks at a temple in Puri, Orissa, ate rice and palm sugar sweetmeats and were very tame, coming to feed when called. The turtles were popularly believed to be descendants of a man named Gopal who offended Juggernaut, and they were summoned by the priest by this name, although they did not always come.

Aspideretes nigricans (Anderson, 1875)
Black softshell turtle

Recognition: The oval to rounded carapace (to 80 cm) is dark brown to olive or black, and, if brown or olive, may bear numerous blackish or rusty brown spots, causing an overall dark appearance. Several longitudinal rows of tubercles occur on the carapace, and a series of enlarged blunt tubercles lies on the anterior rim above the neck. A preneural and a neural bone separate the 1st of the eight pairs of costals. The 8th costals meet at the midline. The gray plastron is heavily spotted with black, especially over the bones. There are well-developed callosities on the posterior bones and, in some, over the epiplastra. The hyoplastra do not touch medially and are strongly divergent posteriorly; their single, short medial process is thick and blunt, and indistinctly terminally bifurcated. The preplastra may touch anterior to the epiplastra. The skull is broad, and its bony snout is slightly longer than the diameter of the orbit. Annandale (1912) reported that the triturating surface of the upper jaw bears low median and internal longitudinal ridges, which is unusual for a trionychid. There is also a strong longitudinal ridge on the symphysis of the lower jaw, with a diameter longer than that of the orbit. Head, neck, and limbs are dark olive to black with greenish spots. Some white may occur on the upper lips and sides of the head, and some light banding may be present about the eyes.

Males have long, thick tails with the vent situated near the terminus.

Distribution: The original geographic range is unknown; the species was already confined to tanks at Chittagong, Bangladesh, when first described by Anderson (1875), and the remaining 150–200 individuals still exist there.

Habitat and Natural History: Annandale (1912) thought *Aspideretes nigricans* to be morphologically intermediate between *Amyda cartilaginea* and *Aspideretes gangeticus*, and probably also in habitat. The only life history data we have on this captive species come from Annandale (in Smith, 1931:169–170) who visited the Mohammedan shrine where they are kept:

They live in a large pond attached to the shrine of Sultan Bagu Bastan (a saint who is said to have lived in the eighteenth century), about five miles from the town of Chittagong. The Mahommedans will neither kill them nor permit them to be killed; they believe they are in some way connected with the saint. The tank is surrounded by steps leading down to a platform a few inches under water, and the turtles are so tame that they come to feed when called, placing their forefeet on the edge of the platform or even climbing upon it and streching their necks out of the water. The largest are tamer than the smaller ones. Some even allowed us to touch them, and ate pieces of chicken from wooden skewers held in our hands. The only sound they emitted was a low hiss. When undisturbed they remained at the bottom of the pond half buried in the mud. A man connected with the shrine told us that they left the water every evening and climbed a small hill, on which they slept. He said that they laid their eggs in the same hill during the rains.

Aspideretes gangeticus (Cuvier, 1825)

Indian softshell turtle

Plate 5

Recognition: The round to oval carapace (to 70 cm) is olive or green, with or without black reticulations; yellow-bordered, dark-centered ocelli are not present, or are poorly developed, in adults, but are often well developed in juveniles. Several longitudinal rows of tubercles occur on the juvenile carapace, but that of large adults is smooth. A preneural and a neural separate the 1st pair of costals. The 8th pair of costals is large and meets at the midline. Carapacial bones have pitted surfaces. The plastron is gray to white or cream. The callosities are large in the region of the hyo-hypoplastra, xiphiplastra, and epiplastra. The preplastra nearly touch in front of the epiplastra, which lie at obtuse or right angles to the midline. The skull is broad and the blunt bony snout is short and broad (shorter than the diameter of the eye) and slightly downturned. The wide triturating surface of the maxilla is ridgeless but somewhat rugose. That of the mandible is wide with a sharp ridge along the inner edge which sends off a short perpendicular process at the symphysis. Mandibular diameter at the symphysis is shorter than the diameter of the orbit. The head is green above with several black oblique stripes running toward the sides, one of which runs backward from the lower edge of the orbit. A black medial longitudinal stripe may extend backward from between the orbits to the nape. The jaws are yellowish and the chin and throat cream to whitish. The limbs are green and usually unmarked.

Males have longer, thicker tails than females, with the vent beyond the carapacial rim near the tip of the tail.

Distribution: *Aspideretes gangeticus* lives in the Ganges, Indus, and Mahanadi river systems in Pakistan, northern India, Bangladesh, and southern Nepal.

Geographic Variation: No subspecies are currently recognized, but variation does exist.

Habitat: This large softshell inhabits deep rivers, streams, and large canals with mud and sand bottoms. It seems to prefer turbid waters (Minton, 1966).

Natural History: Günther (1864) reported the Indian softshell produces low, hoarse, cackling sounds, possibly related to courtship. Nesting occurs in October and November. The hard-shelled eggs are spherical, 23–32 mm in diameter.

This species is omnivorous, eating not only fish, amphibians, and carrion, but aquatic plants as well.

Minton (1966) reported they emerge to bask on sandbanks, and may thermoregulate by resting in shallow waters.

Nilssonia Gray, 1872

Nilssonia formosa is the only species. It is most closely related to the genus *Aspideretes* (Meylan, 1987).

Nilssonia formosa (Gray, 1869a)

Burmese peacock softshell turtle

Recognition: The rounded carapace (to 40 cm) is olive gray to olive brown with blackish reticulations and four dark-centered, light-bordered ocelli in juveniles, which fade with age until the carapace becomes totally olive in large adults. There are several longitudinal rows of tubercles on the juvenile carapace that also disappear with age, but the series of enlarged blunt tubercles above the neck remains on the adult carapace. No preneural bone is present, and only a single neural bone occurs between the 1st of the eight pairs of costals. The 8th pair of costals is well developed and meets at the midline. Surfaces of all the carapacial bones are pitted. The plastron is white and has four callosities (hyo-hypoplastral and xiphiplastral), and all the bones may be seen through the skin. Anterior extensions of the preplastra are long, slender, and separated. The epiplastra lie at obtuse angles to the midline or are straight. The skull is of moderate size, and the bony snout is as long as or slightly shorter than the diameter of the orbit. There is a longitudinal ridge on the mandibular triturating surface at the symphysis; none occurs in the maxilla. Diameter of the mandibular symphysis equals or exceeds the diameter of the orbit. Head, neck, and limbs are olive with numerous yellowish dark-bordered spots; an elongated spot occurs on each side, near the back of the head, and other light spots occur on the temple, at the corner of the mouth, and on the chin. Undersides of chin and neck are cream colored. In adults the head is yellow with black markings.

Males have long, thick tails with the vent situated near the terminus.

Distribution: *Nilssonia formosa* is apparently restricted to the Irrawaddy, Sittang, and Salween rivers of Burma. Smith (1931) reported it was not uncommon in the lower Irrawaddy, and Annandale (1912) thought it ranged to near the Chinese border. Annandale (1912) and Pritchard (1979) reported a colony of *N. formosa* maintained in a tank at the Arrakan Pagoda in Mandalay. Many were deformed (possibly due to inbreeding), and these tame turtles would come when called to be fed chicken and rice.

Amyda Geoffroy, 1809

This is another monotypic genus, with sole living species *Amyda cartilaginea*.

Amyda cartilaginea (Boddaert, 1770)

Asiatic softshell turtle

Plate 5

Recognition: The round to oval carapace (to 70 cm) is olive gray to greenish brown with numerous yellow-bordered black spots and yellowish dots in younger individuals. In adults the yellow spotting tends to disappear and broad black radiating streaks develop. However, many adults show no pattern and have uniformly olive carapaces. Several longitudinal rows of small tubercles are on the juvenile carapace, but these disappear in very large adults, which have smooth carapaces. A series of enlarged tubercles lies on the anterior carapacial rim above the neck. No preneural bone is present, and only a single neural separates the anterior pair of costals. The 8th pair of costal bones is well developed and meets at the midline. Carapacial bones are strongly pitted. The white to grayish plastron has five callosities, which are coarsely pitted. The preplastra touch or nearly touch, and the epiplastra lie at obtuse or right angles to the midline. The moder-

ately sized skull has a bony snout which is much longer than the diameter of the orbit. A well-developed symphysial ridge occurs on the mandible, which is equal in length to the diameter of the orbit. No ridge occurs on the maxillary triturating surface. Head, neck, and limbs are olive with numerous small yellow spots, and slightly larger orange to pinkish blotches may occur on the sides of the head behind the orbit. These light spots fade with age leaving a network of dark lines on the adult's green head.

Males have long, thick tails with the vent near the tip; the tails of females are short. Males have white plastra, females gray.

Distribution: *Amyda cartilaginea* ranges from the Gulf of Tonkin in Vietnam westward through Laos, Kampuchea, and Thailand to southern Burma and southward through Malaysia to Java, Sumatra, and Borneo.

Geographic Variation: No subspecies are currently recognized; however, *Amyda nakornsrithammarajensis* (Wirot, 1979) from Thailand may prove to be valid. Unfortunately, few specimens are available for comparison. It has a more rounded adult carapace which is smooth dorsally but has pointed tubercles along the rim. Photographs presented by Wirot (1979) indicate it retains a heavy pattern of yellow spots on the carapace and head in adults, but lacks the black spots or rays of *A. cartilaginea*.

Habitat: *Amyda cartilaginea* is found in both upland streams and muddy, slow-flowing lowland streams and rivers, and also occurs in ponds, swamps, and oxbow lakes adjacent to large rivers.

Natural History: Nests are dug in mudbanks (Smith, 1931). Wirot (1979) noted that in Thailand three or four clutches of eggs are laid each year; young females lay about 6 to 10 eggs per clutch and older, larger females lay 20 to 30 eggs. Moll (1979) reported the clutch size is 5 to 7, and Bourret (1941) stated a typical clutch is 4 to 8 eggs. The brittle-shelled eggs are spherical with diameters ranging from 21 to 33 mm, and the incubation period is long, 135 to 140 days (Bourret, 1941).

Amyda cartilaginea is highly carnivorous, feeding on fishes, amphibians, crustaceans, aquatic insects, and other water-dwelling invertebrates.

Wirot (1979) reported that this softshell likes to come onto land to burrow into the sand and rest for long periods with only its snout showing. Our captives have seldom basked or come out of the water, although they spent much time buried beneath the sand under water; they also seemed highly nocturnal. *A. cartilaginea* has a fierce disposition and can be expected to bite when handled.

Pelochelys Gray, 1864a

The single species in this genus, *Pelochelys bibroni*, is the largest of the living trionychid turtles.

Pelochelys bibroni (Owen, 1853)

Asian giant softshell turtle

Plate 7

Recognition: The flat, rounded adult carapace may reach 129 cm; it is olive to brown with some darker or lighter spots or lines extending outward from the vertebral area, and a lighter outer border which contains small dark spots. The juvenile carapace bears numerous small tubercles, and a low vertebral keel may be present; however, the carapacial surface becomes smoother with age and the keel disappears. Only a single neural lies between the 1st pair of costals. There are seven or eight neurals forming a continuous series. Of the eight pairs of costals, the last pair or two meet medially. All carapacial bones are pitted. The cream-colored plastron lacks cutaneous femoral flaps. The hyo- and hypoplastra are separate and not fused. The small preplastra are widely separated, and the epiplastra lie at right or acute angles to the plastral midline. There are four or five large plastral callosities. The flattened skull is short and broad, with a very short, rounded, bony snout (shorter than the greatest diameter of the orbit). Its interorbital space is wider than the greatest diameter of the orbit. The proboscis is also short, and that part of the head extending beyond the gular skin fold seems very short, giving the head a small, blunt appearance. The prefrontal contacts the vomer. No ridge occurs on the triturating surfaces of the maxillae. The head is olive with small dark spots, and the jaws are white with dark speckles. Limbs and neck are olive dorsally, but cream colored below. Several small flaps of skin may protrude from the gular region. The toes are webbed.

The smaller males have longer, thicker tails than do the larger females.

Distribution: *Pelochelys bibroni* ranges from Foochow, Canton, Kwangtung, and Hainan Island in southern China southward through Vietnam, Laos, Kampuchea, Thailand, Burma, and Malaysia to Sumatra, Java and other smaller Indonesian islands, New Guinea, and Luzon of the Philippines. It has also been confirmed from the Subarnarekha River, Orissa, India (Moll and Vijaya, 1986).

Geographic Variation: Smith (1931) reported that all specimens he examined from Burma, Thailand, and the Philippine Islands had seven neurals and the last two pairs of costals touching medially, while those from Malaysia, Indonesia, and New Guinea had eight neurals and only the last pair of costals touching medially. Baur (1891), on the basis of skull features, thought specimens of *Pelochelys* from the Philippines differed from mainland specimens. A good quantitative study is needed to determine if subspecies do exist; unfortunately, because of its large size relatively few specimens have been deposited in museum collections.

Habitat: *Pelochelys bibroni* is found in freshwater streams and deep, slow-moving rivers, often far inland. It commonly enters brackish estuaries along the coast and has been caught at sea. Much time is spent under water, and Wirot (1979) reported it can absorb needed oxygen through its pharynx.

Natural History: Mell (1929) reported that *Pelochelys bibroni* lays up to 27 eggs in a single clutch, but that a full complement of eggs is not deposited at one time. Presumably at least two clutches are laid each season. Dr. Edward Moll has told us that Malaysian *Pelochelys* nests in February and March, laying 24–28 eggs per clutch. The spherical eggs are about 35 mm in diameter.

Pelochelys eats fish, shrimp, crabs, mollusks, and some aquatic plants (Wirot, 1979). Although large and potentially dangerous, this turtle seems to have a rather mild disposition (Pope, 1935).

Chitra Gray, 1844

This is one of the three Asiatic monotypic genera of softshells. Its only living species is the huge narrow-headed *Chitra indica*.

Chitra indica (Gray, 1831b)

Narrow-headed softshell turtle

Plate 6

Recognition: The round, flattened carapace may reach a length of 115 cm, and is gray to olive with dark-gray or yellowish, irregularly shaped blotches. The juvenile carapace has numerous small tubercles and a vertebral keel, but the surface of the adult carapace is much smoother. A single neural lies between the 1st pair of costal bones. There are eight neurals in a continuous series. The 8th pair of costals is well developed and touches medially. All carapacial bones are pitted. The cream-colored plastron lacks cutaneous femoral flaps. The hyo- and hypoplastra are distinct and unfused. Preplastra are widely separated and the epiplastra lie at acute angles to the midline. Four, or rarely five, plastral callosities are present. The skull is long and very flattened, with a very short, rounded, bony snout which is much shorter than the greatest diameter of the orbit. The interorbital space is narrower than the greatest diameter of the orbit, and the orbits are anteriorly positioned. That part of the head extending beyond the gular skin fold is short, as is also the proboscis, giving the head a small, blunt appearance. The prefrontal bone does not meet the vomer, and no ridge occurs on the triturating surface of the maxilla. The nasal septum lacks a lateral ridge. The head is olive with dark-bordered yellow streaks; the limbs are olive to brown, paddlelike, and have webbed toes.

Males have longer, thicker tails than do females.

Distribution: *Chitra* ranges from the Ganges, Godavari, Mahanadi, Sutlej, and Indus drainages of Pakistan, India, Nepal, and Bangladesh southward through Burma to western Thailand and, probably, northern Malaysia.

Habitat: *Chitra indica* prefers sandy sections of large rivers, and is highly aquatic, seldom crawling on land except to oviposit. Its large heavy body hinders walking. Minton (1966) thought it probably capable of pharyngeal gas exchange while under water.

Natural History: Wirot (1979) reported 60–110 eggs are laid, and that a 110-cm female laid 107 eggs in captivity. The spherical eggs (34 mm) were white with tough, leathery shells. Hatchlings have brightly marked olive carapaces about 35–40 mm long, and numerous dark-bordered yellow stripes on their head and neck.

Chitra indica eats fish, mollusks, crabs, shrimp, and may take some plants (Wirot, 1979). Smith (1931) and Taylor (1970) commented on the severity of its bites, but Minton (1966) refers to a captive of mild disposition.

Cycloderma Peters, 1854

This is one of the two genera of tropical African softshells with cutaneous femoral flaps. The carapace is oval, and is anteriorly sculptured in one species. No enlarged tubercles occur on the anterior rim above the neck, and no prenuchal bone is present. A preneural and seven or eight neurals in a continuous series are present. There are eight pairs of costals; the 7th and 8th, or only the 8th, are in contact medially. The plastron has the hyo- and hypoplastra fused, the preplastra short and straight, and seven callosities in adults. Callosities are always present on the gular region. Cutaneous flaps occur on the plastron over the hind limb sockets which conceal the limbs when they are withdrawn. No maxillary ridging occurs, and the intermaxillary foramen is small. The prefrontal does not contact the vomer. The jugal is in broad contact with the parietal. The toes are webbed.

Key to the species of *Cycloderma*:

1a. Hyo- and hypoplastral callosities touch xiphiplastral callosities along a long straight suture; jugal does not enter orbit; general coloration brown: ***Cycloderma aubryi***
  b. Hyo- and hypoplastral callosities separated from xiphiplastral callosities or in contact only along a very short suture; jugal enters orbit; general coloration greenish or gray: ***Cycloderma frenatum***

Cycloderma frenatum Peters, 1854
Zambezi flapshell turtle

Plate 6

Recognition: The oval carapace (to 56 cm) is greenish and may contain faint blotches. In adults, it is smooth and keelless; in juveniles, it is roughened by a vertebral keel and numerous longitudinal rows of small tubercles. The carapace is less anteriorly protruding than in its congener, *Cycloderma aubryi*. No prenuchal bone is present. A preneural and a neural separate the 1st pleurals; seven or eight neurals are present, usually forming a continuous series. There are eight pairs of costals; the 7th and 8th pair, or only the 8th, touch medially. All carapacial bones are covered with fine granulations. The adult cream-colored plastron has faded gray blotches. There are seven plastral callosities; those over the epiplastra are very small. The preplastra are large ovals which touch medially. The epiplastra are fused, their posterior prong is inserted into a notch on the anterior border of the fused hyo-hypoplastra. The hyo-hypoplastra are widely separated medially, and the xiphiplastra are short with medially touching callosities. The skull is depressed with a bony snout equal in length to the greatest diameter of the orbit. The interorbital width is about 50% of the height of the orbit. Premaxillae are usually absent; when present, small and scalelike. The prefrontals touch the vomer, if it is present, and the jugal enters the orbital rim. The mandible lacks a symphysial ridge; the symphysial width is less than the greatest diameter of the orbit. The head is gray to green with five longitudinal black stripes beginning on the crown and sides behind the orbits and running backward onto the neck. The chin and throat are either immaculate white or contain some dark streaks. Limbs are grayish green. Four or five crescentic skin folds occur on the upper surface of each forefoot.

Males have longer, thicker tails than do females.

Distribution: *Cycloderma frenatum* occurs in East Africa from Tanzania southward into Zambia, Malawi, and Mozambique.

Habitat: *Cycloderma frenatum* lives in streams, rivers, lakes, and ponds, in both fresh and stagnant waters.

Natural History: Nesting occurs from December to March. Loveridge and Williams (1957) reported females oviposited after rains following an 8-month drought. Clutches normally include 15 to 22 brittle-shelled, spherical eggs about 31 or 32 mm in diameter. Hatchlings have been found from December to February. Their carapace lengths range from 40 to 48 mm (Loveridge and Williams, 1957), and they are colored and patterned like the adults, but brighter.

C. frenatum is carnivorous, feeding on fish, aquatic snails, mussels, and possibly amphibians.

These turtles are shy and withdraw into the shell when disturbed. Much of their lives is spent buried beneath mud or sand.

Cycloderma aubryi (A. Duméril, 1856)

Aubry's flapshell turtle

Recognition: The oval carapace (to 55 cm) is brown with a narrow dark vertebral stripe, smooth and keelless in adults, but with a vertebral keel and numerous scattered tubercles in juveniles. The carapace is sculptured in front causing it to appear anteriorly protruding. No prenuchal bone is present. A preneural and a

neural separate the 1st of the eight pairs of costals; the large 8th pair touch medially. There are seven or eight neurals in a continuous series. All carapacial bones are covered with fine granulations. The yellowish plastron has some faded brown blotches in the adult. Its seven plastral callosities are large and granulated, and cover most of the adult plastron. The preplastra are large and half-moon shaped, and touch medially. The fused epiplastra are also quite large and the fused hyo-hypoplastra almost touch at the midline. Xiphiplastra are totally in contact at the midline. The skull is depressed with a long bony snout (longer than the greatest diameter of the orbit). The interorbital width is about 67% of the height of the orbit. The jugal does not enter the orbit. The mandible lacks a symphysial ridge and its ventral width is less than the diameter of the orbit. The head is brownish with five thin longitudinal lines: a medial one extending backward from the crown to the neck; two, one on each side of the medial line, extending from between the orbits to the back of the head; and two, one on each side, beginning at the nostril and extending backward through the orbit and along the side of the head to the neck. Chin and throat are yellow and speckled with brown; limbs are brown. Six or seven crescentic skin folds occur on the upper surface of each forefoot.

Males have longer, thicker tails than do females.

Distribution: *Cycloderma aubryi* occurs in Africa in the Congo, Cabinda, Gabon, and Zaire.

Habitat: *Cycloderma aubryi* is restricted to water bodies within rainforests.

Natural History: The hatchling has a carapace of about 55 mm, which is orangish to auburn with scattered black spots and a narrow brown vertebral stripe. Its plastron is yellow with a large V-shaped brown mark extending backward from the anterior apex. No plastral callosities occur in the young.

Cyclanorbis Gray, 1852

These freshwater softshells are confined to Africa. The carapace is round to oval, slightly domed, and has large tubercles on the anterior rim. A prenuchal bone may be present or absent. A preneural and up to nine neurals may be present, usually forming an incomplete series. There are eight pairs of costal bones, most of which meet at the midline. The plastron has the hyo- and hypoplastra fused, the preplastra long and angular, and two to nine callosities. Also, cutaneous flaps occur over the hind limb sockets, which conceal the limbs when they are withdrawn. No ridges occur on the maxilla, and the intermaxillary foramen is small, if present. The prefrontal meets the vomer in one species. The jugal is in contact with the parietal. The toes are webbed.

Key to the species of *Cyclanorbis*:

1a. Gular callosities present, xiphiplastra broad and notched posteriorly, prefrontal bones do not touch vomer: ***Cyclanorbis senegalensis***
  b. Gular callosities absent, xiphiplastra pointed posteriorly, prefrontal bones touch vomer: ***Cyclanorbis elegans***

Cyclanorbis elegans (Gray, 1869a)

Nubian flapshell turtle

Recognition: The large, rounded carapace (to 60 cm) is olive to brown in adults with numerous yellow or light-greenish spots on the lateral edges. Longitudinal rows of small tubercles and a low vertebral keel occur in juveniles, but the carapace becomes smoother with age. There are enlarged tubercles on the anterior rim above the neck. A prenuchal bone is usually absent, but a preneural and six to nine neurals are present. These neurals may form a continuous series or have the last two separated. There are eight pairs of costal bones; the 1st pair is separated by two neurals and the last pair is well developed and meets at the midline. All carapacial bones are covered with small granulations. The plastron is yellow with dark spots. Usually a pair of callosities occurs over the hyo-hypoplastra only, but a small pair may also lie over the xiphiplastra. There are never callosities in the gular region. The anterior border of the fused hyo-hypoplastra is straight or concave, but never convex; the posterior border is deeply excavated; its anteromedial process has indistinct prongs, the short medial process barely projects, and the posteromedial process has only three prongs between which the two anterior prongs of the xiphiplastron insert (Loveridge and Williams, 1957). The xiphiplastra do not touch and are pointed posteriorly. The head appears small for the size of the animal. Its bony snout is much shorter than the diameter of the orbit, and the prefrontal touches the slender vomer. The mandible lacks a symphysial ridge; however, there is a strong knob on the inner surface. The symphysial width is less than the diameter of the orbit. In adults the head is brown with light-green or yellow vermiculations. The neck is also brownish, but lighter in color with numerous small yellow spots; the limbs are brown. Four transverse, crescent-shaped folds occur on the skin of each forefoot.

Males have longer, thicker tails than do females.

Distribution: *Cyclanorbis elegans* is confined to Africa, where it ranges from the Sudan westward to Nigeria and Togo.

Habitat: *Cyclanorbis elegans* inhabits slow-moving rivers and marshes.

Natural History: The hatchling is brightly colored with a green carapace covered with large, irregularly shaped yellow spots, and a green to brown head with numerous yellow spots. No plastral callosities are present.

Cyclanorbis senegalensis (Duméril and Bibron, 1835)

Senegal flapshell turtle

Recognition: The oval adult carapace (to 35 cm) is somewhat domed and brown to dark olive gray, with or without small, dark, mottled spots and a light border. Juveniles have several longitudinal rows of tubercles and a low vertebral keel on the carapace, but the adult carapace is much smoother. There are enlarged tubercles on the anterior carapacial rim. A prenuchal, a preneural, and as many as eight neurals are present, but not in a continuous series. The 1st of the eight pairs of costals is separated by two neurals, and the last pair may not touch at the midline. All carapacial bones are covered with small granulations. The plastron is white to cream with several gray or brown spots or blotches. There may be seven to nine

Cyclanorbis senegalensis

well-developed plastral callosities in adults (sometimes absent over the xiphiplastra); small callosities regularly occur in the gular region. The anterior border of the fused hyo-hypoplastra is convex, but never straight or concave; the posterior border is deeply excavated; its anteromedial process has several almost juxtaposed prongs, the short medial process scarcely projects, and the posteromedial process has only three prongs between which are inserted the two anterior prongs of the xiphiplastron (Loveridge and Williams, 1957). Xiphiplastra do not meet and are broad and notched posteriorly. The head is moderate in size with a short, bony snout (shorter than the diameter of the orbit). There is no prefrontal-vomer contact. The mandible lacks a symphysial ridge and has no strong knob on the inner surface; the symphysial width is less than the diameter of the orbit. The adult head is olive to brown above, lighter laterally, and has the chin and throat mottled. Neck and limbs are olive to grayish brown. Five or six transverse, crescent-shaped folds occur on the skin of each forefoot.

Males have longer, thicker tails than do females.

Distribution: *Cyclanorbis senegalensis* ranges in Africa from the Sudan westward through Cameroon to Gabon, Senegal, and Ghana.

Habitat: This poorly known softshell lives in rivers, streams, lakes, and marshes.

Natural History: A female contained six brittle-shelled eggs on 12 April (Loveridge and Williams, 1957). The hatchling has a gray to brown carapace with scattered black vermiculations and yellowish, irregularly shaped spots, a gray to brown finely spotted head, and white throat and chin. No plastral callosities are present.

Lissemys punctata punctata

Lissemys punctata punctata (plastron)

Lissemys punctata scutata

Lissemys Smith, 1931

This genus contains the only Asiatic softshell with cutaneous femoral flaps, *Lissemys punctata*.

Lissemys punctata (Lacépède, 1788).

Indian flapshell turtle

Plate 7

Recognition: The oval, domed, unflared adult carapace (to 27.5 cm) is unique among trionychids in having a series of peripheral bones along the posterior rim (these are not considered homologous with the peripherals of hard-shelled turtles). A prenuchal is also present. Two neural bones separate the 1st pair of costals, and the seven or eight neurals lie in a continuous series. The 7th and 8th pair of costals touch medially. Rib ossification is extensive, and ribs protrude beyond the costals. All carapacial bones are finely granulated. Juveniles have wrinkled carapaces with longitudinal rows of small tubercles, but the adult carapace is much smoother. The adult carapace varies from uniform brown to olive brown or dark green with small dark-brown or large yellow spots; the plastron is cream colored. There is a hinge on the plastral forelobe which allows it to partially close the anterior opening, thus protecting the head and forelimbs. This hinge lies at the point of attachment of the preplastra to the epiplastra. The hyo- and hypoplastra are fused, and posterior extensions are absent from the epiplastra. Seven plastral callosities occur in adults, but none in juveniles. The skull is short and dorsally convex; its bony snout is much shorter than the greatest diameter of the orbit. The interorbital space is also much narrower than the diameter of the orbit. The triturating surface of the maxilla lacks a ridge. There is prefrontal-vomer contact. The fleshy proboscis is short with no lateral ridge on the internasal septum. The head is olive to brown with several elongated and wide yellow stripes: on the snout, one between the orbits, one passing backward on the side from the orbit to the tympanum, and sometimes one from the corner of the mouth backward along the throat. A series of yellow stripes also occurs on the neck. The limbs are olive or brown.

Stock (1972) reported the diploid karyotype to be 66 (16 metacentric and submetacentric, 12 subtelocentric, and 38 acrocentric and telocentric chromosomes).

Males have long, thick tails; females have short tails. Females grow larger than males.

Distribution: *Lissemys punctata* occurs in the Indus and Ganges drainages of Pakistan, northern India, Sikkim, Nepal, and Bangladesh, southward through peninsular India to Sri Lanka, and in the Irrawaddy and Salween rivers of Burma.

Geographic Variation: Three subspecies are recognized (we follow the taxonomy of Webb, 1980a). *Lissemys punctata punctata* (Lacépède, 1788) occurs in peninsular India and on Sri Lanka (it includes the formerly recognized, but no longer valid subspecies *L. p. granosa*). Its carapace is uniformly olive brown to brown, and the 1st peripheral is only slightly larger than the 2d; the head is olive to brown with three oblique, parallel black stripes, at least in the young; and the plastron has an entoplastral callosity that is moderate in size and gray brown in color. *L. p. andersoni* Webb, 1980a (Plate 7) occurs in the Indus, Ganges, and Brahmaputra drainages in eastern Pakistan, northern India, Sikkim, southeastern Nepal, Bangladesh, and northern coastal Burma. Its carapace is gray green with numerous large black-bordered yellow spots, and the 1st peripheral bone is much larger than the rest. The greenish head also contains numerous yellow spots, and the white plastron has a small entoplastral callosity. *L. p. scutata* (Peters, 1868) lives in the Irrawaddy and Salween rivers of Burma. It has an olive-brown to brown carapace with some dark spotting (in juveniles) or reticulations (in adults), and the 1st peripheral is smaller than the 2d. The head is olive to brown with an indistinct dark stripe extending backward from each orbit and another passing backward between the orbits. The entoplastral callosity is large and may contact the hyo-hypoplastral callosities. Webb (1982) suggested that the configuration of its peripheral bones, and the advent of well-developed plastral callosities at a relatively short length, indicating a small maximal size, may mean that *L. p. scutata* is a distinct species.

Habitat: *Lissemys punctata* lives in the shallow, quiet, often stagnant waters of rivers, streams, marshes, ponds, lakes and irrigation canals, and tanks. Waters with sand or mud bottoms are preferred, and *Lissemys* is often forced to estivate in the mud during dry periods. It often basks on sandbars, banks, and floating vegetation.

Natural History: Courtship and mating were observed by Duda and Gupta (1981) during April in wild *Lissemys* and from May to July in captives. The male swims above and around the female, with his neck and limbs extended, periodically stroking the female's carapace with his chin. When receptive, the female faces the male with her neck extended and they bob their heads vertically three or four times in a sequence repeated five to eight times. She then settles to the bottom and the male mounts from the rear. Near the end of copulation, the male releases his grip on the female's carapace and rotates so as to face the opposite direction from her. They remain attached in this position for as long as 15 minutes, during which the female may drag the male about. Finally copulation ends and the pair separates. Nesting occurs throughout most of the year. Two to 14 eggs are laid at a time, and a female lays several clutches each year. The white eggs are spherical (24–33 mm in diameter) with brittle shells. Hatching occurs from May to July, and hatchlings have carapaces 35–44 mm long.

Lissemys is omnivorous. Deraniyagala (1939) reported it feeds on frogs, fishes, crustaceans, aquatic snails, and earthworms in Sri Lanka, but Minton (1966) has observed them eating aquatic vegetation in Pakistan. Both Smith (1931) and Deraniyagala (1939) observed them come on shore at night and feed on carrion or decaying matter.

This flapshell is generally mild mannered, but newly caught individuals, especially large adults, may bite.

Pelusios gabonensis

Podocnemis unifilis

Chelodina steindachneri

PLATE 1

Chelodina expansa

Chelodina rugosa

Emydura australis

Emydura subglobosa

PLATE 2

Phrynops rufipes

Platemys platycephala platycephala

Chelus fimbriatus

Claudius angustatus

PLATE 3

Staurotypus triporcatus

Dermatemys mawii

Carettochelys insculpta

PLATE 4

Amyda cartilaginea

Aspideretes hurum

Aspideretes gangeticus

PLATE 5

Apalone spinifera asper

Dogania subplana

Chitra indica

Cycloderma frenatum

PLATE 6

Pelochelys bibroni

Lissemys punctata andersoni

Dermochelys coriacea coriacea

Chelonia depressa

PLATE 7

Platysternon megacephalum shiui

Callagur borneoensis

Cuora chriskarannarum

PLATE 8

Cuora galbinifrons

Cuora mccordi

Cuora pani

PLATE 9

Cyclemys dentata

Geoemyda spengleri spengleri

Heosemys silvatica

Heosemys spinosa

PLATE 10

Kachuga tecta

Melanochelys tricarinata

Morenia petersi

Notochelys platynota

PLATE 11

Ocadia sinensis

Rhinoclemmys pulcherrima manni

Emydoidea blandingii

PLATE 12

Clemmys guttata

Clemmys muhlenbergii

Pseudemys alabamensis

PLATE 13

Emydidae

Trachemys scripta callirostris

Graptemys oculifera

Graptemys flavimaculata

Graptemys versa

PLATE 14

Acinixys planicauda

Pyxis arachnoides arachnoides

Geochelone carbonaria

Geochelone elegans

PLATE 15

Geochelone pardalis babcocki

Geochelone radiata

Geochelone yniphora

PLATE 16

Dermochelyidae

This family is represented by a single extant genus, *Dermochelys* Blainville, 1816, having only one living species, *D. coriacea*. It is the largest living turtle, reaching 244 cm (Brongersma, 1968) and weighing 659 to possibly 867 kg (Pritchard and Trebbau, 1984), and it ranges widely in tropical and subtropical seas.

Dermochelys coriacea (Linnaeus, 1766)

Leatherback sea turtle

Plate 7

Recognition: The brown to black, smooth carapace is elongated, lyre shaped, and tapers to a supracaudal point above the tail. It lacks horny scutes, but is covered with a ridged, leathery skin. Shell bones, except the nuchal and those at the rim of the plastron, are lost; their place is taken by a mosaic of small bony plates embedded in the leathery skin. The largest of these polygonal bones form seven prominent longitudinal keels, which divide the carapacial surface into eight sections. The nuchal is attached to the neural arch of the 8th cervical vertebra. A series of 10 dorsal vertebrae runs medially along the carapace; the free dorsal ribs, except the last, articulate with the neural arch and two adjacent centra and are embedded in a layer of cartilage. Ribs one and ten are short. The hingeless plastron is cream to white with five longitudinal ridges underlined by small polygonal bones. An entoplastron is absent, but the bones at the rim are apparently homologous to the epiplastra, hyoplastra, hypoplastra and xiphiplastra of other turtles. Head and neck are black or dark brown, with white to yellow blotches, the snout is blunt and nonprojecting,

and there is a toothlike cusp on each side of the gray upper jaw. The skull roof is complete. The vomer meets the premaxillae, separating both the nares and palatines, and the choanae lie posterior to the alveolar surfaces of the vomer. No basisphenoid-palatine contact occurs, and the maxilla is separated from both the pterygoid and quadratojugal. Neither a secondary palate nor a palatine fenestra is present. The quadrate is open posteriorly and does not enclose the stapes. The supraoccipital process is very short. Both jaws are very short behind the coronoid process, and the anterior maxillary ridge is cusped. The neck is short and incompletely retractile; its 4th cervical vertebra is biconvex. Limbs are paddlelike, clawless, and black with some white blotches.

Males have concave plastra and are rather depressed in profile. They are more tapering posteriorly and have tails longer than their hind limbs; the tails of females are barely half as long.

Although no fossils of *Dermochelys coriacea* have been discovered, other leatherbacks are known from as early as the Eocene of Africa, Europe, and North America (*Cosmochelys, Eosphargis, Psephophorus*); and *Dermochelys pseudostracion* has been found in French Miocene deposits (Romer, 1956; Pritchard, 1980).

Distribution: *Dermochelys* ranges throughout the waters of the Atlantic, Pacific, and Indian oceans from Labrador, Iceland, the British Isles, Norway, Alaska, and Japan south to Argentina, Chile, Australia, and the Cape of Good Hope (Pritchard, 1980). It also enters the Mediterranean Sea. It migrates great distances to and from the nesting beaches.

Dermochleys consistently ranges more poleward than other sea turtles; it survives the cold waters off Alaska, Labrador, and Iceland. Frair, Ackman, and Mrosovsky (1972) discovered that the deep body temperature of a leatherback taken from cold water was 18°C above the water temperature, and thought its large size favored heat retention from muscular activity. Their data show that leatherbacks can maintain a considerable differential between their body temperatures and the ambient level in situations where behavioral regulation is not possible. Greer, Lazell, and Wright (1973) demonstrated that a countercurrent circulatory system exists in the fore and hind limbs that permits homoeothermy. Perhaps the high oil content of leatherback flesh also retards heat loss.

Geographic Variation: Two subspecies have been described. *Dermochelys coriacea coriacea* (Linnaeus, 1766) (Plate 7) ranges through the Atlantic Ocean, Gulf of Mexico, and Caribbean Sea, from Newfoundland to Great Britain and south to Argentina and the

Cape of Good Hope. Occasionally it enters the Mediterranean Sea. This race has longer forelimbs in comparison with total body length; has a shorter head; and is darker with less light mottling on the back, lower jaw, and throat. *D. c. schlegelii* (Garman, 1884) ranges through the Pacific and Indian oceans, from British Columbia to Chile, and west to Japan and eastern Africa. It has shorter forelimbs, a longer head, and is paler with more light mottling on the back, lower jaw, and throat.

The subspecies seem to be poorly differentiated (Pritchard, 1971, 1980; Pritchard and Trebbau, 1984), and a detailed study will be necessary to determine their validity. Perhaps sexual differences have been confused with subspeciation.

Pritchard and Trebbau (1984) pointed out that the name *angusta* (Philippi, 1899) is the earliest name applied strictly to leatherbacks from the eastern Pacific, and should these be shown to be a valid race, this would be the proper name of the subspecies.

Habitat: The leatherback is pelagic, but occasionally it enters the shallow waters of bays and estuaries. It shares its major nesting beaches with most of the other species of marine turtles.

Natural History: Courtship and mating have not been described, but they are believed to occur in the water off the nesting beaches at the period of laying. Ovipositing of the Atlantic leatherback occurs from April through November. Nesting females have been reported from Texas, both coasts of Florida, and from Jamaica (Pedro and Morant cays), St. Kitts, Nevis, Barbados, St. Croix, Tortola, Trinidad, Tobago, Venezuela, French Guiana, Surinam, Guyana, Colombia, Costa Rica, Nicaragua, and in Africa from Liberia, Ivory Coast, and Angola (Pritchard, 1980; Pritchard and Trebbau, 1984). The Bahamas, Cayman Islands, Brazil, and Honduras have also been listed but these reports are questionable. The Pacific race nests at various times throughout the year, depending on location. Nesting has been reported from South Africa, Natal, Ceylon, India, Addu Island, Tenasserim, Thailand, Malaya, Indonesia, Borneo, New Guinea, Australia, Mexico (Pritchard, 1980), and Costa Rica (Pritchard, 1982). Leatherbacks usually emerge at night to lay, and many oviposit three to six times a year. Carr and Ogren (1959) observed a female come ashore and nest, and we refer the reader to their lucid description of the process.

The nest may be flask shaped or not, depending on whether the turtle digs down to damp sand. Deraniyagala (1939) reported a nest 100 cm deep, including the body pit. Beaches with a predominance of coarse sand are preferred (Pritchard, 1971). The usual clutch size is 50–170 eggs, but a rather large percentage of the eggs are yolkless. Normal eggs are spherical, have soft, white shells, and are 49–65 mm in diameter. Abnormal eggs may be ellipsoidal, small and spherical, or somewhat dumbbell shaped; they are 15–45 mm in greatest diameter. Incubation lasts 53–74 days. Hatchlings (56–63 mm) are dark brown or black, with white or yellow carapacial keels and flipper margins. Their skin is covered with small scales and the tail is keeled dorsally; both scales and keel soon disappear.

Dermochelys is omnivorous, but seems to prefer jellyfish. This may explain its northward summer migrations, as it follows the drifting jellyfish flotilla. Also, its rather weak jaws are best suited for feeding upon such softbodied prey. Other foods occasionally eaten are sea urchins, squid, crustaceans, mollusks, tunicates, fish, blue-green algae, and floating seaweeds. Several dead, beached leatherbacks have had plastic bags in their stomachs; apparently these were mistaken for jellyfish.

Pritchard (1982) reported the breeding population of *Dermochelys* from the Mexican Pacific states of Michoacán, Guerrero, and Oaxaca is the largest known. He estimated the world population of mature females to be 115,000, but, due to severe stress on all major populations, still thought the species endangered.

Cheloniidae

Four living genera and six species of hard-shelled sea turtles belong to the family Cheloniidae. Most of their feeding and nesting range is in warmer marine waters, and they occur in all tropical oceans with several species ranging well into the temperate zones. Cheloniids are among the oldest fossil turtles. The extinct genera *Allopleuron, Catapleura, Glaucochelone, Glyptochelone, Peritresius* and *Tomochelone* and the living genera *Caretta* and *Chelonia* are all known from Upper Cretaceous deposits of Europe or North America. Morphologically and serologically they are most closely related to *Dermochelys coriacea* (Dermochelyidae) and the extinct sea turtle families Plesiochelyidae, Protostegidae, and Toxochelyidae, but they also show some affinities to the extant terrapin family Emydidae.

The temporal region of the skull is completely roofed. The premaxillae are not fused, meet the vomer, and separate the internal nares and palatines. There is parietal-squamosal contact. A secondary palate is present, but palatine fenestra are lacking. The pterygoids separate the basisphenoid from the palatines but do not touch the maxillae. Bony trabeculae of the basisphenoidal rostrum lie close together or are fused. The quadrate never encloses the stapes, and the maxilla does not contact the quadratojugal. The dentary is confined to the anterior half of the lower jaw. The shell is covered with horny scutes. The oval to heart-shaped carapace retains the embryonic spaces (peripheral fenestrae) between the ribs; thus the neurals and pleurals are greatly reduced. Neurals are hexagonal, shortest anteriorly, and variable in number. The nuchal bone lacks a costiform process but has a ventral area for attachment to the neural arch of the 8th cervical. Carapace and plastron are connected by ligaments, and neither is hinged. Reduction of the plastron is indicated by the small entoplastron. A small intergular scute and small postanal scutes may be present. Inframarginal scutes are usually present. The forelimbs are paddlelike with elongated digits. The deltopectoral crest of the humerus is far down the shaft; the trochanteric fossa of the femur is greatly restricted, and the trochanters are united. The neck is short, and ability to retract the head has been lost. The 4th vertebra is biconvex.

Key to the genera of Cheloniidae:

1a. Four pairs of pleurals; cervical scute not in contact with 1st pleurals   2
  b. Five or more pairs of pleurals; cervical scute touches 1st pleurals   3
2a. One pair of prefrontal scales; lower jaw strongly serrated: *Chelonia*
  b. Two pairs of prefrontal scales; lower jaw only weakly serrated: *Eretmochelys*
3a. Bridge with three inframarginals: *Caretta*
  b. Bridge with four inframarginals: *Lepidochelys*

Chelonia Latreille, in Sonnini and Latreille, 1802

The adult carapace is oval to heart shaped, flattened, lacks a vertebral keel, and is serrated posteriorly. Four pleural scutes occur on each side, and there are five vertebrals. Lateral fontanelles are present in the adult carapace, and there are 11 pairs of peripherals with the 10th pair not touching the ribs, and 9 to 11 hexagonal, anteriorly short-sided neurals. The hingeless plastron has an intergular scute and may have postanal scutes; its bridge has four pairs of inframarginals lacking pores. In the pointed skull, the vomer contacts the premaxillae and separates the maxillae. Triturating surfaces of the maxillae are ridged, but not those of the premaxillae. The descending prefrontal processes touch both the vomer and palatines; the frontal usually enters the orbit. A blunt ridge occurs on the vomer and palatines at the anterior margin of the internal choanae. Only one pair of prefrontal scales and three to four postoculars are present. Each forelimb has one claw.

The karyotype of *Chelonia mydas* is 2n = 56: 14 metacentric or submetacentric macrochromosomes, 10 telocentric or subtelocentric macrochromosomes, and 32 microchromosomes (Bickham et al., 1980; Bachere, 1981).

Key to the species of *Chelonia*:

1a. Carapace heart shaped with sloping marginal scutes; 1st vertebral touches 1st marginal; forelimbs long with large scales overlying phalanges; usually four postocular scales: ***C. mydas***
  b. Carapace oval with upturned marginal scutes; 1st vertebral separated from 1st marginal; forelimbs short with wrinkled skin overlying phalanges; three postocular scales: ***C. depressa***

Chelonia mydas (Linnaeus, 1758)

Green turtle

Recognition: This medium-sized to large (to 153 cm) sea turtle has a single pair of prefrontal scales on the head and a serrated cutting edge on the lower jaw. Its broad, low, heart-shaped carapace lacks a vertebral keel and is serrated posteriorly. All vertebrals are broader than long. There are four pairs of pleurals; the 1st does not touch the cervical. Carapacial scutes are olive to brown and may contain a mottled, radiating, or wavy pattern. The bridge has four inframarginal scutes that lack pores. The formula of the immaculate white or yellow plastron is: abd >< fem > an >< gul > pect > hum > intergul. All skin is brown or, sometimes, gray to black, and many head scales may have yellow margins. The horny inner surface of the upper jaw has well-developed vertical ridges, and the cutting edge of the lower jaw is strongly serrated. Four postocular scales are present.

The male carapace is more tapering posteriorly and the plastral hind lobe is narrower than in females. In males the tail is strongly prehensile in a vertical plane, is tipped with a heavy flattened nail, and extends far beyond the posterior carapacial margin; in females the tail barely reaches the margin. The male forelimb has a large single, curved claw.

The common name of this turtle is derived from the greenish color of its body fat.

Distribution: *Chelonia mydas* ranges through the Atlantic, Pacific, and Indian oceans, chiefly in the tropics. It occurs as far north in the Pacific as Alaska (Hodge, 1981) and in the Atlantic to Great Britain (Brongersma, 1972).

Geographic Variation: Several races have been proposed, and two are commonly recognized as valid. *Chelonia mydas mydas* (Linnaeus, 1758) ranges in the Atlantic Ocean from New England and the British Isles to Argentina and extreme South Africa. It is often abundant about Ascension Island, the Cape Verde and Cayman islands, and off Bermuda. This subspecies is predominantly brown and has an elongated, shallow carapace not markedly indented above the hind limbs. *C. m. agassizii* Bocourt, 1868 ranges from Ethiopia around the Cape of Good Hope to western Africa, and east through the Indian and Pacific oceans to the western coast of the New World from the United States to Chile. It is greenish or olive brown and has a broad, deep shell often markedly indented above the hind limbs. Some individuals are melanistic, becoming slate gray to black in overall color. Caldwell (1962a) described *C. m. carrinegra* from the Gulf of California on the basis of this dark pigment; however, since black turtles also occur in Guatemala and the Galapagos Islands and pale individuals are found in the Gulf of California, there seems to be no geographic distinction between the two color forms and all eastern Pacific green turtles should be considered *C. m. agassizii*.

Hirth (1980b) suggested, however, that *C. mydas* not be split into subspecies until more precise definitions and demarcation of ranges are available. Other authorities would elevate *C. m. agassizii* to specific status, but no thorough study of its relationship to *C. m. mydas* has been conducted, and we feel the two forms are best regarded as subspecies until such a study is completed.

Habitat: The green turtle migrates across the open seas but feeds in shallow water supporting an abundance of submerged vegetation. It shares its nesting beaches with all species of sea turtles (including the Australian flatback, *C. depressa*).

Natural History: Mating occurs in the water off the nesting beaches during the laying season. Although mated pairs have been seen more than 1 km from shore, the greatest concentration is close to shore. Copulation occurs at any hour before or after the first

Chelonia mydas mydas

nesting emergence and may last several hours. Several males (1–5) may simultaneously court and attempt to mate with a single female. Copulating pairs float at the surface, with the male atop the female. The enlarged claws on the front flippers and the strong nail at the tip of the prehensile tail provide a firm three-point attachment; moreover, the male's claws cut into the female's carapace at the third intermarginal seam and leave deep, bleeding wounds. *C. m. mydas* is known to nest on the beaches of Costa Rica, French Guiana, Guyana, Surinam, Veracruz and Quintana Roo (Mexico), Ascension Island, Mona Island, Aves Island, Trinidad, the Dry Tortugas, Mujeres Island, and the Cape Verde Islands. It also may occasionally nest on the coasts of Florida, Georgia, Bermuda, Alta Vela, the Cayman Islands, Cuba, and Brazil. On the beaches of the Caribbean, the Gulf of Mexico, and the Cape Verde Islands the nesting season extends from March to October, with the greatest activity in May and June. On Ascension Island the season extends from December to July with the period from February to April most important. *C. m. agassizii* is known to nest on the western coast of Africa and on the beaches of Ethiopia, Madagascar, the Seychelles Islands, Mauritius, the Maldive Islands, Yemen, Ceylon, Pakistan, Burma, Malaya, Thailand, Indonesia, Borneo, the islands of the Strait of Malacca, Talang Islands, northern Australia, the Caroline Islands, the Marshall Islands, the Hawaiian Islands and other central Pacific islands, the Galapagos Islands, and the Pacific coasts of Central America and Mexico. Carr (1952) reported that the nesting season in Australia extends from late October to mid-February; in the Gulf of Siam the year around, with a July-to-November peak; on Ceylon from July to November; in the Malacca Straits from December through January; in the Seychelles Islands the year around with a March-to-May peak; and in Borneo from May to September. Loveridge and Williams (1957) reported that on the western coast of Africa the laying season is from September to January. Hendrickson (1958) found that the nesting beaches on islands usually were on the leeward side, and in Sarawak (Borneo) the principal ones had fringing coral reefs below mean low tide. Hirth and Carr (1970) compared the characteristics of the nesting beaches at South Yemen, Aldabra Island, Ascension Island, Aves Island, and Tortuguero, Costa Rica. They found that the sand varied from fine to coarse in texture and from olive gray to white in color. The pH ranged from 6.9 to 8.0 and the carbonate content was high in all beaches except at Tortuguero. Organic content was 0.30–1.18%.

Nesting is nocturnal; Carr and Giovannoli (1957) described the nesting process of a female found at Tortuguero, Costa Rica, and we refer the reader to that paper for details.

Females nest in 2- or 3-year cycles, with the latter more frequent. Female *C. mydas* nest several times each season at 10–15-day intervals, and 12–238 (normally 100–150) eggs are deposited in each nest depending on the size of the female. The nearly spherical eggs are 35–58 mm in diameter. Incubation takes 30–90 days, but usually 45–60 days. Hatchlings emerge after dark. Their carapace (35–59 mm) is keeled and dark green to brown; it may have a mottled pattern and light border. The plastron is white or yellow with two longitudinal ridges. Skin is blackish, flippers have white borders, and the upper beak is light. Eggs incubated in warm nests produce a preponderance of females; clutches from cool nests hatch out mostly males (Morreale et al., 1982; Mrosovsky, 1982).

Chelonia mydas is omnivorous, but the juvenile is more carnivorous than the adult. Plants consumed include green, brown, and red algae, mangrove (roots and leaves), and *Zostera, Cymodocea, Thallasia, Halophila, Posidonia, Halodule*, and *Portulaca*. Roots seem to be preferred. Animal food includes small mollusks, crustaceans, sponges, and jellyfish. Deraniyagala (1939) reported that green turtles scavenged near a house, feeding on the kitchen refuse thrown into the water. Captives feed readily on liver, kidney, beef, and fish. The serrated lower jaw probably is an adaptation to grazing. Green turtles are the only marine turtles subsisting mainly on plants—a diet poor in vitamin D. They are also the only marine turtles that come to the shore to bask—perhaps a means of producing the needed vitamin D through the action of the sunlight on skin sterols. Green turtles overwinter on the sea bottom of the Infiernillo Channel in the Gulf of California (Felger et al., 1976), and perhaps this behavior also occurs elsewhere.

Exploitation of the green turtle on both the nesting beaches and the feeding grounds makes its future bleak. It has already disappeared from many beaches where it was once numerous. Collection of eggs on the nesting beaches is particularly injurious. If *Chelonia mydas* is to survive, conservation methods must be applied wherever the turtles breed or feed. In particular, the nesting beaches must be rigorously protected by governmental regulation and must be adequately patrolled during the breeding season.

Chelonia depressa Garman, 1880

Flatback sea turtle

Plate 7

Recognition: The unkeeled oval carapace (to 100 cm) is flattened and only slightly serrated posteriorly in adults (strongly serrated in hatchlings). It is also rather pliable and somewhat slimy to the touch. All vertebrals are broader than long. There are only four pleurals; the 1st is separated from the cervical scute. The lateral marginals may be upturned. The carapace is gray to pale green in adults, darker olive in hatchlings. The wide plastron lacks a hinge and narrows considerably both anteriorly and posteriorly. There is no anal notch, as this area may be filled with one or more postanal scutes. The plastral formula is: fem > pect > an >< gul > abd > intergul > hum > postan. Both plastron and bridge are immaculately cream colored. The head is moderate with a pointed snout. Its lateral lower jaw margins are serrated. As in *C. mydas*, there is only one pair of elongated prefrontal scales, but in contrast there are only three postocular scales. The upper eyelid is covered with numerous small, irregularly shaped scales, the largest of which is much less than 25% of the width of the adjoining prefrontal scale. The forelimb is short; its distal half is covered with single rows of enlarged scales along the phalanges which are separated by rows of smaller scales or wrinkled skin. Head and neck are olive gray dorsally but cream colored ventrally. The limbs are gray.

Adult females are larger than adult males, but have short tails in contrast to the long, thick tail of the male.

Distribution: The total range includes the coastal waters of the continental shelf off north and northeastern Australia.

Habitat: *Chelonia depressa* may occur in the open sea, but is more common in shallow coastal waters.

Natural History: The nesting range of *Chelonia depressa* is predominantly tropical, extending from northern Western Australia eastward around the Australian coast to Mon Repos, Queensland, at 25°S. Most nesting is in November and December, with mating occurring in the waters off the beaches at that time. Most nests are dug in early evening on the tops of sand dunes, or on the steep seaward slopes (Limpus, 1971; Bustard et al., 1975). A shallow body pit is first excavated, and then the nest is dug. Nests measured by Bustard et al. (1975) were oval: 20–22 cm long, 12–15 cm wide, and 30–36 cm deep. The round, white eggs are about 52 mm in diameter with parchmentlike shells, and a clutch averages 50 eggs (7–78). Up to four clutches are laid each season (Limpus, 1971). Incubation takes about six weeks with hatchlings emerging at night. Hatchlings measured by Limpus averaged 61.2 mm (56.6–65.5) in carapace length, much larger than those of *C. mydas*.

C. depressa is more carnivorous than *C. mydas*, feeding on sea cucumbers, prawns, and other invertebrates, and consequently its flesh is less palatable than that of its congener. It is rarely eaten by humans.

Bustard (1972) reported that flatbacks spend much of the day floating on the surface basking in the sun, and that it is not uncommon to see birds resting on their backs.

C. depressa is more numerous in Australian waters than *C. mydas*, and is probably in a less critical state. However, though adults are not eaten, the eggs are, and this could eliminate the species from many of its nesting beaches. Strict protection of the nesting beaches is essential.

Eretmochelys Fitzinger, 1843

The hawksbill sea turtle is the only living member of this genus.

Eretmochelys imbricata (Linnaeus, 1766)
Hawksbill sea turtle

Recognition: This is a small to medium-sized (to 91 cm) sea turtle with four pairs of pleurals (the 1st not touching the cervical) and two pairs of prefrontal scales. The carapace is shield shaped (heart shaped in the young and more elongated and straight sided in the adult), has four pairs of pleural scutes, a keel on the last four vertebrals, and posterior serrations. No lateral fontanelles occur in the adult. All vertebrals are broader than long; neurals 9–11 are six sided and shortest anteriorly. There are 11 pairs of peripherals; the 9th pair does not touch ribs. In the young, the carapacial scutes strongly overlap the next most posterior, but as the turtle matures the overlapping be-

Eretmochelys imbricata imbricata

comes progressively less, until finally the scutes lie side by side. The carapace is dark greenish brown; in the young it shows a tortoise-shell pattern. There are four poreless inframarginal scutes on the yellow bridge. The plastron is hingeless and yellow; in juveniles it may have two longitudinal ridges and a few dark blotches, especially on the anterior scutes. The plastral formula is: an > fem > abd >< hum >< pect > gul > intergul. Head scales are black to chestnut brown at the center and lighter at their margins; the jaws are yellow with some brown streaks or bars. There are two pairs of prefrontals and three postoculars. The chin and throat are yellow, and the neck dark above. The triturating surfaces of the premaxillae and maxillae bear a ridge; the lower jaw is only slightly serrated. The vomer touches the premaxillae separating the maxillae; the descending processes of the prefrontal only touch the vomer. The frontal enters the orbit, and there is a low, blunt ridge on the vomer and palatines. The snout is elongated and narrow, rather like a hawk's beak, and is not notched at the tip. The floor of the mouth is deeply excavated at the mandibular symphysis. Two claws occur on the forelimbs.

The karyotype is 2n = 56 (Bickham, 1981).

Males have somewhat concave plastrons; long thick tails that extend beyond the posterior carapacial margin; and long, heavy claws.

Immunoprecipitation tests on the serum of marine turtles have shown that *Eretmochelys* is more closely related to *Caretta* and *Lepidochelys* than to *Chelonia* (Frair, 1979).

Distribution: The hawksbill ranges in the Atlantic, Pacific, and Indian oceans from California, Japan, the Red Sea, the British Isles, and Massachusetts south to Peru, Australia, Madagascar, northwestern Africa, and southern Brazil.

Geographic Variation: Two subspecies are recognized. *Eretmochelys imbricata imbricata* (Linnaeus, 1766) ranges through the warmer parts of the western Atlantic Ocean, from Massachusetts through the Gulf of Mexico to southern Brazil. It also has been recorded from Scotland and Morocco. This subspecies has a nearly straight-sided carapace that tapers posteriorly, a keel that is continuous on only the last four vertebrals, ridges that converge posteriorly on only the last two vertebrals, and less black on the upper surfaces of the head and flippers. *E. i. bissa* (Rüppell, 1835) ranges through the tropical portions of the Indian and Pacific oceans, from Madagascar to the Red Sea on the east coast of Africa and east to Australia and Japan in the western Pacific, to the Hawaiian Islands in the central Pacific, and from Peru to Baja California in the eastern Pacific. Stragglers occasionally reach California. It has a more heart-shaped carapace, a fully continuous vertebral keel, all vertebrals with ridges that converge posteriorly, and the head and flippers almost solid black. Additional study is needed to determine if these subspecies are valid, and also if other valid subspecies exist.

Habitat: This marine turtle is characteristically an inhabitant of rocky places and coral reefs. It also occurs in shallow coastal waters, such as mangrove-bordered bays, estuaries, and lagoons with mud bottoms and little or no vegetation, and in small narrow creeks and passes. It is occasionally found in deep waters and has been taken from floating patches of *Sargassum* weed. *Eretmochelys* shares its water habitat and nesting beaches with all other marine turtles.

Natural History: Mating occurs in the shallow water off the nesting beaches. According to Hornell (1927), copulation begins soon after the spent females return to the sea, but males have been observed to follow a female onto the beach. Whether mating takes place at any other time or place is not known.

The Atlantic hawksbill nests on beaches in Florida, Jamaica (Pedro and Morant cays), the Cayman Islands, Aves Island, the Virgin Islands, Grenada, Tobago, Trinidad, Guyana, Surinam, French Guiana, Venezuela, Panama, Costa Rica, Nicaragua, Mexico, and the islands and keys off Central America. Nesting lasts from April through November, peaking June and July. The Pacific hawksbill nests on beaches on the eastern coast of Africa (December to February); the Seychelles Islands (throughout the year, with a peak from September to November); islands off Madagascar, Arabia and in the Persian Gulf and Gulf of Aden; Ceylon (November to February); Indonesia, northern Australia; Ryukyu Islands; Samoa; Fiji; Hawaii; and in Honduras, Nicaragua, El Salvador (August to November), and Mexico. Nesting occurs at night and takes about an hour. Females apparently oviposit in 3-year cycles. We refer the reader to Carr, Hirth, and Ogren (1966) for a lucid description of nesting behavior. The white, calcareous eggs usually are spherical, averaging about 38 mm (35–42) in diameter, but a few in each clutch may be slightly elongated. Eggs examined by Deraniyagala (1939) were thinly covered with a mucilaginous secretion which appeared to absorb water and was found to remain moist for 48 hr. The average number of eggs in 57 nests examined by Carr, Hirth, and Ogren (1966) was 161 (53–206). The average incubation period of 13 clutches moved to artificial nests by Carr, Hirth, and Ogren (1966) was 58.6 days (52–74). These clutches contained a total of 2193 eggs that were artificially buried in 9 nests, and 46.7% hatched (range in the individual nests, 12–80%). Probably mortality was influenced by the moving of eggs.

Hatching usually occurs at night or in early morning. Eggs hatch almost simultaneously and the hatchlings follow one another to the surface in quick succession. The hatchling carapace (39–50 mm) is heart shaped with a vertebral keel; the plastron has two longitudinal ridges. Hatchlings are black or very dark brown except for the keels, the shell edge, and areas on the neck and flippers, which are light brown.

Eretmochelys is omnivorous but seems to prefer invertebrates. It is known to consume sponges, coelenterates (Portuguese man-of-war, hydroids, coral), ectoprocts *(Amthia, Steganoporella),* sea urchins, gastropod and bivalve mollusks *(Pinna, Ostrea),* barnacles, crustaceans, ascidians, and fish. Plants eaten are algae, *Cymodocea, Conferva,* and *Sargassum.* Carr (1952) reported that it ate the fruits, leaves, dead bark, and wood of the red mangrove. Captives eat fish, meat, bread, octopi, squid, crabs, mussels, and oysters. Hatchlings seem to be herbivorous, but become more omnivorous as they age.

The translucent carapacial scutes form the tortoiseshell of commerce. These are clear amber, streaked with red, white, green, brown, and black. Usually the turtle is killed before the scutes are removed by the application of heat. There is evidence that if the scutes are removed carefully and the turtle is returned to the sea it can regenerate lost scutes. This may be possible if the epidermal Malpighian cells are not damaged. A single turtle can yield 10–12 lb of tortoise-shell (called "carey" by the Caribbeans). The flesh and eggs are eaten in many parts of the range. This requires caution, for *Eretmochelys* tends to store in its tissues the toxins of various poisonous organisms that it eats. Survival of the hawksbill is questionable. Existing conservation laws must be enforced, and egg collecting stopped.

Caretta Rafinesque, 1814

The one living species in this genus is the loggerhead sea turtle, *Caretta caretta.*

Caretta caretta (Linnaeus, 1758)

Loggerhead sea turtle

Recognition: This is a large (to 213 cm) sea turtle with five or more pairs of pleurals (the 1st touches the cervical), and three poreless inframarginals on the bridge. The elongated carapace has a medial vertebral keel that becomes progressively smoother with age, is

Caretta caretta caretta

highest anterior to the bridge, and is serrated posteriorly. No lateral fontanelles occur in adults. All vertebrals are broader than long; neural bones 7–11 are hexagonal, shortest anteriorly, and may be separated. The marginal scutes average 12 or 13 on each side but vary from 11 to 15. There are 12–13 pairs of peripherals; the 9th and 10th pairs are separated from the ribs. The posterior marginal rim is serrated in juveniles, but becomes smoother with age. The carapace is reddish brown but may be tinged with olive; the scutes are often bordered with yellow. Bridge and plastron are yellow to cream colored. The plastron is hingeless and has two longitudinal ridges, which disappear with age. The plastral formula is: an > abd >< fem > hum > pect > gul > intergul; the intergular scute is very small, and may be absent. The large head is broad posteriorly and rounded in front; its snout is short and broad, and there are two pairs of prefrontals and three postoculars. The head varies from reddish or yellow chestnut to olive brown, often with yellow-bordered scales. No ridge occurs on the triturating surface of the yellowish-brown jaws, and the bony surface of the lower jaw is smooth at the symphysis. The vomer neither separates the maxillae nor touches the premaxillae. The descending prefrontal processes touch both the vomer and palatines. The frontal does not enter the orbit, and there is a blunt ridge on the vomer and palatines. Limbs and tail are dark medially and yellow laterally and below. Two claws occur on the forelimbs. Adults usually weigh about 100–150 kg, but truly massive size is sometimes attained. *Caretta* probably is the largest living hard-shelled turtle; it is exceeded in length and weight only by the leatherback *(Dermochelys coriacea).* Weights up to 450 kg have been recorded and there is a skull in the Bell collection at Cambridge University of such width (28.4 cm) as to indicate a weight of about 540 kg (Pritchard, 1967).

The karyotype is 2n = 56 (Bickham, 1981).

Males have narrow shells gradually tapering posteriorly, and long, thick tails, extending beyond the rear carapacial margin.

Distribution: *Caretta* occurs in the Pacific, Indian, and Atlantic oceans from Washington, Japan, India, Kenya, the British Isles, and Newfoundland south to Chile, Australia, South Africa, tropical western Africa, and Argentina. It also occurs in the Caribbean and Mediterranean seas.

Geographic Variation: Two poorly marked subspecies are recognized. *Caretta caretta caretta* (Linnaeus, 1758) ranges from Newfoundland and the British Isles south to Argentina and the Canary Islands and western coast of tropical Africa; occasionally it enters the Mediterranean Sea. This race usually has 7 or 8 neurals, and averages 12 marginal scutes on each side. *C. c. gigas* Deraniyagala, 1933 occurs from Washington, Japan, India, and Kenya south through the Pacific and Indian oceans to Chile, Australia, and South Africa; there are also records from the eastern Atlantic. It has 7 to 12 neurals in the carapace (the last 5 usually interrupted by costal bones), and 13 marginal scutes on each side.

Pritchard (1979) recently demonstrated that there is much overlap in number of both neurals and marginals between various populations around the world, and has suggested abandoning the trinomial distinctions. Also, Stonebruner (1980) has shown variation in body depth between nesting populations of Atlantic *Caretta*. A thorough study of variation within *Caretta* is sorely needed.

Habitat: The loggerhead wanders widely throughout the marine waters of its range; it has been found as far as 240 km out in the open sea. It enters bays, lagoons, salt marshes, creeks, and the mouths of large rivers. The loggerhead shares its nesting beaches with the other species of sea turtles.

Natural History: Mating has been observed at every hour from dawn to dark but doubtless also occurs at night. Paired loggerheads may copulate for extended periods—more than three hours, according to Wood (1953)—and perhaps the females remate after each nesting. Mating occurs at the surface of the water off the nesting beaches. Although the female is completely submerged, the highest part of the male's carapace usually is out of the water. He grasps the anterolateral margins of her carapace with the claw of each foreflipper and the posterolateral margins with the claw of each hind flipper. He sometimes bites the nape of her neck. His tail is bent down and under hers, so that the cloacal openings touch. Female mating behavior ranges from passive acceptance to violent resistance. Nesting is mostly carried out on temperate zone beaches (except for tropical beaches in the western Caribbean). In the United States nesting regularly occurs in South Carolina, Georgia, and Florida, and occasionally from North Carolina to southern New Jersey. Other nesting sites occur in the western Caribbean in Mexico, Belize, Colombia, Venezuela, Trinidad, the Cayman Islands, Cuba, various islands offshore of the mainland of South and Central America, and possibly on the Bahamas. It also nests in Sicily, Sardinia, Italy, Greece, and Libya in the Mediterranean, and in Mozambique, Tongaland, Natal, the Malagasy Republic, islands off southern Arabia, and Western Australia around the Indian Ocean. In the Pacific Ocean, nesting occurs in Japan, eastern Australia, Fiji, the Solomon Islands, and western Mexico (Pritchard, 1979). Most nesting is in the spring or early summer, and long migrations are often made between the feeding areas and the nesting beaches. Most nests are on open beaches or along narrow bays, if the soil is suitable. The nests are dug at night above the high-tide mark and usually seaward from the dune front. Although the females may come onto the beaches on any night during the season, most nest during periods of high tides. They oviposit up to seven times a season on the same stretch of beach (Lenarz et al., 1981), at intervals of 12 to 15 days (Caldwell, 1962b). Groups of females apparently nest together several times. Both 2- and 3-year nesting cycles occur. Caldwell, Carr, and Ogren (1959) described the nesting behavior in detail. A body pit is first dug, and then a nest cavity 15–25 cm deep. The eggs are spherical (35–49 mm) with leathery shells; usually 64–200 eggs are laid at one time. Incubation takes from 49 to 71 days. Young loggerheads emerge from the nest at night; heat probably controls the time of emergence. They have heart-shaped, yellowish-brown to brown or grayish-black carapaces (38–55 mm) with three longitudinal keels. The plastron also bears several longitudinal ridges. *Caretta* eggs incubated at 32°C or above develop into females, while those incubated at 28°C or below produce males; intermediate incubating temperatures give both males and females (Yntema and Mrosovsky, 1982).

Caretta is omnivorous. It commonly noses about coral reefs, rocky places, and old boat wrecks for food. Animals consumed include sponges, jellyfish, mussels, clams, oysters, conchs, borers *(Natica)*, squid, shrimp, amphipods, crabs, barnacles, sea urchins, tunicates, and various fish; whether the fish are caught fresh or consumed as carrion is not known. It also eats seaweed, turtle grass *(Zostera, Thalassia)*, and *Sargassum*.

Carr, Ogren, and McVea (1980) have recently discovered loggerheads hibernating in the Port Canaveral ship channel off the east coast of Florida.

The loggerhead has been so consistently persecuted at the nesting grounds and so many of them destroyed that the original nesting range cannot be discerned. The rapid development of beaches and coastal islands for home sites and recreational areas probably will destroy most of the North American nesting beaches and perhaps exterminate the Atlantic subspecies. The raccoon is a most efficient nest predator, and the marked expansion of its coastal populations also threatens the future of *C. c. caretta*. Many adults are drowned when caught in shrimp trawls. The main hope for the survival of the Atlantic loggerhead lies in the development of state and national parks and reserves, and the development of turtle-proof shrimp trawls.

Lepidochelys Fitzinger, 1843

The two species of *Lepidochelys* have heart-shaped carapaces with serrated posterior marginals, five to eight pleurals on each side, and, in some individuals, more than five vertebral scutes. Lateral fontanelles are absent in the adult carapace. There are 12–13 pairs of peripheral bones, and 12–15 neurals which are shortest anteriorly and either four sided or six sided. The rigid plastron has at best only a small intergular scute, and there are four pairs of pore-bearing inframarginals on the bridge. In the broad skull, the vomer separates the maxillae and touches the premaxillae. There is a ridge on the triturating surface of the maxilla, but not on that of the premaxilla. The descending processes of the prefrontals touch only the vomer. The frontal enters the orbit. No ridge occurs on the vomer or palatines. There are two pairs of prefrontals and three to four postocular scales on the head. The forelimbs are paddlelike with one to three claws.

The karyotype is 2n = 52 (Nakamura, 1937).

Serological tests by Frair (1969, 1979) suggest that *Lepidochelys* and *Caretta* are closely related, and that both are more closely related to *Eretmochelys* than to *Chelonia*. Frair thought *Lepidochelys* to be the closest relative of ancestral sea turtles.

Key to the species of *Lepidochelys*:

1a. Usually only five pairs of pleurals; color gray: ***L. kempii***
  b. Usually more than five pairs of pleurals; color olive: ***L. olivacea***

Lepidochelys kempii (Garman, 1880)

Kemp's ridley sea turtle

Recognition: This small (to 72 cm) sea turtle has only five pleurals on each side of the heart-shaped, keeled, yellowish to gray carapace. The carapace is wider than long in adults, highest anterior to the bridge, and serrated behind. Vertebrals 1 and 5 are broader than long, but vertebrals 2–4 are often longer than broad. An additional small vertebral may occur between the 4th and 5th. The 1st pair of pleurals touches the cervical, and there are 12 to 14 marginals on each side. The bridge and plastron are immaculate white. The formula for the six large plastral scutes is: gul > an >< fem > abd > pect > hum; however, a small interanal scute may also be present, as well as one or two small intergulars. The head and limbs are gray. The head is wide and somewhat pointed anteriorly; the snout is short and broad. The bony alveolar surface of the upper jaw has a conspicuous ridge running parallel to the cutting edge.

Males have concave plastrons, long tails, which extend beyond the posterior carapacial margin, and a recurved claw on each forelimb.

Distribution: *Lepidochelys kempii* ranges in the western Atlantic Ocean from Nova Scotia and, possibly, Newfoundland south to Bermuda and west through the Gulf of Mexico to Mexico. It also crosses the Gulf Stream to England, Ireland, the Scilly Isles, France, and the Azores, and occasionally enters the Mediterranean Sea. Almost all nesting occurs on the southern coast of Tamaulipas, Mexico, near Rancho Nuevo.

Habitat: Kemp's ridley prefers shallow water. In the Florida Keys it is closely associated with the subtropical shoreline of red mangrove.

Natural History: Courtship and mating occur shortly before nesting, at the surface of the water just off the

Lepidochelys kempii

beaches. The males become quite active at this time; Carr (1967) reported that one even followed a nesting female onto the beach and attempted to mount her on the way. Nesting occurs sporadically from April through mid-August, is usually diurnal, and is on a 1- to 3-year cycle. Although some females come up to nest individually, most do so in large groups (arribadas)—formerly in the thousands, but now only a few hundred. Up to three clutches are laid each season. Nests are flask shaped and about 37–49 cm deep. Eggs are spherical (34–45 mm) with soft shells; 80–140 eggs make up the usual clutch. Incubation takes 50–70 days. Hatchlings have a relatively elongated (38–46 mm), oval carapace with three tuberculate keels and a plastron with four longitudinal ridges. They are dark gray to black, with a white border on both flippers and carapace.

Lepidochelys kempii is predominantly carnivorous, feeding on crabs *(Callinectes, Ovalipes, Hepatus)*, snails *(Nassarius)*, clams *(Nuculana, Corbula, Mulinia)*, jellyfish, fish, and occasionally on various marine plants. It seems to be largely a bottom feeder. Captives readily eat fish.

This ridley's chances of survival are more precarious than those of other sea turtles. Its numbers have been decimated by fishing, nest robbing, oil spills, and the slaughter of nesting females. Fortunately, the Mexican government, in an effort to save this turtle, now protects nesting females on the beaches of Tamaulipas. A group of concerned American conservationists is transplanting eggs and hatchlings from Tamaulipas to Padre Island, Texas, a known natural nesting site, in an attempt to reestablish the breeding colony.

Lepidochelys olivacea (Eschscholtz, 1829)

Olive ridley sea turtle

Recognition: A small (to 71 mm) sea turtle with six to eight (occasionally five to nine) pleurals on each side. The heart-shaped olive carapace is flattened dorsally, highest anterior to the bridge, and serrated posteriorly. The 1st vertebral is usually slightly broader than long, the 2d to 4th are longer than broad, and the 5th is much broader than long. Often the original five vertebrals subdivide to produce up to nine. The 1st pair of pleurals touches the cervical, and there may be more pleurals on one side of the carapace than on the other; in specimens from the eastern Pacific, at least, the higher number usually is found on the left side. Twelve to 14 marginals occur on each side of the carapace. Both bridge and plastron are greenish white or greenish yellow. Two longitudinal ridges are pres-

Lepidochelys olivacea

ent on the plastron, and the plastral formula is: an > fem > gul > abd > pect > hum. The skin is olive above and lighter below. The head is wide, with concave sides, especially on the upper part of the short, broad snout. The bony alveolar surface of the upper jaw may have a gentle elevation extending parallel to the cutting edge, but it lacks a conspicuous ridge.

Males have long, thick tails, which extend well beyond the rear carapacial margin; females' tails usually do not reach the margin. Males have concave plastrons, a more gently sloping lateral profile, and a strongly developed, curved claw on each front flipper.

The pleural scutes of *Lepidochelys olivacea* are clearly divisible into whole and half scutes, the whole scutes being homologous with the five pleurals of *L. kempii*. Displacement of homologues of the seams of *L. kempii* usually is slight, although in cases of extreme splitting (to eight or nine pleurals) the seams become displaced to lessen the size of the small 1st pleural and large last vertebral. In almost every case division takes place in the posterior pleurals; for example, a 6-6 count is produced by division of the 5th pleurals on each side or an 8-8 count by division of the 3d, 4th, and 5th pleurals.

Distribution: *Lepidochelys olivacea* occurs in the tropical waters of the Pacific and Indian oceans from Micronesia, Japan, India, and Arabia south to northern Australia and southern Africa; in the Atlantic Ocean off the western coast of Africa and the coasts of northern Brazil, French Guiana, Surinam, Guyana, and Venezuela in South America; and, occasionally, in the Caribbean Sea as far north as Puerto Rico. In the eastern Pacific it is found from the Galapagos northward to California.

Geographic Variation: No subspecies are recognized; however, the possibility exists that the eastern Pacific and southern Atlantic populations are distinct.

Habitat: Most records are from protected, relatively shallow marine waters, but the olive ridley occasionally occurs in the open sea. Deraniyagala (1939) reported the habitat to be the shallow water between reefs and shore, larger bays, and lagoons.

Natural History: Mating probably occurs offshore from the nesting beaches at the time of nesting. Nesting occurs on beaches in the Marianas, northern Australia, Indonesia, Malaysia, Sarawak, the Bonin Islands, southern Japan, Vietnam, Pakistan, the Seychelles, India, Sri Lanka, East Africa, Mozambique, northern Madagascar, western Mexico to Panama and possibly Colombia, and in the Atlantic from Senegal to Zaire, Brazil, French Guiana, Surinam, Guyana, Trinidad, and Venezuela (Pritchard, 1979). Since these beaches lie in both hemispheres, nesting occurs in almost every month of the year, but usually in the spring and early summer. Females start to emerge from the sea with a rising tide in late afternoon and are finished nesting by the advent of low tide at night. During the peaks of activity many nest on the same stretch of beach; Pritchard (1969) saw 97 nesting simultaneously on a 230-yard stretch of beach in Surinam, and 115 were present on the entire beach. The nesting arribadas may be very large; over 100,000 nest each year at Orissa, India. The complete nesting, from emergence to reentering the sea usually takes less than one hour to complete. Females may nest three times a season but usually do so only twice. Pritchard (1969) found that in Surinam the usual intervals between successive nestings were about 17 days, 30 days, and, possibly, 44 days; one female was recovered after a 60-day interval. Nesting often occurs during periods of strong winds, and Pritchard felt that the interval between nestings was controlled more by environmental factors, such as the tide and weather, than by physiologic factors. Some nest annually, but others apparently are on multiple-year cycles (Hughes, in Bjorndal, 1981). The flask-shaped nests are 48–55 cm deep. Clutches range from 30 to 168 ($\bar{x}$ = 95) eggs. The oval to spherical, white eggs have soft, leathery shells and are 37.5–45.4 mm in diameter. Incubation takes from 49 to 62 days. The relatively elongated, oval carapace (40–50 mm) of the hatchling usually has well-developed pale keels on the vertebrals and pleurals. There are four longitudinal ridges on the plastron. Hatchlings are gray black to olive black, with a small white mark at each side of the supralabial scale, another on the hind part of the umbilical protuberance, and several where the ridges of the plastron cross the abdominal and femoral scutes. Both skin and plastron may be dark grayish brown. A fine white line borders the carapace and trailing edge of the fore and hind flippers.

Lepidochelys olivacea is highly carnivorous, feeding predominantly on fish, crabs, snails, oysters, sea urchins, and jellyfish. Seaweeds are occasionally eaten. Captives eat fish, meat, and bread, and may be cannibalistic.

The taking of nesting females and the gathering of eggs has contributed to decline in *L. olivacea*. Strict conservation laws must be passed to safeguard this species, especially while nesting.

Chelydridae

This semiaquatic New World family ranges from southern Canada to Ecuador, and includes two of the most savage of all living turtles, *Chelydra serpentina* and *Macroclemys temminckii*. They are among the largest of the freshwater turtles, and have big heads, powerful jaws, and vile tempers. Several fossil species are recognized, although only the two above species are extant. Extinct genera include *Hoplochelys* (Paleocene to Eocene, North America), *Gafsachelys* (Eocene, Europe), *Chelydropsis* (Oligocene to Miocene, Europe, Asia, and possibly Africa), *Acherontemys* (Miocene, North America), and *Chelydrops* (Miocene, North America). Of the living genera, *Macroclemys*

dates from the Miocene, and *Chelydra* from the Pleistocene of North America.

The temporal region of the large skull is slightly emarginate, and the frontals do not enter the orbit. There is no parietal-squamosal contact. The maxilla is not connected to the quadratojugal, and only rarely is its crushing surface ridged. The quadrate encloses the stapes. There is no secondary palate, and the tip of the upper jaw is hooked. The well-developed, rough carapace is keeled, strongly serrated posteriorly, has 11 peripherals on each side, and does not become fully ossified until late in life. Seven or eight neural bones and one or two suprapygals are present. The carapace is connected to the reduced, cross-shaped, hingeless plastron by a narrow bridge. Inframarginal scutes are present, and a series of supramarginals also occurs in *Macroclemys*. The abdominal scutes are reduced to the bridge, and do not usually meet at the midline. A T-shaped entoplastron and a median plastral fontanelle are present. The 10th dorsal vertebra lacks ribs, and there is only one biconvex vertebra in the neck. Limbs are well developed, webbed, and heavily clawed. The

---

Key to the genera of Chelydridae:

1a. A row of supramarginal scutes above marginals on each side; upper jaw strongly hooked; carapace with three prominent keels extending entire length: *Macroclemys*

b. No supramarginals; upper jaw not strongly hooked; carapacial keels not extending entire length: *Chelydra*

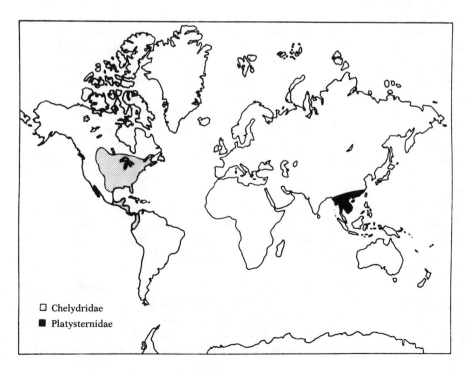

□ Chelydridae
■ Platysternidae

Distribution of Chelydridae and Platysternidae

saw-toothed tail is as long or longer than the carapace. There are only two living genera, each with a single species.

Chelydra Schweigger, 1814

The single species is *Chelydra serpentina*.

Chelydra serpentina (Linnaeus, 1758)
American snapping turtle

Recognition: The carapace (to 47 cm) is massive, strongly serrated posteriorly, and has three low keels, which are composed of knobs located well behind the centers of the scutes. With increasing age the keels become less conspicuous, and old specimens of *Chelydra* are often smooth shelled. Vertebral scutes are broader than long, and the 5th is laterally expanded. There is a short, broad cervical. Beneath the vertebrals is a series of eight quadrilateral or hexagonal neurals, and two suprapygals. Costals are laterally reduced, leaving fontanelles between them and the 11 peripheral bones on each side. No supramarginal scutes are present, but there are 24 marginals. The carapace varies from tan, brown, or olive to black, and there may be a pattern of radiating lines on each scute. The bridge is small and the plastron is reduced, giving a cruciform appearance. The plastron is yellowish to tan, and usually lacks a pattern. Its entoplastron is T-shaped, and there is a medial fontanelle. The abdominal scutes are confined to the bridge, do not usually meet at the midline, and are separated from the marginal scutes by two or three inframarginals. The plastral formula is: an >< hum > pect > fem > gul > abd. The head is large, with a blunt, slightly protruding snout, a slightly hooked upper jaw, and large dorsolateral orbits (the eyes can readily be seen from above). There is little emargination of the temporal region; the squamosal bone contacts the long postorbital but not the parietal. The maxilla does not touch the quadratojugal; the quadrate completely encloses the stapes. The prootic bone forms a portion of the roof of the internal carotid canal. No secondary palate is present, and there is a strongly developed medial ridge on the ventral surface of the vomer. There are no dermal projections on the side of the head, but there are two chin barbels and conical, wartlike projections on the neck. Both temporal region and back of the head are covered with large, flat, over-

Chelydra serpentina serpentina

Chelydra serpentina rossignonii

lapping scales. The powerful jaws are often barred; no medial ridge occurs on the premaxilla or maxilla. The tongue lacks a wormlike process. Skin is gray to brown, with some yellow or white flecks. Lateral rows of tail tubercles are much less conspicuous than the median row, so that there appears to be only one saw-toothed row; the lower surface of the tail is covered with large flat scales.

The karyotype is 2n = 52: 24 macrochromosomes and 28 microchromosomes (Haiduk and Bickham, 1982).

Males grow larger and have longer preanal tail lengths (86% of the length of the posterior plastral lobe) than females (less than 86%). In males the vent is posterior to the carapacial rim.

Distribution: *Chelydra serpentina* ranges from Nova Scotia, New Brunswick, and southern Quebec, west to southeastern Alberta, and southward east of the Rocky Mountains to southern Florida and the Texas coast in the United States, and south through Mexico and Central America to Ecuador. In the United States it occurs to elevations of at least 2000 m.

Geographic Variation: Four subspecies are recognized. *Chelydra serpentina serpentina* (Linnaeus, 1758) ranges from Nova Scotia, New Brunswick, and southern Quebec, west to southeastern Alberta, and southward east of the Rocky Mountains to northern Florida and the coast of Texas. It is distinguished by the following characters: the length of the plastral forelobe (from the level of the hyo-hypoplastral suture to the anterior tip) is less than 40% of the carapace length, the width of the 3d vertebral is much less than the height of the 2d pleural, and the dorsal surface of the neck has rounded wartlike tubercles. *C. s. acutirostris* Peters, 1862 ranges from northern Honduras southward through the Caribbean drainages of Central America to the Pacific lowlands of Colombia and Ecuador. The plastral forelobe is usually longer than 40% of the carapace length, the anterior width of the 3d vertebral is less than 25% of the maximum carapace width, and the neck has low, rounded wartlike tubercles. *C. s. rossignonii* (Bocourt, 1868) occurs in the Atlantic lowland drainages of Mexico from central Veracruz southward across the base of the Yucatán Peninsula and southern Campeche to western Belize, Guatemala, and western Honduras. The plastral forelobe is usually longer than 40% of the carapace width, the anterior width of the 3d vertebral is more than 25% of the maximum carapace width, and the neck has long, pointed tubercles. *C. s. osceola* Stejneger, 1918 is restricted to peninsular Florida. The plastral forelobe is usually less than 40% of the carapace length, the width of the 3d vertebral is the same as or greater than the height of the 2d pleural, and the neck has long, pointed tubercles.

Not all authorities accept this subspecific arrangement. Carr (1952) and Medem (1977) question the validity of the subspecies and feel it more realistic to consider the four forms as only local populations, or demes.

Habitat: The American snapping turtle has been taken in almost every freshwater habitat within its range, and it also enters brackish tide pools. It prefers water with a soft mud bottom and abundant aquatic vegetation or an abundance of submerged brush and tree trunks.

The snapping turtle is one of the more aquatic species. It spends most of its time lying on the bottom of some deep pool or buried in the mud in shallow water with only its eyes and nostrils exposed. The depth of the water above the mud is usually comparable to the length of the neck, for the nostrils must be periodically raised to the surface. This turtle also hides beneath stumps, roots, brush, and other objects in the water and in muskrat lodges and burrows.

Natural History: Mating may occur anytime from April to November. Males often approach females from behind and mount at once; however, Legler (1955) observed a pair facing each other on the bottom with necks extended and noses about 25 mm apart. Each was positioned so that the anterior edge of the plastron touched the bottom and the hindquarters were elevated to a maximum by the hind legs. The turtles made sudden, simultaneous, sideward sweeps of the heads and necks, but in opposite directions; subsequently they slowly brought their necks back to the straightforward position. The sequence was repeated about 10 times at intervals of approximately 10 sec. Taylor (1933) found two *C. serpentina* with their heads close together in shallow water; they appeared to be gulping water and violently forcing it out their nostrils, causing an upheaval of the surface above their heads. Nesting occurs from May through September; June is the peak month. An open site, which may be several hundred meters from the water, is selected and the nest is dug with the hind feet in sand, loam, or vegetable debris. Muskrat lodges are often used as nesting sites. The nest usually is bowl or flask shaped: a narrow opening descends at an angle to a large egg chamber below. The depth of the nest ranges from 10 to 13 cm, and its dimensions vary with the size of the female. The number of eggs in a clutch is 11–83, but usually 20–30. Larger females lay more eggs. Only one clutch is laid each season. The white, tough, spherical eggs are 23–33 m in diameter and weigh 7–15 g. Incubation takes 55–125 days, depending on environmental conditions. Emergence from the nest normally occurs from late August to early October, but may be delayed until the following spring. Hatchlings have nearly round, dark-gray to brown, wrinkled carapaces with three distinct keels. There is often a pattern, consisting of a light spot on the underside of each marginal, and the plastron is black with some light mottling. Skin is dark gray, and head and jaws may be somewhat mottled. Hatchlings are about 24–31 mm in carapace length and 24–29 mm in carapace width. The sex of hatchlings is determined by the incubation temperature: warm (30°C or above) and cold (20°C or below) temperatures cause female development; intermediate temperatures (22–28°C) cause male development (Yntema, 1976; Wilhoft et al., 1983).

Chelydra serpentina is omnivorous. It eats insects, crayfish, fiddler crabs, shrimp, water mites, clams, snails, earthworms, leeches, tubificid worms, freshwater sponges, fish (adults, fry, and eggs), frogs and toads (adults, tadpoles, and eggs), salamanders, snakes, small turtles, birds, and small mammals.

Plants eaten are various algae, *Elodea, Potamogeton, Polygonum, Nymphaeca, Lemna, Typha, Vallisneria, Nuphar, Wolffia,* and *Najas.*

Although sluggish by day, the snapping turtle apparently is quite active at night. It usually moves about by creeping slowly over the bottom; however, when disturbed it can swim rapidly. By day it often floats lazily just beneath the surface with only its eyes and nostrils protruding. It may bask in this fashion, although it may also emerge on logs, for example, to bask.

The snapping turtle has a vicious temper and should be handled carefully. It strikes with amazing speed and its jaws are capable of tearing flesh. The strike often carries the forepart of the body off the ground. When handled, snappers emit a musk as potent as that of musk turtles. Snapping turtles normally are docile when submerged, but even then they can and sometimes do bite viciously.

Macroclemys Gray, 1855

The single living species within this genus, *Macroclemys temminckii,* is one of the unique turtles of the world. It is probably the heaviest freshwater turtle, growing to 66 cm and weighing as much as 80 kg, and is almost completely aquatic: females leave the water only to oviposit, and adult males apparently occasionally emerge only to bask.

Macroclemys temminckii (Troost, in Holbrook, 1836)

Alligator snapping turtle

Recognition: The large, rough, dark-brown or dark-gray carapace is strongly serrated posteriorly, and has three rather high, persistent keels. Knobs on the keels are elevated and somewhat curved posteriorly. Vertebrals are broader than long, and there is a short, broad cervical. Beneath the vertebrals is a series of eight hexagonal neurals and a suprapygal. Costals 1–8 are reduced laterally, leaving small fontanelles between them and the 11 peripherals on that side. A row of three to eight supramarginal scutes occurs between the marginals and the first three pleurals on each side. There are 23 marginals. The bridge is small, and the hingeless, gray to mottled plastron is reduced, giving a cross-shaped appearance. It has a T-shaped entoplastron and a median fontanelle. Abdominals are re-

Macroclemys temminckii

duced, do not usually meet at the midline, and are separated from the marginal scutes by a series of inframarginal scutes. The plastral formula is: fem >< pect > an > hum > gul > abd. The head is large with a pointed snout, a strongly hooked upper jaw, and large, lateral orbits. There is little emargination of the temporal region; the squamosal bone contacts the long postorbital but not the parietal. The maxilla does not touch the quadratojugal; the quadrate bone completely surrounds the stapes. The prootic bone forms little of the roof of the internal carotid canal, which continues within the pterygoid bone. No secondary palate is present, and there is only a weak medial ridge on the ventral surface of the vomer. There are numerous dermal projections on the side of the head and on the chin and neck. The jaws are powerful, but lack medial ridges on the premaxilla or maxilla. There is a wormlike process on the tongue, attached approximately at its center to a rounded muscular base, leaving both ends freely movable. In the young, each end of this elongate process may bear a small branch, but in larger specimens the ends are entire. Skin is dark brown to gray above and lighter below; there may be darker blotches on the head. The tail is about as long as the carapace and has three rows of low tubercles above and numerous small scales below.

The karyotype is 2n = 52: 24 macrochromosomes, 28 microchromosomes (Haiduk and Bickham, 1982).

Males have longer preanal tail lengths than females, and have the vent posterior to the carapacial rim.

Distribution: *Macroclemys* ranges south in the Mississippi Valley from Kansas, Iowa, and Illinois to the Gulf and on the coastal plain from southeastern Georgia and northern Florida to east Texas.

Habitat: The alligator snapper occurs most frequently in the deep water of rivers, canals, lakes, and oxbows. It has also been found in swamp bayous and ponds that are situated near deeper running water, and it sometimes enters brackish coastal waters. Mud bottoms and abundant aquatic vegetation usually are present.

Natural History: Mating occurs in February, March, and April in Florida but probably occurs later in the Mississippi Valley. The male pursues the female and persistently tries to mount her. When successful, he moves his body slightly to one side and pushes his tail beneath her tail as she pulls it upward and to one side. Mating lasts 5–25 min. A pair of large captives copulated on 28 February, and the female laid 44 eggs on 20 April (Allen and Neill, 1950). In Florida, nesting takes place from April through June; in the Mississippi Valley it probably takes place later, perhaps in June or July. The nest cavity may be either flask or funnel shaped, and over 50 m from the nearest water. Eggs are nearly spherical (diameter 30–51 mm) and have a hard, unglazed, smooth shell. Clutches vary from 8 to 52 eggs; apparently only one clutch is laid a year. Natural incubation may last from 100 to 140 days, and most young emerge in September or October. The hatchling (45 mm) is brown, with a roughened carapace and a long, slender tail. Its head is covered with elaborate papillae, and the eye is ringed with conical tubercles. The dark skin may show some lighter mottling. The jaws are long, relatively narrow, and pointed at the tip. The inside of the mouth is light gray brown with black mottling.

Macroclemys probably feeds on any animal it can capture and subdue. In captivity it is known to eat fish, beef, pork, frogs, snakes, turtles, snails, worms, clams, crayfish, and various aquatic plants. Allen and Neill (1950) stated that captives stalked, caught, and ate turtles of the genera *Deirochelys*, *Chrysemys*, and *Kinosternon*. They also reported that a large *Macroclemys* killed but did not eat a smaller example of its own kind. A juvenile we kept killed and partially ate individuals of several species of *Graptemys* and *Chrysemys*. Excrement from newly caught captives often contains numerous fragments of snail and clam shells. Carr (1952) suggested that *Macroclemys* forages actively by night and reserves the use of the "worm" to provide an occasional fish during the more passive daytime period. The "worm"—gray at rest but suffused with blood when active—is the double-ended movable process on the tongue. Captives have been seen to open the jaws, wriggle the process, and lure fish into the mouth.

Platysternidae

The family Platysternidae is represented by a single species, *Platysternon megacephalum*, of southeastern Asia.

Platysternon megacephalum Gray, 1831c

Big-headed turtle

Plate 8

Recognition: *Platysternon megacephalum* has a head so large that it cannot be withdrawn into the shell for protection; it is about half as wide as the carapacial width and completely covered dorsally by an enlarged tough scute. Osteologically, the skull shows a unique combination of features. The temporal region is almost completely roofed over, and the dorsal surface is composed of an enlarged postorbital bone which separates the parietal and squamosal bones. The jugal is prevented from entering the orbit by the meeting of the postorbital and maxilla anterior to it. The frontal bone is also missing from the orbital rim. The jugal does touch both the pterygoid and palatine bones. The quadratojugal and maxilla meet; and the premaxillae medially form a strong hooked beak. Triturating surfaces of the jaw contain at least one well-developed ridge, and the jaw sheaths are large and well developed. In fact, the horny upper jaw sheath is so extensive that it almost extends to the large dorsal scute so that only a narrow band of skin, extending from the orbit to the nares, separates these. A small splenial bone is present, and the quadrate closes behind the stapes. Its carapace (to 18.4 cm) is flat with a slight medial keel (better developed in juveniles), and is indented anteriorly to the small cervical scute. Posteriorly, the carapace contains a slight notch, and the marginals are slightly serrated. Dorsal rib heads are well developed. In color, the carapace ranges from yellow brown to olive and may contain some dark radiations which fan out from posterior to anterior on each scute. Growth annuli may also be present on each scute. The hingeless plastron is well developed, but connected to the carapace only by ligaments at the narrow bridge. It is squared off anteriorly and widely notched posteriorly. Plastron and bridge are yellow and there may be a central pattern of faded brown stipples, or dark-brown blotches, extending along the seams. The plastral formula is: an > hum > fem > pect > abd >< gul. In the pelvic girdle, there is ventral contact between the pubes and ischia from each side. Anteriorly, the two thyroid fenestrae are separated only by a ligament. The head is yellow brown to olive, and dorsally may have some dark-brown or red to orange longitudinal striations or spots. Laterally the head may be darker in color and mottled with yellow or red spots. The chin is yellow to brown and may show either dark or light mottling. The neck is gray brown dorsally and yellowish laterally and ventrally. Its dorsal and lateral surfaces are covered with small scales, and there are large scales ventromedially. The toes are webbed and each digit contains three phalanges. The femoral trochanteric fossa is slightly reduced. Forelimbs are light to dark brown and covered with large scales; hind limbs are brown with a row of large scales along the inner and outer borders and on the heel. The tail is almost as long as the carapace and is covered with large scales.

Males have concave plastra and the anal vent situated beyond the carapacial rim.

Haiduk and Bickham (1982) found the diploid number to be 54 (16 macrochromosomes with median or submedian centromeres, 10 macrochromosomes with terminal or subterminal centromeres, and 28 microchromosomes).

Most authorities recognize Platysternidae as a separate family, but Romer (1956) regarded it as only a

Platysternon megacephalum megacephalum

subfamily (Platysterninae) of the large family Testudinidae, in which he also included tortoises and emydids. More recently, Gaffney (1975a) regarded it a tribe (Platysternini) of Chelydridae on the basis of the decreasing jugal exposure and lateral skull emargination, and the increase in temporal roofing. However, Whetstone (1978) pointed out that *Platysternon* has two biconvex cervical vertebrae, as do only the emydids and testudinids. He also failed to find any derived characters unique to fossil and recent chelydrids which are also shared with *Platysternon*. Additional evidence has been supplied by Haiduk and Bickham (1982), who found the karyotype most similar to the primitive batagurine condition of the Emydidae. The batagurine karyotype can be derived from that of *Platysternon* (or the reverse) by one change, whereas several changes are required to derive *Platysternon* from Chelydridae. Some derived chromosomal characters are also shared between the Emydidae and *Platysternon*. Haiduk and Bickham thought Platysternidae a valid family most closely related to the Emydidae and Testudinidae. Therefore, for the present, it is best to consider *P. megacephalum* representing a separate monotypic family.

Distribution: The big-headed turtle ranges from southern China (Fukien, Kwangtung, Kwangsi, and Hainan Island) southwestward through northern Vietnam, Laos, Kampuchea, and northern Thailand to southern Burma.

Geographic Variation: Five subspecies have been described. *Platysternon megacephalum megacephalum* Gray, 1831c, from southern China, has an unpatterned yellow plastron, a slightly keeled carapace with poorly developed growth annuli, and a slightly serrated posterior rim, the marginals above the bridge flared, a well-developed cephalic shield that often covers the rear of the orbit, yellow mottling on the jaws, and a pattern of narrow radiating lines on top of the head. *P. m. peguense* Gray, 1870b is found from western Vietnam west to southern Burma. It has a dark seam-following plastral pattern, the carapacial keel pronounced and sometimes indications of lateral keels, well-developed growth annuli, the posterior carapacial rim serrated, unpatterned jaws, a strongly

hooked upper jaw, and a black-bordered postorbital stripe. *P. m. vogeli* Wermuth, 1969 occurs in northwestern Thailand. It is similar to *P. m. peguense* in possessing a dark plastral figure, but differs in having a short, narrow, less hooked upper jaw, and a smooth, unserrated carapace. *P. m. tristernalis* Schleich and Gruber, 1984 is from Yunnan Province, China, and similar to *P. m. megacephalum* except it has three additional small scales at the medial junction of the gular and humeral scutes. *P. m. shiui* Ernst and McCord, 1987 (Plate 8) is found in northern Vietnam. Its head, shell, limbs, sockets, and ventral surface of the tail are heavily speckled with yellow, orange, or pink spots; the carapacial surface is smooth and the posterior rim unserrated; the cephalic shield is moderately developed, not entering the orbit; and the upper jaw is strongly hooked. It is likely that further study will prove *vogeli* and *tristernalis* invalid; Ernst is currently studying variation within *Platysternon*.

Habitat: This turtle prefers cool (12–17°C), rocky, mountain streams and brooks.

Natural History: One or two white, ellipsoidal to elongate-tapered (37 × 22 mm) eggs are laid per clutch. The young are more brightly colored with a more serrated posterior carapacial rim, a more pronounced vertebral keel, and a tail longer than the carapace.

Platysternon seems to be nocturnal, spending the daylight hours burrowed within gravel or partially under a rock in the stream bottom. It apparently seldom basks in the wild. When disturbed, *P. megacephalum* bites viciously and retains its grip for some time; the strongly hooked jaws can produce serious bites. However, this savagery is apparently strictly defensive since it is seldom aggressive toward other turtles confined with it.

Big-headed turtles feed on a variety of meats, fishes, and invertebrates in captivity, and are probably carnivorous in nature. At night they probe about the stream bottom for small animals, and may even leave the water to search along the bank and among low shrubs for food. They are accomplished climbers and in captivity have been known to climb out of aquaria and over wire fences.

Emydidae

Semiaquatic pond turtles and their relatives compose the largest and most diverse family of turtles, the Emydidae. Today it is represented on all continents but Australia and Antarctica; however, the fossil record indicates that the family was formerly more widespread in Europe and Asia than at present. The oldest fossil emydids are *Gyremys sectabilis* from the Upper Cretaceous Judith River formation of Montana and *Clemmys backmani* from the Paleocene Ravenscrag formation of Big Muddy Valley, Saskatchewan. Several other genera are known from Eocene deposits in Europe *(Emys, Ocadia)*, North America *(Echmatemys)*, and Asia *(Geoclemys,* and several species of the

Geoemyda complex). Fossils dating from the Oligocene to the Pleistocene have shown increased diversity in this family, and the living forms are represented by 33 genera and 91 species. The literature on many of these species includes little more than the original description, while others are among the most studied turtles.

Species included in Emydidae share several osteological characters. Their skulls are relatively small and similar to those of the tortoises (Testudinidae). The temporal region is widely emarginated posteriorly preventing squamosal-parietal contact. The frontal bone enters the orbit, and the postorbital is wider than that found in Testudinidae. The maxilla and quadratojugal are separated, and the quadrate is exposed posteriorly. The premaxillae usually do not meet to form a hooklike process (but do in *Cuora*). The triturating surface of the upper jaw varies from broad to narrow, and may be ridged or not. There is a tendency toward development of a secondary palate. A vestigial splenial bone may be present. The carapace is usually low arched, but may be considerably domed in some species. A vertebral keel may be present or absent, and some species also may have a pair of lateral keels. Dorsal rib heads are well developed. Carapace and plastron are usually united by a broad bridge. The plastron is well developed, and in several genera may bear a movable hinge between or near the seam separating the pectoral and abdominal scutes. Mesoplastral bones are absent, as also are intergular and inframarginal scutes. Limbs are developed for swimming, and have at least remnants of toe webbing. Usually more than two phalanges occur in the second and third digits, and the femoral trochanteric

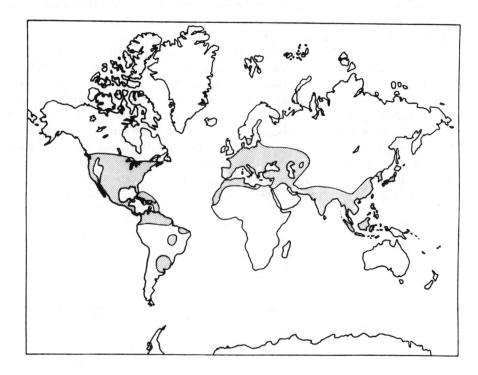

Distribution of Emydidae

136

fossa is reduced. In the pelvic girdle, the pubis normally touches the ischium of the same side; however, cartilage along the ventral midline of the girdle prevents the pubis and ischium from touching those from the other side. The family is divided into two subfamilies.

Subfamily Batagurinae (Old World pond turtles; 23 genera, 56 species). With the exception of the genus *Rhinoclemmys*, which occurs from Mexico to northern South America, batagurines range from southern Europe and northern Africa eastward to southeast Asia, Indonesia, the Philippines, and Japan. In batagurines the lower jaw has the angular bone completely, or nearly, separated from Meckel's cartilage by the prearticular. The basioccipital is broad, in contact with the paracapsular sac and pterygoid, and with a lateral tuberosity forming the floor of the cavity scala tympani. Only a single articulation exists between the centra of the 5th and 6th cervical vertebrae, and the suture between the 12th marginal scutes and last vertebral lies over the suprapygal bone, but not over the pygal.

McDowell (1964), mainly on the basis of skull characters, has further subdivided the Batagurinae into four complexes. (1) The *Hardella* complex includes the Asian genera *Hardella*, *Morenia*, and *Geoclemys*. The triturating surfaces of their jaws are very broad. The upper surface has a pair of ridges that end anteriorly in a pair of cusps (poorly developed in *Geoclemys*); the surface on the lower jaw extends along the midline behind the dentary symphysis. In the lower jaw, the coronoid bone is flattened dorsally and barely rises above the prearticular bone. The vomer is large and forms part of the floor of the cranial cavity posteriorly; anteriorly it contains a shelf that enters the triturating surface of the upper jaw (this shelf hides the praepalatine foramen). The orbito-nasal foramina are large. The squamosal is firmly attached to its surrounding bones. McDowell (1964) also found that in these genera the anterior opening of the Vidian canal lies between the epipterygoid on the outside and the palatine medially, and posterior to the anterior border of the lower parietal process (the anterior border of this process is entirely medial to the Vidian canal). (2) The *Batagur* complex includes the Asian genera *Batagur*, *Callagur*, *Chinemys*, *Hieremys*, *Kachuga*, *Malayemys*, and *Ocadia*. These turtles have jaws with moderately to very broad triturating surfaces, with or without longitudinal serrated ridges, but lacking cusps. The triturating surface of the lower jaw is equal in length to or longer than the dentary symphysis along the midline. The coronoid bone of the lower jaw has a distinct process that rises above the prearticular and surangular bones, but not neces-

sarily above the dentary. The vomer is large and extends backward to the level of the cranial cavity. Anteriorly the vomer has a shelf that enters the upper triturating surface and conceals the praepalatine foramen from below. The anterior opening of the Vidian canal may be in the palatine, lower process of the parietal, or in the parieto-palatine suture. The anterior part of the lower parietal process touches the palatine both lateral and medial to the Vidian canal. The posterior palatine foramen is small. The squamosal bone, when present, is firmly attached to its surrounding bones. The postlagenar hiatus is a narrow slit. (3) The *Orlitia* complex is composed of the two Asian species *Orlitia borneensis* and *Siebenrockiella crassicollis*. In these species the triturating surfaces are moderate to broad posteriorly, but narrow along the midline. Cusps are absent, but there is either an indistinct or a distinct ridge on the upper surface. The triturating surface of the lower jaw does not extend behind the dentary symphysis, and the coronoid bone has a well-developed process that rises above the other lower jaw bones. The vomer is small and is separated posteriorly from the floor of the cranial cavity. Also, its anterior end does not reach the upper triturating surface, and does not hide the praepalatine foramen. The anterior opening of the Vidian canal lies in the palatine. The posterior palatine foramen is somewhat enlarged. The anterior part of the lower parietal process is attached to the palate both medially and laterally to the Vidian canal. The squamosal projects anteriorly and is firmly attached to its neighboring bones. There is a large postlagenar hiatus. Digit 5 contains three phalanges. (4) The *Geoemyda* complex is large and varied. Included are the Old World genera *Annamemys*, *Cuora*, *Cyclemys*, *Geoemyda*, *Heosemys*, *Mauremys*, *Melanochelys*, *Notochelys*, *Pyxidea*, and *Sacalia*, and the New World genus *Rhinoclemmys*. Their triturating surfaces are narrow and lack ridges, except in *Melanochelys* which may have vestigial longitudinal ridges. The lower triturating surface does not extend behind the dentary symphysis. The vomer is small, does not form a part of either the cranial floor or the triturating surface of the upper jaw, and does not cover the praepalatine foramen. The anterior opening of Vidian's canal lies opposite the extreme anterior edge of the lower parietal process. The parietal bone touches the quadrate posterolateral to the trigeminal foramen. Squamosals barely, if at all, touch jugals. Digit 5 of the forefoot contains two phalanges, while the 5th digit of the hind foot contains three phalanges. A plastral hinge is present in adult *Cuora*, *Cyclemys*, *Notochelys* and *Pyxidea*.

More recent studies by Bramble (1974), Hirayama (1984), Sites et al. (1984), and Carr and Bickham (1986) have revealed weaknesses in the arrangement of species in McDowell's complexes. Hirayama performed a cladistic analysis based on 86 characters (47 cranial), Sites et al. reconstructed the batagurine phylogenies on the basis of biochemical characters, Carr and Bickham presented a phylogeny of the Batagurinae based on karyological data, and Bramble concentrated on hinge structure and kineses found in some species of McDowell's *Geoemyda* complex.

Hirayama (1984) found the *Hardella* complex to be artificial. While *Hardella* and *Morenia* were shown to be closely allied by his cladogram, *Geoclemys* appeared more closely related to *Chinemys* and *Malayemys* of the *Batagur* complex than to the former two genera.

Hirayama (1984) and Sites et al. (1984) found the *Batagur* complex more variable than reported by McDowell. Both studies agree with McDowell in the close relationship between *Batagur* and *Callagur*, and in the relationship of these two genera to *Kachuga* (although Sites et al. did find *K. smithi* to be more closely related to other genera in several biochemical characters). McDowell was not able to examine all species of *Kachuga*, and those that he did showed such extremes of divergence in cranial and neural features that he suggested generic-level separation of some species. McDowell also suggested that some species of *Kachuga*, notably *K. trivittata* and *K. kachuga*, were so similar to *Callagur* that additional study might show these three species to form a "single super species." Sites et al. (1984) also considered a *Kachuga-Callagur* clade to be a good alternative to the *Batagur* complex.

Inclusion of *Chinemys*, *Hieremys*, *Malayemys*, and *Ocadia* in the same complex as *Batagur*, *Callagur*, and *Kachuga* has also been challenged by Hirayama (1984) and Sites et al. (1984). Hirayama found *Chinemys* and *Malayemys* most closely related to *Geoclemys*, while Sites et al. found *Chinemys* most closely related to *Heosemys*, *Melanochelys*, and *Mauremys*, and *Ocadia* and *Malayemys* most closely related to *Rhinoclemmys*.

The combination of *Orlitia* and *Siebenrockiella* as a single complex was upheld by Hirayama (1984) and Sites et al. (1984).

The greatest disagreement between the later studies and that of McDowell is in the relationships among the species of the variable *Geoemyda* complex. Hirayama (1984) and Sites et al. (1984) showed this to be a totally artifical grouping with some members more closely related to species in some of McDowell's other complexes. Furthermore, Bramble (1974) suggested that *Heosemys*, *Cuora*, *Cyclemys*, and *Pyxidea* be placed in their own group—the *Heosemys* complex. He showed that *Cuora*, *Cyclemys*, and *Pyxidea* share an identical shell-closing mechanism which may have evolved from *Heosemys*. Of these, *Cyclemys* seems more primitive, *Pyxidea* intermediate, and *Cuora* the most derived. Bramble thought that *Notochelys* has evolved its shell kinesis separately from *Cuora*, *Cy-*

clemys, and *Pyxidea*, and is not closely related to them (but see the comments by Moll, 1985, on sexually dimorphic plastral kinesis).

Hirayama (1984) and Sites et al. (1984) concluded that the subfamily Batagurinae is probably polyphyletic with some species more closely related to the testudinids, and attempts at grouping its species into meaningful complexes should at present be abandoned. Therefore, in the following pages the genera of batagurines are presented in alphabetical order.

Key to the genera of Batagurinae:

1a. Plastron with freely movable hinge between pectoral and abdominal scutes — 2
 b. Plastron rigid, lacking a hinge, or with only a partially movable hinge in adult females — 5

2a. Posterior margin of carapace unserrated; plastral lobes wide, capable of closing shell completely: ***Cuora***
 b. Posterior margin of carapace serrated; plastral lobes narrow, not capable of closing shell completely — 3

3a. Five vertebral scutes — 4
 b. Six or seven vertebral scutes: ***Notochelys***

4a. Carapace very flattened across vertebrals; intergular seam shorter than that between humerals: ***Pyxidea***
 b. Carapace only slightly flattened across vertebrals; intergular seam longer than that between humerals: ***Cyclemys***

5a. Triturating surface of maxilla broad, at least posteriorly — 6
 b. Triturating surface of maxilla narrow — 17

6a. Triturating surface of maxilla very broad throughout its length; if present, medial ridge ends anteriorly as a cusp — 7
 b. Triturating surface of maxilla moderately to very broad, at least posteriorly, without an anterior cusp — 9

7a. Triturating surface of maxilla lacking a medial ridge or, if present, barely noticeable: ***Geoclemys***
 b. Triturating surface of maxilla with a distinct medial ridge — 8

8a. Plastron immaculate yellow; each pleural scute usually with a light-colored ocellus; choanae (internal nostrils) open at level of posterior rim of orbit: ***Morenia***
 b. Plastron with a large dark blotch on each scute; pleural scutes without light-colored ocelli; choanae open behind orbits: ***Hardella***

9a. Triturating surface of maxilla moderate to broad posteriorly, but medially narrowed — 10
 b. Triturating surface of maxilla broad throughout its length — 11

10a. Triturating surface of maxilla with a medial ridge; 2d vertebral mushroom shaped with the stalk pointing posteriorly; 4th pleural scute small: ***Orlitia***

b. Triturating surface of maxilla without a medial ridge; 2d vertebral not mushroom shaped; 4th pleural scute not noticeably smaller than other pleurals: *Siebenrockiella*

11a. Triturating surface of maxilla with a single medial ridge; five claws on forefoot 12
b. Triturating surface of maxilla with two ridges; four claws on forefoot: *Batagur*

12a. Ridge on triturating surface of maxilla sharp and well defined 13
b. Ridge on triturating surface of maxilla reduced or indistinct 15

13a. 4th vertebral much longer than wide: *Kachuga*
b. 4th vertebral wider than long 14

14a. Neck with numerous dark-bordered yellow stripes: *Ocadia*
b. Neck without stripes: *Callagur*

15a. Ridge on triturating surface of maxilla present, but reduced and indistinct 16
b. No ridge on triturating surface of maxilla: *Chinemys*

16a. Carapace posteriorly serrated, and with only a single longitudinal keel: *Hieremys*
b. Carapace not posteriorly serrated (but with a single medial notch), and with three longitudinal keels: *Malayemys*

17a. Carapace with only a single longitudinal keel 18
b. Carapace with three longitudinal keels 19

18a. One or two pairs of light-colored ocelli present on back of head; southeast Asia: *Sacalia*
b. No light-colored ocelli present on back of head; New World: *Rhinoclemmys*

19a. Carapace unserrated, or only slightly serrated posteriorly 20
b. Carapace strongly serrated posteriorly 22

20a. Dorsolateral carapacial keels low, but pronounced, in adults: *Melanochelys*
b. Dorsolateral carapacial keels barely noticeable in adults 21

21a. Plastral buttresses very strongly developed: *Annamemys*
b. Plastral buttresses only moderately developed: *Mauremys* (in part)

22a. Upper jaw medially hooked 23
b. Upper jaw not medially hooked 24

23a. Intergular seam shortest of the medial seams separating plastral scutes: *Geoemyda*
b. Intergular seam not shortest of the medial seams separating plastral scutes: *Heosemys* (in part)

24a. Upper jaw medially notched: *Heosemys* (in part)
b. Upper jaw not medially notched: *Mauremys* (in part)

Subfamily Emydinae (New World pond turtles; 10 genera, 35 species). With the exception of the genus *Emys*, which ranges from Europe and northern Africa into the Middle East, emydines are found from Canada to central South America and the West Indies. In emydines the lower jaw has the angular bone touching Meckel's cartilage. The basioccipital is narrow, separated from the paracapsular sac and pterygoid, lacks a lateral tuberosity, and does not form the floor of the cavity scala tympani. A double articulation occurs between the centra of the 5th and 6th cervical vertebrae, and the suture between the 12th marginal scutes and last vertebral lies over the pygal bone, not the suprapygal. Two complexes seem to exist in this subfamily (McDowell, 1964; Bramble, 1974).

The *Clemmys* complex includes the genera *Clemmys*, *Emydoidea*, *Emys*, and *Terrapene*. The plastron of *Clemmys* is rigid, but that of the other three genera is hinged and movable. *Clemmys* is thought to be ancestral to the hinged forms, of which *Emys* is most primitive, *Emydoidea* intermediate, and *Terrapene* most derived (Bramble, 1974). These turtles have the triturating surfaces of the jaws narrow and ridgeless with the upper triturating surface lacking parts of the palatine or the pterygoid bone. The orbito-nasal foramen is small. The interorbital region is coarsely sculptured, and the postorbital bar is relatively wide. The jugal bone does not touch the palatine. On the plastron, the humero-pectoral seam crosses the entoplastron. The cervical vertebrae are not elongated in *Clemmys*, *Emys*, and *Terrapene*, but are in *Emydoidea*, which also has the cervical extensor muscles hypertrophied. Musk glands are present in all four genera. Formerly, *Emydoidea* was included in a separate complex with *Deirochelys* (Loveridge and Williams, 1957; McDowell, 1964). The two genera share similar neck, jaw, and rib structure, but these are probably convergent features related to similar feeding mechanisms (Bramble, 1974). Also, Jackson's (1978b) studies on fossil *Deirochelys* indicate it is more closely related to *Trachemys* and *Pseudemys*.

The *Chrysemys* complex includes the freshwater genera *Chrysemys*, *Deirochelys*, *Graptemys*, *Pseudemys*, and *Trachemys* and the brackish-water genus *Malaclemys*. All have a rigid plastron with the humeropectoral seam crossing posterior to the entoplastron. The triturating surface is usually broad (narrow in *Deirochelys*), with or without ridges, and with the upper surface containing parts of the palatine or the pterygoid bone (except in *Deirochelys*). The orbito-nasal foramen is usually much larger than the posterior palatine foramen (but not as large in *Deirochelys*). The interorbital region is not coarsely sculptured, and the postorbital bar is not wide. Jugals and palatines touch. With the exception of *Deirochelys*, the cervical vertebrae are not elongated, nor the cervical extensor muscles hypertrophied. Musk glands are absent.

Key to the genera of Emydinae:

1a. Plastron with a well-developed hinge   2
  b. Plastron without a well-developed hinge   4
2a. Neck long (distance from snout to shoulder approximately equal to plastron length); chin and throat bright yellow: **Emydoidea**
  b. Neck short (distance from snout to shoulder approximately equal to half the plastron length); chin and throat not bright yellow   3
3a. Upper jaw medially notched: **Emys**
  b. Upper jaw not medially notched: **Terrapene**
4a. Neck long (distance from snout to shoulder approximately equal to plastron length): **Deirochelys**
  b. Neck short (distance from snout to shoulder approximately equal to half the plastron length)   5
5a. Upper jaw with prominent notch which may or may not be bordered on each side by tooth like cusps   6
  b. Upper jaw without a prominent notch or cusps   8
6a. Carapace serrated posteriorly; vertebral keel present   7
  b. Carapace not serrated posteriorly; no vertebral keel: **Chrysemys**
7a. Crushing surface of upper jaw without a tuberculate ridge extending parallel to its margin: **Trachemys**
  b. Crushing surface of upper jaw with a row of tubercles on ridge extending parallel to its margin: **Pseudemys** (in part)
8a. Crushing surface of upper jaw with a ridge or tuberculate row extending parallel to its margin: **Pseudemys** (in part)
  b. Crushing surface of upper jaw smooth or undulating but not ridged   9
9a. Crushing surface of upper jaw narrow: **Clemmys**
  b. Crushing surface of upper jaw broad   10
10a. Scutes of carapace rough, with concentric ridges or striations formed by growth annuli; head and neck without longitudinal stripes: **Malaclemys**
  b. Scutes of carapace smooth, without concentric ridges or striations formed by growth annuli; head and neck striped: **Graptemys**

Annamemys Bourret, 1939

This poorly known Asiatic genus is often totally overlooked or is designated a synomyn of either *Cyclemys* or *Mauremys*. It contains a single species, *Annamemys annamensis*.

Annamemys annamensis (Siebenrock, 1903a)

Vietnamese leaf turtle

Recognition: The uniformly dark-gray to black carapace (to 17 cm) is low arched, or depressed, and contains three longitudinal keels. The medial vertebral keel is best developed and prominent in adults; the two dorsolateral keels, which run along the dorsal portions of the pleurals, are low and almost nonexistent in older individuals, strongly developed in juveniles. Vertebral 1 is wider anteriorly than posteriorly, but the 5th is the opposite. Vertebrals 2–4 are about as broad as long. The underlying neural bones are hexagonal and short sided anteriorly. Posterior marginals are not serrated; those anterior and lateral are slightly upturned. The plastron is well developed, but does not completely cover the shell opening. Its anterior lobe is truncate, and the posterior lobe contains a deep anal notch. The plastral formula is: abd < pect > fem > an > hum >< gul; however, the seams between the pectorals, abdominals, and femorals are often nearly equal in length. The plastron is sutured to the carapace, and the bridge is well developed (its length is 40–50% that of the plastron). Axillary and inguinal scutes are present and their bony buttresses are very strongly developed, almost touching the neural bones; the axillary buttress touches the 1st rib. Such extensive buttressing is usually associated with

Annamemys annamensis

freshwater turtles that inhabit deep waters (Boulenger, 1889), such as *Callagur, Batagur, Hardella, Kachuga* and *Orlitia*, and forms two extensive lateral chambers housing the lungs. Whether these buttresses indicate close relationships or have evolved several times in turtles is unknown (Savage, 1953). The plastron is yellow to orange with a large black blotch on each scute and on the underside of each marginal. A longitudinal black bar crosses the bridge. The head is somewhat pointed, as the snout is projected forward. Its temporal arch is complete, and the quadratojugal touches both the jugal and postorbital. The upper jaw is medially notched, and the triturating surfaces of the jaws are narrow and lack ridges. Choanae are positioned at the middle of the orbits. The head is dark brown to black with several pairs of yellow stripes. One pair begins dorsal to the nostrils and passes backward on each side above the orbits to the neck. A second wider pair starts at the nostrils and runs backward on each side through the orbit and tympanum to the neck (often broadening on the tympanum). Still a third pair begins on the upper jaw below the nostrils and runs backward along each jaw margin to the side of the neck. The lower jaw is yellow, but the chin varies from yellowish to dark brown. The neck is dark dorsally, but lighter ventrally. Limbs are dark gray or black, and the toes are fully webbed. The tail is not overly elongated, but is thicker at the base in the males.

The taxonomic position of *Annamemys annamensis* has been confused for some time. Siebenrock (1903a) described the species from a 63-mm juvenile, and assigned it to the genus *Cyclemys*, apparently believing the adult to have a hinged plastron, which it does not. Bourret (1939) further clouded the issue when he described a new genus and species, *Annamemys merkleni*, from adult specimens. Savage (1953) showed these two forms to be synonymous; thus the genus became *Annamemys* and the trivial name *annamensis*. The last bit of confusion was caused by McDowell (1964) who thought that the appearance of juvenile *Cyclemys annamensis*, as figured, was so similar to that of juvenile *Mauremys mutica* (= *M. nigricans)*, that *Annamemys* was a synonym of *Mauremys*. However, when adult *A. annamensis* and *M. nigricans* are compared, it is quite apparent they are different species.

Distribution: *Annamemys* is known only from central Vietnam.

Habitat: Lowland marshes and slow-moving streams are inhabited.

Natural History: Its behavior in the wild is unknown, but captives accept both vegetable (lettuce and fruits) and animal (fish) foods.

Batagur Gray, 1855

This monotypic genus consists of the river-inhabiting species, *Batagur baska*, which is one of the largest emydid turtles, reaching a carapace length of at least 60 cm.

Batagur baska (Gray, 1830, in Gray, 1830–1835)
River terrapin

Recognition: The carapace is somewhat domed, has smooth scutes, and is not serrated posteriorly. There is a low, interrupted vertebral keel in juveniles which disappears with age. All vertebrals are broader than long, and the underlying neural bones are elongated, hexagonal, and short sided anteriorly. Vertebral 4 only covers three neurals. The carapace is uniformly olive gray or olive brown. The plastron is well developed, but smaller than the carapacial opening, and has only a shallow posterior notch. The bridge is broad and the plastron is extensively sutured to the carapace. Plastral buttresses are greatly enlarged, extending almost to the neural bones; the axillary buttress is connected to the 1st rib while the inguinal buttress connects between the 5th and 6th costals. On the bridge, the inguinal scute is larger than the axillary. Both the anterior and posterior plastral lobes are shorter than the bridge. The plastral formula is: abd > pect >< fem > hum > an > gul. Plastron and bridge are uniformly yellow or cream in color. For such a large turtle, the head is small to moderate in size with an upturned, pointed, projecting snout, and a medially notched upper jaw. The back of the head is covered with small scales. In the skull, the temporal arch is complete with the quadratojugal touching both the

Batagur baska

jugal and postorbital. The frontal bone may contribute to the orbital rim. The maxilla does not touch either the parietal or the squamosal. The orbito-nasal foramen is large, and there is a dorsal ridge on the palatine bone which may help brace the palate against the cranium. The triturating surfaces of the jaws are broad with two strongly denticulated ridges; the sides of the jaws are serrate. The head is olive gray dorsally, but lighter gray laterally and ventrally with light-colored jaws. Forelimbs are somewhat paddle shaped with four claws. All toes are webbed and there are large transverse scales on the limbs. Skin is olive gray. Interestingly, during the mating (monsoon) season, the male's head, neck, and legs turn black and his iris changes color from yellowish cream to pure white (Moll, 1978). After the breeding season these parts revert to their original colors.

In addition to the sexual dichromatism, males can be told from females by their longer, thicker tails; also adult males are somewhat smaller (to 50 cm) than adult females (to 60 cm).

Distribution: *Batagur baska* is known from the Irrawaddy River in Burma southward through southern Thailand to southern Vietnam, and through Malaysia to Sumatra. Khan (1982) recently confirmed its presence in Bangladesh.

Geographic Variation: Wirot (1979) described the form *Batagur baska ranongensis* from Thailand as differing by its flat, round carapace, as opposed to the elongated, narrower carapace of normal adult *B. baska*. However, this probably just represents an abnormal growth pattern, as all juvenile *B. baska* are rounded, but become more elongated with age. Moll has told us that *B. baska* from the east coast of Malaysia differ in coloration from those of the west coast.

Habitat: *Batagur baska* lives in tidal areas of the estuaries of large rivers, and, during the nesting season, ranges far upstream in these rivers.

Natural History: Presumably, mating occurs just prior to and in the beginning of the monsoon season when the males are in breeding color. Communal nesting occurs from late December to early March on sandbars and banks far upstream from the normal estuarine habitat, and some females may swim over 80 km to reach these sites. A body pit up to 90 cm deep is dug first before the actual 15–30-cm-deep nest cavity is constructed. Digging is slow and deliberate and may take several hours. Females nest up to three times a season, laying 13 to 34 eggs each time. Sometimes these eggs are placed in two or three nest cavities dug during one night. A typical egg is oblong (66–70 × 45 mm), and has either a hard expandable shell or a parchmentlike shell. Natural incubation takes from 70 to 112 days. Hatchlings are olive gray and have round, slightly serrated carapaces with a low medial keel. At the mouth of the Irrawaddy River in Burma, females nest on adjacent sea beaches.

These turtles follow the high tides upstream to forage, retreating back to the estuaries as the tide lowers. They are predominantly herbivorous, preferring certain mangrove fruits, but also take mollusks, crustaceans, and fish (Moll, 1978).

Batagur baska has decreased drastically in numbers owing to the harvesting of its eggs. Fortunately, conservation programs have begun in Malaysia which may insure its existence there. For a good summary of its natural history, see Moll (1978).

Callagur Gray, 1870b

The sole species in this genus, *Callagur borneoensis*, is closely related to *Batagur baska* and the large species of *Kachuga*, and may be the largest emydid turtle. While most individuals are under 60 cm in carapace length, de Rooij (1915) reported one specimen 76 cm in length.

Callagur borneoensis (Schlegel and Müller, 1844)

Painted terrapin

Plate 8

Recognition: The oval carapace is somewhat flattened, has scutes roughened with growth annuli, and is not serrated posteriorly. Juveniles have a well-developed continuous medial keel and an interrupted lateral keel on the pleural scutes of each side. These keels become lower and disappear with age. Vertebrals 1–3 and 5 are broader than long; the 4th is about as broad as long and is smaller than the 3d. A projection on the anterior border of the 4th vertebral fits into a posterior concavity on the 3d. The underlying neural bones are elongated, hexagonal, and short sided anteriorly. The 4th vertebral covers three neurals. The adult carapace is light brown to olive with three broad, black longitudinal stripes. The plastron is well developed, but smaller than the shell opening, and has only a shallow posterior notch. The bridge is broad and the plastron is extensively sutured to the carapace. Plastral buttresses are greatly enlarged, extending almost to the neurals; the axillary buttress is connected to the 1st rib while the inguinal buttress connects between the 5th and 6th costals. On the bridge, the inguinal scute is larger than the axillary. Both the anterior and posterior plastral lobes are shorter than the bridge. The plastral formula is: abd > fem > pect > an >< hum > gul. Both plastron and bridge are uniformly yellow or cream colored. The head is small to moderate in size with an upturned, pointed, projecting snout, and a shallow medial notch on the upper jaw. The back of the head is covered with small scales. In the skull, the temporal arch is complete and the quadratojugal touches both the jugal and postorbital. The frontal bone does not enter the orbital rim. The maxilla does not touch either the parietal or the squamosal. The orbito-nasal foramen is large. Triturating surfaces of the jaws are broad with a single medial denticulated ridge; the sides of the jaws are serrate. The limbs have enlarged transverse scales. There are five claws on the forefeet, and all toes are heavily webbed. Limbs and other soft parts are normally olive to gray.

The karyotype is 2n = 52 (Bickham and Carr, 1983.

The head of the female is olive, that of nonbreeding males charcoal gray. During the mating season the male's head becomes white and a red stripe develops between the eyes (Moll, Matson, and Krebiel, 1981). In addition to the sexual dichromatism, males are shorter (to 39 cm) than adult females (41+ cm) and have longer, thicker tails.

McDowell (1964) thought *Callagur borneoensis* to be most closely related to *Kachuga trivittata*, differing from that species in its elongated neural bones, narrower triturating surfaces, and greater size. He suspected *C. borneoensis* might be the end representative of a single *K. kachuga–K. trivittata–C. borneoensis* superspecies.

Distribution: *Callagur* ranges from extreme southern Thailand southward through Malaysia to Sumatra and Borneo.

Habitat: Adults live mainly in estuaries and tidal reaches of medium to large rivers. Juveniles and hatchlings are less tolerant, and probably inhabit freshwater portions of these same rivers. However, Dunson and Moll (1980) demonstrated that the hatchlings can survive for at least two weeks in 100% sea water. Apparently this allows them enough time to make the migration from their sea-beach nests, some of which may be as far as 3 km from the mouth of a river, to fresh water.

Natural History: Nesting occurs from June to August on the same sea beaches as *Dermochelys coriacea*, *Chelonia mydas*, and *Lepidochelys olivacea* (Dunson and Moll, 1980). The female emerges from the sea, selects a site, and then rapidly digs a 30-cm-deep hole,

lays her eggs, covers them, and returns to the sea. Eggs are ellipsoidal (68–76 × 36–44 mm) with brittle shells; about 12 are laid in each clutch. Hatchlings are rounded with dark spots on each vertebral and one on the posterior border of each pleural scute (de Rooij, 1915).

Adults are almost entirely vegetarian; any animals consumed are probably taken accidentally.

Chinemys Smith, 1931

Chinemys is a genus of semiaquatic batagurines living in semitropical and tropical areas in southeastern Asia. The carapace is low arched and bears either a single medial keel or a medial and two lateral keels. The short neural bones are six sided and shortest anteriorly. The plastron is well buttressed, and lacks a hinge. The entoplastron is crossed by the humero-pectoral seam. The skull is short, with a complete temporal arch, and has shallow inferior, but deep posterior, temporal emargination. Each squamosal touches the postorbital, jugal, and maxilla, but nearly excludes the postorbital from the supratemporal. The maxilla is separated from the inferior parietal process by the palatine, and the orbito-nasal foramen is small. The pterygoid ridge touches the articulating surface of the quadrate. Triturating surfaces are broad and ridgeless. A short angular bone is present on the lower jaw, and the coronoid bone has a very high process extending above the other bones of the lower jaw. Tail length is moderate in adults, but long in juveniles. The toes are webbed.

In *Chinemys reevesii* the diploid chromosome number is 52: 26 macrochromosomes (16 metacentric, 6 submetacentric; 4 telocentric) and 26 microchromosomes (Sasaki and Itoh, 1967; Killebrew, 1977a).

Key to the species of *Chinemys*:

1a. Carapace with three longitudinal keels   2
 b. Carapace with only a single, pronounced medial keel: ***C. kwangtungensis***
2a. Head very broad; vertical profile of snout straight: ***C. megalocephala***
 b. Head not exceptionally broad; snout projecting and presenting an oblique vertical profile: ***C. reevesii***

Chinemys kwangtungensis (Pope, 1934)
Chinese red-necked pond turtle

Recognition: The carapace is elongated (to 20 cm), slightly domed, unserrated posteriorly, and has only a single, pronounced medial keel. Vertebrals 1, 4, and 5 are broader than long; 2 and 3 may be or are as broad as long. The 1st and 4th vertebrals are usually broadest. Lateral marginals may be upturned. The carapace is brown to chestnut brown and usually unmarked; however, some yellowish pigment may occur along the marginal seams. The plastron is elongated and notched posteriorly; the plastral formula is: abd > pect >< fem > gul ≥ an > hum. Axillary scutes of the bridge are much smaller than the inguinals, and the bridge is narrower than the length of the posterior plastral lobe. The plastron is yellowish with brown, irregularly shaped markings on each scute. The bridge is totally brown. The head is broad with a pointed, projecting snout. The medial, anterior profile of the upper jaw is concave, not straight, and the triturating surfaces are broad and ridgeless. The anterior dorsal surface of the head is smooth and scaleless, but the posterior surface is covered with small granular scales. The head is brown to mahogany with narrow yellow vermiculations. A yellow longitudinal stripe extends from the upper edge of the orbit over the tympanum to the neck on each side, and another below it extends from midorbit through the tympanum to the neck. The jaws are yellowish to tan with dark spotting. Limbs and tail are dark brown to black with some yellow scales on the forelimbs.

Males have concave plastra and longer, thicker tails with the vent beyond the carapacial margin.

Distribution: *Chinemys kwangtungensis* ranges from Kwangtung Province, southern China, westward to northern Vietnam.

Habitat: This is a highland species inhabiting mountain streams at altitudes of 300–400 m (Pope, 1935).

Natural History: This little-known turtle lays a clutch of only two eggs (Pope, 1935), which average 51 × 27 mm (Ewert, 1979).

Chinemys megalocephala Fang, 1934
Chinese broad-headed pond turtle

Recognition: This is a large *Chinemys* (to 22.8 cm) with an oblong carapace which is wider posteriorly than anteriorly, moderately domed, unserrated posteriorly, and with three keels. The lateral keels are more pronounced in juveniles, but still present in large adults. Vertebral 1 is about as broad as long; the other vertebrals are broader than long. Lateral marginals are upturned. The carapace is brown to dark brown with a black, longitudinal medial stripe running along the dorsal keel. The posterior plastral lobe is broader than the anterior lobe. There is an anal notch. The plastral formula is: abd > fem >< pect > an > gul > hum. The bridge is about as broad as the length of the posterior plastral lobe, and its axillary scute is only about half as long as the inguinal. Both plastron and bridge are dark brown to black with yellow to green seams and outer rim. The head is large and very broad, with the snout barely projecting and upper jaws not medially notched. Triturating surfaces of both jaws are broad, but ridgeless. The anterior dorsal surface of the head is smooth and scaleless; the posterior surface is covered with small scales. Skin of the soft parts is olive. There are several small yellowish marks on the sides of the head, along with two postorbital stripes running to the neck. Jaws are tan with yellow vermiculations, and the limbs also contain some yellow spots.

Chinemys kwangtungensis

Chinemys megalocephala

Males are much smaller than females, and have concave plastra and longer, thicker tails with the vent beyond the posterior carapacial rim.

Distribution: This species is known only from the Nanking region, China.

Habitat: Fang (1934) reported that this turtle is usually found in ponds and streams near hillsides.

Chinemys reevesii (Gray, 1831b)

Chinese three-keeled pond turtle

Recognition: The elongated (to 23.6 cm) oval carapace is slightly arched, unserrated, but medially notched posteriorly, and has three keels: a pronounced medial vertebral keel flanked on each side by a lower keel extending along the dorsal portions of the pleural scutes. Vertebral 1 is broader anteriorly than posteriorly; it is broader than long in younger individuals, but becomes elongated with age until it may be slightly longer than broad in old turtles. Vertebrals 2–5 are broader than long. The carapace ranges from light to dark brown with darker pigment along the keels, and, in some, light seams. The plastron is notched posteriorly; the plastral formula is: abd >< pect >< fem > gul > an > hum. Both plastron and bridge are yellow with a large brown blotch on each scute; these blotches may be smooth edged, or have extending dark radiations which give the blotch a spiky appearance. The head is moderate in size with a projecting snout and a slight medial notch on the upper jaw. It is dark brown to black with a series of elongated, broken, or curved yellow stripes on the sides. The tympanum is often ringed with yellow, and both chin and lower jaw may contain mottled yellow marks. The neck is grayish brown to black with several solid or broken, narrow yellow stripes. The limbs and tail are uniformly olive to brown.

Males often become more melanistic with age, and may lose all the light markings on the shell, head, and neck.

Females are larger than males, but males have tails with very thick bases and the vent beyond the carapacial rim. The adult male plastron may also be slightly concave; that of the female is always flat. Age and size at attainment of sexual maturity are unknown, but males develop the enlarged tail early at carapace lengths as short as 60 mm.

Distribution: *Chinemys reevesii* occurs in Japan on Honshu and Kyushu, Korea, Taiwan, Hong Kong, and on mainland China south of the Yangtze River west to Canton.

Chinemys reevesii

Geographic Variation: No subspecies are currently recognized, but variation exists (Lovich et al., 1985). Japanese *Chinemys reevesii* average greater carapace lengths than do mainland populations; however, the largest of more than 200 specimens we examined was a 23.6-cm female from Szechwan Province, China. Pritchard (1979) states that Japanese specimens may reach well over 30 cm and possibly to 38 cm, but this is unsubstantiated. Taiwanese specimens often have more neck stripes than turtles from other populations.

Schmidt (1925) described a form of *Chinemys* which he named *Geoclemys grangeri*. He reported it differed from *C. reevesii* in having the axillary scute larger than the inguinal, the occipital scales much smaller, the blotches on the plastron smaller and more sharply defined, the intergular suture more than twice as long as that of the humerals, the 1st marginal broadest, and the bridge a little longer than the posterior plastral lobe. Pope (1935) thought all of these to be within the variance of *C. reevesii*, but gave no proof. Subsequently, Pritchard (1979) resurrected *C. grangeri* on the basis of pet trade animals entering the United States. However, data presented by Lovich et al. (1985) have upheld Pope's (1935) opinion, and this question has presumably been put to rest.

C. reevesii has been reported from the Philippines and the Bering Islands of the USSR. We have examined these specimens, and they are *C. reevesii*; however, since this species is unknown from the Philippines we feel the specimens have wrong locality data. The Bering Island specimen was found washed up onto a beach, and we agree with Nikolskii (1915) that it probably drifted there from Japan.

Habitat: *Chinemys reevesii* lives in shallow, soft-bottomed waters of ponds, marshes, canals and streams, but may also enter rivers.

Natural History: Courtship and mating occur in the spring, with the males pursuing and swimming around the females in an effort to make snout contact (Pope, 1935). Nesting occurs in June and July. Three clutches of four to nine eggs are laid each season (Fukada, 1965). The white eggs are elongated (40 × 23 mm). Fukada (1965) reported an incubation period of 99 days for a clutch in his laboratory, but stated that in the Japanese wild newly hatched young overwinter in the nest and emerge in March or April. Hatchlings have three prominent carapacial keels, long tails, and large heads. Their shells are about 22 mm long.

Chinemys reevesii is fond of basking. An omnivore, it feeds on aquatic plants, fruits, and lettuce, as well as worms, aquatic insects, frogs, and fishes.

This species is a hardy captive with much personality; it makes a good pet.

Cuora Gray, 1855

The southeast Asian box turtles, genus *Cuora*, are tropical and live in semiaquatic habitats. Adult carapaces are either low or high domed, with a medial and two parallel longitudinal keels and a smooth posterior rim. Neural bones are hexagonal and short sided posteriorly. The plastron has a hinge between the pectoral and abdominal scutes (between the hyo- and hypoplastra) that allows the two plastral lobes to move and close the shell. The hind lobe may or may not be notched posteriorly. The entoplastron is intersected by the humero-pectoral seam. There is no bridge and the plastral buttresses have been mostly resorbed; the plastron is connected to the carapace by a ligament. The skull is short, and usually has a complete temporal arch (incomplete in *C. flavomarginata*). The squamosal touches the postorbital but not the jugal; the jugal is separated from the quadratojugal by the postorbital (the quadratojugal is absent in *C. flavomarginata* and *C. galbinifrons*). The anterior edge of the inferior parietal process is not outwardly flared, and is separated from the jugal by the pterygoid, but does touch the palatine. McDowell (1964) noted that the pterygoid fails to grow up and over the medial surface of the quadrate to exclude it from the wall of the cavernosus canal. Triturating surfaces of the jaws are narrow, and the premaxillae meet to form a hook-like process. The toes are at least partially webbed.

The diploid chromosome number is either 50 or 52 (Gorman, 1973).

Bour (1980b) and Hirayama (1984) have reassigned the species *flavomarginata*, *galbinifrons*, and *hainanensis* to the genus *Cistoclemmys* Gray, 1863a, retain-

Key to the species of *Cuora*:

1a. Carapace highly arched or domed, plastron without a posterior notch   2
 b. Carapace slightly arched, domed in one species, plastron with a distinct posterior notch   5
2a. Upper jaw without a medial hook: *C. hainanensis*
 b. Upper jaw at least slightly hooked   3
3a. Carapace with lower 2/3 or 3/4 of pleurals white, cream, or yellow: *C. galbinifrons*
 b. Carapace with lower 2/3 to 3/4 of pleurals dark brown or black   4
4a. A single lemon-yellow stripe passes backward from each orbit to neck; tympanum yellow: *C. flavomarginata*
 b. Three yellow stripes on each side of head, one from tip of snout above orbit to neck and two beginning at nostril and passing through orbit to neck; tympanum brown or olive: *C. amboinensis*
5a. Carapace with three longitudinal black stripes, head with dark-brown or black temporal stripe enclosing lighter olive spots: *C. trifasciata*
 b. Carapace without longitudinal black stripes, head with a yellow or brown stripe on each side   6
6a. Chin with light and dark mottling; plastron light brown with narrow dark seams; head brown: *C. yunnanensis*
 b. Chin without white and black mottling; plastron yellow with extensive black pigment; head yellow or olive   7
7a. Plastron with large medial black mark covering most of surface: *C. mccordi*
 b. Plastron with dark seam-following pattern   8
8a. Plastral seam-following pattern often separated into dark wedges; carapace brown with reddish vertebrals; head lemon yellow: *C. pani*
 b. Plastral seam-following pattern extensive and uninterrupted; carapace olive brown with dark-brown or black keel; head olive: *C. chriskarannarum*

ing the generic name *Cuora* for the other species. They base these designations on cranial features (see Hirayama, 1984, for a descriptive list) and the supposed absence of an interanal seam. In doing this they disregard such important characters as carapacial structure and plastral hinge structure and kinetics (Bramble, 1974). Examination of series of *flavomarginata* and *galbinifrons* by Ernst has shown that loss of the interanal seam (or the low ridge representing it) results from abrasion. Small juveniles had the seam (or ridge) while larger juveniles showed a progressive loss with increased shell length. Also, some adult *amboinensis*, especially those from the Philippines, lose this seam. In addition, Sites et al. (1984) have studied the biochemical relationships among batagurine turtles and have supported the assignment of *flavomarginata* to *Cuora*. The interspecific relationships within *Cuora* are currently being studied by Ernst, and we feel it is premature to split the genus.

Cuora amboinensis (Daudin, 1802)

Malayan box turtle

Recognition: Adults have a high-arched carapace (to 20 cm) with or without a medial keel; the juvenile carapace is depressed and bears three prominent keels. The 1st vertebral is wider anteriorly than posteriorly, the 2d and 5th are longer than broad, and the 3d and 4th are usually broader than long. Posterior marginals are slightly flared, but not serrated, and the carapace also lacks a posterior notch. The carapace is uniformly dark olive or black in color. The large plastron can completely cover the carapacial openings. The posterior lobe is rounded, and usually contains no medial anal notch, or only a slight one. The plastral formula is: abd >< an > pect > gul > fem > hum. Axillary and inguinal scutes are very small or absent. The plastron is yellow to light brown, and there is a large dark-brown or black spot toward the outside of each scute. Undersides of the marginals are yellow, and each has a black spot along its border. The head is small to medium with a protruding snout and a slightly hooked upper jaw; it is olive to dark brown dorsally, becoming black laterally, and yellow to olive posteriorly. On each side, a black-bordered yellow stripe runs anteriorly from the neck passing above the orbit to the tip of the snout above the nostrils, where it meets its counterpart from the other side. Two other pairs of yellow stripes are also present on each side of the head. The first passes posteriorly from the nostril through the orbit and tympanum to the side of the neck. The second passes from below the nostril along the upper jaw backward to the neck. Lower jaw and chin are yellow, limbs are olive to black, and the anterior surfaces of the forelimbs are covered with enlarged horizontal scales.

Males have slightly concave plastra and longer, thicker tails.

Distribution: *Cuora amboinensis* occurs from the Nicabar Islands, Bangladesh (Khan, 1982), and Assam (Moll and Vijaya, 1986), south through Burma, Thailand, Kampuchea, Vietnam, and Malaya, and east in Indonesia to Sulawesi and Amboina (the type locality). It also reaches the Philippines and Celebes.

Geographic Variation: Although *Cuora amboinensis* has an extensive geographic range, no subspecific variation has been described. However, some populations have rather flat carapaces.

Habitat: This turtle inhabits lowland water bodies with soft bottoms and slow currents, such as marshes, swamps, ponds, pools in streams, and manmade flooded rice paddies. Although highly aquatic, it is often found on land far from water. Taylor (1920) reported that the juveniles are entirely aquatic.

Natural History: This is a gentle, shy species. Three to four clutches, usually of only two eggs, are laid between April and the end of June. The brittle-shelled, white eggs are elongated and measure 40–46 mm × 30–34 mm (Smith, 1931). The tricarinate hatchlings are about 46–48 mm in carapace length.

In the wild, *Cuora amboinesis* is decidedly herbivorous, but captives soon learn to take various animal foods.

Cuora amboinensis

Cuora flavomarginata (Gray, 1863a)

Yellow-margined box turtle

Recognition: This is a species of *Cuora* with a high arched shell and a distinct yellow vertebral stripe. The carapace (to 17 cm) is domed, widest behind the center, and smooth bordered. Each vertebral is narrower than its adjacent pleural scute. A prominent vertebral keel is present, as are also two disrupted, low, lateral keels. Well-marked growth annuli may be present on each carapacial scute. The carapace is usually dark brown with a bright-yellow vertebral stripe, a reddish-brown blotch on each pleural and vertebral, and a yellow or reddish-brown blotch along the outer border of each marginal. The dark-brown to black

Cuora flavomarginata

plastron has a narrow yellow band along the outer border of the humerals, pectorals, abdominals, and femorals. Undersides of the marginals are yellow. The posterior lobe is wide, not notched posteriorly, and capable of entirely closing over the hind limbs and tail. The plastral formula is: abd > an > pect > gul > hum > fem. Between the anal scutes, the seam is often reduced or absent in older individuals. The head is gray to light greenish dorsally with a broad yellow stripe running backward from the orbit to the neck. Upper jaws and sides of the head are yellow and the chin is either pink or yellow. Upper jaws are hooked. Limbs are grayish brown on the outside with yellow sockets. Heels of the hind feet are yellow. Forelimbs are covered with large scales. The short tail is gray with a faded yellow dorsal stripe. Numerous blunt tubercles occur at the base of the tail.

In males the plastron is rounded in both front and back; that of females is slightly bent upward in the rear. Also, the male's tail is thicker at its base than that of the female.

Distribution: *Cuora flavomarginata* is known from southern China (Fukien, Hunan, and possibly Szechuan provinces), Taiwan, and the Ryukyu Islands (Ishigaki Shima, Iriomote).

Geographic Variation: Hsu (1930) described a subspecies, *Cuora flavomarginata sinensis*, from Tungting Lake, Hunan, China, as differing from the nominate Taiwanese population on the basis of having the anterior border of the plastron obtusely emarginate; each plastral scute with deeply cut parallel lines; each marginal of the carapace with its posterior angle slightly overlapping the next marginal, and the degree of overlap most pronounced in the posterior third of the marginals so that the posterior border of the carapace is somewhat serrated; a small notch between the anals; and a much shorter tail. Pope (1935) commented that

with the exception of the longer tail, all of the other characters are either common variants to be expected in almost any emydid turtle or differences generally correlated with age, but he thought that until direct comparison between Chinese and Taiwanese *C. flavomarginata* could be made, it was better to consider the forms as distinct subspecies. However, Fang (1934) critically compared specimens from Taiwan and the Ryukyu Islands with Hsu's diagnostic characters and found *C. f. sinensis* could not be differentiated. So, it is perhaps best to regard *C. f. sinensis* as invalid.

Habitat: *Cuora flavomarginata* usually occurs in ponds and rice paddies, but Mao (1971) reported it from an upland shaded stream on a forested hill in Taiwan. The sides of the stream were overgrown with extremely luxurious high-stemmed vegetation, and under the vegetation of the damp bank about 100 specimens were collected in one morning.

Natural History: *Cuora flavomarginata* is fond of basking and spends much time on land, but, when disturbed, quickly pulls into its shell and closes the plastron. In captivity it readily feeds on a variety of foods ranging from fish to bananas. Fukada (1965) reported that a captive female laid two single-egg clutches six days apart in July. The white eggs are elongated, ranging from 37 to 48 mm in length and 21 to 26 mm in width. Hatchlings have carapaces of approximately 40 mm.

Cuora galbinifrons (Bourret, 1939)

Indochinese box turtle

Plate 9

Recognition: The carapace (to 19 cm) of *Cuora galbinifrons* is high domed, widest just behind the middle, and smooth bordered. A vertebral keel is present in juveniles but disappears entirely in adults. The 1st vertebral is wide anteriorly and overall is as long as or a little longer than wide. The 2d to 4th vertebrals are as broad as or broader than long, and the 5th is much broader posteriorly than anteriorly. Pleurals are wider than their corresponding vertebrals; lateral marginals are downturned. A narrow, yellow to cream-colored stripe runs the length of the vertebrals. On each side of this, covering the rest of the vertebrals and upper portions of the pleurals, is a wide, dark-brown to olive stripe often containing a mottled pattern of small dark marks. The lower 2/3 to 3/4 of the pleurals is usually plain white, cream, or yellow, although the extent of this light pigmentation may be reduced or bear dark mottlings in some individuals. Upper surfaces of the marginals are brown to olive with lighter mottlings. This series of alternating light and dark longitudinal areas serves to effectively break up the outline of the carapace. The plastron is large and can completely close; the posterior border of the anals is rounded and unnotched. Axillary and inguinal buttresses are short. The plastral formula is usually: abd > an > pect > gul > fem > hum. The plastron is dark brown or black, but some yellow pigment may occur along the seam connecting the carapace and plastron, and the undersides of each marginal may bear a yellow spot or border. The head is somewhat pointed with a short snout. Triturating surfaces are narrow and smooth, and the upper jaw lacks a strong medial hook. The head is yellow to pale green to gray, and may have dark speckles, especially in juveniles. A narrow dark stripe may occur on each side of the snout. Chin and throat are yellow to cream colored; limbs and tail are olive to gray. The anterior surface of the forelegs is covered with enlarged scales.

There is little sexual dimorphism, although the male has a slightly thicker tail.

Distribution: *Cuora galbinifrons* is known from Tonkin and Annam, Vietnam, and Hainan Island, China.

Geographic Variation: None has been reported, although the carapacial pattern is somewhat variable.

Habitat: Although it readily enters water, *C. galbinifrons* is probably the least aquatic species of *Cuora*. Its habitat appears to be bushy, upland woodlands and forests at rather high elevations (Bourret, 1941).

Natural History: *Cuora galbinifrons* is very shy and withdraws into its shell and closes the plastron when disturbed, remaining so for some time.

Captives feed on a variety of animal foods including earthworms, fish, pieces of beef, and canned dog food.

Cuora hainanensis (Li, 1958)
Hainan box turtle

Recognition: We have not seen this little-known species. The following description is a compilation of the original given by Li (1958) and a more comprehensive translation by Koshikawa (1982).

The high-domed carapace (to 18.6 cm) has a smooth rim. Each vertebral scute is as broad as long and narrower than its adjacent pleural scute. Vertebrals 1–3 each have an anterior medial projection and a posteri-

or medial notch. A medial vertebral keel is present. The carapace is light yellow with the vertebral keel yellow and the rest of the vertebrals and adjacent pleurals and outer border chestnut brown. The yellow area has brown stripes or spots, but the chestnut-brown parts have few yellow stripes. Both anterior and posterior plastral borders are round and unnotched. The interabdominal seam is the longest while that between the humerals is shortest. The anal scute is single, showing no dividing seam or rudiment of one. The plastron is chestnut brown with a few irregularly scattered light-yellow spots; undersides of the marginals are a mixture of chestnut brown and yellow. The head is moderate in size with an obtusely pointed snout slightly projecting in front of the smooth (neither notched nor hooked) upper jaw. The occipital region is covered with small scales. Dorsally the head is olive, and there are chestnut-brown spots on the snout, occipit, sides of the face, and upper jaws. The tympanum is light yellow and the lower jaw and throat are gray white. The neck is yellowish with black-brown stripes on the sides. The hind limbs are gray brown on the outer surfaces and yellow beneath. Forelimbs are covered with large scales. The tail is yellow with blackish-brown blotches.

Distribution: *Cuora hainanensis* occurs only on Hainan Island, China.

Habitat: It seems restricted to mountain streams.

Cuora trifasciata (Bell, 1825)
Chinese three-striped box turtle

Recognition: The carapace (to 20 cm) is elongated, low arched, widest behind the middle, and smooth edged posteriorly (although a shallow medial notch is present). There may be a slight concavity at the seam

Cuora trifasciata

between the first two pleurals. Adults have a low, blunt vertebral keel and indications of lateral keels. Vertebral 1 is wider anteriorly than posteriorly. The carapace is brown with well-marked black stripes along each keel and, in some individuals, dark seams. The dark vertebral stripe is longer than those on the lateral keels, which only extend along the first three pleurals. Schmidt (1927b) has reported that the dorsolateral black stripes are absent in juveniles and first appear in turtles of about 10 cm. The flat plastron is well developed, but the hind lobe is not extensive enough to completely hide the hind limbs. Also, the anal scutes have a wide posterior notch at the midline. The hinge first appears in specimens between 9 and 10 cm in length. The plastral formula is: pect >< abd > an > gul > fem > hum. The plastron is dark brown to black with a yellow lateral border extending from the pectoral to the anals. Some light radiations usually occur along the seams, and Schmidt (1927b) reported that the central part of the plastron may become yellow with rays of black pigment in adults. Undersides of the marginals are bright orange to pinkish yellow to bright yellow with large black spots on at least the anteriormost marginals. The head is narrow and pointed with a slightly projected snout and the upper jaw only slightly hooked. Ground color of the head is olive dorsally, but black laterally. An olive-colored, black-bordered stripe runs backward from the nostril through the orbit; behind this is a wide, black-bordered spot. The upper jaw is yellow, and a yellow stripe extends from the corner of the mouth through the tympanum to the side of the neck. Lower jaw and chin are yellow. Top and sides of the neck are olive and covered with small scales; ventrally there is a wide orange or pinkish-yellow medial stripe. Limb sockets and undersides of the limbs are bright orange or pinkish yellow. Exposed limb surfaces are olive or brown. The front of each foreleg is covered with large horizontal scales, that of the hind limb has small scales. There are enlarged scales along the heel. The top of the tail is olive with two black stripes; orange pigment occurs on the sides and bottom of the tail.

The contrasting bright color pattern of *Cuora trifasciata* makes it one of the more attractive of all turtles.

The male plastron is not very concave, but the tail is longer, thicker at the base, and has the vent beyond the posterior margin of the carapace.

Distribution: *Cuora trifasciata* is known from northern Vietnam and southern China: Kwangsi, Kwantung, Hainan Island, and Hong Kong.

Habitat: Mell (1922) collected *Cuora trifasciata* in clear mountain streams at elevations of 50 to 400 m in southern Kwangtung, but it probably occurs in other aquatic situations. Captives frequently bask and are often quite terrestrial, so this species probably spends much time on land.

Natural History: A captive male we kept repeatedly tried to court and mate with some *Cuora flavomarginata* and *Terrapene carolina* females housed with it. In this respect, its behavior was aggressive and reminiscent of *Clemmys insculpta* and *Mauremys leprosa*. It bit at and pushed the other turtles around the enclosure, and when mounted, held on by hooking the claws of all four feet under the female's marginals. Courtship and mating were attempted both in and out of the water. Nesting occurs in May, and two brittle, white, elongated (48–50 × 26–27 mm) to tapered eggs are laid at one time. Hatchlings have carapace lengths of about 40 mm.

Cuora trifasciata is carnivorous, feeding on a diverse assortment of animals ranging from earthworms and crabs to fishes. Captives do well on a fish diet.

Cuora pani Song, 1984

Pan's box turtle

Plate 9

Recognition: The only publication regarding this species is that of its original designation in Chinese. Fortunately, some individuals of this turtle have now reached the United States via the pet trade, and we are able to supplement the original brief description.

The elongated carapace (to 15.6 cm) is low arched, widest behind the middle at the level of the 8th marginals, and highest at the 2d vertebral. Its marginals are flared and there is a slight posterior notch. A low, rounded medial keel is present. The 1st vertebral scute is flared anteriorly and is the largest; other vertebrals are wider than long, and the 5th is also flared. The cervical scute is long and narrow. The carapace is dark brown, with lighter chestnut to red brown along the vertebrals. The well-developed plastron is yellow with black triangular marks along the seams. A black bar crosses each bridge, and the undersides of the marginals are yellow except at the bridge where their rims are black. The plastral hind lobe is posteriorly notched and broad enough to cover most of the hind limbs when they are retracted. The plastral formula of the few we have examined was: pect >< an > abd > gul > fem > hum; the anal scutes are separated by a seam. The head is narrow and pointed; its snout is not or only slightly projected, and the upper jaw is slightly hooked. Smooth skin covers the dorsal surface of the head, and the iris is greenish yellow. The head is lemon yellow with a narrow brown stripe extending backward from the eye to the neck. Jaws are also yellow. The top and sides of the relatively long neck are brown and covered with roughened scales; the chin and ventral surface are yellow. The outer surface of the unpatterned limbs is olive gray to brown; the sockets beneath are yellow to orange. The anterior surface of each forelimb has large horizontal scales; the hind limbs are covered with small scales, and enlarged scales occur along the heel. The tail is yellow with brown dorsal stripes.

The plastron of those males we have seen was only slightly concave, but we have not seen any larger than 90 mm. The male tail is long and thick with the vent beyond the carapacial rim. Females have smaller tails.

Distribution: *Cuora pani* is known only from Shaanxi Province, China.

Cuora chriskarannarum Ernst and McCord, 1987

Yunnan green box turtle

Plate 8

Recognition: This newly described species has an elongated, flattened carapace (to 16 cm) which is widest behind the center, and has a low medial keel, a slightly serrated posterior rim (there is a small medial notch), and all marginals flared. Vertebrals are wider than long. Vertebral 1 is largest and very flared anteriorly (it reaches the seam separating marginals 1–2 or marginal 2); vertebral 5 is posteriorly flared. The cervical is long and narrow. Growth annuli are present on the scutes. The carapace is olive brown with a yellow rim, black keel, and the seams outlined in dark brown or black. Undersides of marginals are yellow with a black, posteriorly directed wedge at each seam, and some with narrow, black radiating lines. The well-developed plastron is yellow with a broad, black, seam-following pattern which becomes more extensive with age (size). A black bar crosses the bridge. The anterior plastral lobe is truncated; the posterior lobe has the anal scutes tapering toward the midline and a shallow anal notch. The plastral formula is: an > abd > pect > gul > fem > hum in most, but variation occurs; the interanal seam is complete. The head is narrow with a slightly projecting snout and a slightly hooked upper jaw. It is olive (lighter dorsally, darker laterally) with a yellowish-green postorbital stripe and a second yellowish-green stripe extending obliquely downward from the upper jaw to below the tympanum. Both stripes have black borders, and a faint, thin black line may circle the tympanum. The iris is

green, the jaws and chin yellow. The neck is olive dorsally and laterally, yellowish green ventrally; several faint, narrow, yellow lateral stripes are present. Forelimbs have large scales; the outer surface is olive, inner surface and sockets cream to whitish green. Hind limbs have smaller scales, colored as forelimbs. The tail is olive dorsally with two dark-bordered stripes, the venter yellow with an olive tip.

Males are smaller (11.3 cm) with concave plastra, longer, thicker tails, anal vents beyond the carapacial rim, and pointed snouts. Females are more domed with flat plastra, shorter tails, anal vents beneath the carapacial rim, and rounded snouts.

Distribution: *Cuora chriskarannarum* has been collected only at Ta Lau Shan and Chinsha, Yunnan Province, China.

Cuora mccordi (Ernst, 1988)

McCord's box turtle

Plate 9

Recognition: The elliptical carapace (to 13.4 cm) is domed, widest at the level of the 8th marginals and highest at the seam separating the 2d and 3d vertebrals. The marginals are flared anteriorly and posteriorly, laterally somewhat downturned with reverted rims, and only slightly serrated posteriorly. A low, blunt medial keel is present, best developed from posterior vertebral 2 to vertebral 4. The 1st vertebral is flared anteriorly, vertebral 2 is longer than wide, vertebrals 3–5 are broader than long, and the 5th is flared. The cervical scute is longer than broad. Carapacial color is reddish brown with dark seams, a dark wedgelike mark on each marginal, and a yellow border. The well-developed plastron is yellow with an extensive medial black mark and two black blotches on the yellow bridge. Undersides of the marginals are yellow. The plastral hind lobe is posteriorly notched, and broad enough to cover most of the retracted hind limbs. The plastral formula is: abd > pect > an > gul > fem > hum; the interanal seam is complete. The head is narrow and pointed; its snout does not project, and the upper jaw has neither a hook nor notch. Smooth yellow skin covers the dorsal surface of the head, the iris is yellow to greenish yellow, and a bright, narrowly black-bordered, yellow stripe extends backward from the snout through the orbit to the neck. Jaws, chin, and neck are yellow. Large brown to reddish-brown scales cover the anterior surface of the forelegs, and the hind feet are brown. Other limb coloration is yellow. A brown to olive stripe occurs dorsally on the yellow tail.

Males have concave plastra and long, thick tails with the vent beyond the carapacial rim. Females have flat plastra and small tails.

Distribution: *Cuora mccordi* is known only from Kwangsi, China.

Cuora yunnanensis (Boulenger, 1906)

Yunnan box turtle

Recognition: The carapace (to 14 cm) is relatively low arched, widest behind the center, and contains a strongly developed vertebral keel and two low lateral keels. Adult vertebrals are as long as broad, and much narrower than their accompanying pleurals; the 1st and 5th are flared. Posterior carapacial marginals are smooth in adults, but are slightly serrated in juveniles. The carapace ranges from chestnut brown to olive, and in some individuals the rim and keels may be yellow. The plastron is well developed, but the hinge is weak, and the hind lobe does not completely close posteriorly, as it is somewhat narrow and medially notched. The plastral formula is: abd > pect > an > gul > fem > hum; the interanal seam is complete. The plastron is olive to brown with black seams and a yellowish border; a dark bar is present on the bridge. A large red-brown blotch may occur on each scute. The upper jaw is not hooked, but the snout is pointed and slightly projecting. The olive to brown head has a narrow yellow stripe running through the orbit to the neck, and a second yellow stripe extending from the corner of the mouth to the neck. Chin and throat are yellow to orange with olive mottlings. The neck is olive to brown with two orange stripes on each side. Skin of the limbs and tail is olive to brown with orange streaks.

Males have longer, thicker tails with the vent beyond the carapace.

Cuora yunnanensis

Distribution: *Cuora yunnanensis* is known only from Yunnan Province, China, where it has only been collected at altitudes over 1800 m.

Cyclemys Bell, 1834

The oriental genus *Cyclemys* shares a common plastral closing mechanism with *Pyxidea* and *Cuora* and may have evolved from a *Heosemys*-like ancestor (Bramble, 1974). McDowell (1964) has also pointed out its relationship to the genus *Geoemyda*. One widespread species, *C. dentata*, and another poorly known species, *C. tcheponensis*, are currently assigned to *Cyclemys*.

The adult carapace is oval, slightly arched, and longer than wide; that of juveniles is flattened and rounder. A medial and two lateral keels are present, but these become lower with age. The hexagonal neurals are short sided posteriorly, and the posterior border of the carapace is serrated, more so in juveniles. The plastron is large, and a movable hinge develops between the hyoplastron and hypoplastron bones at the seam separating the pectoral and abdominal scutes. This allows the anterior lobe to move, but it cannot entirely close. In the adult, the plastral buttresses are resorbed and the plastron is connected to the carapace only by a ligament. The skull usually has a complete temporal arch, but the squamosal does not normally touch the jugal. Ventrally, the jugal is broad. The parietal touches the quadrate above the foramen for the trigeminal nerve; however, the anterior edge of the inferior parietal process does not touch the jugal or palatine. The ethmoidal fissure is narrowly triangular or shaped like a keyhole. Triturating surfaces of the maxillae are narrow and ridgeless. Cloacal bursae are present. The toes are webbed.

Nakamura (1949), Stock (1972), and Killebrew (1977a) reported the karyotype of *Cyclemys dentata* is 2n = 52; but Kiester and Childress (in Gorman, 1973) found it to vary between 50 and 52.

Key to the species of *Cyclemys*:

1a. Lateral neck stripes rarely reach side of head near corner of mouth: **C. dentata**
  b. Several lateral neck stripes extend through and below orbit to snout and jaws: **C. tcheponensis**

Cyclemys dentata (Gray, 1831b)

Asian leaf turtle

Plate 10

Recognition: The oval carapace (to 24 cm) is slightly arched and contains a single medial keel. Vertebrals are usually broader than long, but the 2d to 4th may be as broad as long. The posterior border of the carapace is serrated, and all of the scutes have smooth surfaces. Carapacial color varies from light to dark brown, black, olive, or sometimes mahogany; narrow black radiations may be present. The adult plastron is narrower than the carapace, and notched posteriorly. The plastral formula is: pect > abd > an > fem >< gul > hum. The plastron varies from yellow or light brown, with dark radiations, to uniformly dark brown or black. The snout is slightly projected, and the upper jaw lacks toothlike projections on each side. Posteriorly, the head skin is divided into large scales; dorsally, it is reddish brown while the sides and jaws are darker brown. Legs are light brown with large transverse scales on the anterior surface.

Females may grow larger than the males, but the males have longer, thicker tails.

Distribution: *Cyclemys dentata* is found from northern India southward through Burma, Thailand, Kampuchea, Vietnam, and Malaya, to Sumatra, Java, and Borneo. It also occurs in the Philippines.

Geographic Variation: Unstudied, but according to Smith (1931) there is considerable variation in coloration that is not correlated with geographic distribution.

Habitat: This semiaquatic turtle lives in shallow streams in both the mountains and lowlands.

Natural History: Wirot (1979) thought several clutches of eggs are laid each year. A typical clutch consists of two or three elongated, white eggs measuring about 57 × 30–35 mm (Smith, 1931; Ewert, 1979). The female's plastron becomes slightly kinetic to allow these large eggs to be laid. Twelve hatchlings measured by Ewert (1979) had an average carapace length of 56.2 mm.

Cyclemys dentata is omnivorous, and feeds on both animals and plants. It is gentle and lively, and does well in captivity.

Cyclemys tcheponensis (Bourret, 1939)

Stripe-necked leaf turtle

Recognition: The carapace is elongated (to 22 cm) in adults, but rounder in juveniles. A single low medial keel is present on adults, and the posterior carapacial border is serrated. Carapacial scutes in juveniles bear some rugosities. The carapace may range from mahogony to olive in juveniles or gray brown to olive brown in adults. The plastron is notched posteriorly, and its formula is quite variable: abd >< pect > gul >< an > fem >< hum. The plastron varies from pinkish or red in juveniles to yellow or light brown in adults. Dark-brown radiations are present in juveniles, but may be lost with age. Head morphology is like that of *C. dentata;* the head is dark olive dorsally with numerous dark mottlings, and four yellow, orange, or pink stripes extend from the neck forward over the side of the head. The most dorsal of these passes above the orbit and onto the snout; a second broad stripe stops at the orbit; below this a third, narrower stripe passes below the orbit and along the upper jaw; and a fourth passes along the lower jaws. The throat is mottled with small dark spots. Skin of the limbs is grayish brown to olive.

Males have longer, thicker tails.

Controversy has surrounded this species ever since it was described by Bourret (1939), who reported it to be intermediate between *Heosemys grandis* and *H. spinosa.* Unfortunately, the holotype is a juvenile, the adult being unknown at that time, and it was unclear whether or not a plastral hinge developed. Consequently it has been assigned and reassigned to several batagurine genera. McDowell (1964) thought the holotype to be merely a variation of *Cyclemys dentata* with the entoplastron abnormally far forward. However, McMorris (1976) studied fifteen specimens from Indochina which matched the description by Bourret (1939), and redescribed and compared it to *C. dentata.* Wirot (1979) and Pritchard (1979) also thought *C. tcheponensis* valid, and we have seen this turtle and agree on its validity.

Distribution: *Cyclemys tcheponensis* ranges westward from Vietnam through Laos and Kampuchea into Thailand.

Habitat: Wirot (1979) reported it lives in streams and brooks in mountainous areas, but spends much time on land.

Natural History: All we know about the life history of this turtle comes from Wirot (1979), who reported it is agile, lively, and becomes tame in captivity. It is omnivorous, and 10–15 eggs are laid per clutch.

Geoclemys Gray, 1855

One living species is assigned to this genus, *Geoclemys hamiltonii.*

Geoclemys hamiltonii (Gray, 1831b)
Spotted pond turtle

Recognition: *Geoclemys hamiltonii* is a relatively large (to 35 cm) freshwater turtle with an elongated oval, three-keeled, domed carapace. It is posteriorly serrated, more strongly in juveniles than adults. The keels are interrupted into a series of elevated protuberances, one on each vertebral and pleural. In adults, vertebral 1 is nearly as broad as long, 2–4 are nearly as long as broad, and the 5th is broader than long. Neurals are short sided anteriorly. Ground color of the carapace is black, and there is a series of dorsally pointing orange, yellow, cream, or white wedge-shaped marks at the base of each pleural and spots or radiations of the same color on the vertebrals, giving this turtle a very striking appearance. These light markings fade with age and old adults may be mostly black. The hingeless plastron is yellow with numerous dark radiations. The bridge is extensive with moderate buttresses; axillary buttresses are attached to the 1st costals and inguinal buttresses attach to the sutures between the 5th and 6th costals. The entoplastron lies just anterior to the humero-pectoral seam. The plastron formula is: abd > fem > pect > gul > an > hum, and the anal scutes are deeply notched. In the skull, the frontal bone is excluded from the orbit, and the postorbital is separated from the squamosal by the jugal. The canal of the carotid artery is enclosed within the pterygoid. The upper jaw is broadly notched medially and its triturating surfaces

are broad and nearly flat, but McDowell (1964) points out that traces of a cusp-and-ridge pattern, much like that of *Morenia*, occur clearly on the rhamphotheca and also on the bony surfaces. No medial ridge occurs on the lower triturating surface. Posteriorly, the large black head is covered with small scales, and it has large white or yellow spots on the top, sides, snout, and jaws. The snout barely projects. Neck and limbs are dark brown or black with large yellow spots. The toes are fully webbed; the tail is short.

Males have concave plastra and thicker tails. Females have flattened plastra and smaller tails.

Distribution: *Geoclemys hamiltonii* is a rare species restricted to the Indus and Ganges river drainages of southern Pakistan, northern India, Assam, and Bangladesh.

Habitat: Little is known of the ecological requirements of this species, but adults apparently enter large rivers and forest ponds. Pritchard (1979) reported it from oxbow lakes and sloughs in quiet, shallow, rather clear waters with considerable aquatic vegetation.

Natural History: Nesting probably occurs in May or June. Hatchlings are more brightly patterned with numerous white spots than are adults, and are about 35–37 mm long.

Geoclemys is entirely carnivorous, feeding on snails and other invertebrates, and probably also fishes and amphibian larvae.

Geoemyda Gray, 1834b

This genus has been used in the past as a "catchall" for various batagurine species, but at present includes only the black-breasted leaf turtle of southeastern Asia.

Geoemyda spengleri (Gmelin, 1789)
Black-breasted leaf turtle
Plate 10

Recognition: This is the nominate species of the *Geoemyda* complex of McDowell (1964). Its carapace is elongated (to 13 cm), slightly wider behind the middle, rather flat topped, and very strongly serrated,

Geoclemys hamiltonii

both anteriorly and posteriorly. Each marginal has an acute, single-pointed corner, with the points of marginals 1–3 and 8–12 flared outward. Present are three well-developed keels; the medial keel is best developed, but the two lateral keels are also prominent. Vertebrals are broader than long, and the underlying neurals are hexagonal with the shortest side posterior. Each carapacial scute has a roughened surface due to growth annuli. Carapacial color varies from yellowish brown or grayish brown to dark brown; usually some dark pigment extends along the keels, especially the medial. The large plastron is elongated, and deeply notched posteriorly. It lacks a hinge, and the scutes are rugose with growth annuli. The plastral formula is extremely variable: abd >< pect >< fem > hum > an > gul. The bridge is about as long as the posterior plastral lobe, and may contain axillary and inguinal scutes. Its buttresses are weak and barely extend beyond the peripheral bones. The plastron is dark brown or black with a yellow lateral border; the bridge is totally dark brown or black. The flat head is not enlarged, and its snout does not project. The upper jaw is medially hooked, and its triturating surface is narrow and lacks ridges. Special skull features noted by McDowell (1964) include: the inferior process of the parietal bones being strongly convergent ventrally and encroaching on the groove housing the trabecular cartilage, so that the cranial cavity is marginally narrowed ventrally, and not contacting the palatines; the anterior edge of the inferior process being very thin and separated from the ventral portion of the jugal bone by both the pterygoid and maxilla; the usual absence of the orbito-nasal foramina (a small pair may be present); and the presence of a keyhole-shaped ethmoid fissure. Skin covering the posterior surface of the head is smooth and undivided. The head is olive to brown with a yellow stripe running backward from the orbit over the tympanum to the neck. Other yellow spotting may occur on the sides of the head and jaws. Anterior surfaces of the forelegs are covered with large, overlapping pointed scales, and large scales also occur on the heels of the hind feet. Small tubercles are present on the thighs and at the base of the tail. The toes are only partially webbed. The limbs and tail are grayish brown.

Nakamura (1949) gave the karyotype as 2n = 52.

Males have concave plastra and long, thick tails with the vent beyond the carapacial rim. Females have flat plastra and shorter tails with the vent under the carapacial rim.

Distribution: *Geoemyda spengleri* ranges from Kwangsi, Kwangtung, and Hainan Island in southern China through Vietnam to the islands of Sumatra, Borneo, and Okinawa.

Geographic Variation: Two subspecies are recognized. *Geoemyda spengleri spengleri* (Gmelin, 1789) (Plate 10) is found in Kwangsi and Kwangtung, China, Vietnam, and on Sumatra and Borneo. It is distinguished by the complete absence of axillary and inguinal scutes. *G. s. japonica* Fan, 1931 occurs only on Okinawa. It has axillary scutes, and sometimes inguinals. Pope (1935) reviewed these bridge characteristics and reported that presence or absence of axillary scutes was a valid character separating the subspecies. He also thought additional study was needed, and that these two forms may represent full species, rather than subspecies.

Habitat: Petzold (1963) reported *Geoemyda spengleri* seldom willingly seeks out water. However, while primarily terrestrial, it apparently occasionally enters freshwater streams (Siebenrock, 1907; Fang, 1930). Pope (1935) thought its apparent abundance in Yaoshan, Kwangsi region, (Fan, 1931) could be taken as evidence that it likes wild, wooded, mountainous country.

Natural History: Ewert (1979) reported that the carapace length of four hatchlings averaged 42.2 mm.

Captives in the Tierpark, Berlin, were omnivorous, feeding on a variety of animal and plant foods (Petzold, 1963).

Hardella Gray, 1870b

The sole species of this monotypic genus is the large (to 53.3 cm) crowned river turtle, *Hardella thurjii*.

Hardella thurjii (Gray, 1831b)
Crowned river turtle

Recognition: The ellipsoidal, moderately domed adult carapace lacks marginal serrations but is notched posteriorly. It also contains a low, blunt medial keel, and is widest just behind the center and highest at the seam separating the 2d and 3d vertebrals. Vertebrals are broader than long, and the keel is reduced to a posterior knob on each. Marginals are slightly flared in adults, but to a greater extent and somewhat serrated in juveniles. Neurals are elongated, six sided, and shortest anteriorly. The carapace is dark gray, brown, or black with a yellow outer rim (at least in younger individuals) and yellow pigment at the junction of the pleural and marginal scutes. The

Hardella thurjii

plastron is firmly and extensively sutured to the carapace and has strong, large buttresses. Axillary buttresses are connected to the 1st ribs, and inguinal buttresses attach midway along the suture between the 5th and 6th costals. The plastron lacks a hinge and has a relatively narrow hind lobe (narrower than the bridge) which is notched posteriorly. The ento-plastron lies anterior to the humero-pectoral seam. The plastral formula is: abd > pect >< fem > hum > an > gul. Both plastron and bridge are yellow with a large dark blotch on each scute. In the skull, the chamber for the paracapsular sac is completely open posteriorly, a feature shared only with cheloniid sea turtles and some Mesozoic turtles. The squamosal bone is attached to the jugal and postorbital bones. The frontal enters the orbital border, but does not pass under the olfactory nerves. The canal for the carotid artery lies between the prootic and pterygoid bones, the orbito-nasal foramen is large, and the choanae open behind the orbits. Triturating surfaces are broad, and that of the upper jaw has a tuberculate medial ridge. The upper jaw is notched medially and the lateral borders of both jaws are serrated. The snout is blunt and only slightly projecting. The head is brown to black with several yellow stripes: one begins over the nostrils and extends backward above the orbit and then dips downward before extending onward to the neck, a second forms a bar across the upper jaw below the nostrils, and a third extends from the corner of the mouth backward along the neck. Occasionally other yellow marks occur on the dorsal surface of the head, or the long stripes may be interrupted. Dorsally, the snout and head are covered by a single large scale; the posterior surface of the head is covered with many small scales. All toes are webbed and the brown limbs have a yellow outer border.

There seems to be extreme sexual size dimorphism in this species: adult males reach about 17.5 cm in carapace length but adult females may be more than three times this length. The male tail is thick and long, and the plastron concave.

Distribution: *Hardella thurjii* occurs in the watersheds of the Ganges, Brahmaputra, and Indus rivers in Bangladesh, northern India, and Pakistan. Minton (1966) also reports a single record for a coastal mangrove swamp near Karachi, Pakistan.

Geographic Variation: No subspecies are recognized, but McDowell (1964) felt that the usually synonymized species, *Hardella indi* Gray, 1870b, may represent a distinct species. This form is restricted to the Indus River system, and was described by McDowell as differing from *H. thurjii*, of the Ganges and Brahmaputra rivers, by having a pair of low lateral keels as well as the medial vertebral keel. This relationship needs reevaluation. Probably, if *H. indi* is a distinct form, it is most likely only a subspecies.

Habitat: This large species prefers water bodies with slow current, mud bottoms, and abundant aquatic vegetation. It has been taken from or observed in oxbows, ponds, canals, and quiet pools in rivers.

Natural History: Hatchlings are more brightly colored than adults. Their carapaces are more sharply serrated, and the vertebral keel more pronounced.

Hardella thurjii is normally a strict herbivore, naturally feeding on aquatic plants, but Minton (1966) observed one small individual eat part of a frog. In captivity they accept a variety of common fruits and vegetables.

These turtles rarely come out of the water to bask, but may do so by floating quietly at the surface. Minton (1966) reported they spend much time resting on the bottom. *Hardella* is of calm disposition, seldom biting when handled.

Heosemys Stejneger, 1902

Heosemys is a genus of semiterrestrial or semiaquatic tropical Asian batagurines. Their carapaces are usually flattened across the vertebrals, but still retain at least traces of a medial and two lateral keels, which are more pronounced in younger individuals. The posterior margin of the carapace is always serrated. Neurals are six sided and shortest posteriorly. The plastron is rigid, well buttressed, and generally hingeless (females may have some kinetic movement to aid the passage of the rather large eggs). The entoplastron is usually crossed by the humero-pectoral seam. A temporal arch is absent from the short skull. The ethmoid fissure is broadly oval, and the small postlagenar hiatus is round. The anterior edge of the inferior parietal process is not flared outward, is separated from the jugal by the pterygoid, but touches the palatine. The orbito-nasal foramen is posteriorly placed at a level with the primitive caroticum foramen and bordered medially by the basisphenoid. Triturating surfaces of the jaws are narrow and ridgeless. Cloacal bursae are present.

Based on skull characters, McDowell (1964) thought *Heosemys* to be an isolated genus, without clear affinity to any other particular genus of his *Geoemyda* complex, although clearly a member of that complex. However, Bramble (1974) suggested that, on the basis of plastral kinesis and closing mechanisms, a *Heosemys*-like ancestor may have given rise to the genera *Cuora*, *Cyclemys* and *Pyxidea*, suggesting all four genera may be united in a *Heosemys* complex.

Key to the species of *Heosemys*:

1a. Medial vertebral keel present on all five vertebrals
 b. Medial vertebral keel absent, or if present, only on posterior vertebrals: ***H. leytensis***
2a. Anterior margin of carapace unserrated 3
 b. Anterior margin of carapace strongly serrated: ***H. spinosa***
3a. Only a medial keel present on carapace 4
 b. Carapace tricarinate, with a prominent medial keel and two lower lateral keels: ***H. silvatica***
4a. Bridge uniformly yellow; carapace high, but flattened across vertebrals; large, to 43 cm: ***H. grandis***
 b. Bridge uniformly dark brown to black; carapace low and flat; medium sized, to 24 cm: ***H. depressa***

Heosemys depressa (Anderson, 1875)

Arakan forest turtle

Recognition: The carapace (to 24.2 cm) is middorsally flattened, but still bears a prominent blunt keel, and is widest behind the middle. Vertebrals are broader than long, and the posterior marginals are strongly serrated. The carapace may be totally light brown or bear either black mottled streaks or a black border. The plastron is notched posteriorly. Its hind lobe is narrower than the length of the bridge. Both axillary

and inguinal scutes are present on the bridge. The plastral formula is: abd > pect > fem >< gul > an > hum. The plastron is yellow to tan with dark-brown or black blotches or radiations on each scute. The bridge is uniformly dark brown or black. The gray to brown head is moderate in size with a slightly projecting snout. A toothlike projection occurs on either side of the medial notch on the upper jaw. Posterior dorsal head skin is divided into large scales. The iris is brown. Neck, limbs, and tail are yellow brown. Forelimbs have large, square to pointed scales on the anterior surface, partially webbed toes, and heavy claws. Hind limbs are flattened, covered with small scales except at the heel where they are enlarged, have little webbing on the toes, and the claws are heavy.

Males have enlarged hind limbs and larger, thicker tails than do females.

Habitat and Distribution: Presumably terrestrial, and possibly a burrower, the species is known only from the hills in Arakan, Burma.

Heosemys grandis (Gray, 1860a)

Giant Asian pond turtle

Recognition: This is one of the largest hard-shelled, semiaquatic Asian turtles, reaching a carapace length of 43.5 cm. The broad, oval carapace is high arched but depressed dorsally and serrated posteriorly. A well-defined, blunt medial keel is present, and all vertebrals are broader than long. In color, the carapace is brown to grayish brown or black, and a pale streak may occur along the keel; in some, light seams appear between the scutes. In younger individuals, the growth annuli are prominent. The plastron is large and notched posteriorly. Both the anterior and posterior lobes are narrower than the medial portion. The plastral formula is: abd > fem > pect > an >< hum

Heosemys grandis

>< gul. The bridge is well developed and longer than the plastral hind lobe. Both axillary and inguinal scutes are present. Plastron, bridge, and undersides of the marginals are yellow. In juveniles a pattern of dark lines radiating outward from a large dark central blotch occurs on each scute, but these disappear with age and the scutes are uniformly yellow in older turtles. Head and neck are broad, snout only slightly projecting, and the upper jaw contains a shallow medial notch flanked by a pair of toothlike cusps. Skin covering the posterior portion of the head is divided into irregularly shaped scales. The head is grayish green to brown with numerous yellow, orange, or pink mottled spots, which disappear with age so that old turtles may have only a few remaining. Jaws are cream to horn colored; forelimbs large with broad scales on the anterior surface. All toes are webbed.

Males have slightly concave plastra and longer, thicker tails than the flat-plastroned females.

Distribution: *Heosemys grandis* ranges southward from Pegu, Burma, through Thailand, Kampuchea, and northern Malaysia.

Habitat: This turtle inhabits various fresh waters, such as rivers, streams, swamps, lakes, and marshes from hill country to sea level. It is not restricted to water, and spends much time on land partially hidden under shrubbery. Captives may spend most of the time out of water.

Natural History: Despite its size, its habits are poorly known. Taylor (1970) and Wirot (1979) reported they feed mainly on aquatic plants in the wild, but captives are generally omnivorous. In zoos, males have been seen biting at the neck and head of females, and possibly this is part of the courtship ritual. Pritchard (1979) reported that juveniles have a large fontanelle or "soft spot" in the middle of the plastron.

This turtle, along with *Hieremys annandalei*, is often placed in the Tortoise Temple in Bangkok.

Heosemys leytensis Taylor, 1920
Leyte pond turtle

Recognition: The carapace (to 21 cm) is somewhat flattened and lacks a keel, or, if one is present, it is only on the posterior vertebrals. Vertebrals are broader than long. Anterior marginals are largest, projecting forward beyond the cervical over the sides of the neck making the anterior rim sharply serrated; posterior marginals are slightly serrated. The carapace is brown to reddish brown and unmarked. The plastron

narrows both anteriorly and posteriorly, and is much smaller than the carapace. It is strongly notched posteriorly. Gulars are strongly projecting and contain a large angular notch between them. The plastral formula is: abd > pect > fem > gul > hum > an. The bridge is longer than the hind lobe of the plastron, and contains an axillary and an inguinal. Plastron and bridge are uniformly yellow to brown or reddish brown. The head appears large, and the snout is pointed and projecting. The eyes are small. The upper jaw is distinctly hooked and slightly bicuspid. Dorsally, the head skin is divided into scales. The head is brown with some darker mottling on the temple. A narrow yellow bar crosses the head just behind the tympanum where it passes downward toward the neck (this bar may be interrupted medially), and a small yellow spot may occur on each side of the lower jaw. The neck is dark brown on top but lighter on the sides and beneath. Limbs are darker brown anteriorly and lighter beneath. Each foreleg contains four enlarged transverse irregular scales on its upper surface, and a large transverse scale at the heel. Each hind limb also has enlarged transverse scales on the anterior surface, but no large scales at the heel. All toes have heavy claws and webbing. Skin on the legs, body, and neck contains very small tubercles giving it a roughened appearance.

Sexual dimorphism has not been described.

Distribution: There are only a few specimens of this species known, collected at Cabalian in southern Leyte and Palawan Province, Philippine Islands (Timmerman and Auth, 1988), but is it really so rare, or is this just a poorly studied area?

Natural History: The biology of this turtle is unknown.

Heosemys silvatica (Henderson, 1912)
Cochin forest cane turtle

Plate 10

Recognition: This rare and little-known species (less than 50 specimens have ever been collected) has an oval, moderately domed carapace (to 13 cm) with three low keels. The medial keel, which extends along all five vertebrals, is better developed than the two lateral keels, which only extend along the upper parts of the first three pleurals. All vertebrals are broader than long, and the posteriormost marginals are slightly serrated. The carapace is uniformly bronze or orangish brown to black. The plastron is relatively wide with a posterior notch. The bridge is fairly long, and both the axillary and inguinal scutes are very small; the inguinals may be absent. The plastral formula is: abd >< pect > hum >< fem > an > gul. The plastron is uniformly yellow to orange with a black blotch on both the pectoral and abdominal scutes at the bridge. The head is of moderate size and the upper jaw has a strong medial hook. The front of the head and the jaws are yellow, and a red spot occurs on the tip of the snout. The iris of the eye is reddish, and red or pink pigment also occurs on the upper eyelid. This pigment may extend to the dorsal surface of the head. Posterior surface of the head and neck are brown to black, limbs and tail are pale brown. Enlarged scales are present on the anterior surface of the forelegs, and the hind foot is somewhat club shaped. No webbing occurs on the toes, but their claws are well developed.

Males have thicker, longer tails with the vent beyond the carapacial rim, concave plastra, and better developed spurlike protuberances on their hind legs. Females are not as colorful as males, and tend to be slightly larger. Mature females develop a movable plastron resulting from erosion of the bony suture between the hypoplastron and carapace (Moll, Groombridge, and Vijaya, 1986).

Heosemys silvatica shares several characters with *Geoemyda spengleri*, and should probably be placed in that genus (Moll, Groombridge, and Vijaya, 1986).

Distribution: *Heosemys silvatica* is restricted to the region of Cochin in southwestern India. It has been collected only rarely since 1911.

Habitat: *Heosemys silvatica* inhabits upland, shady, dense, evergreen forests above 300 m, not necessarily near free water, and lives in short underground burrows (Henderson, 1912), or hides under dead leaves and logs or in rock crevices in spiny cane groves (Vijaya, 1982a; Moll, Groombridge, and Vijaya, 1986). It is nocturnal or crepuscular and apparently rare or very secretive. It is sympatric with *Indotestudo forsteni*.

Natural History: Breeding apparently occurs in October and November, just after the monsoon (Vijaya, 1982a), and males develop bright-red pigment on the head at this time. A female collected in October in Kerala laid two eggs (44–45 × 22.5–23.5) on 22 December at the Madras Snake Park Trust (Whitaker, 1983; Moll, Groombridge, and Vijaya, 1986).

Wild turtles feed on fallen fruits and leaves and invertebrates such as millipedes, mollusks, and beetles (Vijaya, 1982a; Whitaker, 1983); Henderson (1912) reported that captives were entirely herbivorous.

Heosemys spinosa (Gray, 1831b)

Spiny turtle

Plate 10

Recognition: Smaller individuals of this species are like walking pin cushions. The broad, oval carapace (to 22 cm) is flattened dorsally, but has a medial keel. A spinelike projection occurs on the keel at the posterior border of each vertebral. There is also a small spine at the posterior margin of each pleural scute, and each marginal has a pointed, projecting corner making the entire outer rim of the carapace spiny. Many of these spines become worn down and are lost with age, so that larger individuals are much smoother than juveniles. Anterior and posterior marginals are also somewhat upturned. All vertebrals are broader than long. The carapace is brown with a light streak along the medial keel. The plastron is large and the bridge broad. Both anterior and posterior plastral lobes are narrower than the middle section, and there is a deep anal notch. A spinelike anterior projection occurs on each gular of juveniles, but these become more blunt with age and are lost in adults. The plastral formula is: abd > pect > fem > hum > gul > an. Both axillary and inguinal scutes are present. Plastron and undersides of the marginals are yellow with radiating dark lines on each scute. The head is rather small with a slightly projecting snout and a narrow medial depression flanked on each side by a toothlike cusp on the upper jaw. Posterior head skin may be divided into scales. The head is grayish to brown with a yellow spot near the tympanum. Limbs are gray with some yellow spotting. The forelegs have enlarged overlapping scales on the anterior surface, and the hind feet are clublike with enlarged spinelike scales in adults. Thighs and base of tail also contain spiny tubercles; toes are only partially webbed.

DeSmet (1978) reported that the 2n chromosome number is 52 (13 pairs each of macro- and microchromosomes).

Males have slightly longer and thicker tails than do females.

Distribution: *Heosemys spinosa* ranges from Tenasserim, Burma, and southern Thailand southward through Malaysia to Sumatra and Borneo.

Habitat: *Heosemys spinosa* occurs in shallow, clear mountain streams in forests. It frequently wanders about on land in cool, humid, shaded areas, and often hides under plant debris or clumps of grass. Mertens (1971) observed that, at least in captivity, juveniles are more terrestrial than adults.

Natural History: Three white oblong eggs are laid consecutively. Hatchlings are exceptionally large; ten near-hatchling young measured by Ewert (1979) averaged 63.2 mm in carapace length. To accomodate the large eggs housing these young, a hinge develops in the female's abdominal scute allowing the plastron more flexibility during egg laying. This is achieved by an incomplete separation of the abdominal scute into hyoplastral and hypoplastral elements and by fragmentation of the inguinal scute (Waagen, 1984).

Heosemys spinosa apparently is largely herbivorous, preferring fruits, but in captivity it may accept some animal foods, such as canned dog food.

Hieremys Smith, 1916

This is a monotypic batagurine genus; the sole species is *Hieremys annandalei.*

Hieremys annandalei (Boulenger, 1903)
Yellow-headed temple turtle

Recognition: *Hieremys annandalei* is a large semiaquatic turtle, reaching a carapace length of 60 cm. Its uniformly dark-brown or black adult carapace is elongated (much longer than wide), rather low in profile but not depressed, and usually has a serrated posterior border with a pronounced notch between the 12th marginals. A single, rather low medial keel is present along the vertebrals. The 1st vertebral scute is broader anteriorly than posteriorly and the other four are broader than long. The 5th is broader posteriorly than anteriorly. Occasionally a 6th vertebral is present. The 4th vertebral covers portions of four neural bones.

Hieremys annandali

The posterior neurals are not noticeably elongated, and the 4th to 6th are hexagonal and shorter posteriorly. The 2d neural is also hexagonal, but its short side is anterior. Lateral marginals are not downturned. The plastron is connected to the carapace by ligaments, and its buttresses are short. There is a well-developed bridge with large axillary and inguinal scutes. The plastron is well developed, but lacks a hinge and is shorter than the carapace. Its forelobe is angular; the hind lobe is shorter than the bridge and contains a wide anal notch. The humero-pectoral seam lies over the entoplastron. Plastral formula is: abd > fem > pect > gul > an > hum. Plastron and bridge are yellow with a large black blotch on each scute. With age these blotches often grow together so that the entire plastron may become black. The skull has an incomplete temporal arch, as most of the quadratojugal is absent. Squamosals are also absent. The postorbital is widely separated from the supratemporal. Parietals and postorbitals touch, and, in doing so, keep the frontal bone from entering the temporal emargination. The pterygoid is not in contact with the articular surface of the quadrate. The orbito-nasal foramen is large. The triturating surface of the upper jaw is broad, but its ridge is reduced to a slight swelling. The maxilla is prevented from touching the lower parietal process by the palatine. The upper jaw has a medial notch with a toothlike cusp on each side. The lower jaw is denticulated along its margin, and the snout is not overly protruding. The head is blackish to olive with some gray markings, and in younger individuals a series of yellow or cream colored stripes adorns it. One passes anteriorly from each side of the neck over and through portions of the tympanum and orbit to each nostril, a second passes posteriorly from the cusp on the upper jaw along the margin of the jaw onto the side of the neck, and a third begins on the lower jaw and extends posteriorly to the side of the neck. Sometimes several of the stripes join to form a large light-colored blotch behind the eye. In older turtles these stripes fade and may disappear entirely. Chin and throat are cream to white and may contain dark flecking. All toes are heavily webbed, and the fifth contains two phalanges. Forelimbs are covered with large scales, and the limbs are usually darker gray on the upper surfaces and lighter gray below. The tail is short and gray.

The karyotype is 2n = 52 (Bickham, 1981).

Males have concave plastra and thicker tails than do the females.

Williams (in Loveridge and Williams, 1957) reported *Hieremys* to be closely allied to *Geoemyda* (= *Heosemys)* on the basis of the absence of squamosal bones in the skull; the presence of posterolaterally shortened, hexagonal neurals; reduced plastral buttresses; transection of the entoplastron by the humero-pectoral seam; and only two phalanges in the fifth digit. However, McDowell (1964) thought it more closely related to *Kachuga* and other members of his *Batagur* complex than to any member of his *Geoemyda* complex. Its large orbito-nasal foramen; small posterior palatine foramen; triturating shelf on the vomer, which conceals the incisive foramen from below; long angular bone; the extension of the lower triturating surface to behind the mandibular symphysis; and the moundlike swelling along the inner border of the upper triturating surface are all more indicative of *Kachuga, Batagur,* and *Callagur* than of members of McDowell's *Geoemyda* complex.

Distribution: *Hieremys annandalei* occurs in central Thailand, Kampuchea, Vietnam, and northern Malaya.

Habitat: This large species lives in swamps, flooded fields, and rivers with slow current. Smith (1931) purchased many specimens which were caught at the mouth of the Chao Phraya River near Bangkok, so *Hieremys annandali* apparently can tolerate brackish conditions.

Natural History: The nests are excavated in December and January, and the usual clutch is four elongated (57–62 mm), hard-shelled eggs. Hatchlings have carapaces about 60 mm long.

Hieremys is herbivorous and will accept almost any fruit, green vegetable, or aquatic plant.

The common name "temple terrapin" comes from the native custom of placing these turtles in the pools and canals of Buddhist temples, especially those of the Tortoise Temple in Bangkok. The Buddhists believe that saving a life (no matter of what) gains favor for the saver in the hereafter. Consequently, they "save" the lives of the turtles by placing them in the temple waterways. Unfortunately, the obligation ends there and the turtles receive no additional care at the temple.

Kachuga Gray, 1855

This is the largest genus of Asiatic batugurine turtles, with seven species from Pakistan, India, Bangladesh, and Burma. All are semiaquatic, and some reach 50 cm in length. The elongated, oval carapace is high arched with a single, medial vertebral keel, which has posteriorly pointing projections in several species. Each neural is hexagonal in shape and short sided anteriorly. Vertebral 4 is much longer than broad, and

covers parts of four or five neurals. The plastron is rigid, lacking a hinge. Its buttresses are large; the axillary contacts the 1st rib while the inguinal extends upward to about the midpoint of the suture between the 5th and 6th costals. The entoplastron lies anterior to the humero-pectoral seam. On the skull, the temporal arch is usually complete, the quadratojugal touches both the jugal and postorbital. The inferior temporal emargination is deep; the maxilla is excluded from the squamosal. Posterior temporal emargination is also deep; the frontal bone is excluded from the posterior temporal emargination by the parietal and postorbital bones. The squamosal prevents contact between the postorbital and the supratemporal bones. The triturating surface of the maxilla is broad and contains a single median denticulated ridge; the edges of the jaws are also serrated. The maxilla is prevented from touching the inferior parietal process by the palatine. The orbito-nasal foramen is large, and the pterygoid does not extend to the articular surface of the quadrate. The toes are webbed.

Key to the species of *Kachuga*:

1a. 4th vertebral scute narrowly pointed anteriorly, only in slight contact with 3d vertebral; 4th vertebral covering portions of five neural bones 2
 b. 4th vertebral scute not pointed anteriorly, in broad contact with 3d vertebral; 4th vertebral covering portions of four neural bones 5
2a. 26 marginal scutes; posterior rim of carapace strongly serrated: **K. sylhetensis**
 b. 24 marginal scutes; posterior rim of carapace smooth or only slightly serrated 3
3a. Posterior border of 3d vertebral scute straight, not pointed: **K. smithi**
 b. Posterior border of 3d vertebral scute with a projecting spinelike point 4
4a. At least two black blotches on each plastral scute: **K. tecta**
 b. One black blotch on each plastral scute: **K. tentoria**
5a. Vertebral keel low and obscure; carapace with three conspicuous longitudinal stripes in males, but none in females: **K. trivittata,**
 b. Vertebral keel represented as backward projections on 2d and 3d vertebrals 6
6a. Posterior border of 2d vertebral scute pointed backward; a medial vertebral stripe and two poorly marked lateral stripes present on carapace of both sexes: **K. dhongoka**
 b. Posterior border of 2d vertebral scute straight, not pointed; carapacial stripes absent: **K. kachuga**

The diploid karyotype includes 52 chromosomes. Variation occurs in the number of macrochromosomes and microchromosomes, and in the position of the centromere (Stock, 1972; Killebrew, 1977a; De-Smet, 1978; Bickham and Carr, 1983).

Kachuga dhongoka (Gray, 1834, in Gray, 1830–1835)

Three-striped roofed turtle

Recognition: This is a large species of *Kachuga*, reaching a carapace length of 48 cm. Its elliptical carapace is widest behind the middle, depressed, and has a smooth posterior rim. A medial keel is present, but reduced to only a posterior projection on the 2d and 3d vertebrals in adults. Vertebral 1 is broader than long. The 2d is broader than long in juveniles but lengthens to become longer than wide in adults. It also becomes more narrowly pointed posteriorly with its posterior margin fitting into the emargination of the 3d vertebral. Vertebral 3 is broader than long and is shorter than either the 2d or 4th. As in other *Kachuga* species, vertebral 4 is longer than broad and the 5th broader than long. The carapace is olive brown with a dark-brown or black medial stripe along the keel and two indistinct dark lateral stripes. The plastron is long and narrow; both the anterior and posterior lobes are shorter than the broad bridge. There is a posterior plastral notch. The humero-pectoral seam is transverse, and joins the plastral midseam at a right angle. The plastral formula is: abd > fem > pect > hum > an >< gul. On the bridge, the inguinal scute is larger than the axillary. Plastron and bridge are yellow; a brownish-red blotch occurs on each scute in juveniles. The head is moderate in size with a projecting pointed snout and an upper jaw with a shallow notch bordered on each side by a toothlike projection.

Kachuga dhongoka

Lateral jaw rims are serrated. The skin on the back of the head is divided into scales. The head is olive to brown with a yellow stripe on each side of the head extending backward from the nostril over the orbit and tympanum to the neck. The jaws may also be lighter in color. Neck and limbs are olive or yellowish brown. Enlarged transverse scales are present on the limbs.

The males are much shorter than the females, not exceeding 26 cm (Smith, 1931), and have long, thick tails.

Distribution: *Kachuga dhongoka* occurs in Nepal, Bangladesh, and northeastern India in the Ganges and Brahmaputra drainages.

Habitat: This is apparently another turtle of large, deep rivers.

Natural History: Chaudhuri (1912) reported this species nests in sandbanks in March and the young hatch in May and June. A normal clutch comprises 30 to 35 long, oval (55 × 33 mm) white eggs.

Kachuga kachuga (Gray, 1831, in Gray, 1830–1835)

Painted roofed turtle

Recognition: The elliptical carapace (to 49 cm) is widest behind the middle, slightly depressed, and smooth to slightly serrated along the posterior rim. A medial keel is present with posterior projections on vertebrals 2–3; the projection on vertebral 3 is more prominent in juveniles and becomes lower with age. Vertebrals 1, 3, and 5 are broader than long, the 2d is about as long as broad, and the 4th is much longer than broad. The carapace varies from brown to olive in color. The plastron is long and narrow; both the anterior and posterior lobes are shorter than the broad bridge. A posterior anal notch is not always present. The humero-pectoral seam joins the plastral midseam at an obtuse angle. The plastral formula is: abd > fem > hum > pect > an > gul. On the bridge, the inguinal scute is much larger than the axillary. Both plastron and bridge are immaculate yellow. The head is moderate in size with a short, slightly upturned, projecting snout. Its upper jaw contains a shallow medial notch bordered on each side by a toothlike projection. Skin on the back of the head is subdivided into scales. The male's head is red dorsally and bluish gray on the sides; the female's head is olive blue. A pair of red or yellow spots occur on the throat, and the jaws may be brown. The neck is olive brown

Kachuga kachuga, male

in females, but contains a series of seven red or red-brown longitudinal stripes in the male. Red pigmentation in males is most brilliant during the breeding season, fading and disappearing afterward. Limbs are brown to olive with transverse scales on the anterior surface of the forelegs.

In addition to the sexual dichromatism, several other features distinguish adult males from adult females. Females are larger and have short tails. The shorter males have long, thick tails.

Distribution: *Kachuga kachuga* ranges from central Nepal southward through Bangladesh, adjacent northeastern India, and probably northwestern Burma. It resides primarily in the watershed of the Ganges River.

Habitat: Adults are inhabitants of deep-flowing rivers.

Natural History: Chaudhuri (1912) reports that nesting occurs in March on sandbanks with the young emerging in May or June. The eggs are long and oval.

Kachuga smithi (Gray, 1863d)

Brown roofed turtle

Recognition: The elliptical carapace (to 23 cm) is widest behind the middle, arched, and with a smooth or only slightly serrated posterior rim. A medial keel is present, but is low and rather blunt with only slightly raised areas on the posterior portions of the vertebrals. Vertebrals 2 and 5 are broader than long; the 1st, 3d, and 4th are longer than broad. Also, vertebral 4 is tapered and pointed anteriorly. The carapace is brown to tan with a dark medial stripe. The plastron is long and narrow; the anterior lobe is much shorter than the broad bridge, but the hind lobe is only slightly shorter and is notched posteriorly. The humero-pectoral seam joins the plastral midseam at a slight angle. The plastral formula is: abd > fem > hum > pect > an > gul. On the bridge the inguinal scute is larger than the axillary. Plastron and bridge are yellow with a large black blotch on each scute. The head is moderate in size with a projecting short, pointed snout. Its upper jaw is not medially notched. Skin on the back of the head is divided into large scales. The head is yellowish gray or pinkish gray with a reddish-brown spot on the temple and a dark snout. The neck is gray with yellow stripes. The gray forelimbs have enlarged transverse scales.

Males have longer, thicker tails with the vent beyond the rim of the carapace; females have short tails with the vent under the carapace.

Distribution: *Kachuga smithi* lives in the Indus River system of Pakistan and northwestern India. It also occurs rarely in the Ganges River in Bangladesh.

Geographic Variation: Two subspecies have been described (Moll, 1987). The typical race *Kachuga smithi smithi* (Gray, 1863d) occurs in the Indus, Ganges, and Brahmaputra watersheds of Pakistan, India, and Bangladesh. It has a plastral pattern consisting of large dark blotches, and dark pigment on the sides of the head, outer surface of the limbs and feet, and penis. *Kachuga smithi pallidipes* Moll, 1987 is known only from the northern tributaries of the Ganges River in India and Nepal. It lacks a dark plastral pattern, and the pigment on the head, limbs, feet, and penis is reduced.

Habitat: According to Minton (1966), *Kachuga smithi* is common in river channels and larger canals, and is occasionally found in lakes and ponds connected

Kachuga smithi

to rivers. It frequents muddy water with some current where there are logs, bridge abutments, and other protruding objects. It commonly basks in aggregations.

Natural History: Nests are dug in sandbanks, probably in the fall months. Eggs are white and elongated (43–45 × 22–24 mm); a normal clutch is five to eight eggs. Four hatchlings measured by Ewert (1979) had an average carapace length of 39.2 mm.

Kachuga smithi is omnivorous; in captivity it will eat fresh fish, frogs, ground beef, insects, worms, and a variety of fruits and lettuce.

While they love to bask, these turtles are still very wary and slip quickly into the water at the least disturbance. In Pakistan they are quiescent from early December to early March, but some bask during the warmest hours (Minton, 1966).

Kachuga sylhetensis (Jerdon, 1870)

Assam roofed turtle

Recognition: This is the smallest *Kachuga*, attaining a carapace length of only 20 cm. The carapace is oval, arched, and strongly serrated posteriorly. The medial keel is present on vertebrals 3–5, with a backward-pointing projection on the 3d. Vertebrals 1, 2, and 5 are broader than long; 3 and 4 are much longer than broad. Also, the anterior border of the 4th is elongated into a point. There are 26 marginal scutes instead of the normal 24; the extra pair is formed by subdivision of the two supracaudal marginals. The carapace is olive brown, but paler along the vertebral keel. The elongated, oval plastron is notched posteriorly. Both its anterior and posterior lobes are shorter than the bridge. The humero-pectoral seam joins the plastral midseam at an obtuse angle. The plastral formula is: abd >< fem > pect > hum > an > gul. Inguinal and axillary scutes are large, with the inguinal slightly larger. Plastron and bridge vary from yellow, with a large dark brown or black spot on each scute, to totally brown. The head is moderate in size with a projecting snout and hooked upper jaw. Skin on the back of the head is weakly subdivided into scales. The head is brown with a transverse yellow stripe on the back of the head and a yellow stripe along the lower jaw. The neck is brown with pale yellow to cream stripes. The brown limbs have enlarged transverse scales.

Males have longer, thicker tails than females.

Distribution: *Kachuga sylhetensis* occurs only in eastern Bangladesh and adjacent Assam.

Habitat: This turtle apparently lives in flowing foot-hill streams.

Comment: Although this species in some ways resembles *Kachuga tecta* (Theobald, 1876, thought it synonymous with *K. tecta*), it has several features that set it apart from the other *Kachuga* species; study is needed to determine its proper taxonomic position.

Kachuga tecta (Gray, 1831b)

Indian roofed turtle

Plate 11

Recognition: The arched, elliptical carapace (to 23 cm) is widest behind the middle and unserrated posteriorly. The medial keel is prominent with a strong spinelike posterior projection on the 3d vertebral. Vertebrals 2 and 5 are broader than long, while 1, 3, and 4 are longer than broad. Vertebral 3 is posteriorly pointed while the 4th is anteriorly pointed, making the seam between these two scutes very short. The carapace is brown, sometimes yellow or orange bordered, with a red to orange medial stripe. The plastron is long and narrow; the forelobe is much shorter than the broad bridge, and the hind lobe is slightly shorter than the bridge and contains a posterior anal notch. The humero-pectoral seam joins the plastral seam at an obtuse angle. The plastral formula is: abd > fem > an >< hum > pect >< gul. On the bridge, the inguinal and axillary scutes are nearly equal in length, or the inguinal is slightly larger. Plastron and bridge are yellow with at least two black elongated blotches on each scute, except the gulars which usually have only a single blotch. The head is moderate in size with a projecting, short, pointed snout. Its upper jaw is not medially notched. Skin on the back of the head is divided into large scales. Dorsally, the head is black with a large crescent-shaped, orange to yellowish-red blotch on each temple (these may unite posteriorly to form a V-shaped mark). The jaws are yellow, and the neck is black with numerous yellow stripes. Limbs are olive to gray, and spotted and bordered with yellow. They have large transverse scales.

Males have long, thick tails with the vent beyond the carapacial rim. Females have short tails with the vent under the carapace. Females grow larger than males.

Distribution: *Kachuga tecta* inhabits the Ganges, Brahmaputra, and Indus drainages in Pakistan, northern and peninsular India, and Bangladesh.

Habitat: According to Minton (1966), this is a quiet-water turtle. It occurs in quiet streams, canals, ox-bows, ponds, and manmade water tanks. A soft bottom and abundant aquatic vegetation are preferred conditions.

Natural History: Ahmad (1955) reported that a female dissected in March contained nine fully developed oval eggs, 35 mm in length.

This species is predominantly herbivorous. Minton (1966) thought *Kachuga tecta* to be less alert and not as good a swimmer as *K. smithi*.

Kachuga trivittata (Duméril and Bibron, 1835)

Burmese roofed turtle

Recognition: This is the largest *Kachuga*, with a record carapace length of 58 cm. Its elliptical carapace is widest behind the middle and is somewhat arched. A medial keel is present with posterior projections on the first three vertebrals. This keel becomes lower and more blunt with age. Vertebrals 1–3 are as broad as or slightly broader than long, 4 is longer than broad, and 5 is distinctly broader than long. The posterior marginal rim is smooth in adults but somewhat serrated in juveniles. This species shows sexual dichromatism in respect to the carapacial pattern: males have brown to olive carapaces with three distinct black stripes which may be united at the ends; females have uniformly brown carapaces. The plastron is long and narrow; both the anterior and posterior lobes are shorter than the broad bridge. A posterior anal notch is present. The humero-pectoral seam joins the plastral mid-seam at an obtuse angle. The plastral formula is: abd > fem > < pect > an > gul. On the bridge, the inguinal scute is much larger than the axillary. Plastron and bridge are yellow to orange. The head is moderate in size with a slightly projecting snout and a medially notched upper jaw. Laterally, the jaws are slightly serrated. Skin on the back of the head is broken into scales. Head and neck are brown to olive; a large black streak is present on top of the head, and the jaws are lighter in color. Other soft parts are yellowish brown. Transverse platelike scales occur on the anterior surface of the forelegs and at the heel on the hind feet.

Females achieve a larger size than do males, which only reach 46 cm. Males have long, thick tails with the vent beyond the carapacial rim; the tail of females is shorter with the vent under the carapace.

Distribution: *Kachuga trivittata* is found in the Irrawaddy and Salween river systems of Burma.

Geographic Variation: None currently recognized, but Anderson (1878) designated those smaller males lacking the black stripes as a separate species, *Kachuga iravadica*. However, both Boulenger (1889) and Smith (1931) accepted Theobald's (1876) explanation that the black stripes, being a sexual character, do not appear until sexual maturity is attained. Further study of this is needed.

Habitat: Adults are generally inhabitants of large, deep rivers.

Natural History: Smith (1931) reported that nesting occurs in December and January in sandbanks above the tidal limits. A typical clutch includes about 25 elongated (70–75 × 40–42 mm) whitish eggs.

Kachuga trivittata is probably herbivorous.

Kachuga tentoria (Gray, 1834, in Gray, 1830–1835)

Indian tent turtle

Recognition: The arched elliptical carapace (to 26.5 cm) is widest behind the middle and unserrated posteriorly. The medial keel is prominent with a strong spikelike posterior projection on the 3d vertebral scute. Vertebral 2 may be smaller than 1 or 3. Vertebrals 1, 2, and 5 are broader than long; 3 and 4 are longer than broad. Vertebral 3 is posteriorly pointed and 4 is anteriorly pointed, so the seam between these scutes is very short. The carapace is brown to olive and generally paler than that of *Kachuga tecta*, with a narrow, red to orange medial stripe. The plastron is long and narrow; the forelobe is much shorter than the broad bridge; the hind lobe is only slightly shorter

Kachuga tentoria

than the bridge and contains a posterior anal notch. The humero-pectoral seam joins the plastral midseam at an obtuse angle. The plastral formula is: abd > fem > pect >< hum > an > gul. On the bridge, the inguinal is only slightly larger than the axillary. Plastron and bridge are yellow with a single large black blotch on each scute; there is a wide yellow midseam. The head is moderate in size with a projecting, short, pointed snout and an upper jaw that is not medially notched. Skin on the back of the head is divided into large scales. The head is olive to gray with a conspicuous red patch behind the tympanum, but the large crescentlike orange or red temporal blotch found in *K. tecta* is absent. The jaws are yellow or pink. The neck is olive to gray with few, if any, yellow stripes. The uniformly gray limbs lack yellow spots and have large transverse scales.

Males have long, thick tails with the vent beyond the carapacial rim. Females have shorter tails with the vent under the carapace. Males are probably smaller than females.

Kachuga tentoria is very similar to *K. tecta*, and many consider it a subspecies. It is obviously closely related to both *K. tecta* and *K. smithi*, and an intensive study of the relationships of the three is needed.

Distribution: *Kachuga tentoria* is found predominantly in peninsular India, but has also been reported from Bangladesh (McDowell, 1964; Pritchard, 1979).

Geographic Variation: Two subspecies are recognized. *Kachuga tentoria tentoria* (Gray, 1834) occurs in the Mahanadi, Godavari, and Kistna drainages of peninsular India, and also in Bangladesh. It has no bright-reddish band between the pleural and marginal scutes of the carapace, and has, at best, poorly marked stripes on its neck. *K. t. circumdata* Mertens, 1969a was described from the vicinity of Calcutta. It has a reddish band circling the carapace along the seam separating the pleurals from the marginals, and its neck stripes are more clearly marked.

Habitat: The habitat requirements have not been reported, but are probably similar to those of *Kachuga tecta*.

Natural History: Good behavioral and ecological studies are needed. Vijaya (1982b) reported clutches laid at the Madras Snake Park included four to eight eggs, 40–45 × 26–29 mm. Hatchlings were 27–33 mm.

Malayemys Lindholm, 1931

The only species of this genus is the Malayan snail-eating turtle.

Malayemys subtrijuga (Schlegel and Müller, 1844)
Malayan snail-eating turtle

Recognition: This turtle may reach 21 cm in carapace length, but most individuals are much smaller. Its oval carapace is moderately arched, unserrated but notched posteriorly, and contains three discontinuous keels consisting of knobs situated at the posterior border of the scutes on which they lie. The medial keel extends along all five vertebrals, but the lateral keels rarely reach the 4th pleural. Vertebrals are usually broader than long. The underlying neurals are not elongated, and those in the middle of the series are shortest posteriorly. Vertebral 4 covers parts of four neurals. In color, the carapace ranges from light to darker brown, sometimes mahogany, with a yellow or cream-colored border. The knobs of the keels and the marginal seams are darker than the rest of the carapace. The hingeless plastron is narrower than the carapace and notched posteriorly. Its hind lobe is slightly shorter than the bridge. Buttresses are strong; the axillary is attached to the lateroventral surface of the 1st costal and the inguinal extends about a third of the way up along the suture between the 5th and 6th costals. The entoplastron lies anterior to the humero-pectoral seam. The plastral formula is: abd > fem > pect >< an > hum >< gul. The yellow or cream-colored plastron has a pattern consisting of a large

Malayemys subtrijuga

dark-brown or black blotch on each scute. Two dark blotches occur on the bridge. The head is relatively large and the snout projects anteriorly. Posterior emargination of the skull is deep but the ventral emargination is very shallow. The posterior emargination separates the supratemporal from the postorbital and extends nearly to the frontal bone. The squamosal is in broad contact with the maxilla. The orbito-nasal foramen is small. The upper jaw is medially notched; its triturating surface is flat and broad with only a very slight medial ridge (an adaptation for crushing snail shells). In the lower jaw, the angular bone is short and the prearticular extends more anteriorly; the coronoid process is very large. The head is black and is adorned with several yellow or cream-colored stripes. The first begins on the snout above the nostrils and extends posteriorly on each side above the orbits to the neck. A second stripe begins on each side of the snout just behind the nostril and curves downward and then backward passing below the orbit to the neck. Two narrow stripes pass from the nostrils to the medial notch of the upper jaw; when the mouth is closed, these almost touch two similar stripes on the lower jaw. Another narrow stripe begins behind the orbit and crosses the tympanum. Other discontinuous stripes may occur on the lower jaw. Dorsally, the head is covered with a single scale, while the back of the head contains numerous scales. Limbs are gray to black with a narrow yellow outer border.

Killebrew (1977a) reported the karyotype to be 2n = 52 (8 metacentric, 3 submetacentric, and 2 telocentric pairs of macrochromosomes and 13 pairs of microchromosomes), but more recently Bickham (1981) found the diploid number to be only 50 chromosomes.

The plastra of both sexes are flat, so sex determination must be based on the longer, thicker tail of the male.

Distribution: *Malayemys subtrijuga* ranges from southern Vietnam west through Kampuchea and Thailand (and probably southern Burma), and south through Malaya to Java.

Habitat: This turtle is an inhabitant of slow-moving water bodies with soft bottoms and aquatic vegetation. It has been taken from ponds, canals, small streams, marshes, and rice paddies.

Natural History: The elongated white eggs are 40–45 × 20–25 mm (Smith, 1931), and Ewert (1979) reported that 11 hatchlings averaged 35.3 mm in carapace length. *Malayemys subtrijuga* is basically a snail eater, although it also takes worms, aquatic insects, crustaceans, and small fish (Wirot, 1979).

Mauremys Gray, 1869b

Mauremys is a genus of semiaquatic batagurines that occur from southern Europe and northern Africa through the Middle East, in southeastern Asia, and on the Japanese islands. The carapace is low arched and medially keeled. Two lateral keels are also present in young individuals, but are usually obliterated with age. The hexagonal neurals are shortest either anteriorly or posteriorly. The plastron lacks a hinge, and is firmly joined to the pleural bones by moderate to strongly developed buttresses. In the skull, the squamosal always touches the postorbital, but may not touch the jugal. The orbito-nasal foramen is bordered by the basisphenoid anterior to the primitive coroticum foramen. The ethmoid fissure is narrow and shaped like a keyhole. The pterygoid separates the ventral portion of the jugal from the anterior edge of the inferior process of the parietal. This inferior process touches the palatine in the species *caspica*, *leprosa*, and *nigricans*, but is narrowly separated in *japonica*.

The karyotype consists of 52 chromosomes: 28 macrochromosomes, 24 microchromosomes (Nakamura, 1935; Sasaki and Itoh, 1967; Stock, 1972; Bickham, 1975; Killebrew, 1977a).

Key to the species of *Mauremys*:

1a. Sides of head with dark spots, but no longitudinal light stripes: **M. japonica**
 b. Sides of head with at least one longitudinal light stripe ... 2
2a. Sides of head with a broad yellow or orange stripe with a dark dorsal border extending backward from orbit over tympanum to neck (this stripe may be interrupted, and a second stripe may extend downward and backward from lower edge of orbit or corner of mouth to below tympanum): **M. nigricans**
 b. More than two longitudinal stripes occur on side of head ... 3
3a. A round yellow or orange spot lies between orbit and tympanum: **M. leprosa**
 b. Narrow stripes may lie between orbit and tympanum but no round yellow or orange spot: **M. caspica**

Mauremys japonica (Temminck and Schlegel, 1835)

Japanese turtle

Recognition: This is a Japanese semiaquatic turtle with a posteriorly serrated brown carapace. Reaching a length of 18.2 cm, the carapace has a single low medial keel, is heavily serrated posteriorly, and also has a small posterior notch. Vertebrals are generally broader than long with the 1st broadest of all, and each carapacial scute bears growth annuli and well-marked, elevated radiations which diverge anteriorly from the block representing the original hatchling scute. Lateral marginals may be slightly upturned. Older individuals may have almost totally dark-brown carapaces, but those of younger turtles are olive to brown with some yellow along the vertebral keel and yellow and dark mottlings along the seam separating the pleurals and marginals. The brown to black plastron is flat, slightly upturned anteriorly, and has a wide posterior notch. The plastral formula is: abd > fem >< pect > an > gul > hum. The brown bridge is about as wide as the length of the posterior plastral lobe. The head is rather small and the upper jaw is neither hooked nor notched. The triturating surface of the upper jaw is narrow and lacks a ridge. The snout is only slightly projecting. Head skin is light brown, with dark spots on the jaws, chin, and sides. The neck is brown with a series of stripes on raised scales. Limbs and tail are dark brown with some light yellow on the outer border of the limbs and dorsal surface of the tail.

The plastron of the male is only slightly concave, but the lateral margins of the femoral scutes are bent downward. The tail of the male is thicker at the base than that of the female, and has the vent beyond the carapacial margin.

Distribution: *Mauremys japonica* is found on the islands of Honshu, Kyoshu, and Shikoku.

Habitat: Fresh waters with slow currents and soft bottoms (ponds, marshes, swamps, canals, and streams) are preferred.

Natural History: Sexual maturity is reached in three to five years. The mating period extends from April to June and the eggs are laid between mid-May and late June. Two or three clutches of five to eight eggs are laid each year, with the number of eggs decreasing with each subsequent clutch. Normally 10 to 15 days pass between sequential nestings (Fukada, 1965). The white oval egg is brittle shelled and about 25 mm in diameter. Fukada (1965) reported the incubation period to last about 70 days. Hatchlings have carapace lengths of 25 to 33 mm. Their medial keel is prominent and the carapace is very serrate posteriorly. Two low lateral keels are also present on the pleurals, and their carapacial color is brighter than that of adults. The hatchling plastron has either a large dark central blotch or a wide, dark seam-following pattern.

Mauremys japonica is omnivorous, but seems to prefer animal foods. Fukada (1965) reported that it eats fish, frogs, earthworms, insects, and weeds.

Japanese turtles are fond of basking and sometimes wander about on land. They are hearty captives.

Mauremys nigricans (Gray, 1834a)

Asian yellow pond turtle

Recognition: This is a medium-sized (to 19.5 cm) semiaquatic batagurine turtle with a broad yellow stripe extending backward from the orbit over the tympanum to the neck. Its oval carapace is slightly wider behind the center, contains a low medial keel and two rather poorly developed lateral keels (which are more prominent in juveniles), slightly upturned lateral marginals, and slightly flared and serrated posterior marginals. Vertebral 1 is wider anteriorly than posteriorly. The carapace ranges in color from grayish brown to brown and the seams are usually dark. The plastron is large, slightly upturned anteriorly, and notched posteriorly. The bridge is about as wide as the length of the plastral hind lobe. The plastral formula is: fem ≥ abd > pect > hum > an > gul. The plastron is yellow to orange with a large black blotch toward the outside of each scute; the bridge is similarly colored and has two large dark blotches. Head and neck are dorsally gray to olive, and always contain the striped pattern described above. Also, this stripe has its dorsal border darkly outlined. A second yellow stripe may extend diagonally downward and back-

Mauremys nigricans

ward from the lower edge of the orbit or the corner of the mouth to below the tympanum. There is a dorsomedial yellow stripe on the neck. Chin and lower neck are usually yellow, but some individuals may have dark mottles. The snout is conical and slightly projecting, and the upper jaw is medially notched. The triturating surface of the upper jaw is narrow and lacks a ridge or serrations. Limbs and tail are gray to olive above, but yellow ventrally.

The males have slightly concave plastra and longer, thicker tails with the vent beyond the margin of the carapace.

Distribution: *Mauremys nigricans* lives in northern Vietnam, southern China, and on Hainan and Taiwan.

Geographic Variation: *Mauremys nigricans* is quite variable, and a thorough study of its geographic variation is needed, but no subspecies are recognized. Two color phases exist. The unicarinate *"nigricans"* form is melanistic with a dark-brown, gray, or black carapace, the dark plastral blotches fused so yellow pigment is found only along the rim and midseam, and gray to grayish-yellow skin. The tricarinate *"mutica"* form (from *Emys mutica* Cantor, 1842) has a yellow-brown to brown carapace with dark seams, a yellow plastron with a dark blotch on each scute, and yellow skin. Both have an unbroken narrow yellow postorbital stripe that passes dorsal to the tympanum and continues onto the side of the neck, relatively short intergular and interanal seam lengths, and a medially notched upper jaw; many also have a dorsomedial neck stripe.

Habitat: *Mauremys nigricans* inhabits low-altitude water bodies with slow currents such as ponds, marshes, swamps, and creeks.

Natural History: Ewert (1979) reported that the brittle-shelled eggs are 38 × 21 mm, and hatch in an average of 94.2 days when incubated at 25–25.5°C and in 64.7 days at 29.5–30°C. Hatchlings had average carapace and plastron lengths of 32.4 mm and 27.9 mm, respectively.

Mauremys nigricans is fond of basking, and captives do well on a fish diet.

Mauremys leprosa (Schweigger, 1812)
Mediterranean turtle

Recognition: This is a brown, medium-sized (to 18 cm), semiaquatic turtle of the western Mediterranean region. Its low, oval carapace is slightly keeled medially (much more strongly in juveniles than in adults), and has a smooth, unserrated marginal border. Cervi-

Mauremys leprosa

cal and vertebral scutes are usually broader than long. A pair of low lateral keels are present on the pleural scutes of hatchlings but these become lower with age and, at best, are barely visible in adults. The carapace is tan to olive with large, black-bordered, yellow to orange blotches on each scute. A broken medial stripe is usually present, but fades with age, so that older adults may be uniformly tan or olive. The well-developed plastron is notched and somewhat pointed posteriorly. Its formula is: abd > fem > pect > gul > an > hum in males; in females, the pectoral and femoral scutes are reversed. The yellow plastron has a large, central, dark blotch which may contain a yellow median streak. This pattern also fades with age. The yellow bridge is well developed and has two dark blotches which may join medially. The head is not greatly enlarged; it is olive to grayish tan with a series of yellow stripes which begin on the neck, pass dorsal to the tympanum, and extend to the orbit. Often an incomplete yellow ring encircles the tympanum, and a yellow line extends from the neck to the corner of the mouth and continues along the border of the upper jaw to its tip. A round yellow or orange spot lies between the tympanum and orbit; this may touch the tympanic ring. Neck, limbs, and tail are olive with yellow stripes or vermiculations.

The carapace often has infected areas around the scute seams resulting from algae and bacteria invading the soft developing tissues beneath the scutes; this led Schweigger (1812) to apply the trivial name *leprosa* to this turtle.

Females grow larger than males, have flat plastra, wider heads, and shorter tails with the vent beneath the carapace. Males are smaller, have concave plastra, narrower heads, and longer, thicker tails with the vent posterior to the carapacial margin.

Distribution: *Mauremys leprosa* occupies the western end of the Mediterranean region from Spain and Portugal to western Libya, Tunisia, Algeria, Morocco, around northwestern Africa to Senegal, Dahomey, and Niger.

Geographic Variation: No subspecies have been named, but Busack and Ernst (1980) noted that juveniles (to 11-cm plastron length) from Tunisia and Libya usually have the dark pigment reduced on the gulars, humerals, and anterior half of the pectoral scutes, while those from Morocco, Algeria, and Spain have extensive dark pigment on these scutes. Also, Mosauer (1934) reported that Tunisian specimens from Gafsa had uniformly brown carapaces, whereas those from Gabes had a distinct pattern of sharply outlined red spots on their pleurals and some red on each marginal. These red marks faded with age.

Habitat: This turtle lives in almost every type of permanent freshwater habitat within its range: rivers, streams, brooks, ponds, sloughs, marshes, lakes, and the pools of oases. It has been collected in temporary waters such as drainage ditches and latrines, and in brackish waters.

Natural History: Breeding takes place in early spring, March and April, and the usual nesting period encompasses May and June. A wild breeding pair was observed by Ernst on 8–9 March 1976. The 11-cm male trailed the female of the same length, often biting at her flanks and tail. When she was caught, he crawled onto her back, hooked the toes of all four feet under her shell, and held fast. The male then bit the female's gular skin in a bulldoglike grip while manipulating his tail under her to achieve copulation. During this period the totally submerged female was relatively passive with her head withdrawn, although she did crawl about some with the male on top. Mating may also occur on land. Nests are shallow holes, usually in sandy areas or between tree roots. Six to nine elongated (35 m × 21 mm) eggs form a clutch, and the incubation period ranges from 25 days to around 65–75 days, depending on latitude and weather conditions. Hatchlings are about 32 mm in carapace length and are brighter colored than adults, with a medial and two lateral carapacial keels. Hatchlings are extremely shy and remain hidden among bottom debris.

Adults are fond of basking, but are extremely alert, diving into the water at the least disturbance. When handled they may bite, release bladder water, or emit a malodorous secretion from their inguinal glands.

In the wild their diet consists of fish, adult frogs and tadpoles, insects, aquatic invertebrates, algae and aquatic plants, and carrion. Large adults may be cannibalistic. Food can be swallowed both in or out of the water.

The northern populations may hibernate in winter, while those in more southernly, arid situations may be forced to estivate in summer.

For a summary of the biology, see Loveridge and Williams (1957).

Mauremys caspica (Gmelin, 1774)

Caspian turtle

Recognition: *Mauremys caspica* is a tan to blackish, medium-sized (to 23.5), semiaquatic turtle of the eastern Mediterranean region. Its low, oval carapace has a slight medial keel (better developed in juveniles) and a smooth unserrated marginal border, which is slightly upturned and tapered above the tail. A pair of low lateral keels are present on the pleural scutes of hatchlings but these become lower with age and disappear completely in adults. The carapace is tan to olive or black with yellow to cream-colored reticulations patterning the scutes, and some individuals have yellow vertebral stripes. These light lines fade with age but the pleural seam borders become darker. The well-developed plastron is notched posteriorly. The plastral formulae are given in the subspecies descriptions in the section on geographic variation. The plastron is either yellow with variable reddish to dark-brown blotches, or dark brown or black with a yellow blotch along the lateral scute borders. The bridge is either yellow with dark seam borders and dark spots on the corresponding marginals, or almost totally black with a few small yellow marks. The head is not enlarged, and is olive to dark brown with yellow or pale cream-colored stripes. Some stripes extend anteriorly from the neck onto the head. One of these on each side passes above the eye and onto the snout where it meets the stripe from the other side. Several others extend across the tympanum to contact the posterior rim of the orbit, and two additional stripes continue across the snout and pass ventral to the orbit. Neck, limbs, and tail are tan gray to olive or black with yellow, cream, or gray stripes or reticulations.

Females are generally larger than males, have flat plastra and shorter tails with the vent under the rim of the carapace. The smaller males have concave plastra and longer, thicker tails with the vent beyond the rim of the carapace.

Distribution: *Mauremys caspica* ranges from southwestern USSR and central Iran to Saudi Arabia and Israel, northward through Turkey to Bulgaria, and through Cyprus, Crete, and the Ionian Peninsula to Yugoslavia.

Geographic Variation: Two subspecies are recognized. The typical, *Mauremys caspica caspica* (Gmelin, 1774), occurs in Iran to the Zagros Mountains of the central plateau, Iraq, Saudi Arabia, eastern and central Turkey, and the USSR from the eastern Trans-caucasus west to Tbilisi but not farther than Suram Pass. It has wider reticulations on its carapace, a mostly yellow plastron with a dark blotch on each scute, and a yellow bridge with only dark lines or spots. Its plastral formulae are fem > abd > pect > gul > hum > an for males, and abd > fem > pect > gul > hum > an for females. *M. c. rivulata* (Valenciennes, 1833) ranges throughout southeastern Europe (Yugoslavia to Greece, the Ionian Islands, Crete, and Cyprus), Bulgaria, western Turkey, coastal Syria, Lebanon, and Israel. It has narrow or fine reticulations on its carapace, and a totally black plastron and bridge. Its plastral formula is: abd > fem > pect > gul > an > hum in both sexes.

Formerly, *Mauremys leprosa* was considered a subspecies of *M. caspica*, but studies of the electrophoretic properties of its proteins by Merkle (1975), and of its morphology by Busack and Ernst (1980) have shown it to be a separate species.

Habitat: *Mauremys caspica* occurs in large numbers in almost any permanent freshwater body within its range. It also lives in irrigation canals and is quite tolerant of brackish situations. Reed (1957) reported that the turtles at one Iraq site lacked the ability to swim. Instead, they would crawl out of the water periodically to breathe and then slide back in again. A captive from there could not be induced to swim. Reed thought this behavior to be an adaptation to the extreme variability in the supply of surface water in the area.

Natural History: Breeding usually takes place in early spring, but may also occur in the fall (Anderson, 1979). The courtship behavior has not been described, but must be similar to that of captivity. Nesting occurs in June and July. A typical clutch is four to six, elongated (20–30 mm × 35–40 mm), brittle-shelled, white eggs. Hatchlings have round carapaces about 33 mm in length, and are brighter colored than the adults.

Mauremys caspica may occur in large populations in certain areas, especially in permanent water bodies. In temporary waters it is forced to estivate in the mud in summer, and the more northern populations hibernate during winter. It is very fond of basking, but disappears at the least disturbance. Anderson (1979) reported that many are killed each year by humans who obtain their eggs to use in treating ubiquitous eye ailments. Storks and vultures also take a heavy toll of juveniles and adults, respectively.

M. caspica is carnivorous, feeding on small invertebrates, aquatic insects, amphibians, and carrion.

Melanochelys Gray, 1869a

Melanochelys is a genus of dark-colored terrestrial and semiaquatic turtles restricted to Assam, Nepal, Bangladesh, India, Sri Lanka, and Burma. The elongated carapace is moderately to high domed, only slightly serrated (juveniles) or unserrated (adults) posteriorly, and bears three low keels. Neurals are hexagonal and usually shorter posteriorly. The plastron is hingeless. The skull is short and the temporal arch thin and, in some, incomplete. Ventrally, the jugal is narrow and separated from the pterygoid by the maxilla, and the squamosal is loosely attached and often lost during the preparation of the skull. The quadrate does not enclose the stapes and the otic notch is open. The quadrate is well separated from the jugal. The ethmoid fissure is narrow and shaped like a keyhole. The frontals form part of the orbital rim, and the parietal touches the postorbital, but not the jugal. The upper jaw is medially notched, and its narrow triturating surface is ridgeless. The toes are at least partially webbed.

DeSmet (1978) reported that the 2n chromosome number is 50 (19 pairs of macrochromosomes and 6 pairs of microchromosomes), but Carr and Bickham (1986) found 52 chromosomes.

Key to the species of *Melanochelys*:

1a. All toes fully webbed to claws, or nearly so; carapace and plastron dark brown or black: **M. trijuga**
b. Foretoes only half webbed, hind toes barely webbed, if at all; carapace reddish brown; plastron yellow: **M. tricarinata**

Melanochelys tricarinata (Blyth, 1856)

Tricarinate hill turtle

Plate 11

Recognition: This poorly known species has a carapace that is elongated (to 16.3 cm) and rather high domed with three low keels. Vertebrals are broader than long. The carapace is usually reddish brown (described as plum colored by Smith, 1931) to black, and the keels are yellow to brown. The plastron is long and notched posteriorly, and the hind lobe is longer than the bridge. The plastral formula is: abd

> pect > an > gul > fem >< hum. Bridge and plastron are yellowish to orange. Head, limbs, and tail are reddish brown to black. A narrow red, sometimes yellow or orange, stripe is present on each side of the head extending from the nostrils over the orbit and tympanum to the neck. A second similarly colored stripe passes backward from the corner of the mouth. Limbs may contain some yellow spotting, and the forelimbs have large square or pointed scales on their anterior surface. There may be enlarged scales on the heels of the clublike hind feet. The claws are long with the exception of the outermost, which is very small. Foretoes are only half webbed, hind toes barely, if at all, webbed.

The male plastron is slightly concave, and the tail longer and thicker with the vent beyond the carapace. Smith (1931) reported that older males may lose their head stripes. In older females, the hypoplastron is connected to the carapace only by ligaments, allowing the hind lobe of the plastron to move slightly away from the carapace, facilitating the laying of eggs (Moll, 1985).

Distribution: *Melanochelys tricarinata* occurs only in Assam and Bangladesh.

Habitat: *Melanochelys tricarinata* is apparently fully terrestrial, preferring upland woodlands.

Natural History: Theobald (1876) reported a female laid three elongated (44.4 × 25.4 mm) eggs.

Melanochelys trijuga (Schweigger, 1812)

Indian black turtle

Recognition: The carapace is elongated (to 38.3 cm), slightly depressed, and has three keels (a medial keel along the vertebrals and two lateral keels along the dorsal portions of the first three pleurals). The posterior marginals are not serrated (or only slightly so in juveniles) and the lateral marginals are slightly upturned. In adults, the vertebral scutes are variable in width and length. Usually the anterior four are as long as or longer than broad, but the posteriormost is broader than long. Carapacial ground color varies from reddish brown to dark brown or black; often the keels are yellow. The plastron is elongated and well developed. Its anals are notched posteriorly, and the plastral formula is: abd >< pect > fem >< an > gul >< hum. The bridge is about as long as the posterior plastral lobe, and its inguinal scute is short. The plastron is either uniformly dark brown or black, or it may

Melanochelys trijuga coronata

have a yellow border. The head is of moderate size with a relatively short snout and medially notched upper jaw. The head varies in color from brown to black and may contain orange or yellow spots or reticulations, or a large yellow or cream-colored blotch at the temporal region. Limbs and tail are gray to dark brown or black. The anterior surface of the forelegs is covered with large scales. All toes are fully webbed to the claws, or nearly so.

The tail of the male is longer and thicker than that of the female, and he possesses a concave plastron.

Distribution: *Melanochelys trijuga* ranges through most of peninsular India, northern Bangladesh, central Burma, Sri Lanka, the Maldives, and the Chagos Islands. It may also occur in Nepal.

Geographic Variation: Six subspecies have been named, but most are poorly described, and this species is in need of taxonomic revision. *Melanochelys trijuga trijuga* (Schweigger, 1812) transects central peninsular India, extending from just north of Bombay southward to Coorg in the west, to about Bezwada southward, and to Cuddalore in the east. *M. t. thermalis* (Lesson, 1830) occurs on the southeastern coast of peninsular India, the Maldive Islands, and Sri Lanka. *M. t. edeniana* Theobald, 1876 is a Burmese form occurring in Arakan, the Karenni hills, and Moulmein. *M. t. coronata* (Anderson, 1879) occurs on the southwestern coast of peninsular India from Calicut southward through Travancore. *M. t. indopeninsularis* (Annandale, 1913) is found in northern Bangladesh and possibly Nepal (this subspecies differs only in size from *M. t. edeniana*). The last subspecies, *M. t. parkeri* Deraniyagala, 1939 is restricted to lowlands in Sri Lanka. These six races are separated primarily on the

Table 4. Subspecific variation in *Melanochelys trijuga*

Character	*trijuga*	*coronata*	*thermalis*	*parkeri*	*indopeninsularis*	*edeniana*
Carapace						
length	22.0 cm	23.3 cm	28.0 cm	38.3 cm	34.2 cm	28.0 cm
width/length	0.69	0.68	0.72	0.72	0.69	0.71
depth/length	0.36	0.35	0.34	0.42	0.43	0.41
shape	moderately depressed	moderately depressed	moderately flattened	moderately depressed to domed	domed	moderately domed
color	brown to black	brown with dark seams to totally black	dark brown	reddish brown to black	dark brown to black	dark brown to black
pattern	none to yellow keels	none	none to yellow keels	none	none	yellow keels
Plastron						
color	brown to black	black	gray to dark brown or black	dark brown	brown	black
pattern	yellow border	none	none to yellow border	yellow border	yellow border	yellow border
Head						
color	olive to brown	black to olive	black	olive-brown	olive-brown	brown
pattern	yellow, pink, or green streaks or spots	large pale-yellow blotch on temple	yellow, orange or red spots	fine orange	none to a black middorsal stripe	none to olive-brown or yellow reticulations
Iris color	white to chestnut-brown	yellow	brown	yellow	chestnut-brown	?

coloration and pattern of the head, carapacial length, and plastron pattern. Table 4 summarizes subspecific variation. The subspecies can best be identified by a combination of characters and their geographic ranges.

Habitat: Throughout its range *Melanochelys trijuga* is predominantly a freshwater species inhabiting ponds, streams, and rivers with clean water. However, at times it may be found on land far from water. Deraniyagala (1939) reported that *M. t. parkeri* spends much time ashore and even estivates under dead vegetation in the jungle during droughts. He also reported that some *M. t. thermalis* individuals on Sri Lanka live in burrows and have very flattened carapaces. Annandale (1913) thought *M. t. indopeninsularis* visits water during hot weather, but lives habitually, without actually entering water, in damp places.

Natural History: Males of *Melanochelys trijuga* are aggressive during courtship, pursuing and frequently biting the female, especially on the neck. Nesting apparently occurs throughout the year, at least in *M. t. thermalis* (Deraniyagala, 1939), and several clutches of three to eight eggs are laid each year. The calcareous eggs are white and elongated (43–55 × 24–30 mm); incubation lasts about 60–65 days. Hatchlings are about 43–45 mm in carapace length and are more brightly patterned than adults.

M. trijuga is predominantly herbivorous, feeding on aquatic plants such as *Vallisneria*, but Deraniyagala (1939) reported that *M. t. thermalis* is omnivorous and has scavenging habits. This species spends most of the daylight hours basking, but is also active at night, coming on shore and foraging in wet grass and ditches (Annandale, 1913). When first captured, specimens are shy and may emit a strong musky odor.

Morenia Gray, 1870b

The genus *Morenia* is composed of two aquatic species from India, Bangladesh, and Burma. The carapace is slightly depressed and has a medial keel which is more pronounced in the young. The hexagonal neurals are shortened anteriorly. The narrow plastron is well developed, but hingeless. It is firmly sutured to the carapace, but its buttresses are short, extending only to the outer edges of the costals. An important feature is the position of the entoplastron which lies anterior to the humero-pectoral suture. In the skull, the frontal bones touch beneath the olfactory nerves posterior to the ethmoid fissure, and are separated from the orbital border by the prefrontal and postorbital bones. The squamosal is firmly in contact with the postorbital, and the carotid canal lies entirely within the pterygoid bone. In the rear of the skull, the exoccipital is supported by an extension of the opisthotic and drops behind the paracapsular sac to form an aperture for the vagus nerve. The upper jaw is medially notched, and its triturating surface is very wide and has a well-defined tubercular medial ridge. The lower triturating surface is ridgeless. The lateral edges of the jaws are serrated. The choanae (internal nostrils) lie at the level of the posterior orbital rim.

Key to the species of *Morenia*:

1a. Snout strongly projecting, as long or longer than orbital width; cervical scute about half as wide as 1st marginal; India and Bangladesh: **M. petersi**
 b. Snout not strongly projecting, shorter than orbital width; cervical scute about one-fourth as wide as 1st marginal; southern Burma and Tenasserim: **M. ocellata**

Morenia petersi (Anderson, 1879)

Indian eyed turtle

Plate 11

Recognition: The slightly domed carapace (to 20 cm) has a keel of low bulbous projections, and an unserrated posterior rim. Its cervical scute is broader than long (about 50% the width of the 1st marginal). Vertebrals 2–5 are broader than long; the 1st may be longer than broad. The carapace is olive, dark brown, or black with greenish, cream, or yellow borders on the vertebrals and pleurals, and a pale medial stripe. In young individuals, each vertebral has a light-colored horseshoe-shaped mark with an open anterior end, and each pleural contains a pale ocellus near the marginal seam, but frequently these markings are lost with age. Each marginal has a pale vertical stripe. The hind lobe of the plastron is narrow and bears a posterior notch. The bridge is broader than the hind lobe is long, and the axillary and marginal scutes are large. The plastral formula is: abd > pect > hum > fem >< an > gul. The plastron is uniformly yellow, although dark blotches may occur on the bridge and undersides of some marginals. The head is not enlarged, but the snout is pointed and protruding. Dorsal and lateral surfaces of the head are covered with an enlarged scale but its posterior surface has numerous small scales. The head is olive with several yellow stripes on each side: one running laterally along the upper border passing over the orbit, a second below it running backward from the tip of the snout, and a third extending from the orbit backward to the neck. Several other short stripes may occur on the snout. Limbs exhibit a yellow margin; the tail is short.

Males are shorter and have slightly concave plastra, longer tails, and vents beyond the carapacial rim. Females are larger and higher domed, and have relatively shorter tails with the vent under the carapace.

Distribution: *Morenia petersi* is known from northeastern India as far west as Bihar, and from Bangladesh.

Habitat: Slow-moving rivers, ponds, and swamps are the probable habitat of *Morenia petersi*.

Morenia ocellata (Duméril and Bibron, 1835)

Burmese eyed turtle

Recognition: The slightly domed carapace (to 22 cm) has a medial keel of low bulbous projections and an unserrated posterior rim. Its cervical scute is broader than long (about 25% of the width of the 1st marginal). Vertebrals 2–4 are broader than long; vertebrals 1 and 5 are as long as or longer than broad. The carapace is olive, dark brown, or black with a dark-bordered and dark-centered yellow ocellus on each pleural and vertebral (these ocelli fade with age). The plastral hind lobe is narrow and bears a posterior notch. The bridge is broader than the length of the hind lobe, and its axillary and inguinal scutes are large. The plastral formula is: abd > pect > hum >< fem >< an > gul. Plastron and bridge are uniformly yellow. The head is not enlarged and the snout is short and nonprotruding. Dorsal and lateral surfaces of the head are covered with an enlarged scale but the posterior surface has numerous small scales. The head is olive to brown with a yellow stripe on each side running posteriorly from the tip of the snout over the orbit to the neck, and another yellow stripe running backward from the orbit to the neck. Limbs are olive to brown; the tail is short.

Females grow larger than males, and have flat plastra and shorter tails with the vent beneath the carapacial rim. Males are shorter, and have slightly concave plastra and longer tails with the vent beyond the carapace.

Distribution: *Morenia ocellata* is restricted to southern Burma (including Tenasserim).

Habitat: Theobald (in Smith, 1931) reported it to be strictly aquatic, but often occurring in water bodies that seasonally dry up. It probably also lives in permanent rivers, streams, swamps, and ponds.

Natural History: Ewert (1979) reported a hatchling was 42.2 mm in carapace length. The Burmese capture and eat these turtles during the dry season.

Notochelys Gray, 1863a

This genus is unique in normally having more than five vertebral scutes. Usually six or seven are present, as one or two smaller scutes develop between the normal 4th and 5th vertebrals. These additional scutes have symmetrical connections with the adjacent pleural scutes and a definite position in the vertebral series. In this way they differ from the abnormal extra vertebrals that originate through asymmetrical subdivisions of existing vertebrals in other turtles. Only one living species is known, *Notochelys platynota*.

Notochelys platynota (Gray, 1834a)
Malayan flat-shelled turtle
Plate 11

Recognition: The oval to elongated carapace (to 32 cm) is flattened dorsally and serrated posteriorly (more so in juveniles than in adults). A low, interrupted medial keel is present. Six or seven vertebral scutes are usually present; the first four and the most posterior are large and broader than long, but those

between the 4th and last vertebral are smaller and may be as long as broad or a little longer than broad. Neurals are hexagonal in shape and shortest anteriorly. Carapacial color varies from greenish brown to yellowish brown or reddish brown, and each scute contains a dark spot or radiation. Juveniles have two dark spots on each vertebral and one on each pleural. The plastron is bound to the carapace by a ligament and, although there is a bridge, plastral buttresses are usually absent in adults. A poorly developed transverse hinge lies between the hyoplastron and hypoplastron in adults. The entoplastron is intersected by the humero-pectoral seam. The plastral formula is: abd >< pect > an > gul > hum > fem; and the anal scutes are not notched, or only slightly so, posteriorly. Plastron and bridge are yellow to orange with a large dark blotch on each scute. In the skull, the squamosal bone is probably present. Smith (1931) reported that the skull had a complete or incomplete temporal arch depending on the loss of the quadratojugal and postorbital. In several Asian genera, the squamosal (= quadratojugal) is either absent or subject to loss. McDowell (1964) found the squamosal in two skulls of *Notochelys* he examined, and was skeptical of reports of the absence of this bone, since it is loosely attached and easily lost in preparation, leaving little or no evidence on the quadrate of its former presence. When present, the squamosal is separated from the jugal by the postorbital. The orbito-nasal foramen, which houses the palatine artery and an anastomotic artery that joins the artery of the Vidian canal, is on the floor of the cavernosum canal lateral to the border of the basisphenoid and anterior to the primitive caroticum foramen. The ethmoid fissure is broad and the postlagenar hiatus large and rounded. The anterior part of the inferior parietal process flares strongly outward, touching the jugal. Triturating surfaces of the jaws are narrow and ridgeless. The upper jaw has a medial notch bordered by two toothlike cusps. The snout is slightly projecting and the posterior surface of the head is covered with small scales. Head and neck are brown, and young individuals may bear a yellow longitudinal stripe over each eye and another from the corner of the mouth backward to the neck. Limbs are brown and have transversely enlarged scales. The toes are webbed, and the tail is not exceptionally long.

Carr and Bickham (1986) reported a karyotype of 2n = 52.

Males have slightly concave plastra and thicker tails than do the flat-plastroned females.

Distribution: *Notochelys platynota* ranges from peninsular Thailand southward through Malaya, Sumatra, and Java to Borneo. There is also a record from Vietnam.

Habitat: Shallow waters with soft bottoms, abundant vegetation, and slow currents, such as swamps, marshes, streams, and ponds in jungle areas, are preferred.

Natural History: Nothing is known of the reproductive behavior of this species; however, the hatchlings are more brightly colored and more strongly serrated, have a more pronounced keel, and are about 55–57 mm in carapace length.

Notochelys platynota is a herbivore, feeding predominantly on aquatic plants. In captivity it accepts a variety of vegetables, but prefers fruits, such as bananas.

Flower (in Smith, 1931) remarked that when a wild *Notochelys* is disturbed it pulls into its shell with a hiss, and if picked up, defecates on the handler. Unfortunately, captives retain the defecation habit.

Ocadia Gray, 1870b

The genus *Ocadia* Gray, 1870b is represented by only one living species, *O. sinensis.*

Ocadia sinensis (Gray, 1834a)

Chinese striped-neck turtle

Plate 12

Recognition: The elliptical, slightly depressed carapace (to 24 cm) is widest just behind the center, highest at the seam between the 2d and 3d vertebrals, and slightly serrated posteriorly. Three low, discontinuous keels are present in juveniles, but these usually disappear with age. The hexagonal neurals are short sided anteriorly, and not elongated. Vertebral 4 lies over four neurals. Vertebral 1 is wider anteriorly than posteriorly; 2–4 are broader than long, or as long as broad. Vertebral 5 is always wider than long. The carapace is reddish brown to black with yellow seams (especially in juveniles), and occasionally some yellow or orange on the projections of the keels. The hingeless plastron is well developed, sharply bent at the bridge, notched posteriorly, well sutured to the carapace, and with strong buttresses. The axillary buttress is inserted into the carapace well up on the 1st costal, but is not attached to the 1st rib. Its matching inguinal buttress is inserted about half-way up the suture between the 5th and 6th costals. The entoplastron is crossed by

the humero-pectoral seam. The plastral formula is: abd > pect > fem > gul > an > hum. Plastron, bridge, and undersides of the marginals are yellow with a large dark-brown or black blotch on each scute. In the skull, the postorbital bone touches the supratemporal, but the maxilla does not touch the squamosal and is prevented from contacting the inferior process of the parietal by the palatine. The upper jaw is notched medially and its triturating surface has a well-developed, denticulated medial ridge. The frontal bone is separated from the posterior shallow temporal emargination by parieto-postorbital contact. The pterygoid does not touch the articular surface of the quadrate, and the palatine artery is small, but the orbito-nasal foramen is large. In the lower jaw, the coronoid process is only moderately developed and the angular bone is shortened. The snout projects slightly. Smooth skin covers the dorsal surface of the head. Head and neck are olive dorsally and yellow ventrally, and both contain numerous narrow, dark-bordered, pale-green to yellow stripes. The upper jaw is cream colored. Limbs are olive with numerous yellow stripes, and the forelimbs are covered anteriorly with large scales. All toes are webbed, and the 5th anterior toe contains only two phalanges. The tail is long and tapered, even in adults.

Nakamura (1949), Killebrew (1977a), and Carr and Bickham (1986) reported *Ocadia* has a karyotype of 2n = 52 (26 macrochromosomes, 26 microchromosomes), the standard number of chromosomes for a batagurine turtle. However, Stock (1972) reported this turtle has only 50 chromosomes (26 macrochromosomes, 24 microchromosomes).

Males have slightly concave plastra and the vent lies beyond the carapacial margin; females have flat to slightly convex plastra with the vent beneath the carapace.

Distribution: *Ocadia sinensis* is found in Taiwan, southern China (Fukien, Hangchow, Soochow, Kwangtung, Shanghai, and Hainan Island), and northern Vietnam.

Habitat: *Ocadia* inhabits slow-moving lowland waters, such as marshes, swamps, ponds, and canals, with soft bottoms.

Natural History: Smith (1923) reported a captive female laid three eggs in April. The oval, calcareous eggs measured 40 × 25 mm. Hatchlings have tricarinate carapaces of about 35 mm, vertebrals wider than long, and long tails.

Ocadia is herbivorous and will feed on a variety of aquatic plants, as well as lettuce and fruits, in captivity. It is fond of basking and may spend much time in this pursuit.

Orlitia Gray, 1873b

The only living species is *Orlitia borneensis*, which is closely related to the genus *Siebenrockiella* (McDowell, 1964; Hirayama, 1984; Sites et al., 1984).

Orlitia borneensis Gray, 1873b

Malaysian giant turtle

Recognition: This large (to 80 cm) semiaquatic turtle has a narrow, oval carapace which is relatively high peaked in the young at the area of the seam separating the 2d and 3d vertebrals. Adults are somewhat flatter. The vertebrals bear a low medial keel, at least in juveniles, and appear narrow when compared to the adjacent pleural scutes. The 2d vertebral is always somewhat "mushroom" shaped with the stalk pointing backward. The underlying neural bones are elongated, hexagonal, and short sided anteriorly. Posterior marginals are strongly serrated in juveniles, but less so in adults, and are smaller than the lateral or anterior marginals. Pleural 4 is considerably smaller than the first three. All carapacial scutes are somewhat rugose. The carapace is uniformly dark gray, brown, or black. The plastron is long and narrow, with an anal notch. Lateral plastral margins may be keeled. Both the anterior and posterior lobes are much narrower than the pectoral and abdominal scutes. The humero-pectoral seam lies posterior to the entoplastron. The plastral formula is: abd > pect > fem > gul > hum >< an. The bridge contains both axillary and inguinal scutes and is much longer than the posterior plastral lobe. The plastron is sutured to the carapace and the buttresses are strongly developed. Axillary buttresses are attached to the 1st ribs; inguinal buttresses insert over half the distance up the suture between the 5th and 6th costals. Bridge and plastron are yellowish to light brown, and usually unmarked. The head is relatively large and broad with a slightly projecting snout. The upper jaw is slightly hooked and its triturating surface is broad posteriorly, with a well-developed medial ridge, but narrow medially. In the skull, the squamosal is firmly attached to the surrounding bones and extends anteriorly. The orbito-nasal foramen is large (much larger than the posterior palatine foramen). A strip of granular scales lies between the orbit and tympanum, and the skin on the back of the head is broken into small scales. The adult head is uniformly brown or black; that of juveniles may be dark mottled and have a light line extending backward from the corner of the mouth. Neck, limbs, and tail are gray, brown, or black. The forelimbs have large horizontal scales on their anterior surface. All toes are webbed.

Bickham (1981) and Carr and Bickham (1986) gave the diploid chromosome total as 50.

Males have longer, thicker tails than do females.

Distribution: This highly aquatic species is known only from Malaysia, Sumatra, and Borneo.

Habitat: *Orlitia* apparently occurs in large bodies of fresh water, such as lakes and rivers.

Natural History: Little is known of the biology of this apparently rare turtle. Aborigines told Dr. Edward Moll that *Orlitia* nests in piles of debris, but he did not personally observe this. Its eggs are ellipsoidal with brittle shells, averaging about 80 × 40 mm. Hatchlings are about 60 mm long and have very rugose carapaces with strongly serrated posterior marginals.

Mehrtens (1970) reported that a captive accepted a variety of protein foods such as beef heart, horsemeat, canned dog food, and fish. It also accepted overripe bananas, but refused other vegetables. Food was taken as readily on land as in the water.

Orlitia borneensis

Pyxidea Gray, 1863a

This is another monotypic batagurine genus from southeastern Asia. The only living species is *Pyxidea mouhotii*. McDowell (1964), on the basis of skull structure and the lack of cloacal bursae, placed it in the genus *Geoemyda*, but we agree with Bramble (1974) that this is a separate monotypic genus most closely related to *Cyclemys, Cuora,* and *Heosemys,* with which it shares a common mechanism for closing the plastron.

Pyxidea mouhotii (Gray, 1862)
Keeled box turtle

Recognition: The elongated (to 18 cm), oval carapace is flattened dorsally and strongly serrated posteriorly. The anterior margin is also somewhat serrated. Three well-developed keels are present: a medial keel along the center of the flattened vertebrals, and two lateral keels extending along the dorsal portions of pleurals 1–4, forming a border on each side of the flattened dorsal area. Hexagonal neurals are short sided posteriorly. Vertebrals are usually wider than long, but the narrower 1st vertebral may be as long as or slightly longer than wide. The 12th marginal scutes on each side (the supracaudals) are smallest. Uniformly colored, the carapace ranges from yellowish or light brown to reddish brown or dark brown. The plastron is smaller than the opening of the carapace and cannot close completely; the single hinge occurs between the hyo- and hypoplastra (between the overlying pectoral and abdominal scutes). The entoplastron is crossed by the humero-pectoral seam. A posterior notch lies between the anal scutes. The plastron is

Pyxidea mouhotii

united to the carapace by ligaments along a short, but distinct, bridge (approximately one-third as long as the plastron). Its buttresses are weakly developed. An axillary scute may be absent from the bridge. The plastral formula is: abd > an > pect > hum >< fem > gul. The plastron is yellow to light brown with a dark-brown spot on each scute. The head is of moderate size with a strongly hooked, unnotched upper jaw, and a short nonprojecting snout. Triturating surfaces of the jaws are narrow and ridgeless. This species is unique in having the orbito-nasal foramen posteriorly situated at the level of the primitive caroticum foramen, a broad triangular ethmoidalis fissure, and the forward extension of the epipterygoid to the level of the anterior border of the cranial cavity (McDowell, 1964). Posterior head skin is divided into large scales; the head is brown with dark vermiculations. One or two light spots may occur between the orbits and tympanum, and, contrary to some descriptions, these spots may elongate into black-bordered stripes. Limbs are gray to dark brown or black. Anterior surfaces of the forelegs are covered with large scales and the hind legs are somewhat club shaped. The toes are only partially webbed. The tail is moderate in length, and pointed tubercles arise at its base and on the adjoining thighs.

The karyotype is 2n = 52 (Bickham and Carr, 1983). Males have longer, thicker tails than do females.

Distribution: *Pyxidea* is found on Hainan Island, China, and from Vietnam westward to northern Thailand and adjacent Burma.

Habitat: *Pyxidea mouhotii* is a terrestrial species which seldom enters water. Wirot (1979) commented that the only time it becomes wet is when it rains or from the morning dew. It is apparently a forest dweller, and Pope (1935) reported it inhabits mountainous terrain on Hainan. Sachsse (1973) thought it similar in habits to the North American *Terrapene*.

Natural History: The breeding behavior of *Pyxidea* is little known. The eggs are elongated (40 × 25 mm) with brittle shells, and the hatchlings have carapace lengths of about 35 mm (Ewert, 1979).

Pyxidea mouhotii is almost exclusively herbivorous.

Rhinoclemmys Fitzinger, 1835

Members of the genus *Rhinoclemmys* are semitropical and tropical American turtles that frequent aquatic, semiaquatic, and terrestrial habitats. They are the only turtles of the subfamily Batagurinae to occur in the New World. The carapace is low arched to domed, and medially keeled. Each neural is hexagonal and projects behind its associated lateral costal bones. The plastron is rigid, well buttressed, and lacks a hinge. The skull is short, and flattened to slightly convex dorsally with lateral orbits. Frontals form part of the orbital rim. The quadrate does not enclose the stapes; the otic notch is open. The dorsal surface of the tympanic bulla is ridged. The squamosal is only loosely attached; is either separated from the parietal or meets it posterolateral to the trigeminal foramen; barely, if at all, touches the jugal; and is in contact with the postorbital. The parietal is separated from

the jugal but touches the palatine. The jugal is broadened posteriorly. The zygomatic arch is absent or, if present, is narrow and excavated dorsally and ventrally. Triturating surfaces of the maxillae are narrow and ridgeless. Toes may or may not bear webbing.

The karyotypes of the species *areolata*, *funerea*, *melanosterna*, *pulcherrima*, *punctularia*, and *rubida* have been determined (Barros et al., 1975; Bickham and Baker, 1976b; Killebrew, 1977a; Bickham and Carr, 1983, Carr and Bickham, 1986). The species *areolata*, *funerea*, *melanosterna*, *pulcherrima*, and *rubida* have nearly identical karyotypes to the presumed primitive 2n = 52, but the nucleolus organizer regions (NOR) are in a slightly different position than in Asiatic batagurines with 52 chromosomes. *R. funerea* differs from the others in having 8 metacentric or submetacentric macrochromosomes instead of the usual 9. *R. punctularia* is 2n = 56, having 2 extra pairs of heterochromatic microchromosomes, and 8 metacentric or submetacentric macrochromosomes.

Key to the species of *Rhinoclemmys*:

1a. Hind feet heavily webbed   2
  b. Hind feet with little or no webbing   5
2a. Dorsal head stripes from nape to level of orbits or less; no light spots on occipital region   3
  b. Dorsal head stripes from nape to beyond orbits or broken at orbits with a spot anteriorly; light spots on occipital region   4
3a. Snout strongly pointed; chin and lower jaw with dark bars; shell distinctly depressed: ***R. nasuta***
  b. Snout only moderately protruding; chin and lower jaw with numerous large black spots; shell domed: ***R. funerea***
4a. Dorsal head stripes broken, sometimes into numerous spots, with a large light spot anterior to orbit, or stripes unite behind orbits forming a horseshoelike mark: ***R. punctularia***
  b. Dorsal head stripes unbroken and extending anterior to orbits, never united: ***R. melanosterna***
5a. Tip of jaw hooked and unnotched   6
  b. Tip of jaw straight and notched, sometimes with cusp   7
6a. Dorsal head pattern consists of a large, irregular, horseshoe-shaped blotch; carapace depressed: ***R. rubida***
  b. Dorsal head pattern consists of a pair of supratemporal stripes, or no stripes are present; carapace rather high, but flat on top: ***R. annulata***
7a. Head pattern with red stripes, usually two or three, crossing tip of snout and a prefrontal arrow formed where a median sagittal stripe meets two supratemporal stripes on dorsal tip of snout; bridge with extensive dark pigment: ***R. pulcherrima***
  b. Head pattern with only a pair of broad supratemporal stripes posterior to orbit; bridge usually yellow without extensive dark pigmentation: ***R. areolata***

Rhinoclemmys annulata (Gray, 1860b)

Brown wood turtle

Recognition: This is a medium-sized (to 20 cm), brown to black, terrestrial turtle with a medially flattened carapace. The high carapace is flattened across the vertebrals, but still has a low, blunt vertebral keel. It is roughened owing to growth annuli on the scutes, posteriorly serrated, and usually broadest and highest just behind the middle. Carapacial coloration and patterns are extremely variable, ranging from totally black to dark brown with orange pleural and vertebral blotches to tan with yellow blotches on the pleurals and vertebrals. Pleural blotches are often radiations from the dorso-posterior corner; the vertebral keel is usually yellow. This variation is similar to that occurring in the eastern box turtle, *Terrapene*

Rhinoclemmys annulata

carolina carolina, of North America, and provides good concealment among dried leaves and vegetation on the forest floor. The plastron is well developed, upturned anteriorly and notched posteriorly. Its scute formula is: abd > pect > fem > an > hum > gul, and it is black to dark brown with a yellow border and, in some, a yellow midseam. The bridge is black or dark brown. The head is small with a slightly projecting snout and slightly hooked upper jaw which is also laterally serrated. A wide yellow or red stripe may extend from the orbit at a slight angle to the nape, but some individuals lack this stripe. Another stripe runs from the lower posterior orbit to the tympanum where it meets a similar stripe from the upper jaw. There is also a stripe from the upper anterior orbit to the tip of the snout. Forelimbs have large yellowish scales with dark stripes of wide black spots. The toes are not webbed.

Males have concave plastra and longer, thicker tails with the vent beyond the carapacial margins. Females have flat plastra and shorter tails.

Distribution: *Rhinoclemmys annulata* ranges from southeastern Honduras southward through eastern Nicaragua, Costa Rica, and Panama to western Colombia and Ecuador.

Geographic Variation: *Rhinoclemmys annulata* shows much variation in color and patterns in all populations.

Habitat: *Rhinoclemmys annulata* is principally a diurnal, terrestrial, lowland rainforest resident, but it also follows gallery forests onto the highlands to over 1500 m.

Natural History: Courtship has not been fully described, but the male often salivates on the female's head. There are conflicting reports as to whether or not the female digs a nest cavity or just hides the eggs under leaf litter. Egg laying apparently occurs throughout the year with a clutch consisting of only one or two ellipsoidal eggs (70 × 37 mm). The hatchling is about 63 mm in carapace length.

Rhinoclemmys annulata is herbivorous, feeding on ferns, shrubs, and various seedlings. Fruits such as bananas and papaya are also relished. Most activity occurs in the morning. It is also quite active immediately after heavy rains, and then can be found marching along paths and roads. When not active, it scoops out a form in fallen leaves, or retreats beneath tangled vines or root masses. During hot, dry spells it often enters pools of water to cool off.

Rhinoclemmys areolata (Duméril, Bibron, and Duméril, 1851)

Furrowed wood turtle

Recognition: This semiterrestrial turtle is medium sized (to 20 cm) with a high, ovoid carapace which is wider posteriorly than anteriorly, medially keeled, slightly serrated posteriorly, and has flared or laterally upturned marginals. Its surface is smooth in older individuals but rugose in the young. The carapace is usually olive with dark seams and much yellow mottling forming a lichenlike pattern, but may be tan to black. Each pleural has a small yellow or red, often dark-bordered, central spot which disappears with age. The plastron is well developed, slightly upturned anteriorly, notched posteriorly, and has a scute formula of abd > pect > fem > an > gul > hum. It is yellow with a dark central blotch and dark seams. The bridge is yellow. The head is small and has a slightly projecting snout and a notched upper jaw. A yellow or red stripe runs posteriorly from the orbit to the side of the neck, two elongate red or yellow spots lie on the nape, and another stripe runs between the orbit and tympanum. Each eyelid has a light vertical bar, and a light stripe may run from the snout posteriorly along the upper jaw to the tympanum. Lower jaws and chin have black spots or ocelli. The feet are slightly webbed and the forelimbs are covered with large, yellow, black-spotted scales.

Males have concave plastra and slightly longer tails with the vent beyond the carapacial margin; females have flat plastra and tails with the vent beneath the carapace.

Distribution: *Rhinoclemmys areolata* occurs from southern Veracruz, Tabasco, and eastern Chiapas to Yucatán and Cozumel Island in Mexico, southward through Belize and eastern Guatemala. There is a questionable record from eastern Honduras.

Rhinoclemmys areolata

Geographic Variation: No subspecies have been described, but Ernst (1978) reported that *Rhinoclemmys areolata* from Cozumel Island, Mexico, were quite divergent in several characters from mainland populations.

Habitat: *Rhinoclemmys areolata* is generally a savannah inhabitant, but will enter adjacent dense woodlands. It also enters marshy areas, as evidenced by specimens collected with algae-covered carapaces.

Natural History: The elongated eggs are tapered like those of a bird, have brittle shells, and measure about 60 × 31 mm. During egg laying, the carapace and plastron become flexible at the posterior margins allowing the eggs to pass through. At hatching, the young are about 52–55 mm in carapace length.

Rhinoclemmys funerea (Cope, 1876)

Black wood turtle

Recognition: This large (to 32.5 cm), dark, freshwater turtle has a spotted lower jaw and chin. The adult carapace is high, somewhat domed, medially keeled, posteriorly serrated, and usually widest and highest just behind the middle. Its surface is smooth to rugose (due to retention of growth annuli), and it is dark brown to black (some yellow occurs on the juvenile pleurals). The plastron is well developed, upturned anteriorly, and notched posteriorly. The plastral formula is: abd > pect > gul > an > hum. The plastron is black with yellow seam borders and a wide yellow midseam, and the bridge is black to dark brown with yellow seams. The head is of moderate size with a slightly projecting snout and notched upper jaw. It is black with a wide lateral yellow stripe above the tympanum. Two narrower yellow stripes run from the orbit and the corner of the mouth to the tympanum, and there are large black spots on the yellow lower jaw and chin. Skin of the neck and limbs is black with yellow vermiculations. The feet are strongly webbed.

Adult males have concave plastra and longer, thicker tails with the vent posterior to the carapacial margin; adult females have flat plastra and shorter tails with the vent beneath the carapace.

Distribution: *Rhinoclemmys funerea* inhabits the Caribbean drainages of Central America from the Río Coco, on the Honduran-Nicaraguan border, southward to the Panama Canal Zone.

Habitat: The preferred habitats are marshes, swamps, ponds, streams, and rivers in humid forests, where

Rhinoclemmys funerea can often be seen basking on partially submerged logs.

Natural History: Sexual maturity is attained at approximately 20-cm plastron length. Spermatogenesis occurs from April through August and females ovulate from April through July. During courtship, the male chases the female in the water and when she stops, swims to her side, extends his head and neck and rapidly vibrates his head up and down. Up to four clutches of about three eggs are laid each season. The white, brittle-shelled eggs are ellipsoidal (about 68 × 35 mm); however, one female laid an egg 76 × 39 mm. Hatchlings average about 55 mm in carapace length.

In the wild, *Rhinoclemmys funerea* is highly herbivorous, feeding on a variety of fruits, grasses and broad-leaved plants; but in captivity, meats are accepted. It often forages on land at night, and consequently is sometimes parasitized by ticks.

Rhinoclemmys nasuta (Boulenger, 1902)

Large-nosed wood turtle

Recognition: This medium-sized (to 22 cm), dark, freshwater turtle has a strongly projecting snout. Its flattened adult carapace is medially keeled, only slightly serrated posteriorly, widest and highest just behind the middle, and black or reddish brown with black seams. The surface of the adult carapace is usually smooth, but that of juveniles is roughened with small rugosities. The plastron is well developed, slightly upturned anteriorly, notched posteriorly, and yellow with a large reddish-brown to black blotch on each scute. The plastral formula is: abd > pect > fem > an > gul > hum. The bridge is yellow with two dark blotches. The head is of moderate size with a projecting snout and a notched upper jaw. A cream to yellow stripe extends from the tip of the snout to each orbit,

Rhinoclemmys nasuta

another stripe runs posteriorly from the orbit dorsolaterally to the nape, a third light stripe passes from the lower edge of the orbit to the tympanum, and another extends from the corner of the mouth to the tympanum. Dark vertical bars are present on the lower jaws. Skin of the neck and limbs is reddish brown to yellow. The feet are strongly webbed.

Females (to 22 cm) grow larger and wider than males (to 19.6 cm). Males have concave plastrons and longer, thicker tails with the vent beyond the carapacial margin.

Distribution: *Rhinoclemmys nasuta* occurs only in the Pacific drainages of western Colombia (the Quito, Truando, San Juan, Docampado, and Baudó rivers) and northwestern Ecuador, near Esmeraldas.

Geographic Variation: No subspecies have been named, but Ecuadorian *Rhinoclemmys nasuta* have wider carapaces and marginals (Ernst, 1978).

Habitat: *Rhinoclemmys nasuta* lives in large rivers with strong current. It seldom climbs onto land except to nest or bask.

Natural History: What little we know of its biology is from Medem (1962). He reports they lay one, or possibly two, ellipsoidal (67–70 × 35–39 mm) eggs per clutch, usually in January to March, but some may lay throughout the year. Females construct poor, shallow, uncovered nests at best, and many merely lay their eggs among leaves on the ground.

Rhinoclemmys nasuta is principally herbivorous, but Medem reported one had eaten grasshoppers. They do poorly in captivity.

Rhinoclemmys punctularia (Daudin, 1802)
Spotted-legged turtle

Recognition: *Rhinoclemmys punctularia* is a large (to 25.4 cm), dark, aquatic turtle with black-spotted forelimbs. Its domed carapace is medially keeled, posteriorly serrated and notched, and usually widest and highest just behind the middle. The surface is smooth to lightly rugose. It is uniformly dark brown or black in adults, but juveniles may have yellow to bronze radiations on each pleural. The plastron is well developed, upturned anteriorly, and notched posteriorly. Its plastral formula is: abd > pect > fem > an > gul > hum. The plastron is red brown to black with a yellow border and seams. The bridge is yellow with two large dark blotches. The head is small with a slightly projecting snout and a notched upper jaw. It

Rhinoclemmys punctularia punctularia

Rhinoclemmys punctularia diademata

is black with a dorsal pattern ranging from two longitudinal red or yellow stripes that run anteriorly from the nape and touch or pass beyond the orbit, to a broad horseshoelike mark posterior to the orbit. Two light spots may occur on the nape. The eyelids have a light-colored bar, and stripes usually run between the orbit and tympanum and from the snout along the upper jaw to the tympanum. The iris is green to bronze. Forelimbs have large, yellow or red, black-spotted scales, and the hind limbs are gray laterally and yellow with black spotting medially. The toes are strongly webbed.

Males have concave plastra, long, thick tails, and the vent beyond the carapacial margin; females have flat plastra and shorter tails.

Distribution: This turtle ranges from eastern Colombia, the Orinoco drainage of Venezuela, and Trinidad Island eastward through the Guianas and northeastern Brazil.

Geographic Variation: Three subspecies occur in northern South America. *Rhinoclemmys punctularia*

punctularia (Daudin, 1802) ranges from the Orinoco drainage in extreme eastern Venezuela, and Trinidad Island, southeast through the Guianas, to the Amazon drainage of northeastern Brazil. Its dorsal head pattern consists of an oblique yellow or red stripe on each side running posteriorly from above the orbit to above the tympanum; there are two light blotches on the nape, and a light spot on the snout in front of each orbit. The form *R. lunata* Gray represents only a normal pattern variation of *R. p. punctularia* (Ernst, 1978). *R. p. diademata* (Mertens, 1954a) occurs in the Caribbean drainage of eastern Colombia and northwestern Venezuela. Its dorsal head pattern consists of a large, yellow, horseshoe-shaped figure located medially just behind the orbit with the apex pointing anteriorly and the long arms running posteriorly, flaring laterally, and centrally enclosing a dark area; a light blotch occurs on the snout in front of each orbit but no light blotches are on the nape. *R. p. flammigera* Paolillo O., 1985 is found in the region of the confluence of the Ventuari and Orinoco rivers in southern Venezuela. Its dorsal head pattern is characterized by numerous red spots arranged in a radial pattern; loreal, middle lateral, posterior lateral, and parietal spots are always present on each side of the head forming a semicircular pattern. A light spot occurs in front of each orbit, and two light blotches are on the nape.

Pritchard (1979) and Pritchard and Trebbau (1984) considered *R. p. diademata* a separate species based on its allopatric range and lack of intergrades with *R. p. punctularia*, but Ernst (1978) and Paolillo O. (1985) thought it merely a subspecies since only the dorsal head pattern differs, and, since some *R. p. punctularia* have their dorsal stripes united behind the orbits (the *lunata* pattern), the *diademata* head pattern could easily be formed from that of *punctularia*. Determination of the karyotype of *diademata* would help determine its status.

Habitat: *Rhinoclemmys punctularia* inhabits almost all of the freshwater bodies within its range, from ponds, marshes, and coastal swamps to large rivers in savannah areas to deep forests.

Natural History: *Rhinoclemmys punctularia* may reproduce throughout the year, laying several clutches of one to two, elongated, white, brittle-shelled eggs (52–75 × 30–37 mm) under tree roots or vegetation. During courtship, males pursue females, sniffing at their anal vent. The male then moves to the front of the female and proceeds to extend and then withdraw his head on either side of her head; she often bites at his head. When the female is receptive and quiet, the male moves to her rear and copulation follows (Pritchard and Trebbau, 1984). Mating may occur in or out of water.

This turtle is fond of basking, and often roams about on shore. It is omnivorous, accepting a wide variety of plant and animal foods, feeding both in water and on land.

Rhinoclemmys melanosterna (Gray, 1861)
Colombian wood turtle

Recognition: This is a large (to 29 cm) aquatic turtle with an oval, somewhat domed, black to dark-brown carapace widest at the level of marginals 6–7, highest at the level of anterior vertebral 3, and notched posteriorly. A medial keel is usually present, and the carapacial surface may be slightly rugose. The well-developed plastron is upturned anteriorly and notched posteriorly, and has a scute formula of abd > pect > fem > an > gul > hum. It is red brown to black with a yellow border and midseam. The head is small with a slightly projecting snout and a medially notched upper jaw. It is dark brown to black with a dorsal pattern consisting of an oblique pale-green to orange or red stripe on each side running posteriorly from in front of the orbit to above the tympanum and there curving downward toward the posterior side of the tympanum. No light blotches occur on the snout in front of these stripes or on the nape, and the oblique stripes are never united across the forehead. The iris is yellow or bright white. The forelimbs are black spotted, and the toes webbed.

Males are smaller (to 25 cm), have flatter carapaces, slightly concave plastra, and longer, thicker tails with the vent beyond the carapacial rim.

Distribution: *Rhinoclemmys melanosterna* lives in the Caribbean drainages of southeastern Panama and northern Colombia, and Pacific drainages of western Colombia and northwestern Ecuador.

Geographic Variation: No subspecies are recognized; however, Medem (1962; pers. comm., 1974) remarked that those living in a freshwater habitat always had red head stripes, whereas those from brackish waters close to the Pacific Coast of Colombia had green to greenish-yellow stripes. Populations intermediate (intergrade?) between the two possessed orange stripes. There were no morphological differences, although red-striped turtles seemed to be larger. Field study is needed to verify this, since the pigmentation of the head stripes disappears during preservation.

Habitat: *Rhinoclemmys melanosterna* occupies a variety of aquatic sites from ponds and marshes to lakes and large rivers in savannahs to deep rainforests. It also enters brackish coastal waters.

Natural History: Medem (1962) and Castaño-Mora and Medem (1983) reported that nesting occurs throughout the year, but principally in June–August and November. One or two (exceptionally three) ellipsoidal (48–71 × 28–38 mm) eggs form a clutch. No cavity is dug; instead the eggs are covered with rotting leaves. The eggs hatch in 85–141 days, principally in September–November. Hatchlings have carapace lengths of 39–59 mm.

Rhinoclemmys melanosterna is fond of basking. In the wild it is principally herbivorous, but captives will accept a variety of animal foods (Medem, 1962).

Rhinoclemmys pulcherrima pulcherrima

Rhinoclemmys pulcherrima (Gray, 1855)

Painted wood turtle

Plate 12

Recognition: This is a medium-sized (to 20 cm), colorful, terrestrial turtle with a series of red or orange stripes across the snout. Its carapace is rough, owing to growth annuli, medially keeled, posteriorly serrated, notched posteriorly, and usually widest and highest just behind the middle. It is flatter and broader in the northern parts of the range and domed and narrower southward. The carapace is brown with pleurals ranging from solid brown to patterned ones with a single, dark-bordered yellow or red spot to bright yellow or red lines or ocelli. Vertebrals may be unicolored, dark flecked, or with yellow or red radiations. The well-developed plastron is notched posteriorly. Its formula is: abd > pect > fem > an > gul > hum. The plastron is yellow with a narrow to wide, dark, central blotch, and its seams may be dark bordered. The bridge is either completely brown or has a horizontal yellow bar separating the brown pigment from the carapace. A slightly projecting snout and a notched, sometimes cusped, upper jaw are present on the small head. The brown to greenish head bears a series of bright orange to red stripes: (1) a median stripe running forward between the orbits to the dorsal tip of the snout where it meets two other stripes, one from each orbit, to form a prefrontal arrow; the lateral stripes may extend through the orbit to the nape; and any of these stripes may be discontinuous; (2) a stripe running posteriorly from below the nostrils along the upper jaw to the tympanum; (3) a stripe running from each nostril to the corresponding orbit; and (4) several stripes (usually two or three) running from the orbit to the tympanum. Jaws and chin are yellow and the lower jaw and chin may con-

Rhinoclemmys pulcherrima incisa

tain red stripes, large black spots, or ocelli. Other skin is olive to yellow or rufous. Forelimbs are covered with large red or yellow scales with rows of black spots; the toes are slightly webbed, if at all.

Males are smaller (18 cm) with concave plastra and longer, thicker tails with the vent beyond the carapacial margin. Females are larger (20 cm) with flat plastra slightly upturned anteriorly, and shorter tails with the vent beneath the carapace.

Distribution: *Rhinoclemmys pulcherrima* is restricted to the west coast of Mexico and Central America where it extends from Sonora, Mexico, to Costa Rica.

Geographic Variation: Four subspecies are recognized. *Rhinoclemmys pulcherrima rogerbarbouri* (Ernst, 1978) occurs in Mexico from southern Sonora to Colima. Its low, wide, brown carapace has no pleural markings or occasionally only a faint reddish stripe. The underside of each marginal bears a single light bar. The plastron has a wide, often faded, dark central blotch, and the bridge is brown. *R. p. pulcher-*

rima (Gray, 1855) is only found in Guerrero, Mexico. Its carapace is low, wide, and brown with dark flecks and a single dark-bordered, red or yellow central spot on each pleural. There are two or three light bars on the ventral side of each marginal. The narrow, dark central plastral blotch may be forked on the gulars and anals, and its bridge contains a yellow and a black transverse bar. *R. p. incisa* (Bocourt, 1868) ranges from Oaxaca, Mexico, southward to northern Nicaragua. Its brown carapace is medium (in the north) to high domed (in the south) and bears dark flecks, a dark-bordered red or yellow stripe or large ocellus on each pleural, and a light bar on the ventral side of each marginal. Its dark plastral blotch is narrow and unforked, and the bridge is brown. *R. p. manni* (Dunn, 1930) (Plate 12) is found in southern Nicaragua and northern Costa Rica, and is one of the most colorful of all turtles. Its high-domed brown carapace has several large red or yellow ocelli on each pleural and two light bars on the underside of each marginal. The narrow, dark, central plastral blotch may fork on the gulars and anals, and the bridge contains a yellow and a black transverse bar. Recently, Janzen (1980) has suggested the red and yellow ocelli may be a coral-snake mimicry pattern to frighten predators.

Habitat: *Rhinoclemmys pulcherrima* is a terrestrial lowland species, and originally was probably an inhabitant of scrublands and moist woodlands, but now is common in cleared areas, especially those close to streams where it occupies gallery forest. The ornate red terrapin seems, at least in Costa Rica and Nicaragua, to prefer moist situations, and has been observed wading and swimming in streams and rain pools, especially during the dry season. It is very active after rains. When away from water bodies, it usually seeks out moist vegetation.

Natural History: Male head bobbing is a common feature of the courtship, as also is smelling and trailing of the female. In later courtship the female engages the male in nose-to-nose contact and biting (Hildalgo, 1982). Up to four clutches of three to five eggs have been laid by a captive *R. p. incisa* from September through December. The elongated, brittle-shelled eggs measure 24–32 × 37–52 mm, and hatchlings range in carapace length from 35 to 50 mm.

In nature *R. pulcherrima* is probably an omnivore, but with stronger preferences toward plants. Wild foods have not been recorded, but captives readily eat a variety of domestic fruits and vegetables, earthworms, fish, beef strips, and canned dog food. When given a choice they usually choose plant foods over meats.

Rhinoclemmys rubida (Cope, 1870)
Mexican spotted wood turtle

Recognition: This medium-sized (to 23 cm), terrestrial turtle has an irregular horseshoe-shaped blotch on top of its head. Its carapace is flattened, medially keeled, posteriorly serrated, widest and highest just behind the middle, and has a rugose surface due to growth annuli. The carapace is yellowish brown with dark seams and dark mottling on each scute to uniform chocolate brown. A yellow spot is usually present at the center of each vertebral and pleural. The well-developed plastron is slightly upturned anteriorly, notched posteriorly, and has the following formula: abd > pect > an > gul > fem > hum. It is yellow with a brown central blotch and brown bridge. The head is moderate in size with a projecting snout and a hooked upper jaw. A highly variable, broad, horseshoe-shaped red or yellow mark lies on the crown, and usually several light bars cross the snout. A light stripe passes between the orbit and tympanum and another from the corner of the mouth to the tympanum. The yellow jaws and chin contain

Rhinoclemmys rubida rubida

Rhinoclemmys rubida perixantha

dark vermiculations or stipples. Its forelimbs are covered with large yellow or reddish black-spotted scales. The toes are only slightly webbed at best.

Males have concave plastra, longer, thicker tails, and the vent beyond the carapacial margin; females have flat plastra, shorter tails, and vents beneath the carapace.

Distribution: *Rhinoclemmys rubida* is restricted to the lowlands of the west coast of Mexico from Jalisco southward through Michoacán, and from Oaxaca southward through western Chiapas. Although it has not been collected in Guerrero, it probably occurs there.

Geographic Variation: Two subspecies are recognized. *Rhinoclemmys rubida rubida* (Cope, 1870) ranges from central Oaxaca to southern Chiapas. Its carapace is light brown with dark mottling, the gular scutes are approximately twice as long as the humerals, the marginals are slightly flared laterally, and it has an elongated light temporal spot. *R. r. perixantha* (Mosimann and Rabb, 1953) occurs in southern Jalisco, Colima, and Michoacán. It has light-brown marginals without dark mottling, and pleural scutes darker brown than the vertebrals and marginals. Its gular scutes are only slightly longer than the humerals, and the marginals are strongly flared laterally. An oval temporal spot is present.

Habitat: Scrub lowland woodlands and hillsides are the typical habitat.

Natural History: During courtship, the male often bobs his head. The white eggs are elongated (62 × 25 mm) and the hatchlings are 50–52 mm in carapace length.

One was caught while eating a large caterpillar, but captives also eat plant foods.

Sacalia Gray, 1870b

Sacalia is another Asiatic monotypic genus. Its sole living representative is *Sacalia bealei*.

Sacalia bealei (Gray, 1831b)

Four-eyed turtle

Recognition: The elongated, elliptical carapace (to 14.3 cm) is smooth, slightly depressed, with posterior medial keels, and its posterior marginals are unserrated in adults. The medial keel is low and the vertebrals are broader than long. Neural bones are six sided and shortest anteriorly. The carapace is yellow brown to chocolate brown, and may be adorned with dark vermiculations. The plastron is well developed, hingeless, and notched posteriorly. Its buttresses are weak, barely extending beyond the peripheral bones. The bridge is about as long as the plastral hind lobe; its axillary is short and inguinal small or absent. The plastral formula is: abd > pect > an > fem > hum >< gul. Plastron and bridge are yellowish to light olive and may have a pattern of dark vermiculations. The head is of moderate size with its squamosal bones touching the postorbitals but not the jugals. The ethmoid fissure is broad and triangular; the postlagenar hiatus large and round, and the orbito-nasal foramen is bordered by the basisphenoid in front of the primitive caroticum foramen. The anterior edge of the inferior parietal process flares outward to touch the palatine and jugal. The snout does not protrude, and the upper jaw is neither notched nor hooked. Skin on the back of the head is smooth, but small granular scales occur laterally between the orbit and tympanum. The head is yellow brown, olive or dark brown, and its posterior dorsal surface has two or four bright to faded, dark-centered ocelli (the posteriormost pair is always present). Jaws are dark brown, and the chin may contain pink or red pigment. The neck is dark brown dorsally and red ventrally. Three longitudinal yellow stripes extend backward from the head along the dorsal surface of the neck. The forward surface of the forelimbs is covered with large scales, which vary

Sacalia bealei quadriocellata

from yellow, pink or red, to dark brown. The toes are webbed.

The diploid chromosome number is 52: 9 pairs of metacentric or submetacentric macrochromosomes, 5 pairs of telocentric or subtelocentric macrochromosomes, and 12 pairs of microchromosomes (Bickham, 1975).

Males have concave plastra, longer, thicker tails with the vent beyond the carapace, and green ocelli on the head. Females have flat plastra, the vent beneath the carapace, and yellow ocelli on the head.

Distribution: *Sacalia bealei* occurs in southern China (Fukien, Kwangtung, Kwangsi, and Hainan Island) and northern Vietnam.

Geographic Variation: The subspecies *Sacalia bealei quadriocellata* (Siebenrock, 1903a) is based on individuals with four distinct ocelli on the posterior dorsal surface of the head (Sachsse, 1975; Rödel, 1985). Turtles with only two prominent ocelli are placed in the subspecies *S. b. bealei*. The ranges of the subspecies have not been adequately determined.

Habitat: This species prefers streams and small brooks in woodlands of moderate altitude (100–400 m).

Natural History: Two to six (Pope, 1935; Romer, 1978) elongated, brittle-shelled, white eggs constitute a clutch.

In captivity, *Sacalia* sometimes accepts fish but prefers bananas, oranges, apples and grapes.

Siebenrockiella Lindholm, 1929

The black marsh turtle is the sole representative of this genus.

Siebenrockiella crassicollis (Gray, 1831b)
Black marsh turtle

Recognition: This almost totally dark-brown or black turtle has an oval, depressed, tricarinate carapace (to 20 cm) with a strongly serrated posterior margin. It is widest behind the middle. The medial keel is always distinct, but the two lateral keels on the pleural scutes are less pronounced in larger individuals. Vertebrals of adults are dorsally flattened and the first four are usually wider anteriorly and rather narrow posteri-

Siebenrockiella crassicollis

orly, but the 5th is broader posteriorly than anteriorly. The underlying neural bones are not elongated and are shortest anteriorly; those in the middle of the series are almost square. The plastron is well developed, hingeless, and extensively sutured to the carapace. It has a posterior anal notch, and its lateral borders may be slightly keeled, at least as far posteriorly as the femorals. The humero-pectoral seam crosses the underlying entoplastron. The plastral formula is: abd > pec > fem > an > gul > hum. Width of the bridge is about equal to the length of the plastral hind lobe, and it is well buttressed; the axillary buttress attaches midway along the 1st costal and the inguinal buttress about a third of the way along the suture between the 5th and 6th costals. Axillary and inguinal scutes are moderate to large. Bridge and plastron vary from uniformly black or dark brown to yellowish brown with dark blotches or an extensive dark seam-following pattern. The head is large and rather broad with a short, slightly projecting snout. In the skull, the quadratojugal touches both the jugal and postorbital, and the posterior palatine foramen is slightly larger than the small orbito-nasal foramen. The upper jaw is medially notched, with a narrow, ridgeless triturating surface. The head is black to dark gray with a faded white, cream, or yellow spot behind each orbit, and cream to tan jaws. Posteriorly, the head is covered with small scales, and another narrow strip of small granular scales lies between the orbit and tympanum. The black or dark-gray neck is thickened, making the head appear shorter. Limbs and tail are dark gray to black. Toes are webbed, and the anterior surface of the forelegs is covered with large tranverse scales.

Killebrew (1977a) reported the diploid karyotype to be 52, but previously Stock (1972) and Bickham and Baker (1976a) had found it to be 50. Bickham and Baker (1976a) also reported the single individual they examined was heterozygous for a presumed pericentric inversion. Further study by Carr and Bickham (1981) has confirmed heteromorphism in a pair of macrochromosomes in the male, which they interpreted as an XX/XY sex-determining system, the first discovered in the Emydidae and the only such system in turtles other than that found in the kinosternid genus *Staurotypus*.

Males have slightly concave plastra and thicker, longer tails than do the flat-plastroned females. Females retain the light head spots, while these spots fade out in males.

Distribution: This common turtle ranges from southern Vietnam westward through Thailand to Tenasserim in Burma, and southward through Malaya to Sumatra, Java, and Borneo.

Habitat: *Siebenrockiella* lives in water bodies with slow current, soft bottoms, and, often, abundant vegetation; shallow streams, rivers, ponds, lakes, marshes, and swamps may all afford suitable habitats.

Natural History: A captive male we observed courting frequently bobbed his head at the female as he pursued her around the tank. According to Dr. Edward Moll, *Siebenrockiella crassicollis* in Malaysia may lay three or four clutches of one or two eggs each during a nesting season extending from April through June. Ewert (1979) reported the elongated eggs averaged 45 × 19 mm, and that 12 hatchlings had a mean carapace length of 52 mm.

Predominantly carnivorous, *Siebenrockiella* feeds on worms, snails, slugs, shrimp, and amphibians. It also scavenges dead and decaying animals, and Wirot (1979) reported it will eat rotten plants which have fallen into the water. Most food is captured or consumed underwater, but Wirot (1979) stated that it comes onto land at night to forage for food or to mate.

S. crassicollis is a bottom dweller, spending much of the time partially buried in mud; however, at least in captivity, it occasionally basks.

Clemmys Ritgen, 1828

Clemmys is a genus of small to medium-sized (to 23 cm) turtles that frequent aquatic, semiaquatic, or, occasionally, terrestrial habitats in North America. Their carapace is low arched, smooth surfaced or sculptured with growth annuli, and serrated or not posteriorly. A median keel is present in two species, absent in the other two. The hexagonal neural bones

are short sided anteriorly. The plastron is weakly buttressed and hingeless. Its entoplastron is crossed by the humero-pectoral seam. The skull is short, flattened to slightly convex dorsally, and has a complete temporal arch. Its pterygoid does not meet the basioccipital, but the angular bones touch Meckel's cartilage. The jugal is tapered to a point ventrally and does not touch the pterygoid. The frontal bone enters the orbit, and the maxilla forms part of the inferior temporal fossa. Orbito-nasal foramina are small; triturating surfaces narrow and ridgeless. The toes are webbed, and the dorsal surface of the head is covered with smooth skin.

The karyotype is 2n = 50: 16 macrochromosomes with median or submedian centromeres, 10 macrochromosomes with terminal or subterminal centromeres, and 24 microchromosomes (Bickham, 1975).

Key to the species of *Clemmys*:

1a. Hind foot webbed to base of claws, medial keel lacking, found west of Rocky Mountains: *C. marmorata*
  b. Hind foot webbed to base of penultimate phalanges of three middle toes, medial keel may be present, found east of Rocky Mountains 2
2a. Carapace strongly serrated along posterior rim, strong medial keel: *C. insculpta*
  b. Carapace with nearly or quite smooth posterior rim, medial keel weak or absent 3
3a. Head and carapace black with small yellow spots, temporal region with elongated yellow blotch; carapace never keeled: *C. guttata*
  b. Head and carapace brown to black without small yellow spots, temporal region with large, bright yellow, orange, or occasionally red, rounded blotch; carapace weakly keeled: *C. muhlenbergii*

Clemmys guttata (Schneider, 1792)

Spotted turtle

Plate 13

Recognition: This small (to 12.5 cm), blue-black turtle has round yellow spots on its broad, smooth, keelless, unserrated carapace. These spots are transparent areas in the scutes, overlying patches of yellow pigment; they may fade with age, and some old individuals are spotless. The ventral surface of the marginals is yellowish and may have a pattern of black blotches at the outer edge in the young; the bridge is marked with an elongated black mark. Vertebral 1 is usually longer than or as long as broad; the other four vertebrals are broader than long. The yellow or slightly orange plastron has large black blotches on the outer parts, which sometimes cover the entire plastron in older individuals. The plastral formula is: an > abd >< pect > gul > fem > hum; there is a posterior notch on the hind lobe. The black head is moderate in size with a non-projecting snout and a notched upper jaw. A broken yellow band is present near the tympanum, and another may extend backward from orbit; yellow spots may adorn the crown. Other skin is gray to black, and occasional yellow spots occur on the neck and limbs.

Males have tan chins; brown eyes; long, thick tails with the vent near the tip; and slightly concave plastra. Females have yellow chins, orange eyes, and flat or convex plastra, which are slightly longer than those of the males and extend closer to the carapacial margin.

Distribution: *Clemmys guttata* ranges from southern Ontario and Maine southward along the Atlantic Coastal Plain to northern Florida, and westward through Ontario, New York, Pennsylvania, central Ohio, northern Indiana, and Michigan to northeastern Illinois.

Geographic Variation: Although some geographic variation has been recorded, no subspecies have been described.

Habitat: *Clemmys guttata* is equally at home on land or in water. Although most frequently found in bogs and marshy pastures, it also occurs in small woodland streams. It requires a soft substrate and prefers some aquatic vegetation. In a marshy pasture in Lancaster County, Pennsylvania, where all three eastern *Clemmys* occur, *C. guttata* was the most aquatic, followed by *C. muhlenbergii* and *C. insculpta*, in that order.

Natural History: Mating begins in March and the breeding season continues into May. Courtship includes frantic chases of the female by one or several males. The chase may cover 30–50 m and last 15–30 min, and takes place in shallow water and on the adjacent land. The female's hind legs and tail are sometimes bitten during the chase. When the female is finally caught the male mounts her from behind, and, while tightly grasping her carapace, places his tail beneath hers. At times the male slides off and comes to lie on his side; such copulating pairs form an L-shaped figure. Copulation is usually underwater, but may occur on land. Nesting occurs in June, with the nest usually dug in well-drained loamy soil. The nests are flask shaped and about 50 mm in depth. One or two

clutches of two to eight, usually three to four, elliptical, white eggs (31–34 × 15–17 mm) with flexible shells are laid each year. Estimated natural incubation periods are 70–83 days. Hatchlings may overwinter in the nest. Hatchlings are blue black and usually have one yellow spot on each carapacial scute except the cervical, which has none; some hatchlings lack spots. The yellow plastron has a black central figure. The head is spotted and in some the neck is spotted as well. Dimensions are as follows: carapace length 28.0–31.2 mm, carapace width 28.5–33.1 mm; plastron length 25.2–26.9 mm, and plastron width 15.8–16.3 mm. The head and the tail are larger in proportion to shell length than those of adults.

Clemmys guttata is omnivorous. Surface (1908) found animal food in all 27 stomachs he examined and vegetation in three. The animals eaten included worms, slugs, snails, small crustaceans, crayfish, millipedes, spiders, and insects of the orders Ephemerida, Plecoptera, Odonata, Hemiptera, Neuroptera, Lepidoptera, Coleoptera, Diptera, and Hymenoptera. Many of the insects were not aquatic species; perhaps these were captured on land. Conant (1951a) observed *C. guttata* eating frogs, earthworms, grubs, and the grass growing in a flooded meadow. Pennsylvania specimens examined by us contained only filamentous green algae in their stomachs. Our captive specimens feed well on fresh and canned fish, cantaloupe, and watermelon.

Spotted turtles prefer cool temperatures. Ernst found them mating at cloacal temperatures of 8–10°C; that is, at or below the minimum voluntary temperatures of other aquatic turtles. During the warm summer they are not easily found; they estivate in the mud bottom of waterways or in muskrat burrows or lodges. Some also estivate on land. In winter, *C. guttata* hibernates in similar sites.

Clemmys insculpta (LeConte, 1830)

Wood turtle

Recognition: This medium-sized (to 23 cm) turtle has a keeled, sculptured carapace which is broad, low, and rough: each large scute supports an irregular pyramid formed by a series of concentric growth annuli and grooves. Vertebrals are broader than long, and posterior marginals are strongly flared and serrated. The carapace is slightly widened posteriorly and may be slightly indented at the bridge. It is gray to brown, often with black or yellow lines radiating from the upper posterior corners of the pleurals. Undersides of the marginals and bridge often have dark blotches along the seams. The yellow plastron has a pattern of oblong dark blotches on each scute. Its hind lobe is

Clemmys insculpta

notched posteriorly, and the plastral formula is: an > gul > abd >< pect > fem > hum. The blackish head is largest in the *Clemmys*, but is still moderate in size. A nonprojecting snout and a notched upper jaw are present. Other skin is dark brown, often with some orange or red pigment on the neck and forelegs. The tail is rather long.

The male has a long, thick tail, with the vent posterior to the carapacial rim; a concave plastron with a deep end notch; and prominent scales on the anterior surface of the forelimbs.

Distribution: The wood turtle ranges from Nova Scotia south to northern Virginia and west through southern Ontario and New York to northeastern Ohio, Michigan, Wisconsin, eastern Minnesota, and northeastern Iowa.

Geographic Variation: No subspecies have been described, but those from west of the Appalachian Mountains are paler than eastern individuals.

Habitat: Next to the box turtles *(Terrapene)* and the tortoises *(Gopherus)*, this is the most terrestrial North American turtle. It can be found in most habitats within its range. We have observed it in deciduous woods, woodland bogs, and marshy fields (in Pennsylvania, Maryland, Virginia, and New York).

Natural History: Courtship and mating have been observed in the wild on 26 March, 14 May, and 1 October, and in captives during December. The procedure includes a sort of dance, in which the male and the female approach each other slowly with necks extended and heads held high; when the turtles are within about 20 cm of touching each other they suddenly lower their heads and swing them from side to side—a movement that may continue for as long as two hours without stopping (Carr, 1952). Copulation occurs in the water. The male holds the female's shell with all four feet, his forefeet close together around the front edge of her carapace and the claws of his hind feet clamped under the rear edge. Males are usually more aggressive, but females sometimes take the initiative. Nesting occurs in late May and June. The nest is flask shaped and about 100 mm deep. The elliptical eggs are whitish, have smooth, thin shells, and are about 40 mm long and 26 mm wide. Clutch size is from 4 to 18; 7 or 8 is usual. Natural incubation takes 70–80 days. Hatchlings are gray brown and lack any orange or red pigment on the neck or legs. The tail is long: about equal to the carapace length. The keelless shell is low and about as broad as long (30–36 × 27–30 mm).

The wood turtle is omnivorous. Plants eaten include filamentous algae, moss, grass, willow leaves, strawberries, blackberries, and sorrel. Animal foods include a wide variety of insects and mollusks as well as earthworms and tadpoles. Surface (1908) found plant remains in 76% and animal remains in 80% of the 26 specimens he examined. He also found bird remains, indicating scavenging tendencies. Captives readily eat apples and canned dog food, and they relish hard-boiled eggs.

The wood turtle is diurnal and often wanders about on land during the midday hours. It is fond of basking in the morning, on a log in the middle of a large creek, for example. In the dry summer months it often soaks in mud puddles. It is active from late March to mid-October. The winter months are spent in hibernation in the mud bottom of some waterway or in a hole in the bank.

Clemmys muhlenbergii (Schoepff, 1801)

Bog turtle

Plate 13

Recognition: The bog turtle is small (to 11.5 cm) with a large, bright blotch on each side of the head. Its elongated, rough carapace is moderately domed and has an inconspicuous keel. Sides of the carapace are nearly parallel or slightly divergent behind, and its posterior rim is smooth or only slightly serrated. All vertebrals are broader than long; the 5th is longest. Color varies from light brown through mahogany to black; each scute usually has a light center. Lower sides of the marginals and bridge are the same color as the rest of the carapace. The plastron is dark brown to black, with a few irregularly dispersed light marks. The plastral formula is: abd > an > gul >< fem > hum >< pect; there is a posterior notch on the hind lobe. The small head is brown with a large, usually yellow or orange but sometimes red blotch above and

Clemmys muhlenbergii

behind the tympanum. The snout is nonprotruding and the upper jaw medially notched. Other skin is brown and may be mottled with red above and orange or red below.

Males have long, thick tails, with the vent posterior to the carapacial margin; concave plastra; and thick foreclaws. Females have high, wide carapaces and flat plastra.

Distribution: *Clemmys muhlenbergii* has a discontinuous range in the eastern United States. The main range is from western Massachusetts, Connecticut, and eastern New York southward through eastern Pennsylvania and New Jersey to northern Delaware and Maryland. It is also found in northwestern New York, northwestern Pennsylvania, southern Virginia and adjacent western North Carolina, northern South Carolina and Georgia, and eastern Tennessee. There is a doubtful record from Rhode Island.

Habitat: Sphagnum bogs, swamps, and marshy meadows having clear, slow-moving streams with soft bottoms are the preferred habitat. *Clemmys muhlenbergii* occurs from sea level to elevations of more than 1200 m (in the Appalachians of North Carolina).

Natural History: May through early June is the mating period. Females remain secluded in the mating season: the males must search them out. The only data on courtship we have are the observations of Cramer, as reported by Barton and Price (1955). A captive pair was provided with a piece of ground and a pool whose depth was more than twice the height of the turtles. Several days after capture the male mounted the female underwater, hooking the claws of all four feet under her marginals. He then thumped his plastron against her carapace several times, making a noise like two turtles shaken together in a bag. She withdrew her head and he moved forward with-

out losing his footholds; then, putting his head down in front of her, he blew bubbles through his nostrils. This procedure was repeated two or three times, and all the while the male continued thumping lightly against the female's carapace. Next he moved back as far as possible without losing his footholds and attempted to copulate. The entire performance was repeated several times at later dates, and each time both participants were entirely submerged. Nesting occurs in June and July, and the female does not always dig a cavity, but may instead tuck the eggs up under moss or grass tussocks. Probably only one clutch is laid each season, consisting of three to six eggs. The elliptical, flexible shelled eggs are 28–31 × 14–16 mm. Most hatchlings emerge in late August and September; some may overwinter in the nest. Hatchlings are 23–34 mm in carapace length; the width is about 80% of this. They have dark-brown, immaculate carapaces and yellow plastra with a large, dark central figure. The bright head blotches are well developed.

Clemmys muhlenbergii is omnivorous, and it feeds both on land and underwater. Surface (1908) found that the stomach of one individual contained 80% insects and 20% berries. Barton and Price (1955) reported the stomach contents of two adults consisted primarily of insects, with one lepidopterous larva constituting nearly half the total amount in each stomach. Beetles were the next most common food item, followed by the fleshy seeds of pondweed (*Potamogeton*). Large numbers of sedge (*Carex*) seeds had been consumed by these turtles and by nine others subsequently examined. Also represented were caddisfly larval cases and the cocoons of a parasitic hymenopteran or dipteran. The upper intestine of one turtle contained the shells of several young snails (*Succinea ovalis*); that of the other had pieces of a millipede and a cranefly wing. The most common item found in the feces of six turtles caught in mid-August was the exoskeleton of the Japanese beetle (*Popillia japonica*).

Clemmys muhlenbergii is active only during the warmer parts of the day. After emerging from its nocturnal shelter, it basks for a time before foraging. Our observations indicate that the bog turtle requires more heat to initiate and support activity than does *C. guttata, Chrysemys picta,* or *Terrapene carolina;* on many days sufficiently warm and sunny to arouse these turtles, *C. muhlenbergii* remains in seclusion. A muskrat burrow is often used as a hibernaculum. Bog turtles emerge from hibernation in April and return in mid-autumn. They are most active in April, May, June, and September, and they may estivate during the dry months (July and August). An estivating individual was found embedded in hard clay under a board on 8 August. The burrowing habit is well developed: when alarmed the bog turtle rapidly digs into the mucky substrate with which it usually is associated. When handled it retires into its shell, and it rarely makes any attempt to bite or scratch.

Clemmys marmorata (Baird and Girard, 1852)

Pacific pond turtle

Recognition: This is a small to medium-sized (to 19 cm) turtle with a low carapace and, usually, a pattern of spots or lines that radiate from the centers of the scutes. The short, broad carapace is smooth, keelless, widest behind the bridge, and unserrated posteriorly. Vertebrals are broader than long; the 1st touches four marginals and the cervical. The carapace is olive, dark brown, or black, and the pattern may be absent in some individuals. Undersides of the marginals and bridge are marked with irregular dark blotches or lines along the seams. The pale-yellow plastron sometimes carries a pattern of dark blotches along the posterior margins of the scutes. Its hind lobe is notched posteriorly, and the plastral formula is: an > abd >< pect > gul > fem > hum. The head is moderate with a nonprojecting snout and a medially notched upper jaw, it may be plain gray to olive or have numerous black speckles or reticulations, and the jaws and chin are pale yellow. Other skin is gray, with some pale yellow on the neck, forelimbs, and tail.

In males the plastron is concave; in females it is flat. In males the vent is posterior to the carapace margin; in females it is anterior to the margin.

Distribution: *Clemmys marmorata* ranges chiefly west of the Cascade-Sierra crest from extreme southwestern British Columbia to northwestern Baja California. Isolated colonies exist in the Truckee and Carson rivers in Nevada. There is a questionable record from Jerome County, Idaho.

Clemmys marmorata marmorata

Geographic Variation: Two subspecies are currently recognized. *Clemmys marmorata marmorata* (Baird and Girard, 1852) ranges from British Columbia south to San Francisco Bay and also occurs in western Nevada. It has a pair of well-developed, triangular inguinal scutes on the bridge, and the neck and head are well marked with dark dashes. The throat is pale in contrast with the sides of the head. *C. m. pallida* Seeliger, 1945 occurs from San Francisco Bay south into Baja California. It can be distinguished by its poorly developed inguinal scutes (absent in 60% of individuals) and by the uniform color of the throat and neck. The Baja population may prove to be an additional subspecies.

Habitat: This is the most aquatic species of *Clemmys*. It inhabits ponds, marshes, slow-moving streams, and irrigation canals with rocky or muddy bottoms and abundant aquatic vegetation. It also occurs in clear, swift streams to an elevation of 1800 m and has been reported from brackish coastal waters.

Natural History: Nesting occurs from late April through August; the peak period is June to mid-July. Most nests are dug in the morning. Usually they are located along stream or pond margins, and full sunlight seems to be required. Clutch size ranges from 3 to 11. The hard, white eggs are elliptical-oval, measuring 30.0–42.6 mm in length and 18.5–22.6 mm in width. The incubation period is probably about 70–80 days; hatchlings are about 25 mm in carapace length. Their rounded and keeled carapace is brown or olive, with yellow markings at the edge of the marginals. Carapacial scutes have numerous small tubercles, which give them a grainy appearance. The plastron is yellow and has a large, irregular, dark central blotch. Head, limbs, and tail are marked with pale yellow.

Clemmys marmorata is omnivorous but shows a strong preference for animal food. It feeds on the pods of the yellow lily (*Nuphar polysepalum*), adult and larval insects, fish, worms, and crustaceans. It also takes carrion. Bogert (in Carr, 1952) saw a captive pick up a large sow bug on land and carry it to water before attempting to swallow it. Captives feed well on canned and fresh fish, liver, raw beef, and canned dog food.

C. marmorata often can be seen sunning itself on rocks, logs, mudbanks, or mats of vegetation. Shy and wary, it dives into the water at the least disturbance. According to Pope (1939), during the middle of the day those not basking rest on the bottom of pools, but in the early morning and evening they move upstream or downstream, from one pool to another.

Terrapene Merrem, 1820

These New World emydids are characterized by a high domed carapace and a plastron with a single hinge between the pectoral and abdominal scutes which divides the plastron into two movable lobes. A medial keel is present on some species but not others, and the unserrated posterior carapacial rim may or may not be flared. Growth annuli may sculpture the carapacial surface in younger individuals, but become worn smooth with age. Neurals are hexagonal and shortest anteriorly; the 8th is absent so the corresponding costals meet at the carapacial midline. There is no bridge or corresponding buttresses; the plastron is sutured to the carapace. The entoplastron is intersected by the humero-pectoral seam. The skull is short with an incomplete temporal arch. The pterygoid contacts the basioccipital. The jugal is tapered to a point ventrally and does not touch the pterygoid, and the frontal bone barely enters the orbit. Orbito-nasal foramina are small. Triturating surfaces are narrow and lack ridges. The toes are partially webbed, and the dorsal surface of the head is covered with smooth skin.

The karyotype is 2n = 50: 26 macrochromosomes, 24 microchromosomes (Stock, 1972; Killebrew, 1977a).

Key to the species of *Terrapene*:

1a. 1st vertebral scute elevated at a steep angle (50° or more); plastral hinge usually opposite 5th marginal scute — 2
 b. 1st vertebral scute elevated at a low angle (45° or less); plastral hinge usually opposite seam between 5th and 6th marginal scutes or opposite 6th marginal — 3
2a. Carapace height more than 42% of carapace length; *T. carolina*
 b. Carapace height less than 40% of carapace length; *T. coahuila*
3a. Interabdominal seam length 38% or more of plastron hind-lobe length; interfemoral seam length 16% or less of hind-lobe length; interanal seam length 46% or less of hind-lobe length: *T. nelsoni*
 b. Interabdominal seam length 32% or less of plastron hind-lobe length; interfemoral seam length 18% or more of hind-lobe length; interanal seam length 47% or more of hind-lobe length: *T. ornata*

Terrapene coahuila Schmidt and Owens, 1944

Coahuilan box turtle

Recognition: This uniformly colored species is the only truly aquatic American box turtle. Its elongated (to 16.8 cm), narrow carapace is domed, but medially flattened (depth usually less than 40% of the length) with, at best, only vestiges of a medial keel. Vertebral 1 is not straight sided, but instead wedge shaped; 2–4 are very flat dorsally, and all five are broader than long. There may be a hump on the 5th. Posterior marginals are not serrated. The carapace is brown to olive and lacks any pattern. The plastron is large and well developed with no posterior notch on the anal scutes. The plastral formula is: an > abd > gul > pect > hum > fem. Smith and Smith (1979) found that the interpectoral seam length averages 30% and the interhumeral seam length 20% of the length of the anterior plastral lobe. The short interfemoral seam averages only 11% of the posterior plastral lobe length. A large axillary scute is usually present on each bridge. The plastron is yellow to olive with dark seams and some irregular dark flecking on each scute. The skull contains a heavy postorbital bar. The grayish brown to olive head is large with a strongly hooked and notched upper jaw. There may be some dark mottling on the head. Limbs, neck, and tail are also grayish brown to olive. There are five foretoes and four hind toes, all with little webbing. None of the hind toes is capable of medial rotation. *T. coahuila* is the only *Terrapene* with cloacal bursae.

The plastron of the male is concave, that of the female convex or flat. The male iris is brownish and flecked with yellow, that of the female is yellowish and flecked with brown. The female carapace is slightly higher than that of the male, and she has a shorter, less thick tail.

Distribution: *Terrapene coahuila* is restricted to the intermontane Cuatro Ciénegas basin of Coahuila, Mexico.

Terrapene coahuila

Habitat: This aquatic turtle occurs in shallow waters with soft bottoms and abundant sedges, waterlilies *(Nymphea)*, and reeds *(Phragmites)*. It usually is found in water of slow current and has been taken from streams, ponds, and marshes.

Natural History: Mating has been observed in the fall (September and November), winter (December), and spring (March through May) (Brown, 1974). During courtship the male pursues the female with his head extended and frequently bumps her shell with his carapace. Copulation may occur in or out of the water, with the male gripping the female with his claws and at times biting her on the neck and head. The nesting season extends from May to September, and Brown (1974) thought several clutches could be laid. The elongated (33 × 17 mm) eggs have smooth, finely granulated shells, and clutches include one to four eggs ($\bar{x}$ = 2.3; Brown, 1974). Hatching occurs in late summer or early fall. Hatchlings are more brightly colored than adults with yellow and black radiations or spots on the carapace and a yellow postorbital stripe extending backward across the tympanum; the carapace length is about 33 mm.

Terrapene coahuila is omnivorous and can feed on land as well as in water. In the wild it feeds on *Chara* and *Eleocharis*, and various insects, crustaceans, snails, and small fishes. In captivity it accepts a variety of foods such as fish, worms, insects, lettuce and fruits. *T. coahuila* is an active forager, hunting out its prey, but it probably also scavenges.

Brown (1974) estimated the home range of *T. coahuila* to be only 25.6 m in diameter, with population densities of 133–156 adults per hectare of marsh. The turtles spend much of their time buried in the mud bottom or up under overhanging grasses. Although active all year, cooler winter temperatures may temporarily curtail activities.

While other species of box turtles readily float in water and have difficulty submerging or cannot submerge, *T. coahuila* is capable both of submerging with ease and remaining under water for considerable periods. It is interesting, in this respect, that Williams and Han (1964) compared the densities of *T. coahuila* and *T. carolina* and found *T. coahuila* to be much more dense (0.95 and 0.96 g/cm³ for *T. coahuila* versus 0.73 and 0.83 g/cm³ for *T. carolina*). Also, their cloacal bursae may play a role in removal of excess water after diving. Milstead (1967) speculated that *T. coahuila* evolved in response to increasing rigors of terrestrial life as the Cuatro Ciénegas basin, in which it was trapped, became increasingly more arid. *T. coahuila* is endangered owing to its shrinking habitat.

Terrapene carolina (Linnaeus, 1758)

Common box turtle

Recognition: *Terrapene carolina* (to 20 cm) has a keeled, high-domed, elongated carapace highest posterior to the plastral hinge. It is rounded dorsally and not serrated posteriorly; a prominent medial keel is usually present on the 2d to 4th vertebral scutes. Vertebral 1 is elevated at a steep angle (50° or more), and the 1st marginal is usually rectangular. All vertebrals are broader than long. The carapace is brown, usually with an extremely variable pattern of yellow or orange radiating lines, spots, bars, or irregular blotches on each scute. The plastron is often as long as or longer than the carapace, and its lateral rim may be indented at the seam between the femoral and anal scutes. Frequently an axillary scute is present at the 4th marginal. The plastral formula is: an > abd > gul > pect > hum >< fem; there is no posterior notch on the hind lobe. In color the plastron is tan to dark brown; it may be patternless, show dark blotches or smudges, or have a dark central area with branches along the seams. The head is small to moderate in size with a nonprotruding snout and a medially hooked upper jaw lacking a notch.

In most adult males the iris is red; in females it is yellowish brown. The posterior lobe of the plastron is concave in males, flat or slightly convex in females. The claws of the hind foot are, in males, short, stocky, and considerably curved; those of females are longer, more slender, and straighter. Males have longer and thicker tails than do the females.

Distribution: *Terrapene carolina* ranges from southern Maine south to the Florida Keys and west to Michigan, Illinois, eastern Kansas, Oklahoma, and Texas. It has been reported from isolated localities in New York and western Kansas. In Mexico it occurs in the states of Campeche, Quintana Roo, San Luis Potosí, Tamaulipas, Veracruz, and Yucatán.

Geographic Variation: There are six extant subspecies, four in the United States and two in Mexico. *Terrapene carolina carolina* (Linnaeus, 1758) ranges from southern Maine south to Georgia and west to Michigan, Illinois, and Tennessee. It has a short, broad, brightly marked carapace with the marginals nearly vertical or only slightly flared, and four toes on each hind foot. *T. c. major* (Agassiz, 1857) ranges along the Gulf Coastal Plain from the Florida Panhandle to eastern Texas. This, the largest four-toed box turtle, sometimes exceeds 20 cm in carapace length. It has an elongated carapace on which the markings often are absent or are obscured by black or tan pigment; the rear margin flares strongly outward. *T. c. triunguis*

Terrapene carolina carolina

Terrapene carolina yucatana

Terrapene carolina mexicana

(Agassiz, 1857) ranges from Missouri south to Alabama and Texas. It usually has only three toes on the hind foot. The carapace typically is tan or olive with an obscure pattern. Orange or yellow spots usually are conspicuous on both the head and the forelimbs, but in males the head is often totally red. *T. c. bauri* Taylor, 1895 is restricted to peninsular Florida and the Keys. It, too, usually has only three toes on the hind foot, but the carapace has a bright pattern of light radiating lines, and there are two characteristic stripes on each side of the head. *T. c. yucatana* (Boulenger, 1895) is restricted to the Yucatán Peninsula in the Mexican states of Campeche, Quintana Roo, and Yucatán. This four-toed subspecies has a long, high-domed carapace with the 3d vertebral elevated into a small hump, and little flaring of the posterior marginals. The carapace is tan or straw colored with dark radiations or black scute borders. *T. c. mexicana* (Gray, 1849) occurs only in a small area in Mexico, including southwestern Tamaulipas, northeastern San Luis Potosí, and northern Veracruz. It is a three-toed turtle with an elongated, high-domed carapace having the 3d vertebral elevated into a small hump, and only moderate flaring of the posterior marginals. It is often patterned like *T. c. triunguis*, but may also be pale yellow with dark seams.

Habitat: *Terrapene carolina* is predominantly a species of open woodlands, but it also occurs in pastures and marshy meadows. In Mexico, it occurs in scrub forests and brushy grasslands.

Natural History: Courtship and mating usually occur in the spring but may extend through the summer into autumn. Data collected by Evans (1953) from 72 matings of *T. c. carolina* show that courtship is divided into three phases: a circling, biting, and shoving phase; a preliminary mounting phase; and a copulatory phase. In phase 1 the male approaches the female but stops when about 10 cm away, with his legs straightened, his head held high, and often with one leg raised above the ground. The female retracts her head but watches him. He then walks around the female, nipping her shell as he goes, or pushes her shell a few degrees upon its axis, bites it, and then pushes it farther around and bites it again. Up to an hour may elapse during this phase, depending on the readiness of the female to open her plastron. He eventually mounts and almost instantly hooks his toes into the posterior plastral opening, where she holds them tightly. Apparently, titillation of the claws upon the posterolateral edges of the female's carapace is the final stimulus inducing her to open the rear part of her plastron. In phase 2 the male's hind feet follow the edge of the plastron forward and, when near the hinge, the claws hook on and the rear plastron closes upon them. The stimuli that bring about phase 3 are the contact of plastron and carapace; the downward projection of the male's head in front of the female's head as he bites the forward edge of her shell; the touch of his forefeet upon her shell; and the slight motion of his pinioned claws on the plastral edge. For copulation, the male slips backward until his carapace rests on the ground, while the rear ankles of the female press downward and medially upon his feet, which have shifted farther in under her carapace. After several seconds he leans still farther back and then returns to the vertical position, in which intromission occurs. Female *T. carolina* may lay viable eggs for up to 4 years after mating (Ewing, 1943); tubular albumen-secreting glands in the oviduct serve as seminal receptables for sperm storage (Hattan and Gist, 1975). Nesting occurs from May through July. The nest site usually is in an open, elevated patch of sandy or loamy soil; the flask-shaped nests are 75–100 mm deep. The eggs are elliptical and have thin, white shells. They are 24–40 mm in length, 19–23 mm in width. Clutches range from three to eight eggs (four to five is usual), and several are laid each year. Normally about 75–90 days are needed for incubation. Hatchlings have a flat, brownish-gray carapace (30–33 mm) with a yellow spot on each large scute and a medial keel. The plastron is yellow to cream, with a brown central blotch; the hinge is nonfunctional.

T. carolina is omnivorous. When young it is chiefly carnivorous, but it becomes more herbivorous with age. Animals eaten include snails, slugs, worms, insects, crayfish, spiders, millipedes, frogs, salamanders, lizards, snakes, smaller turtles, and small mammals. They often feed on carrion.

In the summer, activity of *T. carolina* is largely restricted to mornings or after rains. Often these turtles avoid the heat of the day by sheltering under rotting logs or masses of decaying leaves, in mammal burrows, or in mud; in the hottest weather they frequently enter shaded shallow pools and puddles and remain there for periods varying from a few hours to a few days. In other words, thermoregulation is accomplished by seeking a more suitable microenvironment. In the northern part of its range *T. carolina* enters hibernation in late October or November. In the deep south it may remain semiactive throughout the winter. When entering hibernation these turtles burrow into loose soil, sand, vegetable debris, the mud of ponds or stream bottoms, or old stump holes; they may enter mammal burrows. The same site may be used in successive winters. They go deeper as the soil temperature drops.

Terrapene ornata (Agassiz, 1857)

Ornate box turtle

Recognition: The carapace (to 14 cm) is generally round or oval and high domed; the highest point is directly above or anterior to the plastral hinge. It is without posterior serrations and is flattened dorsally. Essentially the carapace is keelless, as only rarely does a weakly developed medial keel occur on the posterior half of the 3d and anterior half of the 4th vertebrals. The 1st vertebral is elevated at a low angle (45° or less), and the 1st marginal usually is irregularly oval or triangular. Usually all vertebrals are broader than long. The carapace is dark brown to reddish brown, often with a yellow middorsal stripe; each scute shows radiating yellowish lines. The plastron is often as long as or longer than the carapace, and its lateral margin is not usually indented. The plastral formula is: an > abd >< gul > fem >< pect > hum; the hind lobe lacks a posterior notch. No bridge is present, and there is usually no axillary scute (if present it is at the 5th marginal). The plastron has a pattern of radiating lines on each scute. The head is small to moderate in size with a nonprotruding snout and an unnotched upper jaw. It is brown to green with yellow spots dorsally and yellow jaws. Other skin is dark brown with some yellow spotting. The tail may have a yellow dorsal stripe. There usually are four toes on each hind foot, rarely only three.

In adult males the iris is red; in females it is yellowish brown. In males the first toe on the hind foot is thickened, widened, and turned in. The hind lobe of the plastron is slightly concave in males, flat or convex in females. Females grow larger than males.

Distribution: *Terrapene ornata* ranges from Illinois, Iowa, South Dakota, and eastern Wyoming south to southwestern Louisiana, Texas, New Mexico, southeastern Arizona, and Sonora and Chihuahua, Mexico.

Terrapene ornata ornata

It also occurs locally in northwestern Indiana, and there is a record from Wisconsin.

Geographic Variation: Two subspecies are recognized. *Terrapene ornata ornata* (Agassiz, 1857) ranges from Indiana and eastern Wyoming south to Louisiana and New Mexico. It is distinguished by the five to eight radiating lines on the 2d pleural and by its generally dark appearance. *T. o. luteola* (Smith and Ramsey, 1952) ranges from the Trans-Pecos region of Texas and southeastern Arizona south into Sonora and Chihuahua, Mexico. It has 11–14 radiating lines on the 2d pleural and is generally yellowish in color. Shells of old individuals often lose their pattern and become uniformly pale green or straw colored; this pigment loss does not occur in *T. o. ornata*.

Habitat: This is a "prairie" turtle, inhabiting treeless plains and gently rolling country with grass and scattered low brush as the dominant vegetation.

Natural History: Courtship and mating occur most commonly in spring, soon after emergence from hibernation but sometimes in summer. Brumwell (1940) described mating in *T. o. ornata*. A male pursued a female for nearly 30 min, first nudging the margins of her shell and later approaching rapidly from the rear and hurling himself on her back while emitting a stream of liquid from each nostril. Presumably the liquid was water: both turtles had imbibed in a pond just before courtship began. After the male had achieved intromission the pair remained in coitus for 30 min; however, the act may last as long as 2 hr. In another instance Brumwell saw four males pursuing a single female; they exhibited the same nudging and lunging behavior. Males that attempted to mount other males were repelled by defensive snapping. The female, too, snapped at some of the males that tried to mount her. One male was finally successful and thereafter was unmolested by the other males. In the several matings that Legler (1960b) observed, the male, after mounting the female, gripped her just beneath her legs or on the skin of the gluteal region with the first claws of his hind feet and used the remaining three claws to grip the posterior edges of her plastron. In most instances she secured his legs by hooking her own around them. This coital position differs from that of *T. carolina*, at least in the position of the male's legs. Nesting extends from early May to mid-July and is most frequent in June. Some females that nest early in the season lay a second clutch in July. Nesting sites that are open, well drained, and have a soft substrate are preferred. The nests are flask shaped and 50–60 mm deep. Clutch size ranges from two to eight; four to six eggs are usual. The ellipsoidal eggs have finely granulated but somewhat brittle white shells (21–41

× 20–26 mm). Natural incubation lasts about 70 days. The almost round hatchlings are approximately 30 mm in carapace length. Their carapace is dark brown to black, with yellow spots on the scutes and a yellow dorsal stripe along the vertebrals. The plastron is yellow to cream colored, with a large, dark central blotch; the hinge is not yet developed. The hinge becomes functional in the fourth year.

Under natural conditions *T. ornata* is chiefly carnivorous; captives, however, eat a variety of vegetable matter as well as meat. Insects (chiefly beetles, caterpillars, and grasshoppers) account for approximately 90% of their natural food. Dung beetles constitute the most important staple element of the diet: the disturbance of piles of dung by turtles in the course of their foraging is a characteristic sign of *T. ornata*. Insects form the bulk of the diet most of the year, but certain other foods (for example, mulberries) are eaten in quantity when especially abundant, sometimes to the exclusion of other foods. *T. ornata* also eats carrion.

The daily cycle of activity of *T. ornata* consists of periods of basking, foraging, and rest that vary in length in keeping with environmental conditions. These turtles emerge from their nighttime burrows, forms, or other places of concealment soon after dawn and ordinarily bask for at least a few minutes before beginning to forage. Although foraging sometimes continues in shady spots throughout the day, it usually ceases between midmorning and noon, when the turtles seek shelter. They remain under cover until midafternoon or late afternoon, when they again become active. Activity of all except nesting females ceases at dusk. As temperatures rise in summer the period of midday quiescence is lengthened and the turtles sometimes spend the warmest hours in pools of water. At least in the southwestern part of its range, activity of *T. ornata* seems to be largely controlled by rainfall: it becomes specially active during and after thunderstorms.

Ornate box turtles begin to enter hibernation in October; by the end of November most or all of them are underground. In Kansas, autumn activity is characterized by movement into ravines and other low places and into wooded strips along fields or small streams (Legler, 1960b), places good for basking and burrowing as well as for protection from the wind. The turtles often use animal burrows along the banks of a ravine for temporary shelter, and the overhanging sod at the edge of the ravine provides cover beneath which the turtles can easily dig. Emergence is in March or April.

Terrapene nelsoni Stejneger, 1925b

Spotted box turtle

Recognition: The narrow to oval carapace (to 15 cm) is domed, but somewhat flattened dorsally, with a poorly developed (sometimes absent) medial keel. Vertebrals are broader than long; posterior marginals are unserrated but flared over the hind limbs. The carapace ranges from yellowish to tan to dark or greenish brown and often contains a pattern of scattered small brown to yellow spots. The plastron is large and well developed with no posterior notch on the anal scutes. The plastral formula is: an > abd > gul > pect > fem > hum. Axillary scutes are usually absent, but when present, are at the 5th marginals. The dark brown plastron is yellow bordered and may contain yellow spots and streaks. The head is large with a strongly hooked upper jaw. In the skull, no process arises from the postorbital bone, and the postorbital bar is absent. Skin of the head, neck, limbs, and tail is yellow to brown with brown or yellow spots. Four webbless toes are present on the hind foot.

The plastron is shallowly concave in males and flat or convex in females. The first toe of the male hind foot is capable of medial rotation. Males also have thicker and longer tails than females.

Distribution: *Terrapene nelsoni* occurs in disjunct populations in southern Sonora, Sinaloa, and Nayarit in western Mexico.

Geographic Variation: Two subspecies are recognized. *Terrapene nelsoni nelsoni* Stejneger, 1925b is restricted to the vicinity of Pedro Pablo, Nayarit, Mexico. Its carapace is straw colored to tan or dark brown and the light spots are larger and less numerous. The interhumeral seam and the interpectoral seam lengths average 16 and 35%, respectively, of the ante-

rior plastral lobe length (Iverson, 1982b). Also, its 1st vertebral scute is concave in shape. *T. n. klauberi* Bogert, 1943 occurs in southwestern Sonora and near Terreros in northwestern Sinaloa (Hardy and McDiarmid, 1969). Its carapace is tan to dark brown or greenish brown with numerous, smaller light spots than in *T. n. nelsoni*. Its interhumeral seam and interpectoral seam lengths average 18 and 33%, respectively, of the anterior plastral lobe length (Iverson, 1982b). Its 1st vertebral is flattened. Spotting may be absent from either subspecies.

Habitat: This species inhabits hill country covered with savannah, oak woodlands, or dry scrub forest. *T. n. nelsoni* has been collected only at elevations over 1050 m, while *T. n. klauberi* is known only from elevations below this. Hardy and McDiarmid (1969) found numerous burrows on a hillside where they collected *T. n. klauberi* in Sinaloa, and attributed some of these to box turtles.

Natural History: Nothing is known of courtship and mating in *Terrapene nelsoni*, but the rotating first hind toe is probably used to clasp the female, as in *T. ornata*. Milstead and Tinkle (1967) dissected female *T. n. nelsoni* and found the number of eggs to range from one to four with three the usual number. They found no evidence of second clutches. The elongated eggs averaged 47 × 27 mm.

Captives are omnivorous, feeding on various vegetables, worms, insects, and canned dog food.

Emydoidea Gray, 1870b

The single living species in this genus, *Emydoidea blandingii*, occurs only in North America.

Emydoidea blandingii (Holbrook, 1838)

Blanding's turtle

Plate 12

Recognition: This medium-sized (to 26 cm) turtle has an elongated, smooth carapace that is neither keeled nor serrated. Its vertebrals are broader than long, and the 1st touches four marginals and the cervical. Neurals are six sided and shortest anteriorly. The carapace is blue black; each pleural and vertebral has tan to yellow irregularly shaped spots or slightly radiating lines, and the marginals are heavily spotted. A

Terrapene nelsoni klauberi

movable hinge lies between the pectoral and abdominal scutes. The entoplastron is intersected by the humero-pectoral seam. The plastral hind lobe bears only a slight posterior notch; the plastral formula is: an > gul >< abd > fem >< hum > pect. There are no plastral buttresses, and the plastron is connected to the carapace by ligaments. The plastron is yellow with large, dark, symmetrically arranged blotches, which may be so large as to hide most of the yellow pigment. The flattened head is moderate in size with a nonprotruding snout and a terminally notched upper jaw. The eyes protrude. There is a bony temporal arch. The orbito-nasal foramen is small, but the posterior palatine foramen is large. The inferior process of the parietal touches the palatine. Triturating surfaces of the jaws are narrow and ridgeless; the upper surface has no contributions from the palatine or pterygoid. The skull is similar to that of *Deirochelys* (McDowell, 1964). The top and sides of the head are blue gray with tan reticulations; the chin and throat are bright yellow; and the upper jaw may be marked with dark bars. Other skin is blue gray with some yellow scales on the tail and legs. The neck is very long, and the toes are webbed.

Stock (1972) reported the karyotype is 2n = 50: 20 metacentric or submetacentric, 10 subtelocentric, and 20 acrocentric or telocentric chromosomes.

Sexual dimorphism is not pronounced. In males the vent is behind the carapacial margin and the plastron is slightly concave.

Distribution: Blanding's turtle ranges from southern Ontario south through the Great Lakes region and west to western Nebraska, Iowa, and northeastern Missouri. It also occurs in scattered localities in eastern New York, Massachusetts, southern New Hampshire, Maine, and Nova Scotia.

Habitat: *Emydoidea* prefers shallow water with a soft bottom and abundant aquatic vegetation; it is found in lakes, ponds, marshes, creeks and sloughs.

Natural History: Mating activity occurs in the spring and early summer. There is little formal courtship behavior; males force the females to the bottom, mount almost immediately, grasping the female's shell with the claws of all four feet, and bring the tails together for cloacal contact. Nesting occurs in June and July; the nest is flask shaped. Several clutches of 6 to 17 eggs are laid each season; eggs are ellipsoidal (28–39 × 21–28mm) with parchmentlike shells. Most hatchlings emerge in September after a 65–80-day incubation period; however, some may overwinter in the nest and emerge the following spring. The rounded, keeled, hatchling carapace is dark brown to black, sometimes with spots, and 29–37 mm in length. The

plastron has a large, black central blotch, and the future hinge is suggested by a crease.

Natural food includes crustaceans, fish, frogs, tadpoles, snails, leeches, insects, and some aquatic plants.

Bramble (1974) has shown that *Emydoidea* is most closely related to *Emys* and *Terrapene*, rather than *Deirochelys*, as was previously believed. His conclusions were based on a shared plastral closing mechanism and other morphological similarities. He thought *Emydoidea* convergent with *Deirochelys* in several features related to similar feeding systems.

Emys Duméril, 1806

The single living species of this genus is the semiaquatic *Emys orbicularis*. It is unique in being the only living Old World representative of the subfamily Emydinae, and also the only turtle to occur over most of Europe, especially north of the Alps.

Emys orbicularis (Linnaeus, 1758)
European pond turtle

Recognition: *Emys orbicularis* has an oval (to 20 cm), moderately depressed carapace which is widest behind the middle, and unserrated. A medial keel is present on juveniles, but, as the carapace grows, this becomes progressively lower until it may be lost. Hatchlings may also possess a pair of low lateral keels. Vertebrals are broader than long, with the 5th the widest. The underlying neurals are hexagonal and shortest sided anteriorly. The carapace ranges in color from olive brown to brown or, most often, black, pat-

Emys orbicularis

terned with numerous yellow radiations or dots. The plastron is large, has a movable hinge between the pectoral and abdominal scutes (between the underlying hyo- and hypoplastral bones), lacks buttresses in adults, and has only a ligamentous connection with the carapace. The hinge cannot completely close the shell in adults, but is more flexible in the young. Axillary and inguinal scutes are usually absent, but, if present, are small and poorly developed. The entoplastron is crossed by the humero-pectoral seam. The plastral formula is: an > abd > pect > gul > fem > hum. The plastron varies from totally black or dark brown to yellow with each scute black bordered. The skull has a complete temporal arch. Its frontal bone does not enter the orbit border, and the lower end of the jugal extends inward along the posterior edge of the maxilla to touch the pterygoid. The posterior palatine foramen is large. Triturating surfaces are narrow, unserrated, and ridgeless; the upper jaw is medially notched. Dorsal skin of the head is not divided into scales. Limbs are covered with small to medium-sized scales, not large ones. The toes are webbed, and the tail is relatively long. Skin of the head, neck, limbs, and tail is brown to black, and may contain fine yellow radiations or spots.

Matthey (1931) reported the diploid chromosome number to be 50.

Males have red eyes and longer, thicker tails than do the yellow-eyed females.

Distribution: *Emys orbicularis* ranges from the Caspian Sea in Iran and the Soviet Union westward through Turkey and eastern Europe to the Baltic states of Latvia and Lithuania, and to central Germany. It also occurs in Greece, Italy, southern France, Spain and Portugal in Europe; on Corsica, Sardinia and the Balearic Islands; and in northwestern Africa in Morocco, Algeria and Tunisia. Formerly the range was more extensive, as postglacial remains have been found in Sweden, Denmark, the Netherlands, and England (Loveridge and Williams, 1957).

Geographic Variation: Arnold and Burton (1978) commented that some variation in coloration occurs, but with no obvious geographic trends. However, subspeciation is to be expected as *E. orbicularis* has an extensive geographic range including areas in several climatic regimes. Unfortunately, a good quantitative study of variation in this turtle has yet to be undertaken.

Habitat: *Emys orbicularis* lives in slow-moving water bodies with soft bottoms (mud or sand) and abundant aquatic vegetation, especially overhanging the banks. It has been taken from ponds, lakes, marshes, swamps, brooks, streams, rivers, and drainage canals.

The water in some of its habitats is quite brackish, and Dr. Konrad Klemmer has told us that around Frankfurt, West Germany, a few still survive in highly polluted waters.

Natural History: Breeding occurs from March to May, depending on the latitude. Courtship observed by Ernst consisted of the male swimming behind and trailing the female, occasionally biting at her hindquarters and bumping her with his shell. He climbed upon her shell and held on with his forelegs while nudging her head and forelegs from above with his snout. Biting at this point was not observed, but has been reported by Street (1979). The male also scratched the female's posterior carapace and hind limbs with his hind limbs; possibly he was trying to secure a foothold. Coition eventually was successful, and the entire courtship and mating sequence occurred underwater. Loveridge and Williams (1957) reported that on warm spring nights during the mating season these terrapins emit short piping calls until they find a mate, after which the couple swim about together. No sounds were made by the turtles observed by Ernst. Nesting occurs in May and June, and, contrary to the report by Loveridge and Williams (1957) that the female may use her ridged tail as a borer, the nest is dug entirely with the hind feet. Three to 16 eggs, usually 9 or 10, are laid at one time. The eggs are elliptical (39–30 × 22–18 mm) and have white, pliable, leathery shells. The incubation period varies with latitude, and at the northern extent of the range a long, hot summer is needed for successful hatching to occur; thus these *E. orbicularis* may only successfully reproduce once in four or five years. Hatching normally occurs from August to October, again depending on latitude. Hatchlings have 20–25-mm carapaces, large heads, long tails, and a carapace with a well-developed medial keel and two lower lateral keels.

At incubation temperatures of 24–28°C, only males are produced, but at an incubation temperature of 30°C, 96% of the hatchlings are females (Pieau; 1971, 1972, 1973).

Emys orbicularis is carnivorous, feeding on insects, crustaceans, mollusks, worms, salamanders, frogs, and fishes. Prey is actively stalked, and feeding may occur in or out of water.

The basking habit is well developed, but this is a shy species which dives into the water at any disturbance. These turtles often hide buried in the soft bottom or up under overhanging vegetation along the bank.

Having such an extensive range brings them under diverse climatic conditions. In the northern parts of their range they are forced to hibernate for long periods, but in the more southerly areas, they often estivate to escape the summer's heat.

Chrysemys Gray, 1844

The only living species in this genus is *Chrysemys picta*, the painted turtle of North America.

Chrysemys picta (Schneider, 1783)

Painted turtle

Recognition: The carapace (to 25 cm) is smooth, oval, flattened, and keelless; the highest and widest points are at the center; and the posterior margin is without serrations. Vertebrals are usually broader than long (although, the 1st may be as long as broad or slightly longer than broad), and the underlying neurals are six sided and shortest anteriorly. The carapace is olive to black, with yellow or red borders along the seams and red bars or crescents on the marginals. Some individuals have a well-developed medial stripe, which is red or yellow. The bridge and hingeless plastron are yellow; often there is a black or reddish-brown plastral blotch of varying size and shape. The entoplastron lies anterior to the humero-pectoral seam. The plastral hind lobe bears only a slight posterior notch, and the plastral formula is: abd >< an > gul > pect > fem > hum. Axillary and inguinal buttresses are moderate in length. The head is moderate in size with a slightly projecting snout, and the upper jaw bears a terminal notch bordered on each side by a cusp. The broad triturating surface of the maxilla bears a weak median ridge. Both the palatine and pterygoid contribute to the upper triturating surface. The temporal arch is complete, and the orbito-nasal foramen is larger than the posterior palatine foramen. The lower parietal process does not always touch the palatine, and the pterygoid does not extend backward to the level of the exoccipital. Skin is black to olive. Neck, legs, and tail are striped with red and yellow; the head is striped with yellow. A yellow line extends posteriorly from below the eye and may meet a similar line from the lower jaw. There is a large, yellow dorsolateral spot and a yellow streak on each side of the head behind the eye. The chin is marked with two wide yellow lines, which meet at the tip of the jaw and enclose a narrow yellow stripe. The 5th toe has only two phalanges; the toes are webbed.

The karyotype is 2n = 50: 26 macrochromosomes (16 metacentric, 6 submetacentric and 4 telocentric) and 24 microchromosomes (Stock, 1972; Killebrew, 1977a).

Males have elongated foreclaws and long, thick tails, with the vent posterior to the carapacial margin. Females are larger in all shell dimensions.

Distribution: This is the only North American turtle that ranges across the continent. It occurs across southern Canada, from Nova Scotia to British Columbia, and south to Georgia, Alabama, Mississippi, Louisiana, Oklahoma, Colorado, Wyoming, Idaho, and Oregon. It is also found in scattered localities in Texas, New Mexico, Arizona, Utah, and Chihuahua, Mexico.

Geographic Variation: Four subspecies are recognized. *Chrysemys picta picta* (Schneider, 1783) ranges from southeastern Canada through New England and the Atlantic coastal states to Georgia and thence west into eastern Alabama. This subspecies has the vertebral and pleural carapacial seams aligned, light borders along the carapacial seams, and a plain yellow plastron. The medial carapacial stripe is narrow; it may be poorly developed or absent. *C. p. marginata* Agassiz, 1857 ranges from southern Quebec to Ontario and south in the central United States to Tennessee and northern Alabama. Its range is east of the Mississippi River, and extends eastward into New England, Pennsylvania, West Virginia, Maryland, and Virginia. It has alternating vertebral and pleural seams, dark borders along the carapacial seams, and a variable dark figure on the plastron. This figure is usually no more than half the width of the plastron, and it does not extend out along the seams. The medial stripe is normally absent or poorly developed. *C. p. dorsalis* Agassiz, 1857 is found from southern Illinois and Missouri southward along both sides of the Mississippi River to the Gulf coast of Louisiana and eastward through the northern part of Mississippi into Alabama; there is a relict population in southeastern Oklahoma. It has a conspicuous red or yellow medial stripe, alternating vertebral and pleural seams, and a plain yellow plastron. *C. p. bellii* (Gray, 1831b) ranges from western Ontario across southern Canada to Brit-

Chrysemys picta picta

ish Columbia and south to Missouri, northern Oklahoma, eastern Colorado, Wyoming, Idaho, and northern Oregon; it is also found in many scattered localities in the southwestern United States and in one area in Chihuahua. This is the largest of the painted turtles; it has alternating vertebral and pleural seams; a reticulate pattern of lines on the carapace; and a large, dark plastral figure, which branches out along the seams and occupies most of the plastral surface. Its medial stripe is absent or poorly developed.

Bleakney (1958) offered an explanation for the present distribution of the subspecies of *C. picta*. He suggested that at the time of the latest retreat of the glaciers, the painted turtles were divided into three separate populations, which may well have been separate incipient species: *C. picta* in the southeastern Atlantic coastal region, *C. dorsalis* in the lower Mississippi River region, and *C. bellii* in the southwest. However, the populations did not develop complete reproductive isolation. The retreat of the glaciers was accompanied by northward extensions of the three populations. According to Bleakney, *C. dorsalis* spread up the Mississippi River and met *C. bellii* near the mouth of the Missouri River; hybridization of these two forms produced *C. marginata*, which spread up the Ohio River valley into the eastern Great Lakes Region. Meanwhile *C. picta* spread northward along the Atlantic Coastal Plain and westward along the Gulf Coastal Plain, eventually meeting *C. marginata* in the north and *C. dorsalis* in the west. Wherever these forms met, they eventually interbred, indicating that the whole complex consists of a single species with four subspecies. Intergradation between the subspecies has been well studied in several regions (Bishop and Schmidt, 1931; Johnson, 1954; Hartman, 1958; Waters, 1964, 1969; Ernst, 1967a, 1970a; Ernst and Ernst, 1971; Ernst and Fowler, 1977; Pough and Pough, 1968; and Groves, 1983).

Habitat: *Chrysemys picta* prefers slow-moving shallow water, as in ponds, marshes, lakes, and creeks. A soft bottom, basking sites, and aquatic vegetation are preferred. Along the Atlantic coast it sometimes enters brackish waters.

Natural History: *Chrysemys picta* is the best studied of all freshwater turtles. Courtship and mating usually occur from March to mid-June but sometimes extend into summer and fall. Courtship observed in Pennsylvania always took place in water at temperatures of 10.0–27.8°C. Courtship begins with a slow pursuit of the female; when at last she is overtaken the male passes and turns to face her. He then strokes her head and neck with the backs of his elongated foreclaws. A receptive female responds by stroking his out-stretched forelimbs with the bottoms of her foreclaws. Between the periods of stroking, the male swims away, seemingly trying to entice the female to follow. After this behavior has been repeated several times the female sinks to the bottom, the male swims behind and mounts her, and copulation begins. Nesting occurs from late May until mid-July, with peak activity in June and early July. Most nests are dug in early evening, but morning nestings have also been reported. The flask-shaped nests are dug with the hind feet in loamy or sandy soil in the open. Average dimensions of 14 Pennsylvania nests were as follows: greatest diameter of cavity, 72 mm (65–72); diameter of the neck, 45 mm (41–51); and depth, 104 mm (99–111). The number of eggs per clutch varies from 2 to 20 and differs with the subspecies. The largest race, *C. p. bellii*, lays the most eggs per clutch: 4–20. The medium-sized subspecies *C. p. picta* and *C. p. marginata* lay 2–12 and 3–10 eggs, respectively. Fifteen clutches of intergrade *C. p. picta* × *C. p. marginata* from Pennsylvania averaged 4.73 eggs (4–6). The smallest subspecies, *C. p. dorsalis*, lays only 2–7 eggs per clutch. Apparently in most populations females may lay two or three clutches a year but not all lay each year. The eggs normally are elliptical, white to cream in color, and have smooth, slightly pitted surfaces; they are flexible when first laid but become firmer as water is adsorbed. Eggs in Pennsylvania averaged 32.1 mm (28.8–35.1) in length, 19.2 mm (17.6–22.2) in width, and 7.2 g (6.1–9.1) in weight. Artifically incubated eggs in Pennsylvania took 65–80 days (mean 76) to hatch; those naturally incubated in the nest took 72–80 days (mean 76). The earliest natural hatching took place on 14 August, the latest on 29 August. Hatchlings from eggs laid late in the season (July or August) often are not ready to emerge from the nest before the onset of cold weather; in that case they apparently overwinter in the nest. Overwintering in the nest is well known and has been reported from many localities in the northern part of the range; it seems to be a well-established protective mechanism in painted turtles. Bull and Vogt (1979) discovered that eggs incubated at high temperatures (30.5°C) produced female hatchlings, but eggs incubated at lower temperatures (25°C) produced males—another case of temperature-controlled sex determination. The hatchling *C. picta* is essentially round and has a keeled carapace. The head, eyes, and tail are proportionally larger than in the adult. A deep crease exists across the abdominal plate. The pigmentation and patterns of the shell and skin are brighter and more pronounced than in adults. The hatchling has an external yolk sac 10–25 mm in diameter, and a caruncle that usually drops off by the fifth day.

Painted turtles are omnivorous: most species of plants and animals, living or dead, found in their habitat may be eaten as opportunity arises. Of 56 Pennsylvania adults, animal food was found in 64% of the stomachs (61.2% by volume) and plant remains in 100% (38.8% by volume). Young painted turtles are carnivorous but become more herbivorous as they mature.

C. picta is diurnal; it spends the night sleeping on the bottom, among vegetation, or on a partially submerged object. It becomes active about sunrise and basks for several hours before beginning to forage in the late morning. Another period of basking follows, and foraging is resumed in the late afternoon, to continue into the early evening. The basking habit is well developed: as many as 50 painted turtles can be seen on a log at a time.

Taxonomic Comment: McDowell (1964) revised the New World emydine genus *Chrysemys* on the basis of skull and foot morphology, including in it *C. picta* and the cooter and slider turtles of the genera *Pseudemys* and *Trachemys*, and suggested that three subgenera were involved *(Chrysemys, Pseudemys, and Trachemys)*. Similarities in the choanal structure of *Chrysemys picta* and various species of *Pseudemys* and *Trachemys* upheld both their placement within the genus *Chrysemys* and McDowell's subgeneric distinctions (Parsons, 1968). Zug (1966) found little variation in the penial structure of *Chrysemys picta* and *Trachemys scripta*, *Pseudemys nelsoni*, *P. floridana*, and *P. concinna*, strengthening the inclusion of these turtles within *Chrysemys*. Weaver and Rose (1967) concurred with the inclusion of *Pseudemys* and *Trachemys* in *Chrysemys*, but showed the subgenera to be invalid, based on further examination of skull and shell characters. Ernst and Barbour (1972) and Conant (1975) accepted *Chrysemys* as the generic name for these turtles. However, there remained much disagreement about the generic arrangement of these turtles, and many experts still maintained that *Pseudemys* (including *Trachemys scripta*) was a separate genus.

Holman (1977) expressed doubts about the status of McDowell's (1964) genus *Chrysemys*. He pointed out that under McDowell's concept as many as four congeneric species may occur in the same water body in the southeastern United States and that, although they have similar courtship patterns, there are no records of hybridization between *Chrysemys picta* and other species of *Chrysemys*. However, hybrids are known within the subgenus *Pseudemys: C. floridana* × *C. concinna* (Smith, 1961; Mount, 1975), and *C. floridana* × *C. rubriventris* (Crenshaw, 1965). Holman (1977) urged additional study of the relationships within the genus. Subsequently, the morphological, cytological, biochemical, and parasitological characteristics have been reevaluated (Ernst and Ernst, 1980; Vogt and McCoy, 1980; Seidel and Smith, 1986).

These new studies indicate that *Chrysemys*, *Pseudemys*, and *Trachemys* are best treated as separate genera.

Trachemys Agassiz, 1857

The six species of slider turtles in this genus occur from North America to northern and central South America and on several Caribbean islands and the Bahamas, making this one of the widest ranging of all turtle genera. At times it has been included in the genera *Chrysemys* or *Pseudemys*, but Seidel and Smith (1986) have presented sufficient evidence to show it is a separate genus. The carapace is oval, moderately domed (especially in females), posteriorly serrated, contains a medial keel, and has neurals that are hexagonal and shortest anteriorly. Carapacial pattern often includes yellow stripes or dark-centered ocelli. There is a tendency for older males to become

Key to the species of *Trachemys*:

1a. Snout blunt and rounded   2
  b. Snout elongated and pointed   5
2a. Upper jaw with distinct terminal notch   3
  b. Upper jaw with no terminal notch, or only a shallow one   4
3a. Plastron with an asymmetrical pattern of scattered black spots or ocelli; carapacial scutes smooth or with only shallow longitudinal rugosities: ***T. decorata***
  b. Plastron usually immaculate, but may have faint blotches on gulars and humerals or along seams; carapacial scutes with deep longitudinal rugosities: ***T. terrapen***
4a. Sides of carapace straight or slightly indented at bridge: ***T. decussata***
  b. Sides of carapace elliptical, widest at or near center. ***T. stejnegeri*** (in part)
5a. An elongated, black-bordered yellow spot on chin at corner of mouth; Brazil, Uruguay and Argentina: ***T. dorbigni***
  b. No elongated, black-bordered yellow spot on chin at corner of mouth   6
6a. Plastron immaculate, or with black seam-following pattern; skin gray to brown or olive brown; Hispaniola, Puerto Rico, Bahamas: ***T. stejnegeri*** (in part)
  b. Plastron with either a dark blotch or ocellus on each scute, a dark, elongated, medial blotch, or an extensive pattern covering most of its surface; skin green to olive brown; United States to Colombia and Venezuela: ***T. scripta***

melanistic, and females are usually larger than males. The plastron is hingeless, the axillary and inguinal buttresses are short to moderately long, and the ento-plastron lies anterior to the humero-pectoral seam. The skull is moderate in size with a complete temporal arch, and the orbito-nasal foramen is much larger than the posterior palatine foramen. The cranium is shallow anterior to the basisphenoid (30–34% of the condylobasal length). The lower parietal process touches the palatine, and the posterior pterygoid touches or is very near the exoccipital. Triturating surfaces of the lower jaws are narrow, and an anterior cusp is absent from the median ridge on the upper jaw. No tuberculate denticles occur on the trituating surface. The upper jaw is usually medially notched, the lower jaw rounded. There are three phalanges in the 5th toe, and the toes are webbed.

Barbour and Carr (1941) and Williams (1956) thought the island series to be derived from mainland populations of *Trachemys scripta ornata* which may have been blown to their respective islands during successive hurricanes.

Fifty chromosomes are present: 26 macrochromosomes (16 metacentric, 6 submetacentric, 4 telocentric) and 24 microchromosomes (Stock, 1972; Killebrew, 1977a; Bickham and Baker, 1976a, 1979).

Trachemys scripta (Schoepf, 1792)

Slider

Plate 14

Recognition: *Trachemys scripta* is a medium to large (20–60 cm) turtle with a prominent patch (or patches) of red, orange, or yellow on each side of the head. Because there are several variable subspecies, the following description is highly generalized; see the section on geographic variation for particulars of color and pattern.

The carapace is oval, weakly keeled, and has a slightly serrated posterior rim. Vertebral 1 is longer than broad or as long as broad; the other four are broader than long. The carapace is olive to brown with yellow markings varying geographically from stripes and bars to reticulations and ocelli. Markings on the marginal scutes also are variable but usually take the form of a dark blotch partly surrounded by a light band. Old males often become black. Bridge markings vary from dark blotches to bars. The hingeless plastron is yellow and exhibits a pattern that varies geographically from a single blotch on each scute (or rarely no pattern) to an extensive pattern covering most of the plastron. Its hind lobe is slightly notched posteriorly and the plastral formula is: abd > an

Trachemys scripta venusta

> fem >< gul >< pect > hum. The head is moderate in size, with a protruding snout (more so in the males of some tropical subspecies) and a medially notched upper jaw. Skin is green to olive brown with yellow stripes. Supratemporal and orbitomandibular head stripes are conspicuous; a postorbital stripe of red, orange, or yellow is usually present; and a prefrontal arrow is formed as the supratemporal stripes pass forward from the eyes to meet a sagittal stripe on top of the snout. The neck is marked with numerous stripes and a central chin stripe runs backward and divides to form a Y-shaped mark. Limbs have numerous narrow stripes.

Males are usually smaller than females and have long, thick tails, with the vent posterior to the carapacial rim. In some subspecies, males have elongated, curved foreclaws.

Distribution: In the United States, *Trachemys scripta* ranges from southeastern Virginia south to northern Florida and west to Kansas, Oklahoma, and New Mexico; thence it ranges through Mexico and Central America to Brazil.

Geographic Variation: *Trachemys scripta* is the most variable of all turtles; 14 subspecies have been described and named. *T. scripta scripta* (Schoepff, 1792), to 27 cm, ranges from southeastern Virginia to northern Florida. It has a wide yellow stripe on each pleural scute; a conspicuous yellow postorbital blotch, which may join a neck stripe; and a yellow plastron, which usually has ocelli or smudges only on the anterior-most scutes. *T. s. elegans* (Wied, 1839), to 28 cm, occupies the Mississippi Valley from Illinois to the Gulf of Mexico. It has a wide red postorbital stripe, narrow chin stripes, a transverse yellow stripe on each pleural, and a plastral pattern of one large dark blotch or ocellus on each scute. *T. s. troostii* (Holbrook, 1836), to 21 cm, occurs in the upper parts of the Cumberland and Tennessee rivers, from southeastern Kentucky to northeastern Alabama. It has a narrow yellow postorbital stripe, broad chin stripes, a transverse yellow stripe on each pleural scute, and a plastral pattern of ocelli or small black smudges. *T. s. gaigeae* (Hartweg, 1939a), to 22 cm, is found in the Rio Grande (Big Bend and above) and Rio Conchos drainages of Texas, New Mexico, Chihuahua, and Coahuila). It has a reticulate carapacial pattern, often with small ocelli; an oval, black-bordered, red to orange spot behind the eye and well separated from it; the chin medially striped, with the lateral stripes shortened to ovals that are almost ocelli; and a plastral pattern that varies from a large blotch on each scute to a large, dark central figure spreading out along the transverse seams. *T. s. taylori* (Legler, 1960a), to 22 cm, only occurs in the Cuatro Ciénegas basin of Coahuila. It resembles *T. s. elegans* and has a supratemporal stripe which terminates abruptly on the neck behind an expanded, red, very elongated postorbital stripe; an extensive black plastral pattern with all parts interconnected; small, scattered, elongate or ovoid dark spots on the carapace; and the pectoral midseam longer than that of the gular. *T. s. cataspila* (Günther, 1885), to 22 cm, occurs on the Gulf Coastal Plain of Mexico from northern Tamaulipas to the vicinity of Punta del Morro, Veracruz. Its yellow supratemporal stripe is wide on the temples, and it has dark-centered ocelli on the pleurals and marginals and a medial plastral figure which does not extend along the interanal seam to the rear edge of the anals. *T. s. venusta* (Gray, 1855), to 48 cm, ranges from the city of Veracruz, Mexico, through Honduras (including the Yucatán Peninsula) in the Atlantic and Gulf drainages. The dark-centered ocelli on its pleural scutes are very large, its supratemporal stripe reaches the eye, and the seam-following plastral pattern is extensive. *T. s. yaquia* (Legler and Webb, 1970), to 31 cm, inhabits the lower parts of the Sonora, Yaquai, and Mayo drainages in Sonora, Mexico. Its postorbital mark is yellowish orange and only moderately expanded, the pleural scutes have only poorly defined ocelli with jagged black centers, and the medial plastral mark is extensive but faded in adults. *T. s. hiltoni* (Carr, 1942a), to 28 cm, is restricted to the Rio Fuerte drainage in Sonora and Sinaloa. Its orange postorbital part of the supratemporal stripe is either isolated anteriorly and posteriorly, or is connected posteriorly with a narrow orbital stripe; there are black smudgelike spots on the upper and lower surfaces of each lateral and posterior marginal scute and some pleural scutes; the plastron has a dark central blotch surrounding a narrow yellow medial area. *T. s. nebulosa* (Van Denburgh, 1895), to 37 cm, occurs in freshwater bodies in southern Baja California. The orange or yellow supratemporal stripe does not reach the eye and ends as a large, oval postorbital spot well behind the eye; the carapace usually lacks ocelli, but may have a pattern of black spots and irregular light marks; the plastron bears a series of smudgelike medial blotches. *T. s. ornata* (Gray, 1831b), to 38 cm, occurs on the Pacific coastal plain of Mexico from northern Sinaloa to central Oaxaca and from Guatamala through Central America to Colombia. The orange postorbital stripe usually starts at the orbit, is expanded over the temple, and continues to the neck; the carapace has dark-centered ocelli on the pleurals; the plastral pattern consists of four concentric, faded medial lines which do not extend to the anal notch. *T. s. grayi* (Bocourt, 1868), to 60 cm, occurs from the Pacific coastal plain of Tehuantepec, Mexico, southeastward to La Libertad, Guatemala. The yellow supratemporal stripe reaches the eye; all head stripes are narrow; the carapace has dark-centered ocelli on the pleurals and marginals; and the plastral figure is diffused, fragmented, and faded in adults. *T. s. callirostris* (Gray, 1855) (Plate 14), to 25 cm, lives in the Caribbean drainges of Colombia and Venezuela. It is easily recognized by the large number of ocelli on the underside of the snout and on the upper and lower jaws; its broad, reddish, parallel-sided supratemporal stripe well separated from the orbit; the pattern of ocelli on its carapace; and the extensive pattern of dark lines which cover most of its plastron. *T. s. chichiriviche* (Pritchard and Trebbau, 1984), to 32.5 cm, inhabits the small coastal drainages between the Río Tocuyo and Moron in northern Venezuela. This turtle also has ocelli on its chin, but has a brownish-red, wedge-shaped supratemporal stripe well separated from the orbit, oval or irregular black pleural blotches, and a narrow dark pattern along the plastral midseam. Several additional populations will probably eventually be named as subspecies, as Moll and Legler (1971) reported the following populations that represent unnamed forms: (1) Rio Nazas drainage of Durango and Coahuila; (2) aquadas and cenotes in the northern half of the Yucatán Peninsula; (3) region of Cabo Gracias a Dios, Nicaragua, to the Isthmus of Panama and the Río Atrato of Colombia; (4) Lakes Managua and Nicaragua, and their tributaries; and (5) Río Terraba of Costa Rica to the Río Bayano and Chucunaque drainages of eastern Panama.

It is possible that at least two or more species are involved in this variable complex (see courtship patterns below), and an extensive study of the relationships within this complex is sorely needed.

Habitat: Although most freshwater habitats are occupied, *Trachemys scripta* prefers quiet waters with soft bottoms, an abundance of aquatic vegetation, and suitable basking sites. It is more riverine in the tropics.

Natural History: Courtship and mating occur in both the spring and fall. Two courtship patterns exist, depending on whether or not the males possess elongated foreclaws. If elongated foreclaws are present, such as in subspecies like *T. s. elegans*, the male swims to a position in front of the female, turns to face her, extends his forelegs with palms outward, and strokes her face with his foreclaws (Jackson and Davis, 1972). If the male lacks elongated foreclaws, such as in subspecies like *T. s. taylori*, he does not swim in front of the female, but instead pursues her and vigorously bites the posterior rim of her shell, hind legs, and tail (Davis and Jackson, 1973). In both cases, after the female has been subdued, the male mounts from the rear, and the couple sinks to the bottom to copulate. Nesting in the more temperate zones occurs from April to July but may occur during the dry season (December–May) in tropical subspecies. Nests are usually flask shaped. Several clutches of 2 to 25 oval (30–42 × 19–29 mm) parchment-shelled eggs are laid each season. Incubation occurs in about 65–75 days. Hatchlings are round (30–33 mm) and are more brightly colored than adults. The carapace is green with yellow markings, there are smudges or ocelli on the underside of the marginals, and the bridge and plastron are brightly patterned.

The food preferences of *Trachemys scripta* change with age: juveniles are highly carnivorous, but as they become older they eat progressively larger quantities of vegetable matter. Adults are omnivorous and exhibit no obvious preference for either animal or plant food; they will take almost any food item available. Foods recorded from *T. scripta* include algae, duckweed, and assorted emergent herbaceous plants; animals include tadpoles, small fish, insects (adults and larvae), crayfish, shrimp, amphipods, and various mollusks, mostly snails. Captives readily consume fresh and canned fish, raw beef and hamburger, canned dog food, lettuce, bananas, watermelon, and cantaloupe.

Sliders are quite fond of basking, and may pile up several deep at preferred sites.

Trachemys decorata Barbour and Carr, 1940

Hispaniolan slider

Recognition: The elongated, elliptical carapace (to 30 cm) is slightly domed and has only a weak medial keel. Vertebrals are broader than long. Scutes are relatively smooth, but some may have longitudinal rugosities. Posterior marginals are only slightly serrated. The carapace is gray to brown with dusky ocelli on the pleurals and at the intermarginal seams. The

Trachemys decorata

plastron is well developed, but lacks a deep anal groove. The plastral formula is: abd > pect > an > fem > gul > hum. Plastral ground color is yellow to cream and scattered areoli are present; the bridge contains two or more areoli. The head is of moderate size and the snout is conical and somewhat projecting. The upper jaw is not medially notched, and the ridge on the triturating surface is low. Laterally, the grayish head contains several black-bordered yellow stripes and a wide yellowish-green supratemporal stripe. Stripes on the top of the head are less conspicuous than those on the sides. Limbs and tail are grayish green to brown with conspicuous black-bordered yellow stripes. Melanism apparently does not develop to any great degree in older individuals.

Males are slightly smaller than females, have slightly upturned snouts, elongated foreclaws, and longer, thicker tails with the vent beyond the carapace. Females are slightly larger with straight snouts, short foreclaws, and shorter tails with the vent beneath the carapace.

Distribution: *Trachemys decorata* is restricted to the island of Hispaniola where it occurs in water bodies in the Cul de Sac–Valle de Neiba plain in both the Dominican Republic and Haiti, and on the Tiburon Peninsula of Haiti (Seidel and Inchaustegui Miranda, 1984).

Habitat: This turtle occurs in brackish and freshwater lakes and ponds of moderate depth with soft bottoms and abundant aquatic vegetation.

Natural History: According to Inchaustegui Miranda (1973), courtship and mating occur all year long. The male strokes the female's eyes and interorbital region with his vibrating foreclaws as he swims in front in a face-to-face position. Copulation occurs underwater, and the eggs are deposited from April into July. A typical clutch includes 6 to 18 eggs, and one to four clutches are laid each year. The white eggs are elongated, ranging from 35 to 47 mm in length and 20 to 25 mm in diameter. After an incubation period of 61–80 days (at 30°C) the bright-colored hatchlings (carapace 30–40 mm) emerge.

Adult *Trachemys decorata* are predominantly herbivorous, naturally feeding on a variety of common aquatic plants and algae. In captivity they accept lettuce and various fruits. Juveniles are more carnivorous, accepting fish in captivity and probably eating aquatic insects and other invertebrates in the wild.

This species tames readily in captivity and is fond of basking.

Trachemys decussata (Gray, 1831b)

West Indian slider

Recognition: This slider has an elongated oval carapace (to 39 cm) which is moderately domed, has a medial keel, and is serrated posteriorly. Its scutes bear rugosities in the form of longitudinal or radiating ridges. Vertebral 1 is only slightly elevated and is about as long as broad, vertebrals 2–4 are slightly longer than broad or about as long as broad, and vertebral 5 is much wider posteriorly than long. Ground color varies from brown to greenish or olive brown, and usually there is no pattern except in young individuals. The plastron is notched posteriorly, and the gulars do not contain anterior extensions. The plastral formula is: abd > an > pect > gul > fem > hum. Plastral color is yellow with an extensive, obscure, black seam-following pattern. A series of longitudinal black bars or ocelli occur on the yellow bridge. The head is moderate in size with a bluntly rounded, slightly projecting snout, and a medially notched upper jaw. The head is green to olive brown with yellow stripes. At least two of these stripes run backward along the side of the head from the orbit to the neck, and the dorsalmost expands over the tympanum into a supratemporal stripe. There is usually another stripe below these extending from the corner of the mouth to the neck. Jaws are yellow to tan; neck, limbs, and tail are green with yellow stripes.

Males have elongated foreclaws and long, thick tails; the foreclaws and tail of the female are short. The male carapace is flattened while that of the female is more domed.

T. decussata is another of the species of *Trachemys* in which the males may become progressively melanistic with age. This has caused much confusion, as these dark individuals were once referred to by the specific name *rugosa* (Shaw, 1802), along with the males of several other Antillean terrapins.

Distribution: *Trachemys decussata* occurs on the West Indian islands of Cuba, Isla de Pinos, Grand Cayman, and Cayman Brac.

Geographic Variation: Two subspecies are recognized (Seidel, 1988). *Trachemys decussata decussata* (Gray, 1831b) is restricted to central and eastern Cuba. It has a broad, to oval or elongated, oblong carapace which may be either moderately domed or flattened. Skin of the soft parts is greenish to olive. Its snout is blunt and rounded, and the plastral markings consist of a dark seam-following pattern. Somewhat flattened individuals occur in the Río Jobabo in Oriente Province; these were named *T. d. plana* by Barbour and Carr (1940), but are not presently considered a valid subspecies. *T. d. angusta* (Barbour and Carr, 1940) occurs on western Cuba, Isla de Pinos, and the Cayman Islands. Its elongated carapace is moderately domed, and its skin is grayish brown. Like *T. d. decussata*, it has a bluntly rounded snout and a faded seam-following plastral pattern.

Habitat: This species is an inhabitant of lowland streams, rivers, swamps, and lakes having relatively permanent water, a soft bottom, and abundant aquatic plants.

Natural History: Males stroke the sides of the female's face with their elongated foreclaws during courtship. The nesting season probably extends from April into July. Hatchlings are more brightly colored than adults and their medial keel is more pronounced.

Adults are predominantly herbivorous, feeding on aquatic plants and occasional fruits which may fall into the water. Barbour and Carr (1940) also thought they may eat crayfish. In captivity, juveniles will accept fish.

Captives readily bask and presumably wild *Trachemys decussata* also sun themselves. Barbour and Carr (1940) reported that in Cuba these sliders leave their ponds as they dry up in summer, and estivate burrowed beneath leafmold in woods, scrub areas, or tall grass.

Trachemys stejnegeri (Schmidt, 1928)

Central Antillean slider

Recognition: The oval, slightly domed carapace (to 24 cm) is widest at the midpoint or posterior to the midpoint, serrated posteriorly, and has a low, blunt medial keel. Each scute may contain longitudinal rugosities. Vertebrals are broader than long. The color in adults is gray, brown, olive, or black. In juveniles, yellow streaks occur on the pleurals, vertebrals, and marginals. Melanism develops in older males. The plastron is well developed, and has a slight posterior notch. The plastral formula is: abd > an >< pect >< gul >< fem > hum. The yellow plastron may be immaculate or there may be a black seam-following pattern. Faded olive ocelli may be present on the undersides of the marginals. The head is short and the snout blunt or pointed. The upper jaw is medially notched. The head is gray to olive with cream to yellow stripes; its supratemporal stripe is reddish brown. Neck, limbs, and tail are gray to olive with cream to yellow stripes.

Males have longer, thicker tails with the vent beyond the carapace margin. Females are larger and have shorter tails with the vent under the carapace.

Distribution: *Trachemys stejnegeri* occurs on Hispaniola, Puerto Rico, and Great Inagua, Bahamas, and has been introduced on Marie-Galante (Seidel, 1988).

Geographic Variation: Three subspecies are known (Seidel, 1988). *Trachemys stejnegeri stejnegeri* (Schmidt, 1928) has an elongated, moderately domed carapace, brown to brownish-olive skin, a pointed, somewhat elongated snout, and a plastral pattern which spreads out away from the seams and onto the broad surface of the scutes, especially the gulars. Posteriorly the plastral pattern is less distinct. It is found on Puerto Rico and has been introduced on Marie-Galante, French West Indies. *T. s. vicina* (Barbour and Carr, 1940) occurs only on Hispaniola. It has an elongated, moderately domed carapace, grayish-olive skin, a somewhat pointed, elongated snout, and a seam-following plastral pattern which sometimes has additional ocelli on each scute. *T. s. malonei* (Barbour and Carr, 1938) has an elliptical to oval, high-domed carapace, gray to olive skin, a blunt to rounded snout, and a plastron which is either immaculate yellow or has a dark seam-following pattern or a few dark marks restricted to the upper and lower surfaces of the gulars. It occurs on Great Inagua Island, Bahamas. As can be seen from the descriptions, these subspecies are rather poorly differentiated.

Habitat: *Trachemys stejnegeri* is restricted to freshwater ponds, swamps, streams, and rivers with soft bottoms and abundant aquatic plants.

Natural History: The male strokes the sides of the female's face with his elongated foreclaws during courtship. The nesting season begins in April and probably extends into July. The white eggs are elongated (38–48 × 22–31 mm) and have pliable shells. Three to 14 eggs may be laid in each clutch, and one to three clutches are deposited each season (Hodsdon and Pearson, 1943; Inchaustegui Miranda, 1973). In the Dominican Republic, the incubation period is 57–79 days (Inchaustegui Miranda, 1973). Hatchlings (35 mm) are more brightly colored than adults and their medial keel is more pronounced.

In the Bahamas, the females dig saucer-shaped, symmetrical nests about 6–7.5 cm deep and several centimeters in diameter. There, Hodsdon and Pearson (1943) observed a female scratch away the soil from the top of a nest in August, and two days later young turtles emerged from it. To the best of our knowledge this is the only report of a female turtle aiding its young to escape from the nest, and it is probably erroneous. This type of behavior is common in crocodilians but unknown in other reptiles.

Adults are predominantly herbivorous, but aquatic insects are also taken.

Trachemys terrapen (Lacépède, 1788)

Jamaican slider

Recognition: The oval to elliptical carapace (to 32 cm) is only slightly domed and is wider posteriorly than anteriorly. The medial keel is prominent, and the posterior marginals flared and serrated. Longitudinal rows of shallow rugosities give each vertebral and pleural a wrinkled appearance. Vertebrals 1 and 5 are broader than long, but 2–4 are about equal in width and length. The carapace varies from grayish brown to yellowish olive or dark gray. In young individuals, a yellow bar may occur on each pleural and marginal, but these fade with age. The plastron is widest at the hind lobe, which bears a shallow posterior notch. Its scute formula is: abd > an > pect >< gul > fem > hum. There are usually no markings on the cream to yellow plastron, although faint blotches may occur on the gulars and humerals or along the seams. The head is of moderate size and its snout is short and bluntly rounded. The upper jaw is only slightly notched medially, and the ridge on its triturating surface is low. The head is gray to olive, and any light stripes present are

Trachemys terrapen

Trachemys dorbigni (Duméril and Bibron, 1835)

Brazilian slider

Recognition: This turtle represents the terminal southward extension of the *Trachemys scripta* series in the Americas. Its oval to elongated (to 26.7 cm) carapace is moderately domed (more so in females), contains a low medial keel (more pronounced in juveniles), and is serrated posteriorly. Vertebrals are usually broader than long. The carapace is brown to olive with variously shaped red, orange, or yellow markings on each scute. These marks are usually dark bordered and range from elongated, straight or curved streaks to rounded, light-centered ocelli. Each marginal has a light-colored vertical bar. The plastron is large and contains an anal notch. Its formula is: abd > gul > pect > an > fem > hum. The plastron is yellow or orange with an intricate, broad, dark, seam-following pattern. In juveniles, this pattern contains numerous hollow light areas which become progressively darker with age until the entire plastral surface is almost covered with dark pigment. Males become melanistic until the carapace and plastron are totally black. The head is of moderate size with a somewhat projecting, pointed snout and a slightly notched upper jaw. Skin of the head is green to brown, with numerous yellow to orange, black-bordered stripes. The rather wide supratemporal stripe does not touch, or only narrowly touches, the orbit. Below this on the side of the head are usually three narrower stripes running from the orbit to the neck, and then another wide stripe running from the lower surface of the orbit downward and backward to the neck. The snout contains several narrow stripes on each side. On the chin there are several longitudinal stripes, the middle of which is wide and forks as it passes backward toward the neck. At each corner of the mouth is an elongated black-bordered spot. Limbs and neck are green to brown with yellow stripes.

Males have elongated foreclaws and long, thick tails.

Distribution: *Trachemys dorbigni* has been found at São Luís, Maranhão, Brazil, and in the Rio Guaíba drainage near Pôrto Alegre in Rio Grande do Sul, Brazil. It also occurs throughout Uruguay and in the Paraná and Uruguay drainages of northern Argentina.

Geographic Variation: Two subspecies have been described. *Trachemys dorbigni dorbigni* (Duméril and Bibron, 1835) occurs in Argentina and Uruguay. This turtle is generally brown with orange stripes, and the supratemporal stripe is not enlarged. Ground color of the plastron is orange. *T. d. brasiliensis* (Freiberg,

faded and hardly discernible. Several faded cream to white stripes occur on the chin, and a whitish "mustache" may lie below the nostrils on the upper jaw. Neck, limbs, and tail are also gray to olive, and the forelimbs and neck may contain faded light stripes.

Males have elongated foreclaws and longer, thicker tails than females; however, females grow larger than males. Occasionally, large, white males are found in some of the southern rivers.

Distribution: *Trachemys terrapen* occurs on Jamaica and Cat Island, Eleuthera, and possibly south Andros in the Bahamas. *Pseudemys felis* Barbour, 1935 is synonymous with *T. terrapen* (Seidel and Adkins, 1987).

Habitat: Shallow ponds with mud bottoms and abundant aquatic vegetation seem to be the normal habitat.

Natural History: While courting, the male swims backward in front of the female and strokes the sides of her face with the elongated claws of his outstretched forefeet. Copulation occurs under water. Clutch size is unknown, but Lynn and Grant (1940) dissected a female in April that contained four shelled eggs 39 × 24 mm and four additional large yolks. Nesting probably occurs from May into June. Hatchlings have dark-bordered yellow stripes on the head, neck, limbs, and tail; Williams (1956) showed a photograph of a juvenile *T. terrapen* with an extensive dark blotch on the center of its plastron.

Adult *T. terrapen* are apparently omnivorous. They consume large quantities of aquatic plants, but are still attracted to animal baits.

This species is fond of basking in both the wild and captivity.

1969) inhabits the Rio Guaíba of Rio Grande do Sul, Brazil, and has also been found at São Luís, Maranhão, in northern Brazil. It is green with yellow or pinkish-red stripes and a broad supratemporal stripe. The plastron is yellow to green, and, according to Freiberg (1969), males may lack plastral markings.

Habitat: This species occurs in various bodies ranging from ponds, lakes, marshes, swamps, streams, and rivers. It seems to prefer waters with slow to moderate currents, soft bottoms, abundant aquatic vegetation, and a prevalence of basking sites.

Natural History: Courtship and mating have not been described, but Freiberg (1967) reported that Argentine females lay several clutches each season. Clutches probably include as many as 17 elongated eggs (41 × 27 mm), which have flexible shells. Nesting takes place in December, and the cavity is about 30 cm deep (Freiberg, 1981).

Trachemys dorbigni is probably omnivorous, feeding on both aquatic plants and various invertebrates, fish, and amphibians. It is also an avid basker, often forming large assemblages at good sites.

Comment: Moll and Legler (1971) and Pritchard (1979) considered *Trachemys dorbigni* to be a subspecies of *T. scripta*. Williams (1956) and Wermuth and Mertens (1961, 1977), however, retained it as a full species. It is certainly related to *T. scripta*, especially *T. s. ornata* and *T. s. callirostris*, but so are the West Indian *Trachemys*, which are equally as divergent but still considered separate species. We feel it is best to consider *T. dorbigni* as a full species since it is allopatric and no evidence of reproductive compatability with *T. scripta* has been demonstrated.

Also, the population of *T. dorbigni* at São Luís, Maranhão, Brazil should be critically examined, since this is so far removed from the more southern normal range. Do these turtles show intermediacy between *T. s. callirostris* and *T. dorbigni*, and is this a natural or an introduced population?

Pseudemys Gray, 1855

The five species of basking turtles in the genus *Pseudemys* are restricted to North America. This genus is closely related to the genera *Chrysemys*, *Deirochelys*, *Graptemys*, *Malaclemys* and *Trachemys*, with which it forms a complex. It has at times been included in the genus *Chrysemys* (McDowell, 1964; Zug, 1966; Weaver and Rose, 1967; Gaffney, 1979b), but recent evidence

(Seidel and Smith, 1986) indicates it should be considered a separate genus. Species of *Pseudemys* have oval, elongated, moderately domed carapaces with a serrated posterior rim, and, at least in juveniles, a medial keel. The neurals are six sided and shortest anteriorly. There is a tendency for old males, and sometimes females, to become melanistic, and females are usually larger than males. The plastron is hingeless, with short to moderately long axillary and inguinal buttresses. The entoplastron lies anterior to the humero-pectoral seam. The skull is moderate in size with a complete temporal arch, and the orbitonasal foramen is much larger than the posterior palatine foramen. Contact occurs between the lower parietal process and palatine. Triturating surfaces are broad; that of the upper jaw bears a medial tuberculate ridge. The upper jaw may or may not be medially notched, and if notched may bear cusplike projections. There are four phalanges in the 5th toe; the toes are webbed.

The karyotype is 2n = 50: 26 macrochromosomes (16 metacentric, 6 submetacentric, 4 telocentric), and 24 microchromosomes (Killebrew, 1977a).

Key to the species of *Pseudemys*:

1a. Upper jaw with a prominent notch bordered on each side by toothlike cusps   2
  b. Upper jaw unnotched or with a notch not bordered by toothlike cusps   5
2a. Prefrontal arrow absent; restricted to Texas, New Mexico, and northeastern Mexico: ***P. concinna texana***
  b. Prefrontal arrow present   3
3a. Paramedial stripes end in back of eyes: ***P. nelsoni***
  b. Paramedial stripes continue forward between eyes and onto snout   4
4a. Carapace elevated medially; restricted to vicinity of Mobile Bay, Alabama: ***P. alabamensis***
  b. Carapace flattened medially; restricted to Atlantic Coastal Plain of North America: ***P. rubriventris***
5a. C-shaped mark present on 2d pleural; plastral figure present: ***P. concinna***
  b. C-shaped mark absent on 2d pleural; no plastral figure: ***P. floridana***

Pseudemys concinna (LeConte, 1830)

River cooter

Recognition: This is a large (to 32 cm) freshwater turtle with a C-shaped light mark on the second pleural and a patterned plastron. Its elongated, narrow, flattened carapace is highest at the middle, occasionally restricted in the region of the 6th marginals and slightly serrated posteriorly. Vertebral 1 is as long as broad or longer than broad, vertebrals 2–5 are broader than long. The carapace is brown with yellow to cream-colored markings. Dorsally, each marginal has a dark area, usually in the form of two concentric circles, and some individuals have a light bar on each marginal. Ventrally, each marginal has a dark light-centered spot bordering the seams. The bridge has one or two dark bars. The yellow plastron has a dark pattern that follows the seams, but it may be present only on the anterior half of the plastron and usually fades with age. Its hind lobe bears a posterior notch; the plastral formula is: abd > an > fem > gul >< pect >< hum. Skin is olive to brown, with yellow or cream stripes. The head is moderate in size with a slightly protruding snout and an upper jaw which is neither medially hooked nor has a medial notch bordered by cusps. The supratemporal and paramedial stripes lie parallel, and a sagittal stripe passes anteriorly between the eyes but does not meet the supratemporal stripes. Wide yellow stripes are present on the underside of the neck, and a central chin stripe extends posteriorly and divides to form a Y-shaped mark.

Adult males have long, straight foreclaws and long, thick tails, with the vent behind the carapacial margin. Males may become melanistic.

Distribution: *Pseudemys concinna* ranges along the Piedmont and the Atlantic and Gulf coastal plains from Virginia into Mexico; inland in the Mississippi Valley to southern Illinois, southern Missouri, south-eastern Kansas, and Oklahoma; and west across Texas to New Mexico and south to Coahuila, Nuevo León, and Tamaulipus in Mexico.

Geographic Variation: Five subspecies are recognized. *Pseudemys concinna concinna* (LeConte, 1830) ranges from Virginia to eastern Alabama in streams of the Piedmont and Atlantic Coastal Plain. An isolated population occurs in the New River of West Virginia (Seidel, 1981). Its markings are essentially as described above. *P. c. suwanniensis* Carr, 1937 lacks stripes on the outer surface of the hind foot. Its head stripes are light yellow, and the plastron is yellow to orange, with a well-defined pattern. *P. c. mobilensis* (Holbrook, 1838), occurs in the Gulf coastal streams from the Florida Panhandle to extreme southeastern Texas. It is similar to *suwanniensis* but paler, smaller, and has orange to red head stripes. *P. c. hieroglyphica* (Holbrook, 1836) is the upper Mississippi Valley representative of this species, ranging from southern Illinois and adjacent Missouri south to Tennessee and Alabama and west to Kansas, Oklahoma, and Texas. The C-shaped mark and plastral pattern are well developed, and its shell usually is indented at the bridge. *P. c. texana* Baur, 1893b is the most divergent subspecies: the head pattern is highly variable but always contains broad yellow stripes, spots, or vertical bars; the C-shaped mark is also variable and may be absent altogether; and the upper jaw is notched, with a cusp on each side. It ranges from central Texas and New Mexico south to Coahuila, Nuevo León, and Tamaulipas, Mexico. Ward (1984) elevated *P. c. texana* to specific status, and also described two additional subspecies of *P. concinna*: *gorzugi* and *metteri*. We believe these designations are questionable and feel additional study of the variation within *P. concinna* is needed.

Interbreeding with *Pseudemys floridana* occurs in parts of the range, and some researchers (Fahey, 1980) consider these two turtles to represent one species. However, in other parts of the range they do not interbreed, and we feel that until a study of their relationship over the entire range presents a clearer understanding, *P. concinna* and *P. floridana* are best considered separate species.

Habitat: *Pseudemys concinna* is predominantly a turtle of rivers, preferring those with moderate current, abundant aquatic vegetation, and rocky bottoms. It also inhabits lakes, ponds, oxbows, swamps, large ditches, and cattle tanks, and has been taken in lagoons, brackish tidal marshes, and the Gulf of Mexico.

Natural History: Courtship and mating most commonly occur in the spring. Courtship involves a stroking of the female's face with the male's elongated fore-

Pseudemys concinna hieroglyphica

claws; the male accomplishes this while swimming above the female and extending his forelimbs downward to reach her head. Up to 19 eggs are laid at one time, and perhaps more than one clutch is deposited each year since the nesting season extends from May into June. Sometimes a female will dig several cavities and lay eggs in each until the total clutch has been deposited. The ellipsoidal eggs (33–43 × 22–30 mm) have parchmentlike shells. Hatchlings are 30–36 mm in carapace length, brightly patterned, and have a medial keel.

Adults are predominantly herbivorous, feeding on *Naias*, *Lemna*, *Ceratophyllum*, *Sagittaria*, and filamentous algae, but some animal foods (crayfish, insects, snails, tadpoles, and fish) may also be eaten. The scavenging habit is well developed.

The river cooter is shy and leaves the water only to bask or nest. Basking occurs frequently; large aggregations often are seen on logs or other favored sites. It often shares basking sites with other species of *Pseudemys*, *Graptemys*, *Trachemys* and *Deirochelys*. While basking it is very wary and will dive into the water at the slightest alarm.

Pseudemys floridana (LeConte, 1830)

Common cooter

Recognition: This large (to 40 cm) freshwater turtle has a wide transverse stripe on the second pleural. Its elongated and highly arched carapace is highest at the center and widest behind the center, with a slightly serrated posterior rim. Vertebral 1 may be longer than broad or as long as broad; vertebrals 2–5 are broader than long. The carapace is brown with yellow markings; the wide stripe on the second pleural may be forked at the upper or lower end or both. Each marginal has a central yellow bar on its dorsal surface and,

Pseudemys floridana floridana

on the ventral surface in some individuals, a dark, light-centered spot covers the posterior seam. The plastron is yellow and devoid of any dark pattern. Its hind lobe is posteriorly notched, and the plastral formula is: abd > an > fem >< pect > gul > hum. The head is moderate in size with a slightly projecting snout, and no medial notch or cusps on the upper jaw. Skin is brown or black, with yellow stripes. The supratemporal and paramedial stripes usually are joined behind the eyes. Ventrally, the neck is marked with wide yellow stripes, and a central chin stripe extends backward and divides to form a Y-shaped mark.

Males have elongated foreclaws and long, thick tails, with the vent behind the carapacial rim. Males are slightly smaller and flatter than females, which sometimes show an indentation in the carapace just anterior to the bridge.

Distribution: *Pseudemys floridana* ranges along the Atlantic Coastal Plain from North Carolina to Dade County, Florida, and west across the Gulf Coastal Plain to eastern Oklahoma, southeastern Kansas, southern Missouri, and southern Illinois.

Geographic Variation: There are three subspecies. *Pseudemys floridana floridana* (LeConte, 1830) occurs on the Atlantic Coastal Plain from Virginia to northern Florida and west on the Gulf Coastal Plain to Alabama. This race has numerous head stripes which do not form hairpin markings. *P. f. peninsularis* Carr, 1938a is restricted to peninsular Florida. It has a pair of light lines resembling hairpins on the top of its head; these may be broken or incomplete. *P. f. hoyi* (Agassiz, 1857) inhabits the lower Mississippi Valley and extends west along the Texas coast. It has numerous broken and twisted head stripes and is the smallest of the three races. Ward (1984) placed *P. f. hoyi* in the synonymy of *P. concinna hieroglyphica;* we question this, as the two are usually discernable in the central Mississippi Valley.

In some southern waterways, *P. floridana* and *P. concinna* apparently hybridize (see Fahey, 1980 for a review; also see comment under *P. concinna*).

Habitat: The Florida cooter inhabits almost any aquatic habitat having a slow current, soft bottom, basking sites, and abundant aquatic vegetation.

Natural History: Courtship and mating probably occur throughout the year in Florida, but are probably restricted to the spring farther north. Courtship involves a stroking of the female's head with the elongated foreclaws while the male swims above and slightly behind the female, as in *Pseudemys concinna*. In Florida, nesting occurs throughout the year (Iverson, 1977d), but farther north probably only from late May to mid-July. The nest is flask shaped and about 10–15 cm deep. At least two clutches of 8 to 29 elliptical (29–40 × 22–27 mm) white eggs are laid by a mature female each year. Hatching normally occurs in 70–100 days. Hatchlings are brighter than adults. The carapace is about as wide as long, has a well-developed keel, and is more highly arched and flared. Sixteen hatchlings were 27.0–33.0 mm in length, 25.0–31.2 mm in width, and 16.1–18.2 mm in depth (Carr, 1952).

Adult *P. floridana* are largely herbivorous, consuming a wide variety of aquatic plants, including *Sagittaria*, *Ceratophyllum*, *Myriophyllum*, *Naias*, and *Lemna*, and probably algae. In captivity, they accept lettuce, spinach, cabbage, watermelon, cantaloupe, and bananas. The young take animal food, but become less carnivorous with age.

P. floridana is quite fond of basking and spends many of the daylight hours so occupied. Florida cooters are gregarious, and often as many as 20 or 30 may bask on the same log. Such a group is difficult to approach, but individuals do not seem especially wary.

Pseudemys rubriventris rubriventris

Pseudemys rubriventris (LeConte, 1830)

American red-bellied turtle

Recognition: *Pseudemys rubriventris* is a large (to 40 cm) freshwater turtle having a reddish plastron and a prominent notch at the tip of the upper jaw, with a toothlike cusp on each side. Its elongated carapace is usually highest at the middle, widest behind the middle, flattened dorsally, and slightly serrated posteriorly. Vertebral 1 is as long as broad or longer than broad, the others are broader than long. The carapace is brown to black, with red or yellow markings on the pleurals and marginals. The 2d pleural has a broad, light, central transverse band, which is forked at the upper or lower end or both. Each marginal has a red bar on the upper surface and a dark blotch with a light central spot on the lower surface. Melanism is common in old individuals. The plastron is reddish orange, and in the young has a dark mark that spreads along the seams but fades with age. A wide dark bar crosses the bridge. The plastral hind lobe is only slightly notched posteriorly, at best, and the plastral formula is: abd > an > pect > gul > fem >< hum. The head is moderate in size with a slightly protruding snout and a notched upper jaw (described above). Skin is dark olive with yellow stripes. A sagittal stripe passes anteriorly between the eyes and meets the joined supratemporal stripes on the snout, forming the prefrontal arrow that is characteristic of the red-

bellied group. Five to eight stripes occur beween the supratemporals behind the eyes. The paramedial stripes pass forward from the neck across the occipital region and terminate between the orbits. A supratemporal stripe bends upward from the neck on each side and enters the orbit.

Males have long, straight foreclaws; large, thick tails, with the vent behind the carapacial margin; and lower, slightly narrower shells than females. Females are slightly larger than males.

Distribution: *Pseudemys rubriventris* occurs along the Atlantic Coastal Plain from central New Jersey south to northeastern North Carolina and west up the Potomac River to eastern West Virginia. It also has relict populations in Plymouth County and possibly Essex County, Massachusetts.

Geographic Variation: Two poorly defined subspecies are recognized. *Pseudemys rubriventris rubriventris* (LeConte,1830) ranges on the Atlantic Coastal Plain from central New Jersey south to northeastern North Carolina and west up the Potomac River to eastern West Virginia. *P. r. bangsi* (Babcock, 1937) is restricted to relict populations in Plymouth County and possibly Essex County, Massachusetts. It is recognized on the basis of a higher carapace than that of the nominate race. In *P. r. bangsi* the greatest length of the carapace is 2.4 times its greatest height; in *P. r. rubriventris*, 2.6 times. Studies by Conant (1951b) on Pennsylvania individuals and by us on Potomac River *P. r. rubriventris* indicated a variation encompassing the supposedly diagnostic extremes; this character probably is insufficient to distinguish the populations. Additional study is needed to determine if two different subspecies truly exist.

P. rubriventris, *P. nelsoni*, and *P. alabamensis* form

the red-bellied, or *rubriventris*, group of the genus *Pseudemys*. They share external characters—notched upper jaw, prefrontal arrow, and reddish plastron—and have the skull with the vomer contributing to the triturating surface and with the middle ridge of the lower triturating surface set well to the side of the lingual margin of the surface. The red-bellied group is closely related to the *P. floridana* complex; in fact, *P. rubriventris* and *P. floridana* sometimes hybridize (Crenshaw, 1965).

Habitat: Relatively large, deep bodies of water with basking sites are the preferred habitat. The red-bellied turtle has been found in creeks, river marshes, ponds, and lakes, and has also been taken from brackish water. The New England form is restricted to ponds.

Natural History: Courtship and mating of this shy turtle have not been described, but the elongated male foreclaws suggest some form of stroking is involved. Nesting occurs from late May into July. The nest is flask shaped and about 10 cm deep. Several clutches are probably laid each season, and a single clutch may include from 10 to 17 elliptical (25–37 × 19–25 mm), parchmentlike-shelled eggs. Incubation takes 70–100 days, depending on weather conditions and latitude. The hatchlings are 29–32 mm in carapace length. They are brightly colored and have rounded, keeled carapaces. The pink to red plastron has a large dark figure that spreads somewhat along the seams, a pattern similar to that of *Pseudemys concinna*.

The red-bellied turtle apparently is omnivorous. Known food items include snails, fish, tadpoles, crayfish, and aquatic vegetation. The fact that it is not often lured into traps baited with fish seems to indicate these are not a normal part of the diet. The median ridges on the crushing surfaces of the jaws are tuberculate, like those of *P. floridana* and *P. concinna* (probably an adaptation to a herbivorous diet); it is likely that *P. rubriventris* depends to a substantial degree on aquatic vegetation for nourishment.

The population of *P. r. bangsi* is about 200 individuals restricted to a few ponds. It is considered endangered.

Pseudemys alabamensis Baur, 1893a

Alabama red-bellied turtle

Plate 13

Recognition: The elongated rugose carapace (to 33 cm) is highly arched and elevated along the vertebral scutes; its highest point is often anterior to the middle, and it is widest at the middle. Vertebrals are broader than long; the 1st is the narrowest, the 5th expanded. Posterior marginals are serrated. The carapace is olive to black, with red to yellow bars on the pleurals and marginals. The 2d pleural has a wide, light, centrally located transverse bar, which may be Y-shaped. The plastron has a posterior notch. Its scute formula is: abd > an > pect >< gul >< fem > hum. The plastron is reddish yellow and may show a dark, mottled pattern, which may also occur on the carapace and bridge. The head is moderate in size with a nonprotruding snout and an upper jaw with a prominent medial notch bordered on either side by a toothlike cusp. The skin is olive to black with yellow stripes. Supratemporal and paramedial head stripes are prominent and parallel, but do not join posterior to the orbit. The supratemporals pass forward from the orbit, joining well above and posterior to the nostrils. A sagittal stripe passes anteriorly between the orbits and joins the supratemporal stripes at their junction to form a prefrontal arrow.

Females grow larger and their carapaces are more domed. Males have long, straight foreclaws and long, thick tails, with the vent behind the carapacial rim.

Distribution: Now restricted to the vicinity of Mobile Bay, Alabama; formerly it may have ranged into the Florida Panhandle as far east as Apalachee Bay (Carr and Crenshaw, 1957).

Habitat: Fresh to moderately brackish marshes and backwaters with abundant aquatic vegetation are the primary habitats of *Pseudemys alabamensis*, but it sometimes enters rivers and streams.

Natural History: *Pseudemys alabamensis* nests in May and June, and probably each female lays several clutches of three to six eggs each year (Meany, 1979). Hatchlings are nearly round with keeled, posteriorly serrated carapaces. The carapace is green with a pattern of dark-bordered yellow lines on each scute. Also, each pleural often has a dark-centered yellow ocellus, and there is a black-bordered ocellus over the dorsal and ventral marginal seams. The plastron and bridge are orange, red, or coral with a black seam-following pattern.

This species seems to be primarily herbivorous, consuming many species of aquatic plants. Captives readily eat lettuce, but also take fish.

Owing to its restricted range, water pollution, and predation, *Pseudemys alabamensis* is considered at least threatened.

Pseudemys nelsoni Carr, 1938c

Florida red-bellied turtle

Recognition: *Pseudemys nelsoni* has a high-arched, elongated carapace (to 34 cm) with the highest point usually anterior to the middle, the widest point at the middle, and the posterior rim slightly serrated. The medial keel is lost in adults. Vertebral 1 is longer than broad or as long as broad; vertebrals 2–5 are broader than long. The carapace is variable in color but usually black, with red or yellow markings on the pleurals and marginals. The 2d pleural has a light, central band, which passes dorsoventrally and, just below the dorsal edge, usually bends sharply toward the rear and passes to the upper rear edge of the pleural; often this band is branched, forming a Y-shaped figure. Each marginal has a red bar located centrally on its dorsal surface; the ventral surface exhibits dark smudgelike blotches at the seams. Melanism may occur in older individuals of either sex. The bridge is usually immaculate, but sometimes bears dark blotches. The plastron is reddish orange and may be plain or carry a medial pattern which fades with age. Its hind lobe is only slightly posteriorly notched, and the plastral formula is: abd > an > pect >< fem > gul > hum. The head is moderate with a non-protruding snout and a prominent notch at the tip of the upper jaw, often with a toothlike cusp on each side. Skin is black with yellow stripes, and there is a prefrontal arrow. One to three stripes occur between the supratemporals behind the eyes, and the paramedial head stripes usually are reduced and always end behind the eyes.

Males have elongated, slightly curved foreclaws and long, thick tails, with the vent posterior to the carapacial rim. Females are slightly larger than males.

Distribution: *Pseudemys nelsoni* ranges from the Oke-fenokee Swamp in southern Georgia west to Apalachicola, Florida, and south through peninsular Florida.

Geographic Variation: No subspecies are recognized, but hybridization with *Pseudemys floridana* and *P. concinna* may occur.

Habitat: This turtle has been taken from ponds, lakes, ditches, sloughs, marshes, mangrove-bordered creeks, and slow-flowing rivers. Water containing abundant aquatic vegetation is preferred.

Natural History: Mating apparently occurs throughout the year, and involves the male stroking the female's face with his elongated foreclaws. Nesting also takes place throughout the year, and up to five or six separate clutches of as many as 12 eggs may be laid each year. A favorite nesting site is an alligator's nest of decaying vegetation. The elliptical, parchment-shelled eggs (37–47 × 19–26 mm) probably hatch after 60–75 days. Hatchlings are brighter than adults, and the carapace (28–32 mm) is rounded and slightly keeled. The plastron is orange or red, and the dark plastral markings are solid semicircles with the flat sides along the seams.

Adult *P. nelsoni* are highly herbivorous; the young, like those of other species of *Pseudemys*, probably are more carnivorous. Adults prefer *Sagittaria*, *Lemna*, *Mikania*, *Cicuta*, and *Naias*, but have scavenging tendencies and will eat carrion, such as dead fish. Captive adults eat fish, various meats, and lettuce.

P. nelsoni is a confirmed basker, lying many hours each sunny day on logs or floating mats of vegetation. They are not very wary, and can often be approached quite close.

Deirochelys Agassiz, 1857

This monotypic genus includes only the long-necked, North American chicken turtle.

Deirochelys reticularia (Latreille, in Sonnini and Latreille, 1802).

Chicken turtle

Recognition: *Deirochelys reticularia* is a small to medium-sized (to 25 cm) turtle with an extremely long neck and a reticulate pattern of yellow lines on the tan to olive, oval carapace. The long, narrow, and some-

Pseudemys nelsoni

Deirochelys reticularia chrysea

what depressed carapace is widest behind the middle, and in adults is neither keeled nor posteriorly serrated. Its surface is somewhat rough and sculptured with small longitudinal ridges. Vertebrals are broader than long. The 1st is in contact with four marginals and the cervical, and the undersides of the marginals are yellow and may have a dark blotch at the seam. One or two black blotches may also occur on the bridge. The hingeless plastron is yellow, and in the western race may have a dark pattern bordering the seams. The plastral formula is: abd > an > gul > fem > hum >< pect; the entoplastron is usually slightly anterior to the humero-pectoral seam. Plastral buttresses are moderate. The head is long and narrow with a pointed snout; the upper jaw bears neither a hook nor notch. The narrow triturating surfaces are ridgeless, and the palatine and pterygoid bones do not contribute to the upper surface. The orbito-nasal foramen is small, but the posterior palatine foramen is large. The inferior process of the parietal touches the palatine. The skin is olive to brown, with yellow or white stripes. There is a characteristic pattern of vertical light stripes on the rump, and the foreleg stripe is very wide. Length of the head and neck, measured from the snout to shoulder, is approximately equal to the plastron length and about 75–80% of the carapace length. Cervical vertebrae II to VII are much longer than VIII, and elongation of the neck is probably responsible for modifications to the rib heads for expanded inserts of the retractor neck muscles (McDowell, 1964). Thoracic rib heads are long, slender, and ventrally bowed. The toes are webbed.

The karyotype is 2n = 50: 20 metacentric or submetacentric, 10 subtelocentric, and 20 acrocentric or telocentric centromeres on 26 macrochromosomes and 24 microchromosomes (Stock, 1972; Killebrew, 1977a).

Adult males have long, thick tails, with the vent posterior to the carapacial margin. Females are larger than males.

Distribution: *Deirochelys* occurs in southeastern Virginia and from southern North Carolina south along the Atlantic Coastal Plain to southern Florida, west along the Gulf Coastal Plain to Texas, and northward west of the Mississippi River to southeastern Oklahoma and southeastern Missouri. It has also been found in northwestern Mississippi. There are few records of this species from above the fall line.

Geographic Variation: Three subspecies are recognized. *Deirochelys reticularia reticularia* (Latreille, in Sonnini and Latreille, 1802) occurs along the Atlantic and Gulf Coastal Plains from southeastern Virginia to the Mississippi River. This subspecies has narrow, netlike lines on the olive to brown carapace, a narrow yellow carapacial rim, and often a spot at the juncture of the femoral and anal scutes. Black spots are present on the ventral surface of the marginals at the level of the bridge in about 72% of individuals (Schwartz, 1956a). *D. r. chrysea* Schwartz, 1956a is restricted to peninsular Florida. This race has a network of broad, bright, orange or yellow lines on the carapace and a wide, orange carapacial rim. Black spots are present on the ventral surface of the marginals at the level of the bridge in about 43% of individuals (Schwartz, 1956a). *D. r. miaria* Schwartz, 1956a occurs west of the Mississippi River from southeastern Missouri and central Oklahoma south to the Gulf, and there is a record from western Mississippi. It is flattened, has a network of broad, faint lines on the carapace, and has a plastral pattern of dark markings along the seams. Adults have unstreaked chins and throats.

Habitat: The chicken turtle is fairly common in still-water habitats, such as ponds, lakes, ditches, and cypress swamps; sometimes large aggregations can be found in temporary pools. Apparently it does not live in moving-water habitats.

Natural History: Nesting may occur throughout the year but is concentrated from January to late March. In South Carolina, distinct spring and fall nesting seasons occur (Congdon et al., 1983). Several clutches of 5 to 15 elliptical (32–40 × 20–23 mm), white eggs are laid each year. Incubation takes 70–116 days and the round hatchlings are about 30 mm in carapace length.

Deirochelys probably is omnivorous. Carr (1952) observed free-living chicken turtles eating tadpoles, crayfish, and what appeared to be a bud of the water plant *Nuphar*. Our captive adults readily consume lettuce and fresh or canned fish; juveniles refuse vegetable matter. This species probably becomes more herbivorous with increasing age.

Baur (1889b) first proposed a close relationship of *Deirochelys* and *Emydoidea* on the basis of similar skull and rib specializations, and this was accepted by Loveridge and Williams (1957), McDowell (1964), and Ernst and Barbour (1972), among others. Most authors also noted *Deirochelys* has a close resemblance to *Chrysemys*, *Pseudemys*, and *Trachemys* in its shell features. Bramble (1974) in his study of shell kinesis and other osteological and myological characters indicated that *Emydoidea* is more closely related to *Emys* and *Terrapene* than to *Deirochelys*, and that *Deirochelys* is more closely related to *Chrysemys* and *Pseudemys*. This has been further corroborated by Jackson's (1978b) findings on fossil *Deirochelys*.

Graptemys Agassiz, 1857

The 11 species of *Graptemys* are North American riverine turtles highly adapted to a molluscivorous diet. The genus is a member of the *Chrysemys* complex of emydid turtles, and is most closely related to *Chrysemys*, *Malaclemys*, *Pseudemys*, and *Trachemys*. Their oval carapaces bear a medial keel of varying development from a low, raised mound to well-developed, spiny or blunt protuberances; the posterior rim is strongly serrated. Neurals are hexagonal and anteriorly short sided. The carapace usually bears a pattern of narrow, light, meandering lines or ocelli on the pleurals, vertebrals, and, sometimes, marginals which resembles contour lines on a map; thus, their common name of "map" turtles. The plastron lacks a hinge; axillary and inguinal buttresses are not well developed. The entoplastron lies anterior to the humeropectoral seam. The skull is usually narrow to moderately broad in males, but broader in females. It has a complete temporal arch, and the orbito-nasal foramen is enlarged into a fenestra. The flattened triturating surfaces of the jaws are broad and ridgeless. The anterior border of the inferior process of the parietal is thickened, and the pterygoid touches the exoccipital. The toes are webbed.

Killebrew (1977a) reported the karyotype is 2n = 50: 26 macrochromosomes (16 metacentric, 6 submetacentric, and 4 telocentric or acrocentric), and 24 microchromosomes. McKown (1972), however, found several species to be 2n = 52.

McDowell (1964) revived Boulenger's inclusion (1889) of *Graptemys* in the genus *Malaclemys*. *Graptemys* was not recognized even as a subgenus, because the differences between *M. terrapin* and *G. kohnii* are no greater than those between *G. kohnii* and *G. geographica*. This arrangement was not widely accepted

(see, for example, Ernst and Barbour, 1972, and Dobie and Jackson, 1979). Wood (1977) felt that McDowell had not made a convincing case for the all-inclusive *Malaclemys* and presented other evidence suggesting the origin of many, if not all, *Graptemys* from *Malaclemys*. However, Dobie (1981) showed *Malaclemys* to differ substantially from *Graptemys* in osteology, which raises serious questions about *Malaclemys* as the immediate ancestor of *Graptemys*.

Key to the species of *Graptemys*:

1a. Medial keel low, without prominent spines or knobs — 2
  b. Medial keel well developed, with prominent spines or knobs — 3
2a. Horizontal or J-shaped reddish to orange mark behind eye; scutes of carapace distinctly convex; small size: **G. versa**
  b. Yellowish spot behind eye; scutes of carapace not convex; medium to large size: **G. geographica**
3a. Medial keel with blunt, rounded black knobs: **G. nigrinoda**
  b. Medial keel with sharp, narrow spines — 4
4a. Large solid orange or yellow spot on each pleural: **G. flavimaculata**
  b. Solid orange or yellow spot absent from pleurals — 5
5a. Light ring or oval mark on each pleural: **G. oculifera**
  b. No ring or oval mark on each pleural — 6
6a. Large, solid light mark behind eye — 7
  b. Narrow light lines behind eye — 8
7a. Longitudinal light bar under chin; broad light bars on marginals: **G. pulchra**
  b. Curved or transverse bar under chin; narrow light bars on marginals: **G. barbouri**
8a. Light postorbital stripe originates beneath orbit and continues onto dorsal surface of head, usually preventing neck stripes from reaching the eye — 9
  b. Light postorbital stripe originates behind orbit, does not prevent neck stripes from reaching the eye — 10
9a. Chin with transverse cream-colored bar; no longitudinal yellow mark at symphysis of lower jaw; carapacial scutes appear lumpy: **G. caglei**
  b. Chin lacking a transverse cream-colored bar; a longitudinal yellow mark at symphysis of lower jaw; carapacial scutes smooth, not lumpy: **G. kohnii**
10a. Postorbital mark narrow; no large spots on jaws: **G. pseudogeographica**
  b. Postorbital mark square, rectangular, elongated, or oval; large light spot just under eye and another on lower jaw: **G. ouachitensis**

Graptemys geographica (Le Sueur, 1817)

Common map turtle

Recognition: The olive to brown carapace (to 27 cm) has a distinct but low medial keel, a strongly serrated hind margin, and a reticulate pattern of fine yellow lines. The lower edges of the marginals are yellow with circular markings. Vertebrals are broader than long; the 1st is the narrowest, the 5th flared. The bridge is broad, lacks a hinge, and is posteriorly notched. The plastral formula is: abd > an > fem > pect >< gul > hum. The bridge is marked with light bars. The plastron is immaculate yellow to cream colored in adults, but in juveniles it carries a pattern of dark lines bordering the seams. The broad to moderate head has a nonprotruding snout, and the upper jaw lacks either a medial notch or hook. Skin is olive to brown, with yellow stripes. The postorbital mark is somewhat triangular and variable in size. Frequently the anterior end of one neck stripe turns upward across the tympanum, and a few always reach the orbit. The lower jaw is marked with longitudinal yellow stripes, of which the central stripe is the widest.

Adult males have long, thick tails, with the vent behind the carapacial rim. Females have broad heads, males somewhat narrower heads. In males the carapace is oval, tapering posteriorly; in females it is rounded. Adult males are 10–16 cm in carapace length, adult females 17.5–27 cm.

Distribution: *Graptemys geographica* ranges from southern Quebec and northwestern Vermont west to southern Wisconsin and, west of the Appalachians, south to Arkansas and Georgia. It also occurs in the Susquehanna River drainage of Pennsylvania and Maryland, and the Delaware River.

Habitat: *Graptemys geographica* most often frequents large bodies of water, such as rivers or lakes. Mill ponds, oxbows, and the overflow ponds of rivers often contain many individuals. Abundant basking sites, much aquatic vegetation, and a soft bottom are required. Rapid currents are avoided.

Natural History: Courtship and mating occur in both spring and fall and involve active pursuit of the female by the male. Nesting occurs from late May through mid-July with the peak in June. The nests are flask shaped, and may contain 10 to 16 ellipsoidal eggs (32–35 × 21–22 mm) with parchmentlike shells. Several clutches are laid each season. Eggs incubated at 25°C yield a preponderance of males, while those incubated at 30.5°C yield mostly females (Bull and Vogt, 1979). Hatchlings emerge after about 75 days incubation. They are nearly round (30 mm) with a dorsal keel. The reticulate pattern of the carapace is bright, and the plastral pattern intense.

Graptemys geographica feeds primarily on freshwater snails and clams, but insects (especially the immature stages), crayfish, and water mites are also taken. Fish are occasionally eaten, usually as carrion, and some plant material is consumed.

These turtles are confirmed baskers, often piling up several deep while basking. They are, however, very shy and difficult to approach.

The relationship of *G. geographica* to other map turtles is not well understood; it is the most divergent species. It probably is most closely related to the broad-headed group, which comprises *G. kohnii*, *G. pulchra*, and *G. barbouri*.

Graptemys barbouri Carr and Marchand, 1942

Barbour's map turtle

Recognition: This broad-headed turtle (to 32 cm) has a black-tipped, tuberculate medial keel. The 2d and 3d vertebrals bear slightly concave tubercles; the 1st and 4th bear a low spine or ridge. Vertebrals are broader than long; the 1st is smallest, the 5th expanded. The posterior rim of the carapace is strongly serrated. The carapace is olive to dark brown; pleurals and marginals have a pattern of pale yellow to white oval markings, which open posteriorly (these are often lost in large females). Undersides of the marginals are marked at the posterior seam with a dark, semicircular blotch. The deep plastron is greenish yellow to cream colored, with a narrow black border along the posterior margin of each scute except the anals; the border tends to fade with age. Well-developed spinelike projections are present on the pectoral and abdominal scutes, and the hind lobe is posteriorly notched. The plastral formula is: abd > an > fem > gul >< pect > hum. The head is broad in females,

Graptemys barbouri

but narrower in males. It has a nonprojecting snout, and neither a hook nor notch on the upper jaw. The head is dark brown or black with three broad, light areas, which are interconnected: one on the snout and two behind the eyes. The 12–14 stripes on the nape of the neck are connected with the postocular blotches. The chin exhibits a curved, light bar, which often parallels the curve of the jaw. Other skin is brown or black with light stripes.

Adult males are 9–13 cm in carapace length, adult females 17–27 cm. Males have long, thick tails, with the vent posterior to the carapacial margin. In males the head is small and narrow; in females it is large and short snouted.

Distribution: *Graptemys barbouri* occurs in the Gulf coastal streams of the Apalachicola River system in the Florida Panhandle (including the Chipola) and adjacent Georgia. It apparently occurs only below the fall line.

Habitat: *Graptemys barbouri* lives in clear, limestone-bottomed streams with abundant snags and fallen trees.

Natural History: During courtship the male approaches the female with his neck extended. He swims around her until they are face to face, and after nose contact he strokes her head with the inner side of his forelegs (Wahlquist, 1970). Nesting probably occurs in June, but has not been described. Nests found by Wahlquist and Folkerts (1973) were 8–16 cm deep in moist sand. Clutches found by them consisted of eight and nine eggs (38.3–41.6 × 27.6–30.8 mm). The eggs were ellipsoidal and dull white with flexible shells. Hatchlings emerge in late August and September with 35–38-mm carapaces.

Female foods consist of mussels and snails; males eat smaller snails, insects, crayfish, and fish.

This turtle is rare and threatened by pollution of its waterways and by collection for the pet trade.

Graptemys pulchra Baur, 1893b

Alabama map turtle

Recognition: *Graptemys pulchra* is a broad-headed turtle (to 21 cm) with laterally compressed black spinelike vertebral projections arranged along a black middorsal line. Vertebrals are broader than long, the 1st is smallest. The posterior carapacial margin is slightly serrated. The carapace is olive; each pleural has a pattern of yellow to orange, black-bordered lines in a semicircular or reticulate arrangement, and the

Graptemys pulchra

marginals have wide, semicircular, light markings. The shallow, yellow plastron has a narrow, black border on the posterior edge of each scute. Often both the carapacial and plastral patterns fade with age, especially in large females. The plastral hind lobe has a posterior notch, and the plastral formula is: abd > an > hum >< fem > gul > pect. The snout does not protrude, and the upper jaw lacks both a medial notch or hook. The pattern on the brown head is distinct: the interorbital area is covered with a large yellow blotch extending backward on each side, and a longitudinal, light bar extends backward from the point of the chin. Other skin is dark brown to black with yellow stripes. Females (17–21 cm) are larger and have broader heads than males (8.7–12.5 cm). In males, the tail is long and thick, with the vent behind the carapace rim.

Distribution: This turtle lives in Gulf coastal streams, ranging from the Yellow River in Alabama and the Florida Panhandle west to the Pearl River system in Mississippi and eastern Louisiana.

Geographic Variation: No subspecies are recognized; however, Shealy (1976) has reported that variation occurs in shell depth, coloration, and carapacial and plastral seam alignments. Jeffrey Lovich of the Savannah River Ecology Laboratory is currently studying variation in *Graptemys pulchra*.

Habitat: Deep water with a slow current and a sand or gravel bottom is preferred, and basking sites, such as logs or debris, are necessary.

Natural History: Shealy (1976) observed mating from September to November. The male smells the female's cloacal area, then swims to the front, and, while facing her, rapidly vibrates his head vertically against the sides of her snout, alternating sides about every 5 seconds. The sperm is apparently stored until the next spring; oviducal scrapings in March revealed active and inactive sperm (Shealy, 1976). Nesting occurs from April to July and several clutches of about seven eggs are laid each season. The elongated eggs (38–45 × 24–26 mm) have flexible shells. Incubation takes about 76 days and the young emerge from July to early October; hatchlings are about 30 mm long.

Insects, snails, and clams constitute the diet. Females, with their heavy jaws, can crush mollusks; males and juveniles, with weaker jaws, feed mainly on insects.

Graptemys pseudogeographica

Graptemys pseudogeographica (Gray, 1831b)

False map turtle

Recognition: This is a medium-sized to large (to 27 cm) turtle with a concave anterior profile and a distinct medial keel bearing conspicuous low spines. Its carapace is olive or brown, and each pleural has yellow oval markings and dark blotches. Both upper and lower edges of the marginals have a yellow ocellus at each seam. Vertebral 1 may be longer than broad or as long as broad, and is smallest; other vertebrals are broader than long. The posterior carapacial rim is strongly serrated. The bridge is marked with light bars, and the plastron is immaculate cream to yellow in adults, but with a pattern of dark lines bordering the seams in the young. A posterior notch is present on the hind lobe, and the plastral formula is: abd > an > fem > pect > gul >< hum. The head is narrow to moderately broad with a nonprotruding snout and no medial notch or hook on the upper jaw. Skin is olive to brown, with numerous narrow yellow stripes on the legs, tail, chin, and neck. The postorbital mark is variable but usually consists of a narrow downward extension of a neck stripe behind the orbit; four to seven neck stripes touch the orbit.

Adult males have long, thick tails, with the vent posterior to the carapacial rim, and elongated foreclaws (especially the third). Adult males reach 15 cm in carapace length, females 27 cm.

Distribution: *Graptemys pseudogeographica* occurs primarily in large streams of the Mississippi River drainage from the St. Croix, Wisconsin, and upper Mississippi rivers southward through Louisiana and eastern Texas (Vogt, 1980).

Geographic Variation: No subspecies are recognized. Carr (1952) and Vogt (1980) relegated *Graptemys kohnii* to a subspecies of *G. pseudogeographica*. Recent studies by Vogt (1980) have shown *G. ouachitensis* and

G. pseudogeographica to be separate species. Separation of *kohnii*, *ouachitensis*, and *pseudogeographica* is mostly on the basis of head pattern, but Ewert (1979) has shown that these may be produced by different incubation temperatures and that a single clutch may provide several head patterns. Additional study of the relationships of these three turtles is needed.

Habitat: The false map turtle occurs in lakes, ponds, sloughs, large rivers and their backwaters, and drowned forests in large reservoirs. It prefers water with abundant aquatic vegetation. It seems more common in slow currents, but does occur in the swiftly flowing main channel of the Missouri River.

Natural History: Courtship may take place in autumn, but is most concentrated in spring. During courtship, the male swims over the female, makes an abrupt turn to face her, extends his head, and strokes her head and neck with his elongated foreclaws. When she sinks to the bottom he swims behind and settles over her. The flask-shaped nests are usually dug in late May to July. Several clutches of nine or ten eggs are laid each season. The elliptical (32–36 × 22–24 mm) white eggs have leathery shells. Natural incubation takes 60–75 days (Vogt, 1980); hatchlings are nearly round (25–33 mm), and the black keel is well developed on the first three vertebrals. Incubation temperature influences the sex of the hatchling: eggs incubated at 25°C produced predominantly males, eggs incubated at 35°C produced females (Vogt, 1980).

Natural foods consist of plant material—algae, *Potamogeton*, *Lemna*, and *Vallisneria*—and animals—insects, mollusks, and fish.

This species spends many of the daylight hours basking; for this, it prefers sites remote from shore, and basking *Graptemys pseudogeographica* often pile up several deep. Extremely wary, they slip into the water at the least disturbance, and remain hidden in aquatic vegetation for a long period before resurfacing.

Graptemys ouachitensis Cagle, 1953a

Ouachita map turtle

Recognition: The oval carapace (to 24 cm) has a distinct medial keel bearing conspicuous low spines, and a serrated posterior rim. Vertebrals are usually broader than long, but the 1st, which is smallest, may be as broad as long or longer than broad. The carapace is olive or brown, and each pleural has yellow vermiculations and dark blotches. Upper and lower edges of the marginals have a yellow ocellus at each seam. The bridge is marked with dark bars. The plastron is cream to olive yellow with longitudinal dark lines. Its hind lobe is posteriorly notched. The plastral formula is: abd > an > fem > pect > hum >< gul. The head is narrow to moderately broad with a nonprotruding snout and no medial notch or hook on the upper jaw. Skin is olive to brown or black with numerous narrow yellow stripes on the legs, tail, chin, and neck. The postorbital mark is either square to rectangular or elongate to oval, and usually consists of a downward extension of a neck stripe behind the orbit; one to nine neck stripes reach the orbit. Two large, light spots may occur on each side of the face, one just under the eye and another on the lower jaw, and transverse bars are often present under the chin.

Adult males have long, thick tails, with the vent posterior to the carapacial rim, and elongated foreclaws (especially the third). Adult males are 10–14 cm in carapace length, females 16–24.

Distribution: *Graptemys ouachitensis* ranges from Texas and Louisiana north and eastward to Nebraska, Minnesota, Wisconsin, Indiana, Ohio, and West Virginia.

Geographic Variation: The two subspecies of *Graptemys ouachitensis* were formerly considered subspecies of *G. pseudogeographica* (Cagle 1953a), but Vogt (1980) has compiled sufficient data to show the northern subspecies *ouachitensis* to be a separate species from sympatric *G. pseudogeographica*. *G. o. sabinensis* Cagle, 1953a is restricted to the Sabine River system of Texas and Louisiana. It has an elongate or oval postorbital mark, five to nine neck stripes reaching the orbit, and transverse bars under the chin. *G. o. ouachitensis* Cagle, 1953a ranges from the Ouachita River system of northern Louisiana west to Oklahoma and northward to Kansas, Minnesota, Wisconsin, Indiana, Ohio, and West Virginia. In this subspecies, the postorbital mark is square to rectangular; one to three neck stripes reach the orbit; and there are two large, light spots on each side of the face, one just under the eye and another on the lower jaw.

Habitat: Large rivers, lakes, oxbows, and river-bottom swamps comprise the habitat.

Natural History: Courtship and mating occur in the spring; the male strokes the female's face with his elongated foreclaws. Nesting takes place from late May through June and several clutches of 8 to 17 eggs are laid each year. The elliptical eggs (28–36 × 20–29 mm) have leathery shells. Incubation takes 60–75 days and the round, keeled hatchlings are about 30 mm in length. Eggs incubated at 25°C hatch into males, those incubated at 30.5°C hatch into females (Bull and Vogt, 1979).

Graptemys ouachitensis is omnivorous, feeding on algae and a variety of higher aquatic plants, mollusks, insects, worms, crayfish, and fish. It is fond of basking, but is very wary and difficult to approach.

Graptemys ouachitensis ouachitensis

Graptemys kohnii (Baur, 1890a)

Mississippi map turtle

Recognition: This medium-sized (to 25 cm), broad-headed map turtle has a dark-brown medial keel with tubercles on the 2d and 3d vertebrals pronounced and convex anteriorly. Vertebrals are broader than long; the 5th is flared. The posterior carapacial rim is strongly serrated. The carapace is olive to brown; each pleural has a reticulate pattern of one or more circular markings interconnected by bars, and each marginal has a light figure, which often is open posteriorly. The bridge is marked with light bars. The

Graptemys kohnii

plastron is greenish yellow, with a pattern of double, dark lines extending along the seams; this pattern fades with age. Its hind lobe is notched posteriorly, and the plastral formula is: abd >< an > fem > pect >< gul > hum. The snout is slightly protruding and the upper jaw bears neither a medial hook nor a medial notch. The skin is olive to brown, with yellow stripes. A crescent-shaped postorbital mark extends below the eye on each side preventing the neck stripes from reaching the eye. A straight yellow line extends from the tip of the snout to each crescentic mark, and the lower jaw bears a round spot at the symphysis.

Adult males have elongated foreclaws and long, thick tails, with the vent posterior to the carapacial rim. Females are larger and have proportionately larger heads. Adult females are 15–25 cm in carapace length, adult males 9–13 cm.

Distribution: *Graptemys kohnii* is found in the Mississippi Valley from central Illinois and eastern Nebraska south to the Gulf. It ranges chiefly west of the Mississippi River, but there is a record from as far east as Mobile, Alabama.

Habitat: The quieter parts of silt-bearing rivers, lakes, sloughs, and bayous, always with abundant aquatic vegetation, are the habitats of this turtle.

Natural History: *Graptemys kohnii* is fond of basking and shy and difficult to approach. Apparently it seldom travels overland.

During courtship the male strokes the female's face with his foreclaws. The only observation of nesting was at Natchez, Mississippi, on 1 June.

Apparently *G. kohnii* is omnivorous. Stomach contents have included the aquatic plants *Ceratophyllum*, *Cabomba*, and *Potamogeton* and various species of the duckweed family *(Lemnaceae)*, as well as dragonfly and damselfly nymphs, clams, snails, and a blue-tailed skink *(Eumeces fasciatus)* (Carr, 1952). Captives will eat fish, liver, kidney, and lettuce.

Remarks: Juvenile *Graptemys kohnii* closely resemble juvenile *G. oculifera* in skull characteristics, and these species may be closely allied. *G. kohnii* is also close to *G. pseudogeographica*, with which it was considered conspecific for many years. Ewert (1979) conducted incubation studies on *Graptemys* from Winona, Minnesota, and some clutches yielded both the *kohnii* and *pseudogeographica* patterns of head striping. Further studies using different incubation temperatures resulted in variable numbers of *kohnii*-type hatchlings from each clutch; more *kohnii* hatched at 25°C than at 30°C. The relationship between these two species needs additional study.

Graptemys versa Stejneger, 1925b

Texas map turtle

Plate 14

Recognition: The elliptical carapace (to 18.3 cm) is somewhat flattened, wider posteriorly, and has strongly serrated posterior marginals. The medial keel contains low knobs which are often dark tipped with a yellow area anterior to each. Carapacial scutes are distinctly convex. The carapace is olive, with reticulating yellow lines on each scute. Marginals are patterned dorsally with reticulating yellow lines, and ventrally with fine dark lines surrounding irregular yellow blotches. The plastron is well developed and contains a posterior notch. Its scute formula is: abd > an > fem > pect >< hum >< gul. The plastron is yellow and has a pattern of dark lines along the seams; fine, dark, longitudinal bars cross the bridge. The head is narrow in both sexes, and the snout is somewhat pointed. Triturating surfaces of the jaws are not greatly enlarged. The head is olive with a horizontal or J-shaped, orange or yellow postorbital mark on each side. These marks extend backward at their lower (outer) edge, and may be interrupted. Three to six fine yellow stripes extend up the neck from the tympanum and enter the orbit. The chin often has a pattern of orange or yellow, dark-bordered blotches. Other skin is olive, with many dark lines surrounding yellow areas.

Adult males are 6.5–11.2 cm in carapace length, adult females 9–18.3 cm. Males have long, thick tails, with the vent posterior to the carapace margin.

Distribution: *Graptemys versa* is found in the Colorado River watershed on the Edwards Plateau of central Texas.

Habitat: *Graptemys versa* occupies shallow streams with moderate current.

Natural History: Male courtship apparently involves head bobbing. Captives have eaten fresh and canned fish, chicken, beef, hamburger, dog food, insects, and occasionally lettuce.

Graptemys caglei Haynes and McKown, 1974

Cagle's map turtle

Recognition: This map turtle has a cream-colored transverse bar on the chin and a yellow V-shaped mark on the dorsal surface of its head. Its somewhat flattened carapace (to 16 cm) is elliptical (wider posteriorly), serrated posteriorly, and bears a medial keel of sharp spinelike projections. Vertebrals are also raised in the center, giving the carapace a "lumpy" appearance. The carapace is olive to brown; each scute has yellow, contourlike markings, and the posterior midline of each vertebral is brown or black. The plastron is well developed with a posterior notch. Its scute formula is: abd > an > fem > pect > gul > hum. The cream-colored plastron has a dark seam-following pattern; each scute may also contain black flecks. The cream-colored bridge is crossed by four black longitudinal bars, as also are the undersurfaces of the 5th to eighth marginals. The head is narrow. Its snout is somewhat pointed, and the triturating surfaces of the jaws are not greatly enlarged. The head is black and contains seven cream-colored stripes on the dorsal surface, with the medial stripe widest. Another broad stripe arises on each side of the head below and in front of the orbit and runs backward and upward forming a complete crescent-shaped mark around the orbit and meeting the stripe from the opposite side at the midline just behind the orbits to produce a V-shaped mark (viewed from above). Several other narrower stripes also follow this path, but do not meet. A cream-colored bar extends transversely across the lower jaw. Neck, limbs and tail are black with numerous cream to yellow stripes.

Adult males are 7–11 cm in carapace length, have longer vertebral spines and longer, thicker tails with the vent beyond the carapacial rim. Adult females have larger (to 16 cm) and more rounded carapaces with low vertebral spines. The female tail is shorter with the vent under the carapace.

Distribution: *Graptemys caglei* is restricted to the Guadalupe and San Antonio river drainages in south-central Texas.

Habitat: The streams in which *Graptemys caglei* lives are generally shallow with moderate currents. It has been taken in small and large pools with depths to 3 m, and with both limestone and mud bottoms (Haynes and McKown, 1974).

Natural History: What little we know of the biology of this species has been reported by Haynes and McKown (1974). They collected hatchlings from September through November, indicating a late spring to early summer nesting period. Courtship, mating, and nesting have not been described.

Cagle's map turtle is predominantly insectivorous, although females also eat snails. Plant remains found in stomachs may have been incidentally ingested.

This turtle is a confirmed basker, often being observed on rocks, logs, and cypress knees. Haynes and McKown (1974) commented that logs that had fallen into the river but were still connected to the bank were shunned as basking platforms. When approached, basking *Graptemys caglei* quickly escaped into the water.

Graptemys oculifera (Baur, 1890a)

Ringed map turtle

Plate 14

Recognition: This narrow-headed map turtle (to 22 cm) has laterally compressed, black, spinelike vertebral projections and a slightly serrated posterior carapacial rim. Vertebrals are broader than long; the 1st is smallest. The carapace is dark olive green; each pleural has a broad yellow or orange circular mark, and each marginal is patterned with a wide yellow bar or semicircle. The plastron is yellow or orange, with an olive-brown pattern extending along the seams; this pattern fades with age. Its hind lobe is notched, and the plastral formula is: abd > fem >< an >< pect > gul > hum. The head is small to moderate in size with a nonprotruding snout and an upper jaw which is neither hooked nor notched. The head is black with yellow stripes, as is other skin. The variable postorbital marks are ovoid, rectangular, or rounded and usually not connected with the narrow dorsal longitudinal line. Two broad yellow stripes touch the orbit. The interorbital stripe is wide and is equal to or greater than the width of the broadest neck stripe. The lower jaw is marked with longitudinal yellow stripes as wide as the black interspaces. Ventrally, the neck bears three wide longitudinal stripes.

Adult males have long, thick tails, with the vent posterior to the carapace margin and close to the tail tip. Males also have elongated foreclaws. Adult females are 12–22 cm in carapace length, adult males 7.5–11.0 cm.

Distribution: *Graptemys oculifera* is restricted to the Pearl River system of Mississippi and Louisiana.

Habitat: Sand- and clay-bottomed rivers with rapid currents and abundant basking sites of brush, logs, and debris are preferred.

Natural History: Courtship has not been described, but the elongated male foreclaws indicate it is a "stroker." Eggs are deposited on sandbars in early June. Cagle (1953b) described an incomplete nest as a hole 3 cm in diameter and 3 cm deep leading to a cavity 9 cm in depth. A female collected by him on 4 June had three eggs in the left oviduct and none in the right; two of the oviducal eggs measured 40.3 × 20.6 and 40.0 × 21.0 mm. When found, she was depositing her first clutch of the season; enlarged ovocytes on her ovary indicated a second clutch of four eggs. Hatchlings emerge in late August or September and are 22–33 mm long.

Insects and mollusks are the primary foods. The scissorslike jaws are well adapted for dismembering such animals.

This map turtle is fond of basking, but very shy, and slips into the water at the slightest disturbance. It is endangered through pollution and collecting for the pet trade.

Graptemys flavimaculata Cagle, 1954a

Yellow-blotched map turtle

Plate 14

Recognition: This small (to 17.5 cm), narrow-headed turtle has laterally compressed, black, spinelike vertebral projections and a slightly serrated posterior carapacial margin. Its vertebrals are broader than long; the 1st is smallest. The carapace is olive to brown; each pleural has a broad ring or yellow blotch covering most of its surface, and each marginal contains a wide yellow bar or semicircle. The plastron is light cream colored, with a black pattern extending along the seams; this pattern fades with age. The hind lobe is posteriorly notched, and the plastral formula is: an > abd > fem >< pect > gul > hum. The head is small to moderate with a nonprojecting snout and nei-

ther a hook nor notch on the upper jaw. It is olive with yellow stripes. The postorbital mark usually is rectangular, and joins a longitudinal dorsal neck stripe that is at least twice as wide as the next widest neck stripe. Two to four neck stripes reach the orbit; the interorbital stripe is narrower than the neck stripes. The lower jaw is marked with longitudinal yellow stripes, which are wider than the olive-green interstices; thus yellow predominates. There are about 19 longitudinal yellow stripes on the olive neck; those on the ventral surface are twice as wide as those dorsal. Other skin is olive with yellow stripes.

Adult males have long, thick tails, with the vent posterior to the carapacial rim, and elongated foreclaws. Adult females have broader heads and are larger (10–16 cm) than males (7.5–11.0 cm).

Distribution: *Graptemys flavimaculata* is restricted to the Pascagoula River system in Mississippi.

Habitat: Sand- and clay-bottomed streams with moderate to rapid currents form the habitat. Piles of brush and debris serve as basking sites, and tangled roots are used as shelters.

Natural History: During courtship the male approaches the female with his neck extended. She faces him and extends her neck. He then stretches out his forelimbs and strokes the sides of her head with his claws; she simultaneously attempts to stroke him (Wahlquist, 1970). This behavior is quite similar to that of *Chrysemys picta* and *Graptemys pseudogeographica*.

Almost nothing is known of the behavior of the yellow-blotched sawback. Like other map turtles it is a confirmed basker and extremely difficult to approach. It seldom ventures overland.

The diet consists largely of snails and insects; captives will eat fish.

G. flavimaculata is apparently the dominant turtle species in the Pascagoula River system, but is still endangered by pollution and collection for the pet trade. It is now protected by the state of Mississippi.

Graptemys nigrinoda Cagle, 1954a

Black-knobbed map turtle

Recognition: This is a small (to 15 cm) turtle with broad, rounded, black, knoblike vertebral projections and a strongly serrated posterior carapacial rim. Its vertebrals are broader than long; the 1st is smallest. The carapace is dark olive; each pleural and marginal has a narrow, yellow or orange, semicircular or circular mark. The yellow plastron is often tinted with red and has a black, branching pattern. The plastral hind lobe is notched and the plastral formula is: an > abd > fem >< pect >< gul > hum. The narrow head has a nonprotruding snout and an upper jaw that is neither hooked nor notched. It is black with yellow stripes; the postorbital mark is a vertical crescent connecting dorsally with that of the opposite side to form a Y-shaped mark. Usually two to four neck stripes enter the orbit; the interorbital stripe is narrower than the neck stripes. The lower jaw has longitudinal yellow stripes as wide as the black interstices. Other skin is black, but yellow stripes occur on the limbs.

Adult males have long thick tails, with the vent posterior to the carapacial rim; they also have elongated foreclaws. Adult females are 10–15 cm in carapace length, adult males 7.5–10.0 cm.

Distribution: *Graptemys nigrinoda* occurs below the fall line in the Alabama, Tombigbee, and Black Warrior river systems of Alabama and Mississippi.

Geographic Variation: Two subspecies are recognized. *Graptemys nigrinoda nigrinoda* Cagle, 1954a is restricted to the upper parts of the Tombigbee and Alabama river systems in Alabama and Mississippi. It has a poorly developed plastral figure, which never occupies more than 30% of the plastron. Its postocular mark is, typically, crescentic and strongly recurved. The light lines that reach the eye are seldom interrupted, and its soft parts are predominantly yellow. *G. n. delticola* Folkerts and Mount, 1969 occurs in the interconnecting streams and lakes of the delta of the Mobile Bay drainage, in Baldwin and Mobile counties, Alabama. It differs from the nominate race in having a plastral figure that occupies more than 60% of the plastron. Its postocular mark is neither crescentric nor strongly recurved laterally. In many individuals the light lines that reach the eye are interrupted, and the soft parts are predominantly black.

Habitat: Sand- and clay-bottomed streams with moderate currents and abundant basking sites of brush, logs, and debris are favored. *Graptemys nigrinoda* is found in deeper waters than either *G. oculifera* or *G.flavimaculata*.

Natural History: Male courtship involves stroking of the female's face with the elongated foreclaws. Nesting takes place in June. Hatchlings are brightly marked with knobby vertebral projections, and the entire carapacial rim serrated. *Graptemys nigrinoda* is a basker, and very wary. Snails and insects seem to be its preferred food.

Malaclemys Gray, 1844

The single species in this genus, *Malaclemys terrapin*, is the only truly brackish water North American emydid.

Malaclemys terrapin (Schoepff, 1793)
Diamondback terrapin

Recognition: *Malaclemys terrapin* is small to medium sized (to 23 cm) with markings and concentric grooves and ridges on each large scute; the limbs and head are flecked or spotted. The oblong carapace is widest behind the bridge and slightly serrated posteriorly, and it has a medial keel that varies from low and inconspicuous to prominent and knobby. Vertebrals are all broader than long; neural bones are hexagonal and shortest anteriorly. The posterior peripheral bones may be upturned. The carapace is gray, light brown, or black; if light brown, the scutes are ringed concentrically with darker pigment. Undersides of the marginals and the bridge are often patterned with dark flecks. The oblong, hingeless plastron usually is greenish to yellow with dark blotches or flecks. Its hind lobe is slightly notched posteriorly, and the plastral formula is: abd >< an > fem > gul > pect > hum. The humero-pectoral seam does not cross the entoplastron. The bridge is wide, but its buttresses are relatively weak. The short head is narrow in males but broad in females; it is flat and of a uniform color dorsally. It has a nonprotruding snout and an unnotched or slightly notched upper jaw. The frontal bone often enters the orbit, the maxilla is not or only narrowly in

Malaclemys terrapin centrata

contact with the quadratojugal, and a large posterior process of the pterygoid touches the exoccipital. The triturating surface of the maxillae and palatine is smooth, lacking a ridge, and wide in females but narrow in males. Head and neck are gray to black with flecks or curved markings, but never stripes. Eyes are black, large, and prominent. Jaws are light in color, the chin may be black, and the limbs gray to black.

The diploid chromosome number is 50: 20 metacentric and submetacentric, 10 subtelocentric, and 20 acrocentric and telocentric chromosomes (Stock, 1972).

Adult males are smaller (10–14 cm) than females (15–23 cm). Females have broader, blunter heads, deeper shells, and shorter tails than males, and their vent is closer to the body and under the posterior marginals.

Distribution: *Malaclemys* occurs along the Atlantic and Gulf coasts from Cape Cod to Texas; it is also found on the Florida Keys.

Geographic Variation: Seven subspecies are recognized. *Malaclemys terrapin terrapin* (Schoepff, 1793) ranges along the Atlantic coast from Cape Cod to Cape Hatteras. Its medial keel does not bear terminal knobs on each scute; sides of the carapace diverge posteriorly. Carapace varies from uniform black to light brown with distinct concentric, dark rings; plastron variable, orangish to greenish gray. *M. t. centrata* (Latreille, in Sonnini and Latreille, 1802) ranges from Cape Hatteras south along the coast to northern Florida. Its medial keel bears no terminal knobs on each scute; sides of the carapace are nearly parallel; the marginals curl upward; otherwise similar to *M. t. terrapin*. *M. t. tequesta* Schwartz, 1955 occurs along the Atlantic coast of Florida. Its medial keel bears posteriorly facing tubercles or knobs. The carapace is dark or tan with no pattern of concentric light circles. The centers of large scutes may be slightly lighter than surrounding areas. *M. t. rhizophorarum* (Fowler, 1906) is restricted to the Florida Keys. Its medial keel bears terminal bulbous knobs, and its shell is strongly oblong. The carapacial scutes have no light centers, and the ventral seams of the marginals and plastral scutes are often outlined with black. Spots on the neck often fuse to form a streaked pattern, and the hind legs may be striped. *M. t. macrospilota* (Hay, 1904a) ranges along the Gulf coast from Florida Bay to the panhandle. Its medial keel has terminal, often bulbous, knobs, and the carapacial scutes contain orange or yellow centers. *M. t. pileata* (Wied, 1865) ranges along the Gulf coast from the Florida Panhandle to western Louisiana. Its medial keel has terminal tuberculate knobs. The scutes of the oval carapace lack light centers. Top of the head, upper lip, neck, and limbs are

black or dark brown, the upturned edges of marginals are orange or yellow, and the plastron is yellow and often dusky. *M. t. littoralis* (Hay, 1904a) occurs along the Gulf coast from western Louisiana to western Texas, and has a keel with terminal knobs on its deep carapace. Its carapacial scutes lack distinct light centers, the plastron is pale or white, the upper lip and top of head whitish, and the neck and legs greenish gray with heavy black spotting. Geographic variation is poorly defined; noted differences may represent clines in characters along the linear range of the species. A thorough study is needed.

Habitat: A resident of brackish water, *Malaclemys* lives in coastal marshes, tidal flats, coves, estuaries, and the lagoons behind barrier beaches.

Natural History: Courtship and mating occur from late March to May during daylight hours. Seigel (1980) reported that courtship begins with the female floating at the water's surface; the male approaches from the rear and nuzzles or nudges her cloacal region with his snout. If she remains still, he mounts immediately, and copulation occurs at the surface. If the female swims away, the male pursues her until he catches her. Nesting extends from April through July, depending on latitude. Nests are located above the high-tide mark along the sandy edges of salt marshes and rivers, in the dunes of sea beaches, and on offshore islands. They are flask shaped to triangular, and 100–200 mm deep. At least two clutches may be laid each season, and females may store viable sperm for several years. Clutch size varies from 4 to 18 eggs. The oblong to elliptical eggs (26–36 × 16–22 mm) are pinkish white when fresh and have thin, easily dented shells; the surface is coated with minute, scattered calcareous lumps. The average incubation period is about 90 days; it varies considerably with temperature, and the young may overwinter in the nest. Hatchlings are patterned much like adults but usually brighter. Their shells are round, lengths are about 27–29 mm.

Diamondback terrapins probably are scavengers, but they also take live food. Coker (1906) examined 14 *M. t. centrata* and found remains of the snail *Littorina irrorata* in the stomachs of 12; he also found the snail *Melampus lineatus*, fiddler and other crabs, marine annelids (*Nereis*), and some fragments of marsh plants. Carr (1952) found the stomach contents consisted almost wholly of rather finely crushed fragments of shells, about 90% of which were from the Venus clam *Anomalocardia cuneimeris*. Captives feed readily on chopped fish, crabs, snails, oysters, clams, insects, marine annelids, and beef.

Testudinidae

The terrestrial tortoises compose the family Testudinidae, and that family is second only to the Emydidae in total number of living species (50). Today its members are found primarily in the tropical portions of Africa, Madagascar, India, Southeast Asia, South America, Aldabra Atoll, and the Galapagos Archipelago. However, almost 20% of the species extend into southern Europe (3), western Asia (1), and southern North America (4). The fossil record shows tortoises were much more widespread in the past, occurring in northern Europe and England, central Asia, the West Indies, and in North America to southern Canada. Many islands in the Indian Ocean also had endemic species which have become extinct in modern times through human exploitation. Fossil tortoises date from the early Eocene: *Geochelone (Manouria) majusculus* from New Mexico deposits and *Testudo*(?) *comptoni* from the Isle of Sheppey, England. By the end of the Eocene, tortoises of several other genera appeared in North America: *Geochelone (Manouria) gilmorei (= Hadrianus robustus), G. (M.) utahensis, G. (Cymatholcus) longus, Stylemys uintensis* in Utah; *G. (M.) corsoni* in Wyoming; and *G. (C.) schucherti* in Alabama. In Europe, *Testudo corroyi, T.*(?) *castrensis*, and *T. doduni* appeared in France; *G. (M.) eocaenica* in Germany; *G. (M.) obailiensis* inhabited the USSR (Georgia); *G. (G.) beadnelli, G. (G.) isis* in Africa (Egypt); *G. (M.?) ammon, G. (M.) insolitus, G.*(?) *ulanensis, Indotestudo kaiseni* in Asia (Mongolia); *Kansuchelys*(?) *chiayukuanensis, Sinohadrianus*(?) *sichuanensis* in China; and *S.*(?) *sichuanensis* in Japan. Among living tortoises, *Manouria emys* and *M. impressa* are considered to be most primitive, as they share a number of derived characters with some of the oldest fossil species; thus, they are placed in a separate genus. Additional information on fossil tortoises may be obtained from Williams (1950b, 1952), Mlynarski (1969), and Auffenberg (1974).

Tortoise skulls are short to moderate in length, but very different from those of the Emydidae (Gaffney, 1979b). The temporal region is widely emarginated posteriorly and the squamosal is not in contact with the parietal. The frontal bone may or may not enter the orbit, and the postorbital is reduced (rarely absent) and narrower than that found in emydids. The maxilla rarely touches the quadratojugal and the quadrate is usually closed posteriorly, completely surrounding the stapes. Premaxillae meet dorsally with

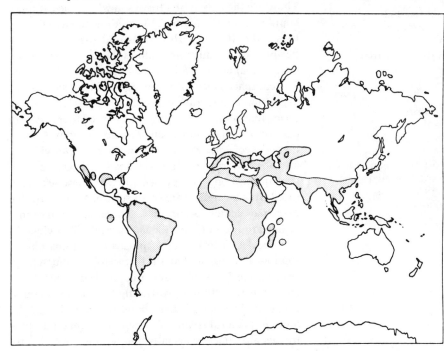

Distribution of Testudinidae

Key to the genera of Testudinidae:

1a. Carapace containing a posterior hinge; submarginal scutes present: **Kinixys**
b. Carapace lacking a posterior hinge; no submarginals present 2
2a. Plastron with hinge 3
b. Plastron rigid, lacking any hinge 4
3a. Plastral hinge lies between humeral and pectoral scutes: **Pyxis**
b. Plastral hinge lies between femoral and abdominal scutes: **Testudo** (in part)
4a. Carapace extremely flat with much reduction of bones so as to be rather flexible: **Malacochersus**
b. Carapace arched with little reduction of bones and rigid throughout 5
5a. Gular single and strongly projected anteriorly 6
b. Gular double, usually not strongly projecting 7
6a. Anal scute small: **Geochelone yniphora**
b. Anal scute large: **Chersina**
7a. Maxilla with medial ridge 8
b. Maxilla lacking any ridge 13
8a. Tail flattened, its dorsal surface covered with enlarged scales: **Acinixys**
b. Tail not flattened but sometimes covered dorsally with an enlarged scale 9
9a. Premaxilla with medial ridge; forelimbs flattened and shovellike: **Gopherus**
b. Premaxilla ridgeless; forelimbs clublike 10
10a. Forefoot with four claws: **Testudo horsfieldi**
b. Forefoot with five claws 11
11a. 5th and 6th marginals touch 2d pleural; humero-pectoral seam does not cross entoplastron 12
b. 5th, 6th, and 7th marginals touch 2d pleural, humero-pectoral seam crosses entoplastron **Indotestudo**
12a. Supracaudal subdivided into two scutes: **Manouria**
b. Only a single supracaudal scute present: **Geochelone**
13a. Carapace arched or domed dorsally; gular scutes as long as or longer than broad; areolae of vertebral scutes raised and conical: **Psammobates**
b. Carapace somewhat dorsally flattened, not arched or domed; gular scutes broader than long; areolae of vertebral scutes flattened, never conical: **Homopus**

decending processes to form a hooklike upper jaw, as in the Kinosternidae. The triturating surface of the upper jaw is moderately developed with or without a medial ridge of varying development on the maxilla, but usually not on the premaxilla (*Gopherus* has a medial premaxillary ridge). No splenial bone is present. The carapace is usually highly arched or domed, and contains a movable posterior hinge in the genus *Kinixys*. A medial keel may be present in juveniles, but disappears with age in all species. Dorsal rib heads are small and often vestigial. Carapace and plastron are firmly sutured at the bridge. The plastron is usually rigid, but is hinged (differently) in *Pyxis* and most *Testudo*. Mesoplastral bones are absent, as also is an intergular scute. Infra- or submarginal scutes are present only in *Kinixys*. The shell is poorly ossified in the genus *Malacochersus*. Limbs are developed for walking on land and supporting much weight and/or for digging. Toe webbing is absent, and there are no more than two phalanges in any digit; the trochanteric fossa of the femur is narrowed by the close apposition of the trochanters. In the pelvic girdle, the pubis normally contacts the ischium of the same side below the thyroid fenestra; ventral cartilages are usually absent in adults and the pubis and ischium of the two sides can meet. The family is composed of 12 genera and 50 species.

Kinixys Bell, 1827

These are the only living turtles in which the carapace contains a movable hinge. This hinge always lies between the 4th and 5th costals and the 7th and 8th peripherals in adults and allows the posterior portion of the carapace to be lowered over the hindquarters. The carapace is usually elongate and may show some expansion and serration of both the anterior and posterior marginals. It is somewhat domed (depressed in one species), flat topped, and has sloped sides. A slight medial keel may be present in younger individuals, and the anterior neurals are six sided. One or two suprapygal bones are present; if two, these are separated by a straight transverse suture. Nine to twelve marginal scutes occur on each side, submarginal scutes are present, and the supracaudal is divided in one species. The hingeless plastron contains only a shallow anal notch, if at all. Its forelobe is much longer than the hind lobe, and the epiplastra are thickened, but usually not protruding beyond the carapacial rim. The entoplastral bone lies anterior to the humero-pectoral seam. Axillary and inguinal buttresses are short, especially the inguinal which supports the anterior carapace at the hinge. The skull is relatively long with a hooked upper jaw. Triturating surfaces of the jaws lack ridges, and the sides of the jaws are not serrated. The maxillae do not contribute to the roof of the palate, and the anterior palatine foramina are large. The temporal arch is relatively weak. The prootic bone is well exposed dorsally and the quadrate bone surrounds the stapes. The limbs are not nearly as club shaped as in other testudinids; each forefoot has four or five claws.

The karyotype consists of 52 chromosomes: 26 macrochromosomes and 26 microchromosomes (Killebrew and McKown, 1978).

Kinixys belliana belliana

Key to the species of *Kinixys*:

1a. Gulars usually twice as broad as long, supracaudal divided, upper jaw tricuspid: **K. natalensis**
 b. Gulars usually less than twice as broad as long, supracaudal not divided, upper jaw unicuspid or bicuspid   2
2a. Posterior portion of carapace sloping or strongly reverted from middle of 5th vertebral   3
 b. Posterior portion of carapace strongly reverted from anterior end of 5th vertebral: **K. homeana**
3a. Posterior rim of carapace strongly serrated; five claws on each forefoot: **K. erosa**
 b. Posterior rim of carapace not serrated, or only weakly so; four or five claws on each forefoot: **K. belliana**

Kinixys belliana Gray, 1831b

Bell's hinge-back tortoise

Recognition: This is the most widespread and best known of the *Kinixys*. Its elongated carapace (to 22 cm) is domed with a flat, sometimes slightly keeled, dorsal surface, sloping sides, the anterior marginals not or only slightly flared, and the posterior marginals not flared and only slightly reverted and serrated. There is, at best, only a shallow indentation at the cervical region, and the posterior portion of the carapace slopes steeply to the marginals. An elongated cervical scute is usually present, although in some it is broad and in others tiny. Vertebrals 2–5 are wider than long and the 5th is flared; the 1st becomes longer with age until it may be longer than wide. A disrupted keel may be present in the form of a narrow longitudinal ridge at the center of each vertebral in younger individuals, but this is usually lost with age.

Normally 11 (9–12) marginals lie on each side, and the single supracaudal is undivided. The center of each carapacial scute is yellow to reddish brown while the surrounding growth annuli are dark brown or black; the carapacial pattern, however, is quite variable. The thickened, paired gulars project at best only slightly beyond the carapacial rim, and may lack a shallow notch. The short plastral hind lobe may not be notched posteriorly. There are two to four small axillary scales and one large inguinal (in contact with the femoral) on each bridge. The plastral formula is: abd > hum > fem >< gul > pect > an. The plastron is yellow with black radiations. The brown or black to yellow or tan head is small to moderate with a nonprojecting snout and an upper jaw which may or may not be hooked. Its large prefrontal scale may not be subdivided or longitudinally divided, the large frontal is occasionally longitudinally divided, and other head scales are small and irregularly shaped. The forelimbs have on their anterior surfaces large, overlapping scutes in five to nine longitudinal rows. Each thigh lacks enlarged tubercles, but the heel may have spurlike scales. Each forefoot has four or five claws. The tail ends in a clawlike tubercle. The limbs and tail are grayish brown.

Males have longer, thicker tails and concave plastra.

Distribution: *Kinixys belliana* occurs from Somalia southward to Swaziland and Natal and westward into Zaire and across West Africa to Senegal. Its range generally skirts the rainforests of tropical Africa.

Geographic Variation: *Kinixys belliana* is quite variable, and many names have been applied to its populations. However, only four subspecies are currently considered valid. *K. b. belliana* Gray, 1831b occurs in eastern Africa, possibly from Somalia southward to

Zululand. It usually has five claws on each forefoot, the pectoral midseam 26–69% of the length of the combined gular and humeral midseams and 21–60% of the abdominal midseam, and a domed carapace (length/height ratio < 2.3) with a pattern of yellow or reddish-brown areolae surrounded by dark pigment. *K. b. spekii* Gray, 1863e ranges in southern and central Africa from Angola and southern Zaire to western Tanzania and Kenya southeastward to Transvaal, Mozambique, Swaziland, and Natal. It usually has five claws on each forefoot, the pectoral midseam as in *K. b. belliana*, but the carapace is distinctly flattened (length/height ratio > 2.3), and may be unicolored brown, or have a pattern of distinct yellow and dark zonary rings. *K. b. nogueyi* (Lataste, 1886) occurs in West Africa from Senegal eastward to Cameroon and the Central African Republic. This subspecies has only four claws on each forefoot, the pectoral midseam 28–39% of the length of the combined gular and humeral midseams and 22–36% of the abdominal midseam, and a domed carapace (length/height ratio < 2.3) which is uniformly brown or with yellow areolae bordered with dark pigment. *K. b. mertensi* Laurent, 1956 occurs in northern Zaire and Uganda. It usually has five claws on each forefoot, the pectoral midseam 0–31% of the length of the combined gular and humeral midseams and 0–32% of the abdominal midseam, and a domed carapace (length/height ratio < 2.3) with a pattern similar to that of *K. b. belliana*.

Habitat: *Kinixys belliana* is a savanna, dry brush, and grassland species, and occurs where there are distinct wet and dry seasons. It is forced to estivate during the dryer periods, and, at such times, has been found in mud bottoms of drying waterholes.

Natural History: Nesting may occur in May. The eggs have brittle shells and are oval to elongated (41–45 × 28–38 mm); two to four compose a clutch. The hatchling carapace is not spinose, and is about 45–47 mm long; in color, hatchlings are either uniformly yellowish-, reddish-, or olive brown, or the areolae may be deep brown surrounded by a yellow border.

K. belliana is omnivorous, feeding on fallen fruits such as bananas and mangos, sugar cane, fungi, grasses, sedges, insects, millipedes, and snails.

Kinixys erosa (Schweigger, 1812)

Serrated hinge-back tortoise

Recognition: The elongated carapace (to 32.3 cm) is domed with a flat, unkeeled dorsal surface, sloping sides, the anterior marginals flared and reverted, and

Kinixys erosa

the posterior marginals only slightly flared but strongly serrated and reverted. A shallow depression occurs in the cervical region, and the posterior portion of the carapace slopes to the rear, but does not drop off nearly as abruptly as in *K. homeana*, and the sloping begins at the middle of the 5th vertebral. A cervical scute is only rarely present. All vertebrals are broader than long; the 5th is flared and widest. There are 11 or sometimes 12 marginals on each side, and the supracaudal is not divided. Carapacial color ranges from totally dark brown to dark brown with yellow to orange centers on all scutes. Occasionally a band of yellow occurs along the base of the pleurals. The thickened, paired gulars project anteriorly beyond the carapacial rim, especially in males, and are notched in front. The short plastral hind lobe may lack a posterior notch or be shallowly notched. There are three or four small axillary scutes and one large inguinal (in contact with the femoral) on each bridge. The plastral formula is: abd > hum >< gul >< pect > fem >< an. The plastron is dark brown to black with some yellow along the seams. The head is small to moderate with a nonprojecting snout and a hooked upper jaw. Its prefrontals are divided longitudinally, the large frontal may be subdivided, and the rest of the head scales are small and irregularly shaped. The head varies from brown to yellow, but usually has some yellow pigment. Each forelimb has on its anterior surface a few large, sometimes overlapping scales in four or five longitudinal rows. Each thigh lacks enlarged tubercles and the heel usually has no spurlike enlarged scales. Five claws are present on each forefoot. Limbs and tail are brown. The tail ends in a clawlike tubercle.

Males have longer, thicker tails and more projecting gulars than do females, and concave plastra.

Distribution: *Kinixys erosa* ranges in West Africa from Gambia eastward to Gabon, Congo, and Uganda.

Habitat: This hinge-back is a forest dweller and seems to prefer moist areas, such as marshes, river banks, and bottom wetlands. It is often found resting in the water and is a fair swimmer. When on land it spends much time buried beneath roots, logs, and plant debris (Schmidt, 1919).

Natural History: Blackwell (1966) observed a mating attempt on 9 April in Nigeria. The male approached from the side and pushed the female's carapace until he could bring the cloacae together. He emitted a series of hissing squeaks near the end of copulation. Eggs are laid in the ground and covered with leaves. One to four oval (40–46 × 31–38 mm), brittle-shelled eggs are laid at one time. A juvenile thought to be a hatchling by Blackwell (1966) was 40 mm long. The young are usually unicolored brown with flattened, very spiny-rimmed carapaces which lack the hinge.

Kinixys erosa is omnivorous, feeding on plants, fruits, small invertebrates and carrion. Eglis (1962) described olfactory, food-exploratory motions as straightforward and not exaggerated.

Kinixys homeana

Kinixys homeana Bell, 1827

Home's hinge-back tortoise

Recognition: The elongated carapace (to 21 cm) is domed with a flat, sometimes slightly keeled, dorsal surface, sloping sides, the anterior marginals flared but only slightly reverted, and the posterior marginals flared and strongly reverted and serrated. Only a shallow depression, at best, occurs in the cervical region, and the posterior portion of the carapace begins to drop off abruptly at the front of the 5th vertebral. An elongated cervical is usually present (rarely absent). Vertebrals 2–5 are broader than long; the flared 5th is widest. Vertebral 1 is broader than long in juveniles, but becomes progressively longer with age until it is longer than broad. A disrupted keel, in the form of a low knoblike projection on each vertebral, may be present on younger individuals; this keel is usually lost with age. There are 11 or rarely 12 marginals on each side, and a single, usually undivided, reverted supracaudal scute. Carapacial color ranges from totally dark brown to tan, or there may be some yellow pigment on the vertebrals and pleurals, or dark borders on these scutes. The thickened, paired gulars project only slightly beyond the carapacial rim, and are widely notched. The short plastral hind lobe is shallowly notched posteriorly. There are two to four small axillary scutes and a large inguinal (in contact

with the femoral) on each bridge. The plastral formula is: abd > hum > gul >< pect >< fem > an. The plastron is dark brown to tan with some yellow at the seams. The brown to yellow head is small to moderate with a nonprojecting snout and a hooked upper jaw. Its prefrontals are divided longitudinally and the large frontal scale may be subdivided. Other head scales are small and irregularly shaped. The forelimbs have on their anterior surfaces a few scattered, large overlapping scutes in five to eight longitudinal rows. The thigh lacks enlarged tubercles, but the heel may have a large spurlike scale. There are five claws on each forefoot. The limbs and tail are brown to yellow. The tail ends in a clawlike tubercle.

Males have longer, thicker tails and concave plastra.

Distribution: *Kinixys homeana* ranges in West Africa from Liberia and Ivory Coast eastward to Zaire.

Habitat: *Kinixys homeana* is another forest dweller, but does not seem to prefer as moist a habitat as *K. erosa*.

Natural History: Blackwell (1968) reported a possible minimum incubation period of five months. The eggs are oval to almost spherical (46 × 35 mm) and have brittle shells. Hatchlings have flattened, brown carapaces, 42–47 mm long, with very spiny rims and no hinge or cervical scute.

Kinixys homeana is apparently omnivorous in the wild, as captives accept both animal and plant foods. Food-searching olfactory movements are usually straight, darting jabs (Eglis, 1962).

Kinixys natalensis Hewitt, 1935
Natal hinge-back tortoise

Recognition: This small species was recently resurrected by Broadley (1981b). Its elongated carapace (to 15.5 cm) is slightly domed with a flat dorsal surface, sloping sides, the anterior marginals not flared or reverted, and the posterior marginals not flared but slightly reverted and serrated. Only a slight depression is present in the cervical region, and the posterior portion of the carapace drops off abruptly. An elongated cervical is present. All vertebrals are broader than long, although the 1st approaches equality, and the 5th is flared. Vertebrals 4 and 5 have raised central protuberances. Twelve or more marginals lie on each side, and the supracaudal is usually divided. Centers of the carapacial scutes are yellow to orange surrounded by dark-brown or black growth annuli. The short paired gulars only slightly project beyond the carapacial rim, and bear a slight notch at best; they are about twice as wide as long. The short plastral hind lobe has only a very shallow anal notch. Three small axillary scutes and one large inguinal (in contact with the femoral) occur on each bridge. The plastral formula is: abd > hum > fem > an > pect > gul. Plastral scutes have yellow centers and seam borders, separated by black areas. The brown to yellow head is small to large with a nonprojecting snout and a hooked tricuspid upper jaw. Its prefrontals are longitudinally divided, the large frontal is undivided, and other head scales are small and irregular. The clublike forelimbs have on the anterior surface seven or eight longitudinal rows of large overlapping scales. Five claws occur on each forefoot, and the tail ends in a clublike tubercle. Both limbs and tail are brown to yellow.

Males have longer, thicker tails and concave plastra.

Kinixys natalensis

Distribution: *Kinixys natalensis* ranges from the Lebombo Range on the Mozambique-Swaziland border, south through western Zululand to Greytown, Natal (Broadley, 1981b).

Habitat: *Kinixys natalensis* is an inhabitant of dry shrubby slopes and thickets and of grasslands to elevations of 1000 m.

Acinixys Siebenrock, 1903b

This is a monotypic genus; its sole species is the Madagascar flat-shelled spider tortoise.

Acinixys planicauda (Grandidier, 1867)
Madagascar flat-shelled spider tortoise

Plate 15

Recognition: The elongated carapace (to 12 cm) is somewhat flattened dorsally, with descending almost parallel sides, a small notch in the cervical region, and unserrated (or slightly in juveniles), downturned posterior marginals. The cervical scute is short and broad. Vertebrals are broader than long with slightly raised centers surrounded by prominent growth annuli. Neurals are hexagonal or squared, but never octagonal; two suprapygals are present. Eleven marginals lie on each side, and the supracaudal is undivided. No submarginals are present. Each vertebral and pleural has a light-brown or yellow center surrounded by a wide dark-brown or black border, and in older tortoises an additional yellow border surrounds this dark border. Yellow rays extend outward from the centers of these scutes across the dark border (usually 4–9 on each vertebral, 2–4 on each pleural). Each marginal is dark with a yellow stripe. The hingeless plastron is well developed; its forelobe is longer than the hind lobe, narrower, and slightly tapered toward the front. The hind lobe is short and broad, but lacks a median anal notch; the anals are somewhat rounded posteriorly. The plastral formula is: abd > hum >< fem >< pect > an > gul; the paired gulars are thickened and extend slightly beyond the carapacial rim. One or two small axillary scutes and larger inguinals occur on the broad bridge. Axillary and inguinal buttresses are short but strong. The plastron is yellow with some scattered dark blotches or radiations along the sides. The head is moderate in size with a nonprotruding snout and a slightly hooked upper jaw. The maxillae have a slight median ridge but the sides of the jaws lack serrations. The quadrate encloses the stapes and the anterior palatine foramina are small. The head is dark brown or black with some variable yellow marks. Limbs are club shaped with five claws on each forefoot. Anterior surfaces of the forelegs have enlarged yellow, nonoverlapping scales in seven to nine longitudinal rows. Spurlike scales occur on the heel, and some small blunt tubercles on the thighs. On the anterior surface of the hind leg there is also a longitudinal row of large nonoverlapping scales. The tail is flattened and ends in a large naillike scale.

Males have longer, thicker tails with the vent near the tip.

Obst (1978, 1980) and Bour (1981) placed *Acinixys* within the genus *Pyxis*, designating it as a subgenus. Characters used by Obst (1980) and Bour (1981) in support of this new alignment are, unfortunately, juvenile traits which *Acinixys* and *Pyxis* share also with the young of *Homopus*, *Psammobates*, *Chersina*, *Kinixys*, and even some *Testudo* and *Geochelone*. Therefore we feel it is premature to combine these two genera, and urge additional study of their relationships.

Distribution: *Acinixys planicauda* occurs on the southwestern coast of Madagascar, in the vicinity of Morondava.

Habitat: Presumably this tortoise inhabits tropical deciduous forests and dry woodlands, as does *Pyxis arachnoides*.

Natural History: Pritchard (1979) reported the egg is relatively large and slightly ovoid.

Pyxis Bell, 1827

The single species, *Pyxis arachnoides*, is the only tortoise with an anterior plastral hinge.

Pyxis arachnoides Bell, 1827

Malagasy spider tortoise

Plate 15

Recognition: The domed, hingeless carapace (to 15 cm) has descending, almost parallel sides, a notch in the cervical region, and unserrated, downturned posterior marginals. A cervical scute may be absent but, when present, is long and narrow. Vertebrals are broader than long with slightly raised centers surrounded by prominent growth annuli. Neurals alternate in shape between octagonal and square; two suprapygals are present. Eleven marginals lie on each side; the supracaudal is undivided. No submarginal scales are present. Each dark-brown or black vertebral and pleural contains a starlike pattern consisting of a yellow center from which extend several rather broad rays (usually 6–8 on each vertebral, 4–6 on each pleural). The plastron is well developed, and its forelobe is longer but narrower than the hind lobe, and tapered toward the front. A movable hinge lies between the humeral and pectoral scutes (between the epiplastra and hyoplastra, either crossing the posterior portion of the entoplastron or behind it) allowing the forelobe to be raised to almost close the anterior portion of the shell. The hind lobe is broad and very short with a posterior notch. The plastral formula is: abd > gul > pect > hum >< an > fem; the paired gulars are not greatly thickened, and only extend slightly past the carapacial rim when the forelobe is at its lowest position. A single small axillary and one or two larger inguinal scutes lie on the broad bridge. Axillary and inguinal buttresses are short and do not extend to the costals. The plastron varies from totally yellow to having some dark blotching at the bridge. The head is moderate in size with a nonprotruding snout and a slightly hooked upper jaw. Its premaxillae and maxillae lack a median ridge and the sides of the jaws have serrations. The quadrate does not enclose the stapes and the anterior palatine foramina are large. The head is black with some yellow speckles. The neck is odd in having either totally procoelous cervical vertebrae or the second biconvex, only a rare anomaly in other turtles (Williams, 1950a). Limbs are clublike with five claws on each forefoot. On the anterior surface of the foreleg are seven or eight longitudinal rows of enlarged, nonoverlapping scales. Spurlike scales occur on the heel, and some blunt tubercles on the thighs. A longitudinal row of large nonoverlapping scales also lies on the anterior surface of the hind leg. The tail ends in a spinelike scale. Limbs and tail are yellowish brown.

Males have longer, thicker tails.

Distribution: *Pyxis arachnoides* occurs only along the southern coast of Madagascar from the Mahajamba River southward around Cape Sainte-Marie almost to Fort-Dauphin.

Geographic Variation: Three subspecies are currently recognized. *Pyxis arachnoides arachnoides* Bell, 1827 occurs along the southwestern coast of Madagascar in the vicinity of the Onilahy River, which virtually splits its range into northern and southern portions. Its uniformly yellow plastron has a movable hinge, forward projecting gulars, and axillary scutes only a little longer than broad when viewed from beneath. *P. a. oblonga* Gray, 1869a ranges along the southern coast of Madagascar from La Linta to Lake Anony. Its plastron has a movable anterior hinge, some black pigment or blotches, and only slightly projecting gulars. Its axillary scutes are broader than long when viewed from below. *P. a. brygooi* (Vuillemin and Domergue, 1972) occurs in the northern portion of the range, southwest of the Mangoky River between Morombe, Lake Ihotry, and Fanemotra Bay. It lacks mobility in its anterior plastral hinge, has some black pigment on the plastron, slightly projecting gulars, and axillary scutes longer than broad when viewed from beneath.

Habitat: The spider tortoise lives in tropical deciduous forests and dry woodlands (Auffenberg and Iverson, 1979).

Chersina Gray, 1831a

This monotypic genus includes one of the small South African tortoises, the angulated or bowsprit tortoise.

Chersina angulata (Schweigger, 1812)
South African bowsprit tortoise

Recognition: The elongated, hingeless carapace (to 26.7 cm) is domed with the sides dropping abruptly. The cervical scute, if present, is narrow, and there is a

Chersina angulata

deep anterior notch in the cervical region. Anterior marginals are expanded but smooth bordered, lateral marginals not upturned or flared, and posterior marginals somewhat expanded and downturned. A single suprapygal is present; submarginals are absent. The vertebral scutes are wider than long; the 1st is largest and narrows to an anterior point, the 2d is shortest, and the 5th is flared posteriorly. Anterior neural bones are usually hexagonal while the others are variable and approach octagonal and quadrilateral shape (Loveridge and Williams, 1957). The carapace is yellowish brown to olive with wide dark borders and dark centers on each vertebral and pleural, and a narrow dark triangle at the anterior seam of each marginal. The unhinged plastron has the epiplastral portion of the forelobe thickened and somewhat protruding, and covered only by a single gular. Its hind lobe has a posterior notch. The plastral formula is: abd > gul > hum >< an > fem >< pect. The plastron is yellow to reddish with a broad central black blotch. Undersides of the anterior marginals are yellow, and those posterior to the inguinal have a black spot. The head is moderate in size; its snout does not protrude and the upper jaw is hooked (sometimes bi- or tricuspid). The prefrontal scale is divided longitudinally, and sometimes separated from the large, single frontal by a small scale (Loveridge and Williams, 1957). Premaxillae and maxillae lack ridging, but the edges of the jaws are weakly serrated. The quadrate usually encloses the stapes, and the anterior palatine foramina are large. The face is usually black or dark brown; the top of the head is often yellow. Forelimbs are anteriorly olive to brown with large yellow scales along the outer border and yellow posteriorly. Hind limbs are dark on the outer surface and yellow beneath. The toes bear strong claws.

Males have plastrons with a shallow posterior concavity and a stronger gular projection, and longer, thicker tails.

Distribution: *Chersina angulata* is found along the coastline of Cape Province, South Africa, from East London westward to the mouth of the Orange River at elevations to 900 m. South-West African records are questionable (Greig and Burdett, 1976).

Habitat: *Chersina angulata* is best associated with the dry grass and brushlands of coastal plains and upland plateaus, but does enter the coastal forest zone.

Natural History: During courtship, males pursue and bite at the legs and tail of the female. Males also bite, ram, push, and hook other males with their gular projections while attempting to overturn them during dominance combat bouts. Nesting occurs in August when the female digs a 75–100-mm-deep hole to deposit her single egg (occasionally two eggs are laid). The egg is usually oval, 34–35 × 37–42 mm, and brittle shelled. Natural incubation may last from 180 days to 12 or 14 months (Loveridge and Williams, 1957), and hatchlings have flattened shells 30–32 mm long.

Like other tortoises, *C. angulata* is basically herbivorous, probably eating a variety of native plants. Captives do well on lettuce and other greens, bean and pea pods, and various fruits, and they can sometimes be induced to eat raw beef and canned dogfood. Eglis (1962) has described the olfactory movements of this turtle while smelling food as one or several basically straightforward motions with or without sideways oscillation at completion.

Homopus Duméril and Bibron, 1835

This genus of small tortoises is endemic to South Africa. The hingeless carapace is only weakly domed with ascending sides and a flattened top. Its rim may or may not be reverted and serrated. Eleven or twelve marginals lie on each side, a single supracaudal is present, and submarginals are absent. A small but broad cervical scute is present, and a low medial keel may occur, at least on younger individuals. Neurals are either hexagonal or four sided; none are octagonal. An enlarged suprapygal encloses a smaller one between its ventral rami. The hingeless plastron contains an anal notch. Its paired gulars are slightly thickened but not greatly projecting, and each is broader than long. The entoplastron lies anterior to the humero-pectoral seam. Both axillary and inguinal buttresses are short but strong. The upper jaw is usually hooked. Triturating surfaces of the jaws lack ridges, and the maxilla does not contribute to the roof of the palate. The anterior palatine foramina are large. The temporal arch is relatively weak. The prootic bone is narrowly exposed dorsally and the quadrate bone encloses the stapes. Limbs are clublike, with four or five claws on each forefoot. See Loveridge and Williams (1957) for a discussion of variation within the genus.

Key to the species of *Homopus*:

1a. One inguinal present; five foreclaws   2
 b. Two or more inguinals present; four foreclaws   3
2a. Usually only 11 (11–12) marginals on each side; 5–6 longitudinal rows of large imbricate (overlapping) scales on forelimbs: *H. signatus*
 b. Usually 12 (11–13) marginals on each side; 3–5 longitudinal rows of large imbricate scales on forelimbs: *H. boulengeri*
3a. Upper jaw strongly hooked; posterior rim of carapace only slightly serrated at best; no small scales above nostrils: *H. areolatus*
 b. Upper jaw weakly hooked at best; posterior rim of carapace strongly serrated; several small scales lie above nostrils: *H. femoralis*

Homopus boulengeri Duerden, 1906

Donner-weer or Boulenger's cape tortoise

Recognition: This is a small (11 cm) tortoise with a slightly domed carapace which is flattened dorsally. A slight anterior cervical indentation is present, and the anterior and posterior marginals are only slightly upturned and serrated. A small cervical scute is also present. Vertebrals are wider than long, the 5th flared. Centers of the vertebrals and pleurals are raised and surrounded by growth annuli. There are 11–13 marginals, usually 12, on each side, and the supracaudal scute is undivided. Color varies from uniform tan to reddish brown or even olive brown (especially in juveniles). The vertebrals are often dark bordered. The plastron is large, and its forelobe is anteriorly truncated; the hind lobe bears a posterior notch; and the plastral formula is variable: abd > hum > pect >< fem >< an > gul. Only one inguinal occurs on the bridge, and it contacts the femoral scute. Plastral coloration varies from immaculate yellow, tan, or olive to having a large, medium-dark blotch or dark-bordered scutes. The head is moderate in size with a nonprojecting snout and a nonhooked or hooked, often bi- or tricuspid, upper jaw. The prefrontal scale is small and longitudinally divided, and is followed posteriorly by several other small scales. The head is light green to yellow or tan with brown jaws. Each forelimb is covered anteriorly with large overlapping scutes in three to five longitudinal rows, and a large conical tubercle usually lies on the posterior surface of the thigh. Normally, there are five claws on each forefoot. The limbs and tail are yellow to light green.

Males have posteriorly concave plastra with deeper anal notches, and longer, thicker tails.

Distribution: The range of *Homopus boulengeri* is disjunct; it is found on the Karroo Plain of Cape Province, South Africa, and verified localities are also known from near Aus, South-West Africa.

Habitat: *Homopus boulengeri* is most commonly found in dry woodland and scrub savanna.

Natural History: Eglis (1962) reported olfactory movements, which may be associated with courtship, consisted of a double or triple swinging sideways motion of the head with some additional short, straightforward motions. Apparently, a clutch comprises only a single, rather large (39 × 22 mm), pointed egg; to lay an egg this large, some flexibility of either the plastral hind lobe or the egg must occur.

Diet in nature has not been recorded, but probably consists of grasses and other small herbaceous plants.

Homopus areolatus (Thunberg, 1787)

Beaked cape tortoise

Recognition: The slightly domed carapace (to 11.5 cm) is dorsally flattened, scarcely indented in the cervical region, and has the anterior marginals only slightly expanded and the unexpanded posterior marginals not, or only slightly, serrated. A small but broad cervical scute is present; the 1st and 4th vertebrals are longer than broad, and the others are broader than long. A slight medial keel may be present, especially in younger specimens. Vertebrals and pleurals have broad areolae surrounded by raised growth annuli. There are usually 11, but occasionally 10 to 13, marginals on each side, and the supracaudal is undivided. Areolae of the carapacial scutes are reddish brown with yellow, olive, dark-brown, or black borders. A dark bar lies along the anterior seam of each marginal. The yellow plastron usually has some dark pigment toward the center. Its forelobe is anteriorly truncated, the hind lobe notched posteriorly. The plastral formula is: abd > hum > an > fem >< pect >< gul. Each bridge has one or two (sometimes to five) axillary scutes and three or four inguinals, the innermost touching the femoral scute. The head is moderate in size with a nonprojecting snout, and a strongly hooked, tricuspid upper jaw. Usually, no small scales lie above the nostrils, and the large prefrontal scale may be divided or partially (posteriorly) divided longitudinally. The frontal scale may be subdivided. Other dorsal head scales are small. The head varies from yellow to tan or reddish brown, the

Homopus areolatus

jaws are tan. The neck varies from yellowish brown to reddish brown, as do also the limbs and tail. Each forelimb is covered anteriorly with large, overlapping scales in three or four longitudinal rows. A large conical tubercle may be present on the thigh. There are four claws on each forefoot.

Males have posteriorly concave, usually uniformly colored plastra and longer, thicker tails; females have flat, usually medially dark plastra and short tails.

Distribution: *Homopus areolatus* occurs in Cape Province, South Africa, north to 31°S latitude.

Geographic Variation: Some variation in color exists between populations, but this has not been adequately studied.

Habitat: The beaked tortoise lives in deciduous woodlands, dry scrub and sourgrass areas where it seeks out the shade of bushes. It usually occurs from sea level to over 700 m elevation, but rarely is found up to 1300 m (Greig and Burdett, 1976).

Natural History: Eglis (1962:6) described courtship in this species as follows:

> The tortoises form a more or less right angle with their shells, their heads meeting at the confluence. Always responsible for the tête-à-tête, the male managed to "infiltrate" close to the wary female by using a strange method of locomotion: the hind legs, stiff and in stiltlike hyperextension under the steeply upraised posterior carapace, pushed what may have appeared to the female to be an uninterested, passive anterior part into view. The forelegs were dragging. At one point, the female appeared to be intrigued enough by the sliding maneuver to sidle toward the male. At other times, especially after an attempt to mount, the male would rush around the female for nose contact, first from one side and then from the other, and the female would agitatedly skitter aside.

A posture on the part of the female that appears to be restricted to courtship and is tied in with the sliding approach is to tilt the head downward and to the side; ordinarily, when induced to look back around the shell, the female would tilt her head upward and to the side.

The dominating feature of the second phase was continual vicious biting by the male, who would often check the extremities of the female, probably to make sure of their being withdrawn.

Two to five eggs are laid in each clutch; the eggs are elongated (27–33 × 20.5–23 mm) with hard shells; incubation may take up to eight months (Loveridge and Williams, 1957). Hatchlings have 30–32-mm carapaces. *Homopus areolatus* eats grasses and other small herbaceous plants.

Homopus femoralis Boulenger, 1888a

Karroo cape tortoise

Recognition: This is the largest species of *Homopus*, reaching 15 cm in carapace length. Its carapace is flattened dorsally, scarcely indented in the cervical region, and has the anterior and posterior marginals expanded, reverted, and serrated. A small, broad cervical scute is present, and the 1st vertebral is longer than broad, or at least as long as broad, while the others are broader than long. Eleven marginals lie on each side, and the supracaudal is undivided. The carapace is yellowish brown to dark brown or olive with the scutes dark bordered in younger individuals. The scutes of some are orange or red tinged. The plastron is yellow to olive, with dark pigment on the anterior of each scute in the young, but immaculate in older tortoises. Its forelobe is anteriorly truncated and scarcely notched; the hind lobe has an anal notch. The plastral formula is: abd > hum > an >< fem > gul > pect. Each bridge has a single axillary and two or three inguinal scutes, the innermost touching the femoral scute. The head is moderate in size, with at best a weakly hooked, tricuspid upper jaw, and a nonprojecting snout. Several small scales lie above the nostrils. The prefrontal scale is large and divided longitudinally; the frontal is also large or is subdivided; other head scales are small. Head and neck are yellow to tan with some pink or orange pigment; the jaws are brown. The forelimbs are anteriorly covered with large imbricate scales in three or four longitudinal rows, and a large conical tubercle is present on the thigh. The heels have large spurlike tubercles. Four claws occur on each forefoot. Limbs and tail are yellow to tan with tinges of pink or orange.

Males have posteriorly concave plastra with deeper anal notches, and longer, thicker tails.

Distribution: *Homopus femoralis* is restricted to South Africa where it occurs in the eastern Cape Province, southwest Orange Free State, and possibly extreme southwest Transvaal (Greig and Burdett, 1976).

Habitat: This tortoise inhabits arid woodland and scrub savanna to elevations near 1800 m.

Natural History: Nesting is at intervals with a clutch comprising a pair of oval to spherical (29–33 × 25 mm) eggs.

Grasses and other small herbaceous plants form the diet.

Homopus signatus (Schoepff, 1792)

Speckled cape tortoise

Recognition: This is one of the smallest of all turtles, reaching a carapace length of only 10 cm. The slightly domed carapace is dorsally flattened, scarcely indented in the cervical region, and has the anterior and posterior marginals expanded, reverted, and serrated. The cervical scute is small; most vertebrals are broader than long, but the 1st may be as long as broad. Additional vertebrals are sometimes present, raising the total from five to six or seven; and the young have a medial keel (Loveridge and Williams, 1957). There are 11–12 marginals, usually 11, on each side, and the supracaudal is undivided. The carapace is cream colored to yellowish green with numerous black spots or radiations. The plastron is also cream to yellow, and has faded brown blotches and radiating lines. Its forelobe is anteriorly truncated; the hind lobe has an anal notch. The plastral formula is: abd > hum > an > gul >< pect >< fem. Each bridge has a single inguinal in contact with the femoral scute. The head is moderate in size, with a nonprojecting snout and a weakly hooked upper jaw. The small prefrontal scale is longitudinally divided, and is followed by several other small scales. Head and neck are yellow and dorsally spotted with black. The forelimbs are anteriorly covered with large overlapping scales in five or six longitudinal rows. Each thigh bears a very large conical tubercle. Five claws occur on each forefoot, and the limbs and tail are yellow to tan.

Males have posteriorly concave plastra with deeper anal notches, and longer, thicker tails.

Distribution: *Homopus signatus* is confined to Little Namaqualand, from Kleinzee and Springbok southward to Graafwater and Clanwilliam; South-West African records need confirmation (Greig and Burdett, 1976; Boycott, 1986).

Geographic Variation: Two subspecies seem to be valid (Boycott, 1986). *Homopus signatus signatus* (Schoepff, 1792) has a strongly grooved carapace with depressed scute areolae, black radiations, strongly serrated posterior marginals, and the cervical scute wider than long, and occupies the northern and eastern portions of the range. *H. s. peersi* (Hewitt, 1935) has a slightly grooved carapace with flattened or raised scute areolae, fine black dashes on an orange to salmon ground color, weakly serrated posterior marginals, and the cervical scute longer than wide. It is found in the southwestern portion of the range about Clanwilliam and Goergap.

Habitat: *Homopus signatus* occurs in arid woodlands and scrub savanna, and Loveridge and Williams (1957) reported they were fairly numerous on granite koppies near Bitterfontein, Cape Province.

Natural History: Eglis (1962) reported olfactory movements, possibly associated with courtship, consisting of stiffly executed, left-to-right pendulous sweeps of the head.

Psammobates Fitzinger, 1835

These colorfully patterned small tortoises are restricted to southern Africa. The hingeless carapace is domed with ascending sides, and, in two of the three species, the vertebral areolae are conically raised. The anterior and posterior carapacial rims vary from smooth to slightly or strongly serrated. There are 12 or 13 marginals on each side, only a single supracaudal scute, and no submarginal scutes. A small cervical scute is present. The anterior neural bones are hexagonal. There are either one or two suprapygal bones; if two, they are separated by a straight transverse suture. The hingeless plastron contains an anal notch. The paired gulars are broader than long, and neither greatly thickened nor projected. The entoplastron lies anterior to the humero-pectoral seam. Axillary and inguinal buttresses are short but strong. The skull is short with a hooked upper jaw. Triturating surfaces of the jaws lack ridges, but the jaw rims are serrated. Maxillae form part of the lateral roof of the palate; the anterior palatine foramina are small or moderate. The temporal arch is relatively weak. The prootic bone is narrowly exposed dorsally and the quadrate bone en-

closes the stapes. Limbs are clublike; each forefoot has five claws. See Hewitt (1933, 1934) and Loveridge and Williams (1957) for discussions of variation within the genus.

Key to the species of *Psammobates*:

1a. A single axillary scute fused with humeral; cervical scute large; carapacial rim strongly serrated: **P. oculifera**
 b. One to three axillary scutes, not fused with humeral; cervical scute small; carapacial rim only slightly serrated at most   2
2a. A single axillary scute; forelimb with large nonoverlapping scales: **P. geometricus**
 b. Two or three axillary scutes; forelimb with large overlapping scales: **P. tentorius**

Psammobates geometricus (Linnaeus, 1758)

Geometric tortoise

Recognition: The carapace (to 24 cm) is elliptical (widest posteriorly), domed with abruptly descending sides, deeply notched at the small cervical, and rarely serrated posteriorly. Vertebrals are broader than long. Each carapacial scute is covered with raised growth annuli, and these often cause the center (areolus) of the vertebrals to be raised in a conical or pyramidal fashion. There are 11 or 12 marginals on each side, and the supracaudal is not divided. The carapace is dark brown or black, and each vertebral and pleural scute has a yellow center with yellow stripes radiating outward from it (8–15 rays on each vertebral, 9–12 on each pleural). The plastron is large and well developed. Its forelobe tapers to the front, is narrower than the hind lobe, and bears a shallow anterior notch as the gulars are slightly divergent. The hind lobe tapers toward the rear and has a deep posterior notch. Single axillary and inguinal scutes are found on the bridge. The variable plastral formula is: abd > gul >< hum >< fem >< an < pect. The plastron is yellow with some brown or black pigment along the seams. The head is moderate in size with a convex forehead, a nonprojecting snout, and a hooked upper jaw. Its prefrontals are longitudinally divided and followed by subdivided frontal scales. Other head scales are small. Head and neck are dark brown or black with irregularly shaped yellow reticulations. The black or dark-brown forelimbs are covered anteriorly with six or seven longitudinal rows of large, nonoverlapping, irregularly shaped yellow scales separated by smaller scales. Hind limbs are also dark colored, and lack large conical tubercles on their thighs.

Males have concave plastra and longer, thicker tails.

Distribution: *Psammobates geometricus* is restricted to the Tulbagh, Paarl, and Malmesbury districts of Cape Province, South Africa (Loveridge and Williams, 1957). Apparently it was once more widespread, but loss of habitat to farming has caused a decline until it is now one of the most endangered tortoises in the world.

Habitat: *Psammobates geometricus* exists only in small populations on isolated patches of uncultivated land with acidic sandy soil; its present range almost follows the contour of the Rhenosterbushveld. Vegetation in this area consists of grasses and short dry shrubs.

Natural History: Eglis (1962) reported olfactory movements, which may be associated with courtship, consisting of stiffly executed, single, left-to-right sideways motions of the head. A single clutch comprises 12–15 eggs (Loveridge and Williams, 1957); the eggs are spherical, about 30 cm in diameter, and have brittle shells. Rau (1976) reported that the eggs hatch in April and May. Hatchlings are 30–32 mm in carapace length, and many show a pattern consisting of a single yellow X-shaped mark on each vertebral and pleural. Their plastrons have a broad yellow-bordered central black blotch with a yellow midseam and various yellow reticulations.

Rau (1969) observed *Psammobates geometricus* feeding on sedges and leaves of *Crassula ciliata* and the iris, *Romulea*.

Nomenclatural Comment: Wallin (1977) showed that the Linnaean type of *P. geometricus* was actually a specimen of *Geochelone elegans*. Since the type species of *Geochelone* is *G. elegans* and the type of *Psammobates* is *geometricus*, following the rules of priority would require that all of *Geochelone* be changed to *Psammobates*, whose type species would be *geometricus*. In order to prevent this chaos, Hoogmoed and Crumly (1984) have designated an appropriate lectotype, thereby avoiding an ICZN decision.

Psammobates oculifera (Kuhl, 1820)

African serrated star tortoise

Recognition: The carapace (to 13.3 cm) is widest behind the center, domed with abruptly descending sides, shallowly notched at the large cervical, and strongly serrated along both the anterior and posterior rims. Vertebrals are broader than long. Each carapacial scute is covered with concentric growth annuli, but the centers of the vertebrals are not raised in a conical or pyramidal fashion, as in *P. geometricus*.

Psammobates oculifera

There are 10 to 12, but usually 11, marginals on each side, and the supracaudal is not divided. The carapace varies from yellowish brown to tan with numerous yellow, dark-brown, or black radiations on each scute (6–10 rays on each vertebral and pleural). The plastron is large and well developed. Its forelobe tapers to the front, is about the same width as the hind lobe, and bears a shallow anterior notch; the gulars are slightly divergent. The hind lobe tapers toward the rear and has a deep posterior notch. Single axillary and inguinal scutes occur on each bridge; the axillary usually is fused to the humeral. The plastral formula is: abd > hum > gul >< fem >< an > pect. Each plastral scute usually has a yellow center from which extend yellow and dark-brown or black rays. The head is small to moderate in size with a convex forehead, a nonprojecting snout, and a hooked, often tricuspid, upper jaw. Its prefrontal scales are subdivided or divided longitudinally, and are followed by subdivided frontal scales. Other head scales are small. Head and neck are tan to brown with some yellow markings; jaws are yellow. The brown forelimbs have on their anterior surfaces a few large, irregularly shaped, nonoverlapping scales in two to four longitudinal rows. The brown hind limbs have a large and occasionally also a few smaller tubercles on the thighs.

Males have concave plastra and longer, thicker tails; Loveridge and Williams (1957) reported the supracaudal scute is not curved in males but downwardly directed in females.

Distribution: *Psammobates oculifera* ranges in southern Africa from extreme western Transvaal and the western Orange Free State northwestward through the Kalahari Desert of Botswana and South-West Africa almost to Angola.

Habitat: The habitat is brush or savanna areas with sandy soils. In the Transvaal, *Psammobates oculifera* has been found estivating from March to September up to half buried in red sand, under the scanty shelter of fallen *Acacia* branches (Milstein, 1968).

Natural History: Courtship has been observed in late November and consisted of the male scuttling around the female, butting her shell, and periodically emitting short, low, grunting coughs (Loveridge and Williams 1957). The egg has a brittle shell and is somewhat elongated (39.5 × 31 mm).

Like other *Psammobates*, *P. oculifera* feeds exclusively on grasses, sedges, and other short herbaceous plants.

Psammobates tentorius (Bell, 1828)

African tent tortoise

Recognition: The carapace (to 14.1 cm) is oval to slightly elliptical, domed with abruptly descending sides, usually shallowly notched at the small cervical, and only slightly serrated anteriorly and posteriorly. Vertebrals 3–5 are broader than long, the 2d may be as long as broad, the 1st (which is somewhat pointed anteriorly) is often longer than broad, the 5th is flared. Each carapacial scute is covered with growth annuli causing the centers of the vertebrals to be raised in a conical or pyramidal fashion. On each side lie 11 to 13 marginals, and the supracaudal is not divided. The carapace varies from yellow, orange, or reddish to yellowish brown. Each scute usually has light-yellow and black or dark-brown radiations (4–14 black rays on the vertebrals and pleurals, 1–4 on the marginals), but some *P. t. verroxii* have uniformly reddish-brown or tan carapaces or yellow areolae (the *bergeri* form, a synonym of *P. t. verroxii*). The plastron is large and well developed. Its forelobe is tapered to the front and about the same width as the hind lobe, and has only a shallow anterior notch with slightly divergent gulars. The hind lobe tapers toward the rear and bears a deep anal notch. There are two or three (rarely one) axillaries and an inguinal on each bridge; the axillaries do not fuse with the humeral scutes. The plastral formula is varied: abd > hum >< gul >< an >< pect >< fem. The plastron is yellow to orange, sometimes uniform, but usually with some pattern of dark pigmentation (see geographic variation). The head is small to moderate with a slightly convex forehead, a nonprojecting to slightly projecting snout, and a hooked, bi- or tricuspid, upper jaw. Its prefrontal scales may be divided longitudinally or they may be subdivided, the frontals are subdivided, and other head scales are small. Head, neck, and limbs are grayish brown to yellowish or reddish brown to tan. There may be some dark pigment on the head and the snout may be yellow. Large irregularly shaped, overlapping scales lie in two to four longitudinal rows on the anterior surface of the forelegs. Usually one or more large tubercles occur on the thigh, and there are spurlike scales at the heel of the hindfoot.

Males have concave plastra and longer, thicker tails.

Distribution: *Psammobates tentorius* ranges from the Great Namaqualand of South-West Africa southeastward to Cape Province, South Africa.

Geographic Variation: *Psammobates tentorius* is highly variable, and a great number of names have been applied to its various pattern forms (see Loveridge and Williams, 1957, for a discussion of these names); however, currently only three subspecies are considered valid. *P. tentorius tentorius* (Bell, 1828) occurs at elevations below 900 m in southeastern Cape Province, South Africa, mostly south of the 32d parallel (Loveridge and Williams, 1957). It has a well-defined plastral pattern consisting of a dark central blotch which may extend outward along the seams, but lacks yellow radiations. The carapace often has 13 marginals on each side, and the vertebral centers are conically raised; 8–12 black rays occur on each vertebral, 12–14 on each pleural, and 3–4 on each marginal. *P. t. trimeni* (Boulenger, 1886) occurs at elevations below 900 m in extreme western Cape Province from Lambert's Bay north to beyond the Orange River in Great Namaqualand in South-West Africa (Loveridge and Williams, 1957). It has a well-defined plastral pattern in which the central dark blotch is crossed by yellow radiations or is indented by a series of light encroachments from the ground color. The carapace has no more than 12 marginals on each side, and the vertebral centers are conically raised; four to eight black rays occur on each vertebral and pleural, and three or four on each marginal. *P. t. verroxii* (A. Smith, 1839) ranges at elevations over 900 m from the northern Cape Province northwestward into the Great Namaqualand of South-West Africa (Loveridge and Williams, 1957). DeWaal (1980) did not find it in the southern Orange Free State, although Loveridge and Williams (1957) reported it occurred there. This subspecies has a yellow plastron which may totally lack dark pigment, but when dark markings are present they are random in manner, not presenting a uniform outline. The carapace has no more than 12 marginals on each side, and the vertebral centers are flattened and not conically raised; there are five or six black rays on each vertebral and pleural of which the anterior pairs meet the posterior pairs from the next forward scute to form ocelli; one to three rays occur on each marginal.

Habitat: The tent tortoise lives in various habitats within its range: sandy desert, savanna, scrub brush, and dry woodlands, both on the flats and rocky outcrops. In the cooler parts of its range, it may hibernate from June to September, while others in warmer parts of the range may estivate in summer.

Natural History: Eglis (1962) described olfactory motions, possibly associated with courtship, in the various subspecies as follows: *Psammobates tentorius tentorius*, a well-pronounced, swinging, figure-eight-shaped sideways motion; *P. t. trimeni*, a swinging, hook-shaped sideways motion; and *P. t. verroxii*, a double or triple swinging or a single hook-shaped sideways motion, sometimes a double-jointed movement consisting of a forward motion with a subsequent sideways motion to either side. Nesting occurs in September and possibly the season extends to December. Only one *(P. t. trimeni* and *P. t. verroxii)* or two or three *(P. t. tentorius)* eggs are laid in the single yearly clutch. Eggs are almost ellipsoidal (27–34.5 × 21–24 mm) and have brittle shells. Hatching occurs in April or May, and the hatchling carapace is about 25.4 mm long.

Feeding occurs in the morning and late afternoon. Like other tortoises, it is a vegetarian, feeding on grasses, sedges, and other short herbaceous plants.

Manouria Gray, 1852

The Asian tortoises of this genus lack hinges on either the carapace or plastron. The oval carapace may be either domed or flattened dorsally, and may have or lack a cervical indentation. Both the anterior and posterior marginals may be serrated. Usually 11 marginals lie on each side, and the supracaudal is divided into two scutes. The cervical scute is broad and triangular, and no submarginal scutes are present. Neurals are hexagonal, and there are usually two suprapygals. The plastron is well developed and has a deep anal notch. The entoplastron lies anterior to the humero-pectoral seam. Axillary and inguinal buttresses are short, and barely touch the costals. The skull is moderate to long with only a slightly hooked upper jaw. The triturating surface of the maxilla is strongly ridged, but the premaxilla lacks a ridge. The maxillae do not contribute to the roof of the palate, and the anterior orbito-nasal foramina are small, but the lateral caroticum foramen, containing the palatine artery, is large. The prootic bone is well exposed dorsally and anteriorly, and the quadrate bone usually encloses the stapes. There is no supraangular pro-

cess. Limbs are club shaped with five claws on each forefoot.

The two living species are the least derived of extant tortoises, sharing primitive mental glands with batagurines (Winokur and Legler, 1975). See Obst (1983) and Hoogmoed and Crumly (1984) for a review of the taxonomic history and a discussion on the validity of the name *Manouria*.

Key to the species of *Manouria*:

1a. Carapace uniformly dark brown, olive, or black; posterior marginals only slightly serrated; pectoral scutes may not meet at plastral midline; several very large pointed tubercles (spurs) on each thigh: ***M. emys***
 b. Carapace yellow brown to brown with dark seams; posterior marginals strongly serrated; pectoral scutes always meet at plastral midline; a single large, conical tubercle surrounded by very much smaller scales on each thigh: ***M. impressa***

Manouria emys (Schlegel and Müller, 1844)

Asian brown tortoise

Recognition: This tortoise is the largest in Asia. Its oval carapace (to 60 cm) is domed with descending sides; however, it may be somewhat flattened across the 2d and 3d vertebrals. There is little indentation in the cervical region; instead both the anterior and posterior marginals are upturned and slightly serrated. A rather broad cervical scute is present. Vertebrals are wider than long; the 5th is expanded. Well-defined growth annuli surround the flat areolae of the vertebrals and pleurals. Eleven marginals lie on each side, and the supracaudals are divided both dorsally

Manouria emys phayrei

and ventrally. The carapace varies from olive or brown to black; vertebral and pleural areolae may be tan in young individuals. The plastron is well developed and has both an anterior and a posterior notch. Plastral lobes are almost equal in length and width. The plastral formula is: abd > hum > gul >< fem > an > pect; the pectoral scutes may or may not extend to the midseam. The gulars are thickened and extend beyond the carapacial rim. The bridge is wide; the two or more inguinal scales are larger than the single axillary. The plastron is yellow with black shading, usually around the periphery. The head is moderate to large with a nonprojecting snout and a slightly hooked upper jaw. Its prefrontal is divided longitudinally, and followed by a single large frontal scale; other head scales are small. The head is black with some pink, bronze, or brown pigment. Limbs and tail are black. The anterior surface of each forelimb is covered with large, pointed, overlapping scales. Several very large pointed tubercles (spurs) occur on each thigh, giving rise to the colloquial name of "six-footed" tortoise. The tail ends in a horny scale.

Males have longer, thicker tails and more concave plastra than females.

Distribution: *Manouria emys* ranges from Assam, Bangladesh, and Burma southward through Thailand and Malaysia to Sumatra, Borneo, and the Indo-Australian Archipelago.

Geographic Variation: *Manouria emys emys* (Schlegel and Müller, 1844) ranges from southern Thailand (Ranong, Nakorn, and Si Thammaraj provinces) through Malaysia to Sumatra, Borneo, and some of the smaller islands of the Indo-Australian Archipelago. It has a flattened brown carapace (to 48 cm), with lighter vertebral and pleural areolae; the pectorals are widely separated. *M. e. phayrei* (Blyth, 1853) occurs in Assam, Burma, and from Tak Province in northern Thailand to Kanchanaburi Province in the central region. It is almost totally black with a domed, not conspicuously flattened, carapace which may reach 60 cm, and its pectorals meet at the midline of the plastron. Wirot (1979) was apparently unaware of Blyth's earlier description and attributed the name *Geochelone nutapundi* Reimann to this race.

Habitat: This is another highland monsoon forest dweller. It lives in tropical evergreen woodlands, and prefers moist situations. It sometimes forages in shallow mountain streams (Wirot, 1979). Much time is spent burrowed into moist soil or under leaf litter.

Natural History: McKeown, Juvik, and Meier (1982) reported the only observations of reproduction. Courtship of *Manouria emys phayrei* is generally simplified,

wherein the male utilizes shell ramming to immobilize the female prior to mounting. Male head bobbing during courtship and vocalizations while mounted also occur. Nesting behavior of *M. e. phayrei* is quite bizarre. The female constructs a large leaf-litter mound by "back sweeping" ground litter for up to 4 m from the nest site. After ovipositing in the leaf mound, she guards the nest for up to three days. Clutches laid in these nests at the Honolulu Zoo contained 23 to 51 eggs (larger than that reported for other tortoises; Wirot, 1979, reported the clutch size for *M. e. emys* to be 5–8); nestings occurred in April, May, and September, and the female usually laid two clutches a year. The spherical, white eggs have brittle shells and range from 51 to 54 mm in diameter. Hatchling *phayrei* have uniformly brown to gray-brown carapaces and black plastrons; carapace length ranges from 60 to 66 mm.

Wirot (1979) reported the foods eaten by *M. e. phayrei* include aquatic plants such as tubers and lotus.

Manouria impressa (Günther, 1882)

Impressed tortoise

Recognition: The oval carapace (to 27 cm) is flattened dorsally, has an indentation at the broad cervical, and is strongly serrated around its entire rim. Posterior marginals are somewhat upturned, and pleurals somewhat concave. Vertebrals are wider than long, and the 5th is expanded. Günther (1882) reported a slight indication of a medial keel on the 4th and/or 5th vertebrals. Well-defined growth annuli surround the flat vertebral and marginal areolae. Eleven marginals lie on each side, and two supracaudals. The carapace is yellowish brown to brown with dark seams, but some have dark radiations along the outer border of each scute. Large black blotches occur on the marginals. The plastron is well developed with a deep anal notch and a broad anterior notch which somewhat separates the gulars. Its forelobe is longer but narrower than the hind lobe. The plastral formula is: abd > hum > fem > gul >< an > pect; the gulars are thickened and extend slightly beyond the carapacial rim. The bridge is wide; the inguinal is large and often subdivided, and the axillary is small to moderate. The plastron is yellowish brown with darkened seams, and some dark streaking may be present. The head is large with a nonprojecting snout and an upper jaw which lacks a hook, or is only slightly hooked. Its large prefrontal is longitudinally divided, and followed by a large undivided frontal scale; other head scales are small. The maxillae are ridged, but the premaxillae are not. The head is yellow to tan with pink pigment about the snout on some. Forelimbs are black; hind limbs and tail dark brown. The anterior surface of the somewhat flattened forelimbs is covered with large, overlapping pointed scales. A single large conical tubercle lies on each thigh, and the tail may end in a horny scale.

Males have longer, thicker tails than do females.

Distribution: *Manouria impressa* ranges from the Karenni Hills in Burma southward to Malaysia and Kampuchea, and recently has been discovered in Hunan Province, China (Zhao, 1986).

Habitat: This tortoise inhabits forests on hills and mountains. McMorris (1975) thought the natural habitat seems to be fairly dry, and not usually associated with water bodies. The tortoises rely on heavy dew or rain-drenched vegetation for water. They do poorly in captivity if kept in humid situations.

Natural History: Wirot (1979) reported the mating season coincided with the rainy period, and a captive female laid 17 eggs between 16 and 29 May (McMorris, 1975). Hatchlings are yellowish to light brown with rounded, medially keeled, heavily serrated carapaces.

They seem to be more active at twilight. Wirot (1979) reported they feed on plants, grass, and bamboo shoots, and we suspect fallen fruits are also consumed in season.

Since this rare species is declining in the wild and does poorly in captivity, its sale in the pet trade should be prohibited.

Indotestudo Lindholm, 1929

Indotestudo is usually regarded as a subgenus of *Geochelone* (Auffenberg, 1974); however, considerable evidence has accrued that indicates it should be considered a separate genus. Charles R. Crumly, who has acted as our advisor on tortoise systematics, has completed morphological studies which indicate nine major differences (Table 5), and we follow his suggestion to elevate *Indotestudo* to full generic rank.

The two medium-sized Asiatic tortoises belonging to *Indotestudo* occur in southern peninsular India, northeastern India, Nepal, and Bangladesh southward to Vietnam and Thailand, and on Celebes and Halmahera, Indonesia. The hingeless, low carapace is usually elongated, flattened dorsally, has a cervical notch, and the posterior marginals may be somewhat serrated. The anterior neurals are alternately octagonal or four sided. There are usually two suprapygal bones separated by a transverse suture; the anterior is larger than the posterior and partially surrounds it just as in most *Geochelone* and *Gopherus*. Usually, 11 marginals lie on each side of the carapace; the supracaudal is single and undivided. No submarginal scutes are present. A narrow cervical is present in one species. The hingeless plastron is well developed with a posterior notch so deep it may completely separate the anal scutes. The paired gulars are somewhat thickened, but not greatly projecting. The humero-pectoral

Table 5. Differences between the tortoise genera *Geochelone* and *Indotestudo* (from Charles R. Crumly)

Character	*Geochelone*	*Indotestudo*
Cervical scute	Usually absent, but present as a broad, short element in *G. radiata*, *G. yniphora*, and some *G. gigantea*	Present in one of three species, and here it is long and narrow.
Marginal–second-pleural contact	Two marginals (rarely three, but if so, just barely)	Three marginals
Position of humero-pectoral seam	Does not cross entoplastron	Crosses entoplastron
Skull arches	Usually broad	Narrow
Palate	Narrow, and deeply extending posteriorly	Broad, and only shallowly extending posteriorly
Postmaxillary scales	Absent	Sometimes present
Trachea	Long	Short
Long terminal tail scale	Absent (except in *G. platynota*)	Present, and not surrounded by large scales
Color	Radiating pattern or dark brown with or without light areolae	A light cream-yellow ground color with dark-brown irregular blotches more abundant on carapace than on plastron

sulcus usually crosses the entoplastron. Axillary and inguinal buttresses are short but strong, just extending to the costals. The skull is short to moderate with a hooked, usually tricuspid upper jaw. Sides of the jaws are somewhat serrated. Triturating surfaces of the maxillae usually have a weak to moderate medial ridge, but no ridge occurs on the premaxillae. The maxillae do not contribute to the roof of the palate, and the anterior orbito-nasal foramina are small and concealed in ventral view. The temporal arch is relatively weak. The prootic bone is typically well exposed dorsally and anteriorly; the quadrate encloses the stapes. Limbs are clublike; five claws occur on each forefoot. The tail ends in a large horny scale.

DeSmet (1978) reported the diploid karyotype was 52 (34 macrochromosomes, 18 microchromosomes).

Indotestudo elongata

Key to the species of *Indotestudo*:

1a. Cervical scute absent, interpectoral seam usually less than 70% as long as interhumeral seam: *I. forsteni*

b. Cervical scute present, interpectoral seam as long as or longer than interhumeral seam: *I. elongata*

Indotestudo elongata (Blyth, 1853)

Elongated tortoise

Recognition: The elongated carapace (to 27.5 cm) is domed, but flattened dorsally with descending sides, has a shallow cervical notch, a long narrow cervical scute, and the posterior marginals somewhat flared and serrated (more strongly serrated in juveniles). Vertebral 1 is about as broad as long, but 2–5 are broader than long; the 5th is expanded. Well-defined growth annuli surround the flat vertebral and pleural areolae. Usually 11 marginals lie on each side, and the undivided supracaudal scute is downturned between the somewhat expanded marginals. The carapace is yellowish brown or olive, with black blotches on the vertebrals and pleurals. The well-developed plastron has a deep anal notch. Its forelobe tapers anteriorly and is shorter and narrower than the hind lobe. The plastral formula is: abd > fem > pect ≥ hum > gul > an. The gulars are somewhat thickened, and the bridge is wide with a small axillary and a larger inguinal scute. The plastron and bridge are yellow and usually unpatterned. The head is moderate with a non-protruding snout and a weakly hooked, tricuspid upper jaw. Its large prefrontal scale is longitudinally divided, and followed by a large frontal scale which is

often subdivided; other head scales are small. The head is pale cream to yellowish green without dark spots or blotches. Limbs are brown to olive. The anterior surface of the forelimb is covered with moderate to small overlapping scales (the outermost largest).

Males have longer, thicker tails, and deeper anal notches than do females.

Distribution: *Indotestudo elongata* ranges from Nepal, Bangladesh, and northeastern India (Jalpaiguri, East Bengal, and Singhbhum in Bihar) southward through Burma, Laos, Thailand, Kampuchea, and Vietnam to Penang, Malaysia.

Habitat: The elongated tortoise lives in tropical evergreen and deciduous forest on hills, mountains, or high plateaus. Although partial to cooler humid forests, it has been taken in open areas during the warmest parts of the day, and there is evidence (Swindells and Brown, 1964) that *Indotestudo elongata* can withstand air temperatures to 48°C.

Natural History: During the breeding season (July, Biswas et al., 1978) the skin around the eyes and nostrils becomes bright pinkish red; a clutch comprises two to four eggs.

Fruits and flowers are the usual foods, but Wirot (1979) reported they also eat fungi and slugs.

Indotestudo forsteni (Schlegel and Müller, 1844)

Travancore tortoise

Recognition: The elongated carapace (to 30.9 cm) is domed, but flattened dorsally, with descending sides; a cervical notch is due to the absence of any cervical scute; the posterior marginals are serrated (more strongly so in juveniles). Vertebrals are broader than long, the 5th is expanded. Well-defined growth annuli surround the vertebral and pleural areolae. There are usually 11 marginals on each side, and the single supracaudal scute may be downturned between the somewhat expanded posterior marginals. The carapace is either unpatterned gray to brown, or yellow brown to brown or olive with dark blotches on each scute. The plastron is well developed. Its forelobe tapers anteriorly and is shorter and narrower than the hind lobe. The plastral formula is: abd > fem >< hum > gul >< pect > an. The gulars are somewhat thickened. The bridge is wide, and the axillary and inguinal scutes are moderate in size. Plastron and bridge are yellowish to brown with small black blotches. The head is moderate in size with a nonprotruding snout and a weakly hooked, tricuspid upper jaw. Its large prefrontal is divided longitudinally, and the single frontal scale is almost as large; other head scales are small. The head is yellow with some brown or orange pigment. Neck, limbs, and tail are brownish gray to olive. The anterior surface of the forelimbs is covered with large, yellow, irregularly shaped, overlapping scales (the outermost largest).

Males have longer, thicker tails with the vent nearer the tip than do females.

Distribution: *Indotestudo forsteni* occurs in the states of Karnataka and Kerala in southwestern India, and has been introduced on Celebes and Halmahera, Indonesia.

Habitat: The prime habitat is probably upland tropical mesic forest. In India, it is found at elevations to 1000 m.

Natural History: Males recognize females by olfaction. When the male approaches the female, his neck is extended and the head is moved through a short vertical arch followed by a small rapid circle (Auffen-

Indotestudo forsteni

berg, 1964). The female is then rammed and pushed with the gulars to immobilize her (no biting occurs) and mounting follows. Breeding takes place from November through January, and the pinkish pigmentation around the orbits and nasal area intensifies in both sexes, but to a greater extent in males. Hatchlings have humpbacked shells with strong posterior serration.

Like other tortoises, *Indotestudo forsteni* is a vegetarian. During fruiting of particular trees, small aggregations may form to feed on the fallen fruits (Auffenberg and Iverson, 1979). Males may vocalize during rainstorms.

Comment: Hoogmoed and Crumly (1984) have shown that *Testudo forsteni* Schlegel and Müller, 1840 is a senior synonym of *T. travancorica* Boulenger, 1907, and supposed differences in the plastral pattern, gular measurements, and general shell morphology do not distinguish *travancorica* from *forsteni*. They believe *forsteni* was introduced onto the islands east of Wallace's Line from mainland Indian stock.

Geochelone Fitzinger, 1835

This is the largest genus of tortoises, both in number of living species (21) and in size attained (to 130 cm, and over 185 kg). It is pantropical with living species in Africa, Madagascar, India, Sri Lanka, Southeast Asia, Indonesia, South America, and on the oceanic Aldabra and Galapagos Islands. Formerly the genus was even more widespread, as fossil species are known from North America as far north as southern Canada, the British Isles, western Europe, northern Africa, the Middle East, Central Asia, and additional islands in the Indian Ocean (Auffenberg, 1974). Fossil *Geochelone* have even been found on Ellesmere Island (78°N) (Estes and Hutchison, 1980).

No adults are less than 25 cm in carapace length. The hingeless carapace varies in shape from oval to elongated, and from domed to somewhat dorsally flattened (especially in males of some species) to saddlebacked (narrower and elevated anteriorly, wider and lower posteriorly). Some form of cervical indentation or notch is usually present, and the anterior marginals may be flared or downturned. Both anterior and posterior marginals may be serrated. Usually 11 marginals lie on each side of the carapace, and the supracaudal is usually single and undivided. There are no submarginal scutes, but a cervical scute is present in most species. The anterior neurals are alternately octagonal or quadrilateral in shape, and usually two suprapygal bones are present: the larger anterior suprapygal bifurcates posteriorly and partially surrounds the smaller posterior. The plastron lacks a hinge and varies from narrow to broad. Its posterior notch may be deep, shallow, or nonexistent. The thickened gular scutes are paired and broader than long in all species but *Geochelone yniphora*, which has them fused into a long single element; the entoplastron is almost always anterior to the humero-pectoral seam. Axillary and inguinal buttresses are short, and barely touch the costals. The skull is short to moderate with a hooked, sometimes bi- or tricuspid upper jaw. Sides of the upper jaw are usually serrate. The triturating surface of the maxilla is strongly ridged, but no medial ridge occurs on the premaxilla. The maxillae do not contribute to the roof of the palate, and the anterior orbito-nasal foramina are small and concealed from ventral view. The temporal arch is weak to moderately wide. The prootic bone is well exposed dorsally and anteriorly, and the quadrate bone usually encloses the stapes. The limbs are club shaped with no toe webbing; each forefoot has five claws. In a few species, the tail ends in a large horny scale.

The diploid chromosome compliment is 52: 26–30 macrochromosomes, 22–26 microchromosomes (Sampaio et al., 1971; Stock, 1972; Goldstein and Lin, 1972; Dowler and Bickham, 1982; Bickham and Carr, 1983).

Living species have been separated into several subgenera. Auffenberg (1974) recognized six: *Aldabrachelys* Loveridge and Williams, 1957 *(G. gigantea); Asterochelys* Gray, 1873a *(G. radiata, G. yniphora); Chelonoidis* Fitzinger, 1835 *(G. carbonaria, G. chilensis, G. denticulata,* and the *G. elephantopus* complex); *Geochelone* Fitzinger, 1835 *(G. elegans, G. pardalis, G. platynota, G. sulcata); Indotestudo* Lindholm, 1929 *(G. elongata, G. forsteni, G. travancorica);* and *Manouria* Gray, 1852 *(G. emys, G. impressa).* However, we feel *Indotestudo* and *Manouria* are sufficiently different to warrant full generic rank, and Crumly (1982a) feels *Indotestudo* is monophyletic (see discussion and Table 5 under *Indotestudo*). Crumly (1982a) has also shown that, based on 26 selected cranial features, *Asterochelys, Chelonoidis, Geochelone* and *Manouria* do not share derived features and are not monophyletic. *Aldabrachelys* (including several fossil forms, as well as *G. gigantea*) is probably monophyletic since its members share the derived vertically elongated external narial openings. Cladistic analyses by Crumly have indicated possible relationships between *G. (Geochelone) pardalis,* and *G. (Asterochelys) yniphora,* and among *G. (Geochelone) elegans, G. (Aldabrachelys) gigantea,* and *G. (Asterochelys) radiata,* but final decision must await further anatomical studies, inclusion and analysis of other tortoise genera, and consideration of functional

Key to the species of *Geochelone:*

1a. Gular scutes fused into one elongated element: **G. yniphora**
 b. Gular scutes paired — 2

2a. External narial opening an elongated vertical slit: **G. gigantea**
 b. External narial opening more or less rounded, not higher than wide — 3

3a. Cervical scute present; **G. radiata**
 b. Cervical scute absent — 4

4a. Carapacial pattern of light radiating lines — 5
 b. Carapacial pattern of light blotches, or no pattern — 6

5a. Plastral pattern of dark blotches: **G. platynota**
 b. Plastral pattern of dark radiating lines: **G. elegans**

6a. Pleural scutes very narrow — 7
 b. Pleural scutes usually not appreciably narrowed — 8

7a. Carapace uniformly tan or brown; frontal scale large: **G. sulcata**
 b. Carapace yellow to olive with black or dark-brown markings; frontal scale usually absent or small and divided: **G. pardalis**

8a. Tail ends in a large terminal scale: **G. chilensis**
 b. Tail lacking a large terminal scale — 9

9a. Carapace uniformly black or dark brownish gray; forelimbs black or gray; carapace very large, to 130 cm: **G. elephantopus** complex
 b. Carapace with yellow, orange or red vertebral and pleural areolae; forelimbs with large yellow or red scales; carapace to 72 cm — 10

10a. Lateral sides of carapace straight; carapace with yellow or orange vertebral and pleural areolae; large foreleg scales yellow or orange; interseam length of femoral scutes less than that of seam separating humeral scutes; gular scutes do not reach entoplastron: **G. denticulata**
 b. Lateral sides of carapace distinctly concave; carapace with yellow or red vertebral and pleural areolae; forelimbs with large orange or red scales; interseam length between femoral scutes equal to or greater than that separating humeral scutes; gular scutes overlap entoplastron: **G. carbonaria**

hypotheses. The subgenera, as now recognized, are often no more than convenient geographical assemblages, and until additional analyses are completed we agree with Crumly (1982a) that the subgenera of living *Geochelone* should be abandoned.

A contrary opinion is that of Bour (1980a) who elevates the subgenera to full generic status, but we reject this for the same reasons, and also because Bour, while stating reasons for these changes, did not provide the data upon which he arrived at these decisions.

Geochelone elegans (Schoepff, 1795)

Indian star tortoise

Plate 15

Recognition: The oval carapace (to 28 cm) is domed with a very convex dorsal surface and abruptly descending sides. A deep cervical indentation is present, but no cervical scute. Posterior marginals are serrated and, in some, slightly upturned. Vertebrals are usually broader than long, although the 1st may be longer or as long as broad; the 5th is expanded. Well-defined growth annuli surround the raised vertebral and pleural areolae, causing the carapace to appear lumpy. There are usually 11 marginals on each side, and a single, undivided, downturned supracaudal scute. The carapace is dark brown or black; a series of 6–12 yellow stripes radiate outward from the yellow or tan vertebral and pleural areolae. Each marginal has one to three yellow stripes beginning at a yellow spot in the lower posterior corner and extending upward toward the pleurals and vertebrals. The well-developed plastron is slightly upturned anteriorly. Its forelobe is longer but narrower than the hind lobe, which bears a deep anal notch. The plastral formula is: abd > hum > gul > fem > pect >< an; the paired gulars are thickened but do not greatly protrude anteriorly. The bridge is wide with single axillary and inguinal scutes of moderate size. Plastron and bridge are yellow with black radiations. The head is moderate with a nonprojecting snout and a weakly hooked, sometime bi- or tricuspid, upper jaw. Its large prefrontal scale is divided longitudinally and followed by a single, rather narrow frontal scale which may extend forward to partially separate the two halves of the prefrontal. Other head scales are small. The head is yellow to tan with brown jaws; limbs and tail are also yellow or tan. The anterior surface of each forelimb is covered with large and small irregularly shaped to pointed scales in five to seven longitudinal rows. Several small to moderate conical tubercles oc-

cur on the thigh. The tail usually lacks an enlarged scale.

Males have longer, thicker tails than do females; females are larger and broader.

Distribution: The star tortoise ranges in peninsular India from Orissa in the east and Sind and Kutch in the west southward to the tip (Minton, 1966). It also occurs on Sri Lanka, and on other small offshore islands where its only fresh water comes from the monsoons.

Geographic Variation: No subspecies are recognized, but a size gradient exists from north (larger) to south (smaller) through peninsular India (C. R. Crumly, pers. comm.). This is reversed in Sri Lanka where large individuals occur.

Habitat: *Geochelone elegans* lives in a variety of habitats ranging from tropical deciduous forests to dry savannas, but seems to require an ample supply of water. Seasonally they are most active during the monsoon rains when they may wander about all day; in dry seasons they are active only during early morning and late afternoon or evening. They soak in puddles and pools of water.

Natural History: The mating season corresponds to the rainy season from mid-June to mid-October. While courting, males pursue females, and, when they are caught, ram and push the females with their thickened gular scutes. When she is immobilized, the male mounts from the rear while resting his forefeet on her 3d pleural scutes. He may then raise his hind feet free of the ground and smash his anal scutes against the lower, rear portion of her carapace. During this phase and after intromission the male may utter short grunts. Aggressive males ram and push rival males during the mating season to drive them away from potential mates. Nestings have been observed in the wild and in captivity from April to November, and three to nine (Whelan and Coakley, 1982) clutches are laid each year. A typical nest is flask shaped and about 10–15 cm deep. Eggs are elongated (38–52 × 27–39 mm) with brittle shells; 3–10 are laid at one time (Deraniyagala, 1939; Whelan and Coakley, 1982). Incubation may last 147 days, and the hatchlings may remain in the nest until rains soften the soil allowing them to escape. Hatchlings have rounded, slightly domed carapaces (35 mm), with or without yellow radiations.

Like other tortoises, *G. elegans* is predominantly herbivorous, feeding on grasses, fallen fruits, flowers, and the leaves of succulent plants. Occasionally carrion is eaten.

Geochelone platynota (Blyth, 1863)

Burmese star tortoise

Recognition: This rare tortoise is seldom seen outside of Burma, and, unfortunately, there it is usually eaten by the natives soon after being caught. The oval carapace (to 26 cm) is domed but flattened dorsally with descending sides and a slight cervical indentation; the posterior marginals are only slightly expanded and weakly serrated. No cervical scute is present. Vertebral 1 is about as long as broad or occasionally longer, vertebrals 2–5 are broader than long; the 5th is expanded and the 4th relatively small. Well-defined growth annuli surround the flat vertebral and pleural areolae. There are usually 11 marginals on each side, and the supracaudal scute is undivided and downcurved. The carapace is dark brown or black with six or fewer radiating stripes extending from the yellow areola of each vertebral and pleural. Two yellow stripes form a V-shaped pattern on each marginal. The well-developed plastron is notched both anteriorly and posteriorly. Its forelobe is longer, but narrower than the hind lobe. The plastral formula is: abd > hum > fem > gul > an > pect; the pectoral scute is extremely narrow. The gulars are somewhat thickened, broader than long, and do not greatly extend anteriorly. The bridge is wide, and its axillary scute is smaller than the inguinal. Plastron and bridge are yellow, and each plastral scute has a dark-brown or black blotch. The head is moderate with a nonprojecting snout and a weakly hooked, tricuspid upper jaw. The large prefrontal is divided longitudinally and is followed by a large single frontal scale; other head scales are small. Skin of the head, limbs, and tail is yellow to tan. The anterior surface of the forelimbs is covered with large pointed to rounded scales. The tail ends in a large horny scale.

Males have longer, thicker tails with the vent nearer the tip.

Distribution: *Geochelone platynota* is restricted to Burma where it ranges from northern Burma southward to Moulmein (Smith, 1931).

Habitat: *G. platynota* is a forest dweller.

Natural History: According to Smith (1931), nesting occurs at the end of February. He also stated that the eggs are large (55 × 40 mm) and few in number.

Geochelone pardalis (Bell, 1828)

Leopard tortoise

Plate 16

Recognition: The carapace (to 68 cm) is domed with abruptly decending (almost vertical) sides and somewhat convex vertebral scutes. There is a prominent cervical indentation, but no cervical scute. Anterior and posterior marginals are only slightly expanded and somewhat upturned. Vertebral 1 is as long as or longer than broad, the rest are broader than long, and the 5th is expanded. Well-defined growth annuli surround the vertebral and pleural areolae; areolae may be raised or flattened. There are usually 11 marginals on each side, and a single, undivided supracaudal which may be downturned between the posterior marginals. Dark brown or black pigment occurs along the growth annuli of each pleural and vertebral surrounding the yellow, tan, brown, reddish-brown, or olive areolae. The seams are usually light colored and black smudges occur on the marginals. These marks often fade with age and old individuals may be almost entirely tan or brown; the young, however, are strikingly marked with the characteristic leopardlike spotted pattern. The plastron is well developed with a deep anal notch. Its two lobes are about the same length, but the hind lobe is slightly wider. The variable plastral formula is: abd > hum >< fem >< gul >< an > pect; the paired gulars are thickened, but only slightly protruding. The bridge is wide with two axillary scutes (one large, one minute) and a single inguinal which touches the femoral scute. The head is moderate with a nonprotruding snout and a hooked, often tricuspid upper jaw. The large prefrontal scale may be single or divided longitudinally; the frontal scale is usually absent, but may be small and subdivided; other head scales are small. The head is uniformly yellow or tan. Limbs and tail are also yellow to brown. The anterior surface of each forelimb is covered with three or four longitudinal rows of large, irregularly shaped, nonoverlapping (or rarely overlapping) scales. Two or more large conical tubercles occur on the hindside of each thigh (these are much smaller than those of *G. sulcata*), but the tail lacks a large terminal scale.

The sexes are hard to differentiate, but Loveridge and Williams (1957) found that males had longer, thicker tails and the posterior third of their plastra slightly concave while that of the female is flat.

Distribution: *Geochelone pardalis* is restricted to Africa where it ranges from southern Sudan and Ethiopia southward through eastern Africa to Natal and South Africa and westward to southern Angola and South-West Africa.

Geographic Variation: Two subspecies are known. *Geochelone pardalis pardalis* (Bell, 1828) is restricted to southwestern Africa where it ranges northward to near Luderitz Bay and Keetmanshoop, and possibly as far as Rehoboth (Loveridge and Williams, 1957). It formerly occurred to the Cape, but is now extinct in the immediate proximity of the Cape, and rare over the rest of its range. It has a relatively low, flat-topped carapace (carapace length/carapace depth is 2.02–2.62). *G. p. babcocki* (Loveridge, 1935) (Plate 16) occurs in eastern and southeastern Africa from Sudan and Ethiopia southward to Natal and west through Cape Province to South-West Africa and southern Angola (Loveridge and Williams, 1957). It has a high, domed or convex carapace (carapace length/carapace depth is 1.61–2.07). Pritchard (1979) remarked that *G. pardalis* from the highlands of the Graaff-Reinet area of South Africa are more distinctly marked and larger than those from surrounding lowland populations.

Habitat: These tortoises shun heavily forested areas, but inhabit savannas, plains, and kopjes with dry woodland, thorn scrub, and grasslands from sea level to over 2900 m. It is possible that in parts of the range, they may bury themselves to estivate or hibernate.

Natural History: During the mating period both males and, to a lesser degree, females become aggressive, butting and ramming would-be competitors for their mates. During courtship, the male trails the female, often for some distance, and may butt her to immobilize her. When finally mounted, he extends his neck and utters a gruntlike bellow. Nesting occurs from May until October, depending on location and latitude. The female digs a flask-shaped hole about 100–300 mm deep, and up to 5–7 clutches of 5–30 eggs may be laid each season (the internesting period is usually about three weeks). The white eggs are spherical (36–40 mm) with brittle shells. Incubation may take over a year. Hatchlings are brightly patterned with round, flattened, serrated carapaces (45–50 mm).

They feed on a variety of wild plants, fungi, grasses, and fallen fruits: crussulas, spekboom, thistles, and the introduced prickly pear. Succulents are preferred, possibly because of their higher water content.

Geochelone sulcata (Miller, 1779)

African spurred tortoise

Recognition: This is the largest continental tortoise (76 cm), and is surpassed in size only by the island species from Aldabra and Galapagos. The oval carapace is flattened dorsally with abruptly descend-

Geochelone sulcata

ing sides, a deep cervical notch, both the anterior and posterior marginals serrated, and the posterior marginals upturned. No cervical scute is present. Vertebrals are broader than long; the 5th is smallest and somewhat expanded. Well-defined growth annuli surround the flat vertebral and pleural areolae. There are usually 11 marginals on each side, and a single undivided supracaudal scute which extends downward between the somewhat expanded posterior marginals. The carapace is uniformly brown. The plastron is well developed with a deep anal notch. Its forelobe tapers to the front, and the two bifurcated gulars project forward beyond the carapacial rim. Both plastral lobes are about the same length, but the forelobe is slightly broader. The plastral formula is: abd > hum > fem > gul > pect >< an. The wide bridge has two axillaries (the inner very small) and two inguinals (the inner small). Plastron and bridge are uniformly cream or yellow. The head is moderate with a nonprotruding snout and a weakly hooked upper jaw. The large prefrontal scale is divided longitudinally and is followed by a large single frontal scale; other head scales are small. The head is brown with darker jaws; limbs and tail are also brown. The anterior surface of the forelimbs is covered with large, irregularly shaped, knobby, overlapping scales in three to six longitudinal rows. The hind surface of the thigh bears two or three large conical tubercles. The tail lacks a large terminal scale.

The larger males have slightly longer, thicker tails and more concave plastra than do females, but the external sexual differences are slight.

Distribution: *Geochelone sulcata* ranges from Ethiopia and Sudan westward through the dry regions of Chad, Niger, and Mali to southern Mauritania and Senegal. Its range generally lies along the southern perimeter of the Sahara Desert.

Habitat: *Geochelone sulcata* lives in areas varying from desert fringes to dry savannas. Most habitats have standing water for only limited periods at best, and this turtle relies heavily on metabolic water and the moisture in its food. To avoid desiccation and the unbearable heat, it digs burrows (some very long) or pallets. Most activity is at dusk or dawn, and basking may occur in the early morning to raise temperatures after the night chill.

Natural History: Cloudsley-Thompson (1970) reported that mating may occur from June to March, but most frequently from September through November. During courtship, the male circles the female and occasionally rams her with his shell (Grubb, 1971). Mounting follows during which the male vocalizes; Grubb (1971) described the mating call as a short grunt or ducklike quack. Nesting takes place in autumn or winter and the incubation period is long, 212 days (Cloudsley-Thompson, 1970). Up to 17 eggs are laid at a time; the white eggs are almost spherical (41–44 mm) and have brittle shells. Hatchlings are yellow to tan with rounded, serrated carapaces (45–53 mm).

Geochelone sulcata is a vegetarian, relying on succulent plants for food. In captivity it consumes grass, lettuce, berseem *(Medicago sativa)*, morning-glory leaves *(Ipomaea)*, *Idigofera linnei*, and *Euphorbia hirta*.

Geochelone gigantea (Schweigger, 1812)

Aldabra tortoise

Recognition: Among the tortoises, this species is second only in size to the slightly larger species found in the Galapagos Islands. *G. gigantea* reaches a record carapace length of 105 cm and may weigh 120 kg. It has a thick, elongated, domed carapace with descending sides, only a slight cervical indentation at best, and posterior marginals that are flared, slightly upturned, and unserrated. A cervical scute is usually present. Vertebrals are broader than long, the 5th is smallest and expanded, and the 1st is narrower than the 2d. Growth annuli surround the raised vertebral and pleural areolae, but these are generally low, and the carapace may be worn very smooth in old individuals. Usually 11 marginals lie on each side of the carapace, and the supracaudal is single, undivided, and downturned between the posterior marginals. The carapace is uniformly brownish gray. In comparison to other tortoises, the plastron is somewhat short and lacks a posterior notch. Its forelobe tapers anteriorly and is longer, but much narrower, than the hind lobe. The plastral formula is: abd > hum > fem

Geochelone gigantea

> pect >< gul >< an; the paired gulars are short and thick and scarcely project beyond the carapacial rim. The bridge is moderate in width (about one-third the carapace length), and bears a single small axillary scute and a larger inguinal. The plastron and bridge are also brownish gray. The head is narrowly pointed and somewhat wedge shaped (when compared to Galapagos tortoises), and with a convex forehead. The snout is nonprojecting, and the upper jaw is either weakly hooked or bi- or tricuspid. The large prefrontal scale is divided longitudinally, its sides are parallel and do not diverge posteriorly; a relatively small frontal scale follows, and other head scales are small. Head, neck, limbs, and tail are gray. The anterior surface of the forelimbs is covered with large, slightly rounded, nonoverlapping scales. No conical tubercles occur on the thighs, and the tail lacks a large terminal scale.

Males are larger and have longer, thicker tails than do the smaller females.

Distribution: *Geochelone gigantea* occurs naturally only on the coral-limestone Aldabra Atoll in the Indian Ocean. Population estimates range to 150,500 individuals, with most occurring on Grande Terre where the density is about 27/ha (Picard, 5/ha; Malabar, 7/ha; Bourn and Coe, 1978). A small colony was established on the island of Curieuse in the granitic Seychelles as a tourist attraction in 1978, and a hatchling was found there in 1980 (Stoddart et al., 1982).

Arnold (1976) proposed that giant tortoises may have first reached Aldabra by oceanic drifting from Madagascar, and, since the atoll has apparently been completely submerged twice since the tortoises settled on it, at least three such colonizations may have occurred.

Geographic Variation: No subspecies are recognized, but some individuals are domed while others are more saddlebacked. Also the tortoises living in some areas are larger. However, Arnold (1979) found the total variation to be less than that found in the Galapagos tortoises.

Habitat: *Geochelone gigantea* occurs in a variety of habitats, including grassland, scrub areas, and mangrove swamps. Sheltering trees or bushes are necessary to escape the extreme midday tropical sun, and some tortoises make use of freshwater and/or saline pools and mudholes to cool off. Heavy grazing by tortoises has resulted in a dwarfed vegetation over much of the area.

Natural History: Mating takes place from February to May. The male approaches the female and smells her carapace, then attempts to mount with neck outstretched, forefeet on the anterior portion of her carapace, and hind feet on the ground. His tail is then laid next to hers and intromission follows. Most copulatory attempts are unsuccessful. Males emit stereotyped deep-pitched groans or bellows while mounted. Nests are dug in the dry season from June through September. Where the soil is deep enough, well-camouflaged flask-shaped cavities about 25 cm deep are excavated; elsewhere the eggs may be laid in shallow, scraped-out, unfilled depressions. Patches of grass or scrub in open areas seem to be preferred sites (Bourn, 1977). Nesting occurs at dusk or in the night, allowing the females to avoid overheating. Females in high-density populations lay fewer eggs (4–5) in a single clutch every few years, while those from low-density populations may lay several clutches of 12–14 eggs each year. Eggs are almost spherical (48–51 mm) with brittle shells. Hatching occurs from early October to mid-December during the rainy season, some 98 to almost 200 days later.

A factor aiding survival is the tortoises' ability as herbivores to both graze and browse. In the drier open areas they are grazers, feeding primarily on sedges, grasses, and small herbs which form the distinctive "tortoise turf" plant community. Usually over half of the tortoises observed eating will be feeding on this turf. So minute and low-growing are these plants that the tortoises inevitably ingest soil as they feed, allowing much erosion. Dry conditions and heavy grazing prevents formation of a continuous plant cover, and in many places dry earth is exposed. Along much of the coastal dunes, *Sporobolus* grass is dominant to the exclusion of other grasses and sedges and is the favorite food of tortoises there.

In the wooded and scrub areas, tortoises browse on many types of woody plants. A number of species are readily eaten, and some show a conspicuous browse line about a meter above the ground, or as high as the tortoises can stretch their necks. Many woody plants are eaten, but apparently unpalatable species are ignored. Although the branches of the screwpine (*Pandanus*) are out of reach, they are readily eaten if cut down, as are also dry coconut fronds. Tortoises have considerable influence on regeneration of the browse plants, and the saplings of *Ficus* and *Flacourtia* are especially stunted by repeated cropping.

Although primarily vegetarians, the tortoises eat many other foods if the opportunity arises, including feces and the decaying flesh of dead land crabs and other tortoises. They are probably attracted to these by the stench, and possibly these foods provide additional sources of water. There also seems to be a predilection for red-colored foods.

Drinking water is generally scarce on the atoll, though in places pools are abundant. During the wet season, rain puddles provide drinking water, but when free water is unavailable, tortoises probably gain most water through metabolism.

Taxonomic Comment: Bour (1982) has recently proposed and defined a separate genus, *Dipsochelys*, for some giant tortoises of the Aldabra-Seychelles group (including *G. gigantea*), but we feel this designation is premature, as it was based on the examination of relatively few specimens.

Geochelone radiata (Shaw, 1802)

Radiated tortoise

Plate 16

Recognition: The elongated carapace (to 40 cm) is high domed with abruptly descending sides, only a slight cervical indentation, and the posterior marginals upturned and serrated. A rather broad cervical scute is present. Vertebrals are broader than long, and the 5th is expanded. Well-defined growth annuli surround the raised vertebral and pleural areolae. There are usually 11 marginals on each side and a single, undivided supracaudal which is downturned between the posterior marginals. The carapace is dark brown or black; vertebrals and pleurals have yellow or orange areolae from which extend 4 to 12 yellow or orange radiating stripes. Each marginal has 1 to 5 yellow or orange radiations beginning from a light spot at the lower center of the scute and extending upward toward the pleurals. These light stripes may fade with age, but most individuals are strikingly marked. The plastron is well developed. Its forelobe tapers to the front and is longer but narrower than the hind lobe, which bears a deep anal notch. The plastral formula is: abd > hum > fem > gul > an > pect; the paired gulars are thickened and may protrude beyond the carapacial rim, especially in males. The bridge is wide with a single small axillary and a large inguinal scute. The plastron is yellow with a large black triangle at the outer edge of the humerals, pectorals, abdominals, and femorals; the gulars are usually unpatterned and the anals have black radiations. Black radiations may also occur on the abdominals. The head is moderate in size with a nonprotruding snout and a slightly hooked upper jaw. Its large prefrontal is longitudinally divided and followed by a single large frontal scale; other head scales are small. Crumly (1982b) found that 59% of the adult *G. radiata* skulls he examined contained an irregular parietal foramen, previously known only from some *Gopherus* individuals and *Geochelone carbonaria*. The head is usually yellow with a darkened area on the dorsal surface, and the limbs and tail are yellow. The anterior surface of the foreleg is covered with overlapping scales, only a few of which are large. No enlarged tubercles occur on the thigh, and the tail lacks a large terminal scale.

Males have slightly longer tails, deeper anal notches, and more protruding gular scutes.

This species appears most closely related to the other Madagascar endemic, *Geochelone yniphora*. Auffenberg (1974) thought *G. radiata* and *G. yniphora* to be probably conspecific, but this seems hardly the case.

Distribution: *Geochelone radiata* is endemic on Madagascar where it is restricted to the Antandroy Territory in the south between the Mandrare and Menarandra waterways.

Habitat: *Geochelone radiata* lives in dry, *Didierea* woodland, stands of euphorbs, and scrub and thorn brush.

Natural History: Courtship and mating in Florida captives observed by Auffenberg (1978) occurred from spring to early summer. Tactile and olfactory cues seemed more important than visual or auditory cues. Chemical cues are largely from cloacal scents (pheromones?). The male sniffs the female's cloacal region while moving the head vertically. This is often accomplished while walking behind (trailing) the female with head extended. The male then pushes the female's carapace with his extended gulars, and tries to lift or overturn her by hooking his gulars under her shell and pushing upward. He then mounts from the rear and, placing his forefeet on her carapace as a pivot, swings the posterior part of his body downward while simultaneously lifting his hind feet off the ground. This results in his violently striking the lower rear of her carapace with the thickened anal region of his plastron. At this instant the male usually emits rhythmic grunts. The female's cloacal region is probed by the male's tail and intromission follows. Cap-

tive nestings have occurred in March, July, August, and September. Eggs are almost spherical (36–42 × 32–39 mm) with brittle shells; clutch size varies from 3 to 12. Incubation is long, 145–231 days. Hatchlings are brightly marked and have flattened, rounded carapaces (32–40 mm).

G. radiata is herbivorous, and feeds on grasses, introduced *Opuntia* cacti, fruits, and other succulents in its natural habitat. Red foods seem to be preferred.

Geochelone yniphora (Vaillant, 1885)
Northern Madagascar spur tortoise
Plate 16

Recognition: The oval carapace (to 44.6 cm) is extremely domed with descending sides, a broad cervical indentation, and the posterior marginals slightly flared and serrated. A cervical scute, small dorsally but large ventrally, is present. Vertebrals are broader than long, or as broad as long, and the 5th is expanded. Well-defined growth annuli surround the raised vertebral and pleural areolae. There are usually 11 marginals on each side, and a single, undivided supracaudal. The carapace is yellowish brown and the outer edges of each vertebral and pleural are darker brown. Each marginal has a dark-brown triangle at its anterior seam. The well-developed plastron is usually immaculate yellow, but may contain some brown pigment. Its forelobe is much larger than the hind lobe, and the thickened elongated gular scute is upturned and projects well beyond the carapacial rim. The hind lobe has a broad anal notch. The plastral formula is: abd > gul >< hum > fem > an >< pect; the gular scute is longer in males than in females, which may have the interhumeral seam longer than the gular. The bridge is wide with a small axillary and a larger inguinal scute. The head is moderate with a nonprojecting snout and a slightly hooked upper jaw. Its large prefrontal scale is divided longitudinally and followed by a single large frontal scale; other head scales are small. The head varies from totally black or dark brown to brownish with some large, yellow, lateral spots near the tympanum. Neck, limbs, and tail are yellow to tan. The forelegs are covered anteriorly with large, yellow, overlapping scales. No enlarged tubercles occur on the thigh, and the tail lacks a large terminal scale.

Males have concave plastra, longer, thicker tails, and longer gular scutes than females.

Distribution: The former range probably included more of the western desert of Madagascar, and possibly some nearby islands, but today *Geochelone* *yniphora* is restricted to a small area about Baly Bay in northwestern Madagascar, where only a few hundred individuals still exist (Juvick et al., 1981).

Habitat: *Geochelone yniphora* lives along the tropical coastal plain in bamboo forests. This is an area of distinct seasonal precipitation, and the tortoises seem to prefer the more xeric microhabitats.

Natural History: Knowledge of the reproductive biology of *Geochelone yniphora* is based on studies by McKeown, Juvik, and Meier (1982) at the Honolulu Zoo. Courtship and breeding behavior occurred from late May to mid-September, and courtship behavior was similar to that of *G. radiata*. Males trailed and circled females while sniffing at them. This led to subsequent ramming, pushing, and hooking with the elongated gular scute to overturn the female. One male frequently bit at the female's head and forelegs while circling. Mounting and copulation followed, during which the males vocalized. A single female laid six clutches (September, October 1979; March, October, November 1980; and September 1982) totaling 25 eggs (3–6, $\bar{\chi}$ = 4.2). The white eggs were spherical (42–47, $\bar{\chi}$ = 43.6 mm) with brittle shells. The flask-shaped nests were about 11 cm deep. Unfortunately, to 1982, no successful hatchings have occurred at Honolulu.

Juvik et al. (1981) reported that the tortoises remain mostly inactive during the cool dry season from May to October, and do not dig burrows, but instead push under surface litter. During the day, the tortoises are most active in the morning and late afternoon; other times are spent in the dense thickets. The herbivorous tortoises feed mostly on the leguminous shrub *Bauhinia* cf. *pervillei*, but also eat the grass *Heteropogon contortus* and other sedges and grasses.

This is one of the most endangered of all turtles; currently less than 400 individuals survive in the wild (Curl et al., 1985). Its habitat must be preserved, and captive breeding programs should be undertaken.

Geochelone carbonaria (Spix, 1824)

South American red-footed tortoise

Plate 15

Recognition: This handsome tortoise has an elongated carapace (to 51 cm) with a shallow cervical indentation, the lateral sides distinctly concave when viewed from above, and a smooth posterior rim. There is no cervical scute. Vertebrals are broader than long, and the 1st and 5th are laterally expanded. Well-defined growth annuli surround the raised vertebral and pleural areolae. There are usually 11 marginals on

each side, and the single supracaudal is undivided and downturned. The carapace is black, with the vertebral and pleural areolae yellow to reddish orange; a light spot of the same color occurs at the base of each marginal. The plastron is well developed. Its upturned forelobe tapers toward the front and is about as long and broad as the hind lobe, which bears an anal notch. The plastral formula is: abd > fem ≥ hum > gul > an >< pect; the paired gulars are thickened, but do not extend much beyond the carapacial rim, if at all. The dorsal surface of each gular scute is usually not subdivided. The bridge is wide with a moderate axillary and a moderate to large inguinal which is in broad contact with the femoral scute. The plastron is yellowish brown with some dark pigment along the mid- and transverse seams. The head is moderate in size with a nonprojecting snout and a slightly hooked upper jaw. Its prefrontal scale is short and divided longitudinally, and followed by a large undivided frontal scale. Other head scales are small. Head scales are yellow, red, or orange; the jaws are dark. The anterior surface of each forelimb is covered with large, red, slightly or nonoverlapping scales. No enlarged tubercles occur on the thighs, and the tail lacks a large terminal scale.

Males have concave plastra, lower flattened carapaces with deep lateral concavities, and longer, thicker tails. Females have more domed, shallow, laterally concave carapaces, flat plastra, and shorter tails.

Distribution: *Geochelone carbonaria* occurs in southeastern Panama and west of the Andes in Chocó of Colombia, but its main range is east of the Andes in eastern Colombia, Venezuela, and the Guianas to eastern Brazil, south to Rio de Janeiro, and west to eastern Bolivia, Paraguay, and northern Argentina. It seems absent from almost all but the eastern parts of the Amazon Basin. This tortoise may occur naturally on Trinidad, and has been introduced on quite a few Caribbean Islands, including St. Croix in the Virgin Islands.

Geographic Variation: No subspecies have been described, but Pritchard and Trebbau (1984) suggested that pattern and size differences exist between various populations.

Habitat: Where it is sympatric with *Geochelone denticulata*, *G. carbonaria* is more prevalent in moist savannas (which are seldom entered by *denticulata*), but where *denticulata* does not occur, *carbonaria* seems to be a humid forest dweller. Perhaps there is competitive exclusion by *denticulata*.

Natural History: Mating occurs, at least in captivity, year round. As in *Geochelone denticulata*, males distinguish other males from females by their responsive head movements; and, if a male, some ramming and pushing follows. If no responsive head movement is given, the male walks behind the other tortoise and sniffs its cloacal region, presumably for odors confirming it a female of the proper species. The male may then mount the female or he may ram and push to subdue her. During copulation the males utter clucking sounds. Nesting occurs from June to September and the female digs a flasklike cavity about 20 cm deep. Two to 15 eggs are laid in one clutch, and probably several clutches are deposited each season. Eggs are elongated (40–59 × 34–48 mm) with brittle shells. Hatchlings are rounded and flat (39–45 mm) and have no toothlike projections on the rims of the anterior marginals, as occur in hatchling *G. denticulata*.

Geochelone carbonaria feeds on grasses, succulents, fallen fruits, and carrion.

Geochelone denticulata (Linnaeus, 1766)
South American yellow-footed tortoise

Recognition: This is the largest tortoise (to 82 cm) living on the mainland of South America. It has an elongated carapace with a shallow cervical indentation, the lateral sides parallel, and posterior marginals slightly serrated. There is no cervical scute. Vertebrals are broader than long, and the 1st and 5th are laterally expanded. Well-defined growth annuli surround the slightly raised areolae of the vertebral and pleural scutes. There are usually 11 marginals on each side, and the single supracaudal is undivided and downturned. The carapace is brown with yellow to orange vertebral and pleural areolae; yellowish or orange pigment also occurs at the lower edge of each marginal. The plastron is well developed. Its upturned forelobe tapers toward the front and is about as long as, but slightly narrower than, the hind lobe, which bears

Geochelone denticulata

an anal notch. The plastral formula is: abd > hum > fem > gul >< pect >< an; the paired gulars are thickened, but do not extend much beyond the carapacial rim, if at all. Each gular scute is dorsally subdivided, ultimately producing four scales. The bridge is wide with a moderate-sized axillary and a smaller inguinal that barely touches the femoral scute. The plastron is yellowish brown with darker pigment along the mid- and transverse seams. The head is moderate in size with a nonprojecting snout and a slightly hooked upper jaw. Its large prefrontal scale is divided longitudinally and followed by a subdivided frontal scale; other head scales are small. Head scales are yellow to orange with dark borders; the jaws are dark brown. The anterior surface of each forelimb is covered with large yellow or orange, non- or only slightly overlapping scales. No enlarged tubercles occur on the thighs, and the tail lacks a large terminal scale.

Males have concave plastra; the carapace expanded over the hind limbs; low, flattened, elongated profiles; and longer, thicker tails. Females have domed, nonexpanded carapaces, flat plastra, and shorter tails.

Distribution: *G. denticulata* ranges from southeastern Venezuela through the Caribbean lowlands of the Guianas to Brazil, where it occurs throughout the Amazon Basin to eastern Ecuador and Colombia, northeastern Peru, and northern and eastern Bolivia. It also occurs in an isolated range in eastern Brazil and on Trinidad.

Habitat: *Geochelone denticulata* is a denizen of tropical evergreen and deciduous rainforests.

Natural History: Mating probably occurs throughout the year. Males identify each other through characteristic head movements not made by females. The male then sniffs the cloacal region of the female, perhaps to determine if she is his species. Some pushing and ramming and biting at her limbs may follow to immobilize her before copulation, or mounting and intromission may occur immediately. Clucklike vocalization may be emitted by the male during courtship and while mounted. Nesting probably also occurs throughout the year (it has been observed in August to February in Colombia), and it is likely several clutches are laid each year. A normal clutch comprises 1 to 12 (usually 4 to 8) eggs; eggs are elongated (40–60 × 35–44 mm) with brittle shells. Incubation takes four or five months. Hatchlings are rounded and flat (47–60 mm) and have distinct toothlike projections on the rims of the anterior marginals (hatchlings of *G. carbonaria* have smooth anterior rims),

Geochelone denticulata feeds on grasses, succulent plants, fallen fruit, and carrion.

Geochelone chilensis (Gray, 1870a)

Chaco tortoise

Recognition: The oval carapace (to 43 cm) is flattened dorsally with descending sides, a cervical indentation, and a serrated marginal rim. The posterior marginals are slightly upturned, and no cervical scute is present. Vertebral 1 is usually broader than long, but may be only as broad as long. Other vertebrals are broader than long, and the 5th is expanded. Well-defined growth annuli surround the slightly raised vertebral and pleural areolae. There are usually 11 marginals on each side, and a single, undivided supracaudal scute which is downturned between the posterior marginals. The carapace may be either totally yellowish brown or have dark-brown or black growth annuli surrounding tan areolae, and a dark wedge of pigment at the posterior seam of each marginal. The well-developed plastron has a deep anal notch. Its forelobe tapers toward the front, and the paired gulars may be slightly divided by a notch. The forelobe is longer but slightly narrower than the hind lobe. The plastral formula is: abd > hum >< fem > gul > pect >< an; the paired gulars are thickened, but do not project much past the carapacial rim. The bridge is broad with an axillary about half as large as the single inguinal. The plastron varies from uniformly yellow brown to having a dark triangular wedge along the seams of each scute. The head is moderate in size with a nonprojecting snout and a hooked, bi- or tricuspid upper jaw. Its large prefrontal scale is divided longitudinally, and followed by a large frontal scale which may be either entire or subdivided; other head scales are small. Head, limbs, and tail are yellowish brown. The anterior surface of each foreleg is covered with large, angular, non- or slightly overlapping scales, and several enlarged tubercles occur on each thigh. The tail ends in an enlarged scale.

The sexes are difficult to distinguish, but males are smaller and have slightly longer tails.

Distribution: *Geochelone chilensis* ranges from southwestern Bolivia, western Paraguay, and northwestern Argentina southward to about 40°S in northern Patagonia (Auffenberg, 1974).

Geographic Variation: At this time, we do not recognize any subspecies. However, Freiberg (1973) described two other tortoises from Argentina, the large *Geochelone donosobarrosi* and the small *G. petersi*, which are probably only variants of *G. chilensis*. The situation is confusing since *donosobarrosi* (south) and *petersi* (north) occur at the opposite ends of the range of *G. chilensis*, and possibly the size differences are clinal. Also, sexual dimorphism has been poorly stud-

Geochelone chilensis

ied in *G. chilensis*, and the small, elongated, thick-shelled *petersi* may represent males, whereas the larger, oval, thinner shelled *donosobarrosi* may be typical of females. Unfortunately, *G. chilensis* has been one of the least collected of continental tortoises, so comparisons are difficult.

Habitat: *Geochelone chilensis* lives in a variety of dry lowland habitats: savanna, thorn and scrub brush, and desert. In northern Patagonia it survives the winter in open burrows or pallets.

Natural History: Freiberg (1981) reported courting males gently push the females about. Nesting occurs from November to January. From one to six oval to spherical (42–49 × 32–38 mm) brittle-shelled eggs are laid at a time, and two clutches are laid each season. Hatching occurs after about 125 days to a year, and hatchlings are rounded (50–65 mm), flat topped, and yellowish brown.

Natural foods include grasses, succulents, and cacti.

The Galapagos tortoise complex

One of the most confusing situations in turtle systematics and biology involves the status of the giant tortoises inhabiting the Galapagos Archipelago. All are members of the genus *Geochelone* Fitzinger, 1835, but are they separate species or merely subspecies of a single species? Most recent systematic accounts have treated them as subspecies of *Geochelone elephantopus* (Harlan, 1827) (Wermuth and Mertens, 1961, 1977; Hendrickson, 1966; MacFarland, Villa, and Toro, 1974a; Pritchard, 1979). However, most are allopatric with no evidence of interbreeding, and, when cross breeding occurs in captivity, a large proportion of abnormal individuals result (Robert Reynolds, pers. comm.). It is probably best to follow Fritts (1983), who considers each geographically isolated population a separate species, until more detailed analysis is completed. Further confusing the issue, the name *G. elephantopus* has been misapplied to tortoises occurring on the islands of Albemarle (Wermuth and Mertens, 1961, 1977), Abingdon and Duncan (Garman, 1917), when it really belongs to the extinct saddlebacked form of Charles Island. Here we treat all 12 possibly extant island forms (MacFarland, Villa, and Toro, 1974a) as full species, following a slightly modified version of Van Denburgh's (1914) species list for the names: *abingdonii* (Abingdon Island), *becki* (Albemarle Island), *chathamensis* (Chatham Island), *darwini* (James Island), *ephippium* (Duncan Island), *guntheri* (Albemarle Island), *hoodensis* (Hood Island), *microphyes* (Albemarle Island), *nigrita* (Indefatigable Island), *phantastica* (Narborough Island), *vandenburghi* (Albemarle Island), and *vicina* (Albemarle Island). Variances from Van Denburgh (1914) include substitution of *nigrita* (Duméril and Bibron, 1835) for *porteri* (Rothschild, 1903), over which it has priority, and the dropping of the form *wallacei* (Rothschild, 1902) (Jervis Island; Van Denburgh, 1914) which is based on a single specimen considered an artificial introduction (Snow, 1964; MacFarland, Villa, and Toro, 1974a).

The 12 species differ in a number of morphological characters such as the color, thickness, and overall shape of the carapace, maximum size, and the lengths of the neck and limbs; however, there is much variation within each taxon, and the differences between some taxa are small (MacFarland, Villa, and Toro, 1974a). The major differences occur between those with domed carapaces (*nigrita* and *vandenburghi*) and the saddlebacked species (*abingdonii, becki, ephippium, hoodensis,* and especially *phantastica*), in which the carapace is strongly flared and upturned above the neck, an adaptation which allows the tortoise to stretch its neck and head upward to a greater extent, allowing feeding on high vegetation or for territorial displays. The saddlebacked shell probably evolved from the domed form, and Fritts (1983) has reported some differences between the saddlebacked species that suggest this type of shell may have evolved more than once. Pritchard (1979) presents good profile drawings of the various saddlebacked species. Recently, in a electrophoretic study of blood proteins from seven island species, Marlow and Patton (1981) found that genetic similarity is unrelated to carapace shape, suggesting that the shape has been independently derived several times on separate islands.

Most natural history data reported for Galapagos tortoises has also been published under the blanket name "*elephantopus*," and can only be properly applied to species if the island of origin is given. Unfortunately, this has invalidated several good reproductive

Key to the species of Galapagos tortoise complex:

1a. Carapace saddlebacked, anteriorly narrowed with an elevated rim, wider and lower posteriorly with a rounded rim ... 2
 b. Carapace domed or intermediately shaped, not greatly narrowed or elevated anteriorly, high in center and rear ... 6
2a. Plastron length more than 72% of carapace length; height of carapace at cervical indentation usually less than 53% of straight carapace length ... 3
 b. Plastron length less than 72% of carapace length; height of carapace at cervical indentation more than 53% of straight carapace length: *Geochelone phantastica*
3a. Plastron length more than 88% of carapace length: *G. hoodensis*
 b. Plastron length less than 88% of carapace length ... 4
4a. 8th marginal reduced, its dorsal border narrow and wedge shaped: *G. abingdonii*
 b. 8th marginal little reduced, its dorsal border relatively long and not wedge shaped ... 5
5a. Adult carapace length to 84 cm, height to marginals (vertical distance from surface when shell is laid upon a flat surface to lower edge of marginal scutes at midbridge level) 7–12% of carapace length: *G. ephippium*
 b. Adult carapace length to 105 cm, height to marginals 5–9% of carapace length: *G. becki*
6a. Carapace domed, higher in middle than at either end ... 7
 b. Carapace intermediate in shape between saddlebacked and domed ... 8
7a. Anterior carapacial rim slightly upturned; vicinity of Volcán Alcedo, central Albemarle Island: *G. vandenburghi*
 b. Anterior carapacial rim horizontal, not upturned; Indefatigable Island: *G. nigrita*
8a. Pectoral scutes usually reduced and not extending to plastral midseam: *G. chathamensis*
 b. Pectoral scutes not reduced, extending to plastral midseam ... 9
9a. Bridge length greater than 44% of carapace length ... 10
 b. Bridge length less than 44% of carapace length: *G. darwini*
10a. Sum of straight width, curved width, and half the height to marginals is less than twice the straight carapace length: *G. microphyes*
 b. Sum of straight width, curved width, and half the height to marginals is more than twice the straight carapace length ... 11
11a. Posterior marginals serrated, 1st marginals with prominent points: *G. vicina*
 b. Posterior marginals not serrated, 1st marginals lacking prominent points: *G. guntheri*

studies of captive tortoises. In addition, the only karyotypes described (2n = 52: 26 macrochromosomes, 26 microchromosomes; Golstein and Lin, 1972; Dowler and Bickham, 1982) were reported as *"elephantopus"* with no locality data.

The Galapagos Archipelago lies 965 km west of Ecuador on the Equator in the eastern Pacific Ocean. There are 24 named volcanic islands and numerous islets and rocks. The islands are bathed in the cold, southern Humboldt current, thus the air temperatures, while hot, are not excessive. There is little rainfall, and the low parts of the islands are very dry; however, some of the higher (over 300 m) of the numerous craters are often cloud covered and have a damp climate and more lush vegetation, especially on the windward sides of the islands. Tortoises probably first reached these islands by rafting from mainland South America. Marlow and Patton (1981) have shown in their electrophoretic studies that the Galapagos species share no greater similarity with mainland *Geochelone carbonaria, G. chilensis,* or *G. denticulata* than they do among themselves, so the most direct common ancestor of the Galapagos species is undoubtedly extinct.

Captive Galapagos tortoises are often confused with the Aldabra tortoise, *Geochelone gigantea,* but there are several easily recognized pertinent differences. The Aldabra tortoise has a thicker, domed carapace usually with a cervical scute; that of the Galapagos tortoise is thinner, lacks a cervical scute, and often is saddlebacked. Also, some of the Galapagos species reach a greater size. The heads are much different. That of the Aldabra tortoise is more pointed and wedge shaped with the two halves of the large, divided prefrontal scale lying parallel and not diverging posteriorly; the frontal scale is relatively small. The head of the Galapagos species is not wedgeshaped, but instead more blocklike or squared off. Its prefrontal scale is divided into two short, broad scales which diverge posteriorly, and the following frontal scale is about the same size.

The key below is based on examination of the specimens in the United States National Museum of Natural History, measurements of specimens in the California Academy of Sciences provided by its herpetological staff, and literature descriptions; not all species were examined by the authors.

Geochelone abingdonii (Günther, 1877)

Abingdon Island tortoise

Recognition: The black carapace (to 98 cm) is shaped like a saddle, very narrow, compressed, and slightly upturned anteriorly, and wider and lower posteriorly with a rounded margin. The shell is rather thin and surprisingly light in weight. Only a slight cervical indentation is present. The height at the cervical indentation is 46% or more of the carapacial length. Vertebrals are usually broader than long, but the 1st is narrowed and may be as long as or longer than broad; the 5th is expanded. The surface of the vertebral and pleural scutes ranges from smooth to pitted. Eleven marginals lie on each side. Those anterior are serrated and slightly upturned, but their ventral surfaces are never completely vertical; the posterior marginals are slightly flared and downturned; the single, undivided supracaudal scute is downturned. Lateral marginals are vertical or downturned, and the 8th is wedge shaped with a very short, reduced, dorsal border. The black plastron is well developed, but shorter than the carapace, and tapered (narrowed) anteriorly and posteriorly. At best, only a slight anal notch is present. The plastral formula is: abd > hum > fem > gul >< an > pect; the paired gular scutes do not project beyond the carapacial rim. The bridge is narrow (43–45% of carapace length) with single small axillary and inguinal scutes of the same size. The head is small with a nonprojecting snout and a weak bicuspid upper jaw. Its divided prefrontal scale is small, as is the single frontal scale. Head, neck, limbs, and tail are gray to black. The neck is very long with a biconvex 4th cervical vertebra. The anterior surface of each forelimb is covered with large, slightly overlapping to nonoverlapping scales. The short tail lacks a large terminal scale.

Males have concave plastra, and yellow jaws and throats.

Distribution: *Geochelone abingdonii* is restricted to Abingdon (Pinta) Island, Galapagos.

Habitat: Extremely rugged, as most of the island is covered with bare volcanic rock with numerous crevices and pits. Vegetation consists of scattered brushy thickets and *Opuntia* cacti.

Natural History: Specimens collected in 1906 had eaten grasses and cacti (Van Denburgh, 1914). Günther (1877) reported the anterior margins of the carapace were more or less broken, possibly as a result of male combat.

The future of this species looks grim. Its population was apparently always small, and in 1959 goats were introduced to the island which destroyed much vegetation and drove the remaining tortoises from their lowland breeding areas. Females were seemingly also scarce, as only one has ever been collected (Pritchard, 1979). Today, only one individual, "Lonesome George," a captive male at the Charles Darwin Station on Indefatigable Island, is known to exist; but Mac-

Farland, Villa, and Toro (1974a) thought a few may still remain in the wild. For this species to survive, males and females will have to be brought together for captive breeding and the young raised to a sufficient size to withstand predation before being returned to Abingdon Island. Also, hunting of feral goats must continue until they are brought under control or eliminated.

Geochelone becki (Rothschild, 1901)

Volcán Wolf tortoise

Recognition: The grey, saddlebacked carapace (to 105 cm) is relatively thick with little or no cervical indentation, the anterior carapacial rim upturned, and the posterior marginals flared and slightly serrated. The carapace is compressed or narrowed anteriorly, but not nearly as much as some other saddlebacked species, and the height at the cervical indentation is 44% or more of the carapace length. Vertebrals are broader than long, and the 5th is somewhat expanded. In adults, the carapacial surface is nearly smooth. There are 11 marginals on each side and the single, undivided supracaudal is bent downward between the flared posterior marginals. Anterior marginals are not greatly expanded, or greatly upturned, and their ventral surfaces are not upturned to become vertical. Lateral marginals are vertical or downturned, and the 8th is not reduced. The gray, well-developed plastron is shorter than the carapace, and is tapered both anteriorly and posteriorly. At best, only a shallow anal notch is present. The plastral formula is: abd > hum > fem > an >< gul > pect; the paired gular scutes do not project beyond the carapacial rim. The bridge is narrow (37–41% of carapace length) with single axillary and inguinal scutes of about the same size. The head is moderate in size with a nonprotruding snout and a slightly hooked, bicuspid upper jaw. Its divided prefrontal scale and the following frontal scale are small. Head, neck, limbs, and tail are gray. The neck is long with a biconvex 4th cervical vertebra. Anterior surfaces of the forelimbs are covered with large, nonoverlapping scales. The tail is short and lacks a large terminal scale.

Males are larger, have slightly longer, thicker tails and yellowish lower jaws and throats.

Distribution: Restricted to the area about Volcán Wolf, Bank's Bay, and Cape Berkeley at the northern end of Albemarle (Isabela) Island, Galapagos; MacFarland, Villa, and Toro (1974a) estimated the population to be between 1000 and 2000 individuals, and apparently not endangered.

Geographic Variation: Pritchard (1979) reported that *Geochelone becki* has a variety of carapacial shapes ranging from domed to flattened or saddlebacked.

Habitat: *Geochelone becki* lives on very rugged, steep slopes with brushy thickets.

Natural History: The population is reproducing, but details of their reproductive biology are lacking; Pritchard (1979) presented a photograph of a nesting female.

Wild foods consist of coarse grasses, but apparently not cacti (Van Denburgh, 1914).

Geochelone chathamensis (Van Denburgh, 1907)

Chatham Island tortoise

Recognition: *Geochelone chathamensis* has a wide, black shell, its shape intermediate between the saddlebacked and domed species: adult males are rather saddlebacked, but females and young males are wider in the middle and more domed. The carapace (to 89 cm) has only a slight cervical indentation; the anterior rim is little upturned, if at all, and the posterior marginals are downturned. Height of the shell at the cervical indentation is less than 40% of the carapace length. Vertebrals are broader than long, and the 5th is expanded. Surface of the scutes varies from smooth to rough with growth annuli in adults. There are 11 marginals on each side, and a single, undivided, downturned supracaudal scute. Anterior marginals are little serrated, and usually only slightly upturned, at best; their ventral surfaces are never vertical. Lateral marginals are vertical or downturned, and the 8th is not reduced. The black plastron is well developed, shorter than the carapace, and tapered (narrowed) on both lobes; an anal notch is usually absent. The plastron formula is: abd > hum > fem > gul >< an > pect. This is the only species of the Galapagos complex in which the pectoral scutes do not always meet at the plastral midline (in 9 of 18 examined by Pritchard, 1979, and 3 of 4 examined by Ernst). The paired gulars extend almost to the carapacial rim in adult males. The bridge is broad (about 42–45% of carapace length) with the inguinal scute slightly larger than the axillary. The head is small with a nonprojecting snout and a bicuspid upper jaw. Its divided prefrontal scale and the frontal scale are small. Skin of the head and soft parts is dark gray or black; white or cream-colored pigment may occur around the mouth, chin, and nostrils. The neck is long with a biconvex 4th cervical vertebra. Anterior forelimb surfaces are covered with large nonoverlapping scales. The short tail has no large terminal scale.

Distribution: *Geochelone chathamensis* is restricted to the northeastern part of Chatham (San Cristóbal) Island of the Galapagos Archipelago.

Habitat: *Geochelone chathamensis* lives in brushy thickets, woods, and grassy areas, and often soaks in mud puddles.

Natural History: Nesting begins in September and four to six eggs are deposited in each nest (Pritchard, 1979). Natural foods include cacti, grasses, forbs, and the leaves of bushes.

MacFarland, Villa, and Toro (1974a) estimated the population to be between 500 and 700 individuals; however, while mating and nesting naturally occur, all hatchlings are eaten by dogs or the nests destroyed by feral donkeys, and recent attempts at captive breeding at the Charles Darwin Station have not been very successful. Originally the tortoises were found throughout Chatham Island, but, due to easy accessability, mariners collected more tortoises there than on any other islands during the 1800s, extirpating them from all but the rarely visited northeastern part of the island.

Geochelone darwini (Van Denburgh, 1907)

James Island tortoise

Recognition: The grey to black carapace (to 96.5 cm) is intermediate in shape between the saddlebacked species and those with domed shells. It has only a shallow cervical indentation; the anterior carapacial rim is not appreciably upturned, and the posterior marginals are flared, slightly upturned, and slightly serrated. The carapace is wide and not compressed anteriorly; height at the cervical indentation is 42% or more of the carapacial length. Vertebrals are broader than long, and the 5th is expanded. In adults, the carapacial surface is smooth. There are 11 marginals on each side, and the single, undivided supracaudal is downturned between the posterior marginals. Anterior marginals are not greatly expanded or upturned; their ventral surfaces are nearly horizontal. Lateral marginals are vertical or downturned, and the 8th is not reduced. The well-developed gray plastron is shorter than the carapace, and tapered (narrowed) anteriorly and posteriorly; there is only a very shallow anal notch. The plastral formula is: abd > hum > fem > an >< gul > pect; the paired gular scutes do not protrude beyond the carapacial rim. The bridge is 40–43% of the carapace length, with single axillary and inguinal scutes. The head is of moderate size with a nonprojecting snout and a slightly hooked, bi- or tricuspid upper jaw. Its divided prefrontal scale is small,

as is the single frontal scale. Head, neck, limbs, and tail are gray, but the jaws and throat are darker with some yellow markings. The neck is long with a biconvex 4th vertebra. Anterior surfaces of the forelimbs are covered with large, irregularly shaped, nonoverlapping scales. The short tail lacks a large terminal scale.

Males are larger, have thicker tails and concave plastra.

Distribution: *Geochelone darwini* lives only on James (San Salvador) Island of the Galapagos Archipelago, where, according to Pritchard (1979), it is found in four distinct, though interconnected, zones in the west-central area of the island. Each of these zones has a separate nesting area at low altitudes. MacFarland, Villa, and Toro (1974a) estimated the remaining population to be between 500 and 700 tortoises, mostly adults; but few hatchlings survive due to predation by introduced pigs and goats. An artificial incubation and hatchling raising program has been initiated at the Charles Darwin Station.

Habitat: The tortoises occur at altitudes between 200 and 700 m in rugged, rocky, brushy, wooded areas.

Natural History: Pritchard (1979) reported that nesting occurs from August to October, and that the nests contain 4–10 eggs.

Foods consist of grasses, forbs, and cacti.

Geochelone ephippium (Günther, 1875)

Duncan Island tortoise

Recognition: This saddlebacked species is one of the smallest of the Galapagos tortoises. Its brownish-gray, oblong carapace (to 84 cm) has only a very shallow cervical indentation, the anterior marginals little to much upturned, and the slightly serrated posterior marginals flared and upturned. The carapace is usually compressed or narrowed anteriorly, and its height at the cervical indentation is usually more then 40% of the carapacial length. Vertebrals are broader than long, and the 5th is expanded. Although growth annuli may be present on old tortoises, the surfaces of the vertebral and pleural scutes are generally smooth. There are 11 marginal scutes on each side, and the single, undivided supracaudal is downturned between the posterior marginals. Anterior marginals are not greatly expanded, and, while often upturned, their ventral surfaces are not totally vertical. Lateral marginals are vertical or downturned, and the 8th is not reduced. The well-developed brownish-gray plastron is tapered (narrowed) toward both front and back, and shorter than the carapace. Its hind lobe bears only a slight posterior notch, at best. The plastral formula is: abd > hum > fem > gul > an > pect; the paired gulars do not project beyond the carapacial rim. The bridge is narrow (38–42% of carapace length) with single axillary and inguinal scutes of about the same size. The head is small with a nonprotruding snout and a weakly hooked, bicuspid upper jaw. Its divided prefrontal scale and the following single frontal scale are small. Head, neck, limbs, and tail are dark gray. The neck is long with a biconvex 4th cervical vertebra. Anterior forelimb surfaces are covered with large, rounded, nonoverlapping scales. The tail is short and lacks a terminal scale.

Males have yellowish lower jaws and throats, and slightly longer, thicker tails.

Distribution: *Geochelone ephippium* is restricted to Duncan (Pinzon) Island in the Galapagos Archipelago.

Habitat: The surviving population of about 150–200 tortoises lives on the southwestern slopes of a volcanic crater in thickets and grassy areas.

Natural History: Nesting occurs between August and December; two to eight spherical (58–60 mm), brittle-shelled eggs are laid at one time, and several clutches are laid each year. Natural incubation may take from 85 to 120 days. MacFarland, Villa, and Toro (1974a) estimated that approximately 7,000–19,000 tortoises hatched on Duncan Island in the 10 years prior to 1974, but no juveniles could be found, as all were apparently eaten by introduced black rats. A program of egg incubation and hatchling rearing has been established at the Charles Darwin Station, and 4–5-year-old tortoises have been returned to Duncan Island (Metzer and Marlow, 1986). At this age they are apparently large enough to deter rat attacks, and those released seem to be doing well.

Natural foods include grass, cacti, moss, and forbs.

Geochelone guntheri (Baur, 1890b)

Sierra Negra tortoise

Recognition: The gray-brown to black carapace (to 120 cm) is intermediate in shape between domed and saddlebacked. There is a slight cervical indentation, and the carapace is rather broad anteriorly. Anterior marginals are not bent upward or only slightly; posterior marginals are downturned; the height at the cervical indentation is not more than 44% of the carapacial length. Vertebrals are all wider than long, and the 5th is expanded. Surfaces of the vertebral and

pleural scutes are rather smooth, especially in large specimens, but growth annuli are present. Eleven marginals occur on each side, and the single, individual supracaudal scute is downturned. Lateral marginals are small, but the 8th is not reduced and has a long dorsal border. The well-developed gray-brown plastron is shorter than the carapace. Its forelobe tapers anteriorly, and the hind lobe is notched posteriorly. The plastral formula is: abd > hum > an > fem > gul > pect; the paired gulars do not project beyond the anterior carapacial rim. The bridge is wide with a short axillary and a larger inguinal. The head is moderate in size with a nonprojecting snout and a slightly bicuspid upper jaw. Its short prefrontal scale is longitudinally divided and followed by a larger frontal scale; other head scales are equal in size and rather small. Females have gray heads, and the lower jaw and throat may be yellow in males. The gray neck is long with a biconvex 4th cervical vertebra. Limbs and tail are gray. Anterior surfaces of the forelimbs are covered with large, pointed, slightly overlapping scales. No large terminal scale occurs on the tail.

Males have longer, thicker tails and lighter colored throats and chins.

Distribution: *Geochelone guntheri* is restricted to the Vilamil Mountain (Sierra Negra) area of southeastern Albemarle (Isabela) Island.

Geographic Variation: Some individuals are more saddlebacked than others. MacFarland, Villa, and Toro (1974a) thought that *Geochelone vicina* and *G. guntheri* may eventually be combined. Their morphological features are extremely similar, and until 1925 no known physical barriers separated the two populations; they are now partially separated by an extensive lava flow, but a large area along the southern coast of Albemarle Island remains open as a potential pathway for genetic exchange. Male *G. vicina* have more elevated shells than the flatter male *G. guntheri*.

Habitat: The habitat is relatively dry with cacti, grassy areas, brush, and some trees. Beck (in Van Denburgh, 1914) reported that these tortoises push under shrubs to avoid the heat, and that they frequently bathe in mudholes.

Natural History: Nesting probably occurs from July to November, and clutches contain about 9 (8–17) eggs (Pritchard, 1979). Eggs are almost spherical; 10 measured by Van Denburgh (1914) ranged from 60 × 58 mm to 57.6 × 57.1 mm. Foods consist of grasses and cacti.

MacFarland, Villa, and Toro (1974a) estimated the population of *G. guntheri* to be 300–500 individuals, mostly small to medium-sized animals, and reported natural reproduction. However, few adults remain, as most have been killed by settlers, and continued patrolling will be necessary to protect the remaining tortoises.

Geochelone hoodensis (Van Denburgh, 1907)

Hood Island tortoise

Recognition: *Geochelone hoodensis* is the smallest (to 75 cm) member of the Galapagos complex. Its black, saddleback carapace has a deep cervical indentation, the anterior rim only weakly upturned, and posterior marginals downturned and slightly serrated. It is narrow anteriorly and wider posteriorly. Height at the cervical indentation is 42% or more of the carapacial length. Vertebrals are broader than long; the 5th is expanded. Vertebrals and pectorals are roughened with growth annuli, and their centers are raised. There are 11 marginals on each side, and the single, undivided supracaudal is downturned. Anterior marginals are not expanded or much upturned; their ventral surfaces are never vertical. Lateral marginals are vertical or downturned, and the 8th is reduced with a narrow anterior border. The black, well-developed plastron is shorter than the carapace and narrow at each end; no posterior notch, or only a shallow one, is present. The plastral formula is: abd > hum > fem > an >< gul > pect; the paired gulars may project to slightly beyond the carapacial rim. The bridge is about 37% of the carapace length, and has single axillary and inguinal scutes. The head is small with a nonprojecting snout and a bicuspid upper jaw. Its divided prefrontal and frontal scales are small. The head is dark gray to black with yellowish or white pigment on the jaws, chin, and throat; elsewhere, the skin is gray or black. The neck is long with a biconvex 4th cervical vertebra. Anterior surface of each forefoot is covered with large, nonoverlapping scales. The short tail lacks a large terminal scale.

Males are larger than females, are more saddlebacked, and have slightly thicker tails.

Distribution: *Geochelone hoodensis* lives on Hood (Espanola) Island, Galapagos.

Habitat: Rocky, brushy areas are inhabited.

Natural History: Mating occurs from December to August and nesting from late June to December. Possibly two to four clutches of three to seven spherical, brittle-shelled eggs are laid each year.

Natural foods include grass and cacti.

The estimated population on Hood Island is very small, 20–30 individuals (MacFarland, Villa, and Toro, 1974a), and the tortoises are now so dispersed they seldom find each other to reproduce. There is no evidence of recent copulations in the wild, although there seems to be a preponderance of females in the population. Seafarers in the 1800s exploited this species almost to extinction, and the large population of feral goats on the island now destroy any nests and hatchlings, as well as compete with adult tortoises for food plants.

A breeding colony has been established at the Charles Darwin Station, and through August 1972, 25 young *G. hoodensis* had been raised to be used to restock Hood Island (MacFarland, Villa, and Toro, 1974b).

Geochelone microphyes (Günther, 1875)

Volcán Darwin tortoise

Recognition: The brownish-gray, oval carapace (to 103 cm) is intermediate between saddlebacked and domed and rather flattened, with a shallow cervical indentation, the anterior marginals not greatly upturned, and the posterior marginals slightly downturned. The carapace is not appreciably narrowed anteriorly, as in the saddlebacked species, and its height at the cervical indentation is not more than 44% of the carapacial length. Vertebrals are broader than long, and the 5th is expanded. Surfaces of the vertebrals and pleurals are rather smooth in adults. Eleven marginals lie on each side, and the single, undivided supracaudal is downturned between the posterior marginals. Anterior marginals are not expanded and are almost horizontal; lateral marginals are vertical or downturned, and the 8th is not reduced. The brownish-gray, well-developed plastron is shorter than the carapace, with narrow lobes. Its forelobe tapers toward the front, and the hind lobe is only slightly notched posteriorly. The plastral formula is: abd > hum > fem > gul > an > pect; the paired gulars do not project beyond the carapacial rim. The bridge is wide (45–46% of carapace length) with a small axillary scute and a larger inguinal. The head is moderate in size, with a nonprotruding snout and a weakly hooked, bi- or tricuspid upper jaw. Its divided prefrontal is small, as is the following frontal scale. Head, neck, limbs, and tail are gray. The neck is long with a biconvex 4th cervical vertebra. Anterior surfaces of the forelimbs are covered with rounded, slightly overlapping scales of varying sizes. The tail is short and lacks a large terminal scale.

Van Denburgh (1914) reported that males have low shells with flat backs, the front of the carapace being only a little lower than the middle of the back. The male tail is also somewhat longer and thicker.

Distribution: *Geochelone microphyes* occurs in the vicinity of Volcán Darwin in north-central Albemarle (Isabela) Island of the Galapagos Archipelago.

Habitat: This tortoise lives mostly on the volcanic slopes in dry thickets, cacti, and grass bunches.

Natural History: Nesting occurs in the lowlands and slopes along the western side of Volcán Darwin; about eight eggs are laid in each clutch (MacFarland, in Pritchard, 1979).

Natural foods include mostly grasses, cacti, and other succulents. MacFarland, Villa, and Toro (1974a) estimated the endangered population to be 500 to 1000 individuals.

Geochelone nigrita (Duméril and Bibron, 1835)

Indefatigable Island tortoise

Recognition: This is the best known of the Galapagos tortoises. Its black, oval carapace (to 130 cm) is domed, higher in the center than in the front, and broad anteriorly instead of narrow and compressed. A shallow cervical indentation is present; height at this point is less than 40% of the carapacial length. The anterior rim is slightly serrated but not upturned; the posterior marginals are serrated, flared, and may be upturned. Vertebrals are broader than long; the 5th is laterally expanded. Surfaces of the vertebrals and pleurals are rough with growth annuli. There are 11 marginals on each side, and the single, undivided supracaudal is downturned between the posterior marginals. Anterior marginals are expanded but not upturned; their ventral surfaces are horizontal; lateral marginals are vertical or downturned with the 8th not reduced. The black plastron is rather large, but still shorter than the carapace. Its two lobes are narrowed or tapered, and an anal notch is present. The plastral formula is: abd > hum > fem > an >< gul > pect; the paired gular scutes do not extend to the carapacial rim. The bridge is broad (47–49% of carapace length) with the inguinal scute larger than the axillary. The head is small with a nonprojecting snout and a weakly hooked, bicuspid upper jaw. Its divided prefrontal scale is large, as is also the frontal scale which follows. Head, neck, limbs, and tail are dark gray or black. The neck is long with a biconvex 4th cervical vertebra. Anterior surfaces of the forelimbs are covered with rounded, overlapping scales of varying sizes, and the short tail lacks a large terminal scale.

Males are larger and flatter with more concave plastra and slightly thicker tails.

Distribution: *Geochelone nigrita* occurs on Indefatigable (Santa Cruz) Island of the Galapagos Archipelago.

Habitat: The habitat on Indefatigable Island is more moist than that of several of the other islands with tortoises. There are many open grass glades, tree stands, and thickets. Mosses and lichens grow on many rocks, and the tortoises often soak in the numerous mud holes.

Natural History: Most mating occurs between February and April in the inland areas; mating is rare in the arid coastal zone (Carpenter, 1966; Rodhouse et al., 1975). Carpenter (1966) reported that as the male approaches, the female extends her head and he smells it. The female then turns and the male smells her tail and mounts, stretching his neck and tail to the fullest. Intromission follows and males emit bellows or grunts. During the mating season males are quite aggressive toward other males, charging and butting them. Nesting occurs from April (Carpenter, 1966) to December (MacFarland, Villa, and Toro, 1974a). The feeding areas of the southwestern population lack suitable soil for nesting and the females are forced to migrate to the "campos" areas of silty soil and little vegetation in the lowlands. The flask-shaped nests are dug in areas of full sunlight to a depth of 175–300 mm; one or two clutches of 3–16 eggs (usually 9–10) are laid each year. The brittle-shelled eggs are almost spherical (56–63 × 56–58 mm). Incubation may take more than 200 days, and the young usually emerge during the wet season between January and May. Hatchlings have rounded, domed carapaces about 60 mm long with rough, granulated surfaces. Each vertebral and pleural may have a light-brown or gray ring surrounding a darker area.

Indefatigable tortoises may graze on grasses or browse on *Opuntia* cacti or numerous other succulent plants; see Rodhouse et al. (1975) for a list of food plants. When free water is available, they frequently drink; during the dry season they depend on metabolic water from their food.

The tortoises are active throughout the daylight hours. At night they sleep under bushes, beside trees, or under overhanging rocks (Carpenter, 1966).

MacFarland, Villa, and Toro (1974a) estimated the main population of *G. nigrita* in southwest Indefatigable to be 2000–3000 with another 50–100 in a separate population in the east. However, predators (pigs, black rats, cats, goats) destroy large numbers of nests and young in the main population, and MacFarland, Villa, and Toro (1974a) felt that recruitment was too low to replace naturally lost adults. The smaller population is subject to poaching and is declining. A program of nest protection, artificial incubation of eggs, and hatchling raising has begun. The young will be released to repopulate Indefatigable when large enough to avoid predation.

Geochelone phantastica (Van Denburgh, 1907)
Narborough Island tortoise

Recognition: This species is known from only a single, old adult male collected in 1906. It has an extremely saddle-shaped, gray-black carapace (86 cm) with a shallow cervical indentation, and the anterior and posterior marginal scutes strongly upturned. The carapace narrows anteriorly; height at the cervical indentation is more than 53% of the straight carapacial length. Vertebral 1 is longer than broad, the others are broader than long, and the 5th is expanded. Some radiations and growth annuli are visible on the vertebrals and pleurals. Eleven marginals lie on each side, and a single, undivided, upturned supracaudal scale is present. The 1st marginals are expanded and more upturned than in any other Galapagos species (their ventral sides are almost vertical). Lateral marginals are small, but the 8th is not reduced. The gray-black, well-developed plastron is shorter than the carapace with both lobes tapered; the hind lobe bears a slight posterior notch. The plastral formula is: abd > hum > fem > an > gul > pect; the paired gulars do not project beyond the anterior carapacial rim. The bridge is very narrow (approximately 35% of carapace length) with single short axillary and inguinal scutes. The head is of moderate size, with a nonprotruding snout and a weakly hooked upper jaw. Its prefrontal scale is small and divided longitudinally, followed by a single, somewhat larger frontal. The head is gray with a yellow lower jaw and throat. The gray neck is long with the 4th cervical vertebra biconvex; limbs and tail are also gray. Large scales cover the anterior surfaces of the forelimbs. The tail is short and lacks a large terminal scale.

Distribution: *Geochelone phantastica* is known only from Narborough (Fernandina) Island of the Galapagos Islands.

Habitat: The single specimen was collected on the side of a volcano in a rocky area with scattered grass clumps, cacti, bushes, and vines.

Natural History: All we know of the life style of this tortoise was reported by the collector, R. H. Beck (in Van Denburgh, 1914). When collected, on 3 April 1906 in the late afternoon, the tortoise was eating grass, and there was evidence that tortoises on the island had also been feeding on *Jasminocereus* cacti, which has rather strong spines. Beck also remarked that he found a rock along the trail that the male had apparently mounted in an attempt to copulate in the absence of females.

Status of this species is uncertain. Although it has not been seen since 1906, Hendrickson (1965) did find some old droppings and partially eaten *Opuntia* cacti, and a few individuals may still exist on some of the rather inaccessible parts of the island. MacFarland, Villa, and Toro (1974a) pointed out that, regardless of its present status, its fate has been determined by natural causes (probably volcanism), since humans never exploited Narborough Island, and no introduced mammals occur there.

Geochelone vandenburghi (DeSola, 1930)
Volcán Alcedo tortoise

Recognition: The domed, black carapace (to 125 cm) has a shallow cervical indentation, the anterior rim serrated and slightly upturned, and the posterior marginals serrated and flared. It is broad and low anteriorly; height at the cervical indentation is about 36% of the carapacial length. Vertebrals are broader than long with the 5th expanded. Vertebral and pleural scutes are smooth to striated in adults. Each side bears 11 marginals, and the single, undivided supracaudal scute is downturned between the posterior marginals. Anterior marginals are not greatly expanded and their ventral surfaces never vertical. Lateral marginals are vertical or downturned, and the 8th not reduced. The black, well-developed plastron is shorter than the carapace and tapered both anteriorly and posteriorly with no anal notch. The plastral formula is: abd > hum > fem > an >< gul > pect; the paired gulars do not extend beyond the carapacial rim. The bridge is narrow (30–32% of carapace length) with single axillary and inguinal scutes. The head is moderate in size, with a nonprotruding snout and a slightly hooked or bi- or tricuspid upper jaw. Its divided prefrontal and single frontal scales are small. Head and other skin are gray to black. The neck is long with a biconvex 4th cervical vertebra. Anterior surfaces of the forelimbs are covered with large, nonoverlapping scales. The short tail lacks a large terminal scale.

Males have concave plastra and slightly thicker tails.

Distribution: *Geochelone vandenburghi* is restricted to the higher slopes and crater of Volcán Alcedo on central Albemarle (Isabela) Island, Galapagos.

Habitat: Grassy slopes and thickets; *Geochelone vandenburghi* seems to enjoy soaking in mud pools formed from geyser runoff. It is extremely difficult to get through the thick vegetational band near the rim of the outer slope, thus these tortoises have been little disturbed, and 3000–5000 individuals exist (MacFarland, Villa, and Toro, 1974a), the largest population of tortoises on the archipelago.

Natural History: Copulation probably occurs in almost every month, and heavy mating activity has been observed in March (Pritchard, 1979) and April (DeSola, 1930). Males are quite vocal while courting or mounted, emitting deep, basal roars (DeSola, 1930). DeSola (1930:79–80) described the mating act as follows:

Making his advances he carefully approaches and observes her and if she shows any signs of response, i.e., as approach toward him, he will quicken his pace and commence the deeply resounding guttural tortoise shout. He collides against her heavily in a manner that appears fierce, bumping her carapace with his own for about five to eight minutes and often for longer periods. During this time he often nips at her legs but in so doing she never retracts her limbs, however brutal his attack may seem. Again he crashes against her while she views his antics unheeding.

This constant concussion, appearing painful to the quiet observer, continues and then another hoarse bellow follows. Slowly but persistently he cleaves behind her and awkwardly mounts her from the posterior extremity. Inserting his penis (which before had been concealed and now protrudes from the cloacal vent) into her dilating cloaca, he stretches forth his long thick neck with its heavy head and straddles forward over the hinder neural and costal shields of her carapace. Stretching and tensely holding this equilibrium so difficultly obtained, as he is now fully mounted in a semihorizontal, somewhat slanting, spread-eagle position, the first spasmodic tupping action of their congress begins. Its preliminary jerky motion almost takes the observer unawares. Opening his strangely peculiar and diabolical face he gives vent to another yell, which sounds more shrill and piercing than the others.

The female all the while is crouched with forelegs retracted and hinder limbs stretched strongly outward, uplifting and supporting his great weight and bulk.

Natural foods include mostly grasses and sedges (Fowler de Neira and Johnson, 1985), but forbs, cacti, and fruits are also eaten when available. MacFarland and Reeder (1974) observed two species of Darwin's finches, *Geospiza fuliginosa* and *G. fortis*, remove ticks from *Geochelone vandenburghi*, *G. ephippium* and *G.*

nigrita. They also showed a photograph indicating aggressive behavior between *G. vandenburghi* males.

Geochelone vicina (Günther, 1875)
Iguana Cove tortoise

Recognition: This is another species with a thick, heavy shell intermediate between saddlebacked and domed, and not appreciably narrowed anteriorly. Its black carapace (to 125 cm) has a shallow cervical indentation, little upturning of the anterior rim, and the posterior marginals somewhat serrated and flared. Height at the cervical indentation is less than 45% of the carapacial length. Vertebrals are broader than long; the 5th is expanded. Surfaces of the vertebrals and pleurals are roughened with growth annuli, but may be very smooth in old tortoises. Eleven marginals lie on each side, and the single, undivided supracaudal scute is downturned between the posterior marginals. Anterior marginals have prominent anterior points, are not greatly expanded and never upturned; their ventral surfaces are horizontal. Lateral marginals are vertical or downturned, the 8th is not reduced and has a well-developed dorsal border. The black plastron is well developed, but shorter than the carapace, and narrowed (tapered) at both ends with a posterior notch. The plastral formula is: abd > hum > fem >< an >< gul > pect; the paired gulars do not reach the carapacial rim. The bridge is broad (47–49% of carapace length), and its axillary is smaller than the inguinal. The head is moderate in size with a nonprojecting snout and a bicuspid upper jaw. Its divided prefrontal and single frontal scale are large. Head, neck, limbs, and tail are blackish gray. The neck is long with a biconvex 4th cervical vertebra. The anterior surface of each forelimb is covered with large, irregularly shaped, nonoverlapping or slightly over-

Geochelone vicina

lapping scales. The short tail lacks a large terminal scale.

Males are larger and more saddlebacked; females are more domed.

Distribution: *Geochelone vicina* is found near Cerro Azul on southern Albemarle (Isabel) Island, Galapagos. Its range may overlap that of *G. guntheri*.

Habitat: *Geochelone vicina* lives in various habitats such as grass patches, brush, and woodlands.

Natural History: Natural nesting occurs from late June into November, but has also been observed in captivity from January to April (Throp, 1976). Three to 19 ($\bar{x} = 9$) eggs are laid at one time (Throp, 1976; Pritchard, 1979), and the spherical eggs (diameter 59–65 mm, Van Denburgh, 1914) have brittle shells. Natural foods include grasses and cacti.

The remaining population of *Geochelone vicina* is only about 400–600 tortoises, and, while mating and nesting occur, virtually all nests and hatchlings are destroyed by black rats, pigs, dogs, and cats (MacFarland, Villa, and Toro, 1974a). Eggs are now being brought to the Charles Darwin Station where they are incubated and the resulting hatchlings raised (MacFarland, Villa, and Toro, 1974b).

The species *vicina* and *guntheri* are very closely related and may be sympatric. Van Denburgh (1914) had difficulty distinguishing between them, and MacFarland, Villa, and Toro (1974a) thought they may eventually be combined. Perhaps they represent two subspecies of a slightly variable species, *G. vicina* (by priority).

Testudo Linnaeus, 1758

These five small to medium-sized tortoises are found predominantly around the Mediterranean, but range through the Middle East to the Caspian Sea, Afghanistan, and Pakistan. The hingeless carapace is usually domed with descending sides; its posterior rim may or may not be flared and serrated. Anterior neural bones are alternately octagonal or four sided. There is usually only a single suprapygal bone, but if two are present, they are separated by a straight transverse suture. Normally 11 marginals lie on each side; the supracaudal scute may be single or divided. No submarginal scutes are present, but there is a narrow cervical scute. The adult plastron of both sexes has a weak hinge between the abdominal and femoral scutes in four of the five species, and a shallow anal notch is present. The paired gulars are not greatly thickened and projecting. The entoplastron is usually anterior to the humero-pectoral seam. The short axillary and inguinal buttresses are strong, but barely, at best, reach the costal bones. The skull is short with a hooked, usually tricuspid, upper jaw. Triturating surfaces of the maxillae usually have a weak ridge, but no ridge occurs on the premaxillae. The maxillae do not contribute to the roof of the palate, and the anterior orbito-nasal foramina are usually small (large in one species). The temporal arch is relatively weak. The prootic bone is typically concealed dorsally and anteriorly by the parietal, and the quadrate encloses the stapes in four of the five species. Limbs are clublike with unwebbed toes; four species have five foreclaws, one has four. See Loveridge and Williams (1957) for discussions of variations within the genus.

The diploid chromosome number is 52 (20 metacentric-submetacentric chromosomes, 10 subtelocentric, and 22 acrocentric or telocentric; Falcoff and Fanconnier, 1965; Stock, 1972).

The genus has been subdivided into three subgenera (Loveridge and Williams, 1957; Auffenberg, 1974). (1) Subgenus *Testudo* Linnaeus, 1758 includes the living

Key to the species of *Testudo*:

1a. Adult plastron lacking movable hinge between abdominal and femoral scutes; carapace flattened, about as broad as long; four claws on each forefoot: **T. horsfieldi**

 b. Adult plastron with movable hinge between abdominal and femoral scutes; carapace domed with descending sides, much longer than broad; usually five claws on each forefoot 2

2a. Supracaudal usually divided; no enlarged tubercles on thigh; 5–10 longitudinal rows of small scales on anterior surface of foreleg; tail ending in horny, enlarged scale: **T. hermanni**

 b. Supracaudal usually single; enlarged tubercles usually present on thigh; 3–6 longitudinal rows of large scales on anterior surface of foreleg; tail with or without enlarged, horny scale 3

3a. Hindside of thigh with a large conical tubercle; no enlarged scale on tail: **T. graeca**

 b. Hindside of thigh lacking enlarged tubercle; tail usually with an enlarged terminal scale 4

4a. Supracaudal and posterior marginals greatly flared; carapace elongated (> 20 cm); usually 4–5 longitudinal rows of enlarged scales on anterior surface of foreleg: **T. marginata**

 b. Only supracaudal scute flared to rear; carapace short (< 14 cm); usually only 3 longitudinal rows of enlarged scales on anterior surface of foreleg: **T. kleinmanni**

species *T. graeca*, *T. hermanni*, and *T. marginata* and several fossil tortoises. These have a weak hinge on the plastron, a single or divided supracaudal scute, maxillae with a weak ridge, orbito-nasal formina small and partially concealed, the quadrate surrounding the stapes, and usually five foreclaws; a horny, enlarged scale may be present at the tip of the tail. (2) Subgenus *Pseudotestudo* Loveridge and Williams, 1957 includes only *T. kleinmanni*, which has a hinged plastron, the supracaudal scute usually single, no ridge on the maxillae, the orbito-nasal foramina large and not concealed, the quadrate not completely surrounding the stapes, usually five foreclaws, and no horny tail scale. (3) Subgenus *Agrionemys* Khozatsky and Mlynarski, 1966 includes only *T. horsfieldi*, which lacks a movable plastral hinge (but has the suture/sulcus relationships as in other *Testudo*), has a single supracaudal scute, a moderate ridge on the maxillae, small and concealed orbito-nasal foramina, the quadrate enclosing the stapes, four foreclaws, and a horny tail scale.

Testudo graeca Linnaeus, 1758

Spur-thighed tortoise

Recognition: The rounded carapace (to 30 cm) is domed, highest behind the center, with abruptly descending sides, a broad cervical notch, and the posterior rim slightly to greatly flared and slightly serrated. The juvenile carapace may be slightly flattened. The cervical scute is generally long and narrow. Vertebrals are broader than long; the 5th is expanded, but narrower than the 4th (the 5th vertebral is much broader than the 4th in *T. hermanni*). Vertebral and pleural areolae are rarely raised, and the shell appears smooth compared to the lumpy *T. hermanni*. Eleven marginals lie on each side, and a single undivided supracaudal scute is present. Carapacial color varies from yellow or tan with black or dark-brown blotching to totally gray or black. The plastron is well devel-

Testudo graeca graeca

oped. Its forelobe is not or only slightly upturned and gently tapers toward the front; it is shorter and narrower than the hind lobe, which bears a posterior notch. The plastral formula is: abd > hum >< gul >< an >< pect > fem; the paired gular scutes are slightly thickened, but do not or only slightly extend beyond the carapacial rim. The bridge is broad, with a small axillary and one or two small or moderate inguinals that do not touch the femoral. Plastron and bridge are yellow to greenish yellow, brown, or gray, and usually with some dark-brown or black spotting on the plastron. The head is moderate in size with a nonprotruding snout and a weakly hooked, tricuspid upper jaw. Its prefrontal is rarely divided longitudinally, but the large frontal may be subdivided; other head scales are small. The head varies from yellow to brown, gray, or black, with or without dark spotting. Three to seven longitudinal rows of large, overlapping scales lie on the anterior surface of each foreleg. The heel may lack spurlike scales, but each thigh has a very large, conical tubercle. The tail lacks a large terminal scale. Neck, limbs, and tail are yellowish brown to gray.

Males have longer, thicker tails with the vent beyond the carapacial rim.

Distribution: *Testudo graeca* ranges from southern Spain eastward across coastal northern Africa to Israel and Syria, and from northern Greece, Romania, and Bulgaria eastward through Turkey to Transcaucasus, Russia, and Iran.

Geographic Variation: Four subspecies have been described. *Testudo graeca graeca* Linnaeus, 1758 lives in northern Africa from the Cyrenaica Peninsula of Libya westward along the Mediterranean Coast to Tunisia, Algeria and Morocco, and also on the Pityusae Islands of the Balearics, and in southern Spain. It is small (to 20 cm), flattened dorsally, lacks much expansion and flaring of the posterior marginals, and has a dark head. *T. g. terrestris* Forskål, 1775 ranges from eastern Libya and northern Egypt eastward through the Sinai Desert and Israel to Syria. It is small (to 20 cm), elevated dorsally, has some expansion but little flaring of the posterior marginals, and yellow pigmentation on the side and top of the head. *T. g. ibera* Pallas, 1814 occurs from northeastern Greece, Romania, Bulgaria, and Turkey eastward to the Transcaucasus of Russia and central Iran. It is larger (to 30 cm), is slightly flattened dorsally, has some expansion and slight flaring of the posterior marginals, and a unicolored tan or gray head. *T. g. zarudnyi* Nikolskii, 1896 is found only on the eastern part of the Central Iranian Plateau and adjacent Afghanistan. It is large (to 27.5 cm), elongated, and elevated dorsally, with much expansion, flaring, serration, and slight upturning of the posterior marginals, and a plain gray-brown head.

Habitat: *Testudo graeca* lives from sea level to over 3000 m on plateaus and mountains. It most often occurs on dry open steppes, barren hillsides, and wastelands where vegetation varies from sea dune grasses to scrub thorn brush or dry woodlands.

Natural History: Mating takes place from April to July; during courtship males pursue, ram, and push females about, often biting at their heads and limbs. Males emit whistlelike noises while copulating. Nesting may also occur from April to July, but usually in May or June. The nest is a flask-shaped cavity 10–12 cm deep, and two to seven eggs compose a typical clutch. The white eggs are ellipsoidal (30–42.5 × 24.5–35 mm) with brittle shells. In the southern parts of the range, the young may emerge from August to October, but Nikolskii (1915) reported they overwinter in the nest in Russia. Hatchlings are yellow with dark-brown or black pigment along the seams of their rounded carapaces (30–35 mm). Tortoises seem to be most active in the morning but seek refuge during the midday heat.

Food in the wild includes many plants (grasses, sedges, clover, trefoil, *Echweria, Sedum*), and, according to Loveridge and Williams (1957), yellow flowers such as buttercups and dandelions are relished. In the northern parts of the range, *T. graeca* is forced to hibernate, and in the southern warmer areas it may estivate in the summer.

This species is one of the most popular turtles in the European pet trade, and the hundreds of thousands sold on the market have severely depleted its numbers in the wild, causing extinction of some local populations. Fortunately, new regulations have curtailed or slowed much of this trade, assuring *T. graeca* will survive in the wild.

Testudo hermanni Gmelin, 1789

Hermann's tortoise

Recognition: The rounded carapace (to 20 cm) is domed, highest behind the center, with abruptly descending sides, a slight cervical notch, and the posterior rim downturned and slightly serrated. A long narrow cervical scute is present. Vertebrals are broader than long; the 5th is expanded. Vertebral and pleural areolae are often raised and surrounded by growth annuli, giving the shell a lumpy appearance. The carapace is variable in color, ranging from yellow, olive, or orange to dark brown; lighter individuals contain variable amounts of dark blotching. The plastron is well developed with a forelobe that is slightly upturned, tapered toward the front, and shorter and narrower than the hind lobe. The hind lobe bears a

Testudo hermanni hermanni

posterior notch. The plastral formula is: abd > hum > an >< gul > fem > pect; the paired gulars are thickened, but extend only slightly past the carapacial rim. The yellow bridge is broad, with a small axillary and small inguinal scute (which may be absent) that does not touch the femoral. The plastron is dark brown or black with a yellow border and midseam. The head is moderate in size with a nonprotruding snout and a hooked upper jaw. Its large prefrontal and frontal scales may be longitudinally divided or subdivided; other head scales are small. The head is brown or black; the chin may be lighter in color. Five to ten longitudinal rows of small, nonoverlapping scales lie on the anterior surface of each foreleg. The large scales on the heel are not greatly elongated, and there are no tubercles on the thighs. The tail has a large terminal scale.

Males have longer, thicker tails with the vent near the tip.

Distribution: *Testudo hermanni* ranges across southern Europe from northeastern Spain to the Balkans, Bulgaria, and eastern Turkey. It also lives on the Ionian Islands, Balearics, Corsica, Sardina, Elba, Pianosa, and Sicily.

Geographic Variation: Two subspecies are recognized. *Testudo hermanni hermanni* Gmelin, 1789 is found in western Turkey, Bulgaria, southern Romania, Albania, Greece, Yugoslavia, southern Italy, and Sicily. It has a long (males 19, females 20 cm), flat, dull-colored carapace, and lacks a light spot behind and beneath each eye. *T. h. robertmertensi* Wermuth, 1952 lives in northwestern Italy, southern France, northeastern Spain, and on the Balearic Islands, Corsica, and Sardina. It is small (males 14 cm, females 16.5 cm), high domed, brightly colored with contrasting areas of yellow and black, and has a bright spot behind and beneath each eye.

Habitat: *Testudo hermanni* occurs in a variety of dry meadows, scrub hillsides, woodlands, and on rocky slopes, and seems to prefer dense vegetation; however, moist places are avoided.

Natural History: In the wild, most courtship and mating occur in the spring, but in captivity, copulation may continue through the summer. During courtship the male pursues the female, rams her shell, and bites at her head and legs. While mounted, males often emit squeaklike grunts. Nesting occurs in May and June, and the flask-shaped nests are 7–8 cm deep. Two to 12 eggs may be laid in a clutch, and possibly several clutches are laid each season. Eggs are pinkish white, and elongated to spherical (28–35 × 21–26 mm) with brittle shells. Incubation takes about 90 days and the young emerge in August and September. Hatchlings have rounded gray-brown or black and yellow carapaces (31–33 mm).

Although it feeds on a variety of plants, it seems to prefer legumes (beans, clovers, lupines), which formed 90% of the foods taken by tortoises examined by Meek and Inskeep (1981). Ranunculaceae (buttercups, etc.) amounted to 7% and Graminaceae (grasses) 3% of their sample. Animal foods sometimes eaten include earthworms, snails, slugs, insects, carrion, and feces. Most feeding is done in the late afternoon and evening. Aggression may occur between rival males, especially during the breeding season, resulting in ramming contests.

Testudo horsfieldi Gray, 1844

Central Asian tortoise

Recognition: The rounded carapace (to 22 cm) is almost as broad as long and flattened dorsally (about twice as long as high). The highest point is near the center, and the sides are not abruptly descending. A weak cervical notch is present, and the posterior rim is not greatly flared, but may be slightly serrated. The cervical scute is usually long and narrow, but the vertebrals are all broader than long; the 5th is expanded. Areolae of the vertebral and pleural scutes are slightly raised and surrounded by growth annuli. There are usually 11 marginals on each side, and the supracaudal is undivided. The carapace ranges in color from totally light brown to yellowish brown with extensive dark-brown pigment on each scute. The plastron is well developed, but lacks the movable hinge between the abdominal and femoral scutes that occurs in other *Testudo* species. This has been considered a primitive character, but is more likely a derived condition. The plastral forelobe is not upturned, tapers to the front,

Testudo horsfieldi

and is slightly longer and narrower than the hind lobe, which bears a posterior notch. The plastral formula is: abd > gul >< hum >< an > fem > pect; the paired gulars are thickened and extend slightly beyond the carapacial rim. The broad bridge has a small axillary and usually only a single small inguinal that does not touch the femoral. The bridge is usually yellow, but the plastron is black with yellow seams. The head is moderate in size with a nonprotruding snout and a hooked, often tricuspid upper jaw. Its large prefrontal is divided longitudinally; the frontal is large and usually undivided, and the other head scales are small. The head is yellowish brown with dark jaws. Limbs are also yellowish brown. Five or six longitudinal rows of large overlapping scales lie on the anterior surface of each foreleg. Spurlike scales are present on each heel, and blunt tubercles on each thigh. The tail ends in a horny claw.

Males have longer, thicker tails with the vent close to the tip.

Distribution: *Testudo horsfieldi* ranges from the Caspian Sea in southeastern Russia southward through Iran, Afghanistan, and Pakistan to northern and western Baluchistan.

Habitat: The Central Asian tortoise is found in a variety of dry habitats, ranging from rocky deserts, sandy and loamy steppes, to rocky hillsides. In these habitats, they most frequently occur in oases, along brooks, or near springs, where vegetation is more lush. In the north they hibernate in plugged burrows during the winter, and in the south they may estivate in the warmer months.

Natural History: In Russia, mating takes place in the spring following emergence from the winter retreats. Nests are dug in June with hatching in August or September in the south; the young may overwinter in the nest and not emerge until the next June in Russia (Nikolskii, 1915). Possibly up to four clutches of three to five eggs are laid each year. Eggs are elliptical (41–50 × 26–35 mm) with a hard smooth shell. The yellow and black hatchlings have rounded carapaces (29–34 mm) with posteriorly serrated rims.

Natural foods consist solely of vegetation: grasses, flowers, fleshy leaves of plants, and fruits. These tortoises may not drink, but instead subsist on metabolic water produced from food. During the hot periods of the day and at night, they remain in burrows or under boulders or rock outcroppings.

Testudo kleinmanni Lortet, 1883
Egyptian tortoise

Recognition: This species closely resembles *Testudo graeca*, and has in the past been considered a dwarf race of it. Its domed carapace (to 13.5 cm) is highest behind the center with abruptly descending sides, a cervical notch, and a flared supracaudal. Juveniles are dorsally flattened. A cervical scute is present which is usually long and narrow, but may be broad and triangular. Vertebrals are broader than long; the 5th is expanded. Vertebral and pleural areolae may be raised slightly and surrounded by growth annuli. There are 10 to 12 marginals on each side; the supracaudal is usually undivided and projects beyond the marginals. The carapace is greenish yellow to yellowish brown, and usually has dark seam borders. The plastron is well developed. Its forelobe is upturned, tapered anteriorly, and shorter and narrower than the hind lobe, which has a posterior notch. The plastral formula is: abd > an > pect >< hum > gul > fem; the paired gulars are thickened, and rarely project beyond the carapacial rim. The broad bridge has an axillary and one or two small inguinal scutes, which may touch the femoral scute. Plastron and bridge are yellow with a central dark blotch on each abdominal scute and along the seams. The head is moderate in size with a nonprotruding snout and a weakly hooked, tricuspid upper jaw. Its large prefrontal scale may be longitudinally divided or subdivided; the frontal may be entire or subdivided; other head scales are small. The quadrate almost, but not completely, encloses the stapes, and the orbito-nasal foramina are large. Head and limbs are yellowish brown; the jaws tan. Three (occasionally four) longitudinal rows of large, nonoverlapping scales lie on the anterior surface of each foreleg, and there are five claws on each forefoot. Spurlike scales occur on each heel, but no tubercles on the thighs. The tail lacks a terminal claw.

Males have long tails with the vent toward the tip.

Distribution: *T. kleinmanni* lives in the Cyrenaica Plateau of northeastern Libya, northwestern deserts of the Nile Delta, and the Sinai Desert of Egypt.

Habitat: The Egyptian tortoise inhabits deserts, dry woodlands, and brushy areas of scrub thorn. Loveridge and Williams (1957) remarked that it is active throughout the Egyptian winter, though more sensitive to cold than either *T. graeca* or *T. marginata*.

Natural History: Basically, what we know of the biology of *T. kleinmanni* has been reported in Loveridge and Williams (1957). Courtship and mating occur in September and October; the male emits vocalizations. Nesting takes place from May into July; hatching occurs from September to February. Clutch size has not been determined, but the eggs are spherical to ellipsoidal (28–30 × 22–23 mm) with brittle shells. Hatchlings are yellow with slightly elongated, oval carapaces about 32 mm long.

Food in the wild consists largely of saltwort and sea lavender.

Testudo marginata Schoepff, 1792
Marginated tortoise

Recognition: The elongated carapace (to 30.5 cm) is domed and highest just behind the center with abruptly descending sides, a weak cervical notch, and the serrated posterior rim greatly flared. A narrow cervical scute is present. Vertebrals are wider than long; the 5th is expanded. Areolae of the vertebral and pleural scutes may be raised slightly and are surrounded by growth annuli. Eleven marginals lie on each side; the undivided supracaudal is very flared, as also are the posterior marginals. The carapace is black to dark brown with yellow vertebral and pleural areolae, and a yellow bar on each marginal. The plastron is well developed, and its forelobe upturned, tapered toward the front, and shorter and narrower than the hind lobe, which bears a posterior notch. The plastral formula is: abd > an > hum >< pect >< fem > gul; the paired gulars are thickened and may project slightly beyond the carapacial rim. A single axillary and one or two moderately sized inguinals which may touch the femoral scute occur on the broad bridge. Plastron and bridge are yellow, with some dusky pigment on the bridge and two rows of large black triangular blotches along the plastron. The head is of moderate size with a nonprotruding snout and a hooked upper jaw. Its large prefrontal and frontal scales may be longitudinally divided or subdivided; other head scales are small. The head is black, the chin yellow or gray, and the jaws black or brown.

Limbs are brown to yellowish brown. Four or five longitudinal rows of large overlapping scales lie on the anterior surface of each foreleg; each forefoot has five claws. Spurlike scales lie on each heel, and weak tubercles may occur on the thighs. The tail may lack a terminal claw.

Males have longer, thicker tails with the vent near the tip.

Distribution: This tortoise ranges in Greece from Mt. Olympus southward. It also occurs on some small offshore Greek islands, Skyros and Poros in the Aegean, and Sardinia (introduced?) (Arnold and Burton, 1978).

Habitat: *Testudo marginata* seems to prefer dry scrub woodlands often on hillsides, but it also enters olive groves. It probably hibernates in winter at some localities.

Natural History: Little is known of this tortoise, and most data come from captive individuals. Mating usually occurs in the spring (April to early June), but Clark (1963) observed a wild pair in courtship on the island of Spetsai in the fall (mistakenly reported as *Testudo graeca ibera*, R. J. Clark, pers. comm.). Courtship is aggressive and consists of chasing, butting of the female's shell, and biting of her head and limbs (Hine, 1982). Hine (1982) reported that females respond to the male's gutteral utterances with deliberate head movements from side to side. Eggs are laid during June and July in excavations about 10 cm deep. In captivity, some females fail to dig a nest and instead deposit the eggs at random about their enclosures. Clutches range from 3 to 11 eggs; hatching occurs from late August into October, depending on soil temperatures. Eggs are nearly spherical to ellipsoidal (31–37 × 27–36 mm) with brittle shells, Hatchlings have rounded carapaces (about 35 mm) that lack posterior flaring. The plastral hinge is also nonfunctional. Each carapacial scute has a yellow areolus surrounded by a dark border.

Testudo marginata is herbivorous and will accept a wide variety of grasses, flowers, and fruits in captivity.

Gopherus Rafinesque, 1832

The four living species of North American gopher tortoises are closely related to the extinct genus *Stylemys* (Eocene–Miocene) and show numerous adaptations for burrowing. They have a hingeless, oblong, dorsally flattened carapace which lacks a cervical indentation and may have flared and serrated posterior marginals. The cervical scute is about as broad as long, and there are usually 11 marginals on each side, with a single, undivided supracaudal scute. No submarginal scutes are present. The nuchal bone lacks lateral costiform processes, and there are 11 peripheral bones on each side. Anterior neurals are usually alternately tetragonal or octagonal, but sometimes hexagonal. Two suprapygal bones are present. The hingeless plastron is well developed and has a deep posterior notch. Its gulars are paired and project anteriorly. The entoplastron is anterior to the humero-pectoral seam. Axillary and inguinal buttresses are short and barely touch the costal bones. The skull is short and broad with a nonprojecting snout and a slightly hooked upper jaw. A medial ridge lies on both the premaxillae (a feature shared only with *Stylemys*) and maxillae. The maxillae do not contribute to the roof of the palate, and the anterior orbito-nasal foramina are small and concealed from ventral view. The temporal arch is moderately developed. The prootic bone is well exposed dorsally and anteriorly, and the quadrate bone usually encloses the stapes. In *G. flavomarginatus* and *G. polyphemus*, the sacculus contains a large otolithic structure and the inner ear chambers are hypertrophied, but in *G. agassizii* and *G. berlandieri*, the sacculus contains only a small otolithic mass and the inner ear chambers are only slightly inflated. In *flavomarginatus* and *polyphemus*, the cervical vertebrae are short and stout with enlarged, closely joined pre- and postzygapophyses, and a specialized, interlocking joint is present between cervical 8 and the 1st dorsal vertebra, but in *agassizii* and *berlandieri*, these vertebrae are not appreciably shortened, their pre- and postzygapophyses not enlarged, and the 8th cervical vertebra does not form a specialized, interlocking joint with the 1st dorsal vertebra (Bramble, 1982). The 1st dorsal vertebra is also attached to a distinct, bony strut on the nuchal bone in *flavomarginatus* and *polyphemus*, but not in *agassizii* and *berlandieri*. Instead, there are a small zygapophysis and a neural arch sutured to the 1st neural (Bramble, 1982). Forelimbs are flattened for digging with two to four subradial carpals and five claws; hind limbs are club shaped. No toe webbing occurs, and the tail lacks a large terminal scale.

The diploid chromosome number is 52 (26 macrochromosomes, 26 microchromosomes): 20 metacentric and submetacentric, 10 subtelocentric, 22 acrocentric and telocentric chromosomes (Stock, 1972; Killebrew and McKown, 1978; Dowler and Bickham, 1982).

Key to the species of *Gopherus*:

1a. Distance from base of 1st claw to base of 4th claw on forefoot approximately equal to same measurement on hind foot 2
  b. Distance from base of 1st claw to base of 3d claw on forefoot approximately equal to distance from base of 1st claw to base of 4th claw on hind foot; restricted to southeastern United States: ***Gopherus polyphemus***
2a. Marginals of carapace lighter than rest of shell; carapacial scutes with dark areolae: ***G. flavomarginatus***
  b. Marginals of carapace not lighter than rest of shell; carapacial scutes with light areolae 3
3a. Paired axillary scutes; 3d vertebral scute broadest: ***G. berlandieri***
  b. Single axillary scute; 5th vertebral scute broadest: ***G. agassizii***

Gopherus flavomarginatus Legler, 1959

Bolson tortoise

Recognition: The oblong carapace (to 37 cm) is rather flat topped, low arched, widest behind the center, and lacks a cervical indentation. The posterior marginals are flared and slightly serrate. The cervical scute is about as broad as long. Vertebrals are broader than long; the 1st is usually the narrowest, the 3d broadest, and the 5th laterally expanded. Well-defined growth annuli surround the slightly raised vertebral and pleural areolae, but become worn smooth with age. There are usually 11 marginals on each side, and the single supracaudal scute is undivided and downturned. The carapace varies from pale greenish yellow or lemon yellow to straw colored or brown. Carapacial areolae are dark brown or black; lateral marginals contain much yellow pigment and are always

Gopherus flavomarginatus

paler than the rest of the carapace. Some individuals may also have some dark radial stripes on the carapaces. The plastron is well developed and covers most of the carapacial opening. Its forelobe is as broad as or slightly narrower than the hind lobe, which bears a well-defined posterior notch. The plastral formula is: abd > gul > fem >< hum > pect >< an. The paired gulars are thickened and may project beyond the carapacial rim, especially in males. The bridge is broad with a large axillary scute and one or two inguinals. The plastron is yellow with a dark-brown or black blotch on each scute; these blotches fade with age. The head is broad to moderate with a nonprojecting snout, and the upper jaw may be slightly hooked. The scales on the top of the head are large and irregular; the prefrontal is longitudinally divided, and the following frontal scale may also be subdivided. The iris is yellow to brown. Head, limbs, and tail are yellow to brown; the jaws are tan to brown. The anterior foreleg surface is covered with large, slightly overlapping, brown-centered scales in seven or eight longitudinal rows. The hind limbs contain two large dark-tipped spurs on each thigh.

Males have more concave plastra; longer, more projecting gular scutes; and slightly longer, thicker tails than do females. Also, they have larger integumentary chin glands, especially during the mating season.

Distribution: *Gopherus flavomarginatus* is restricted to an internally drained basin (Bolson de Mapimi) which borders on southeastern Chihuahua, southwestern Coahuila, and northeastern Durango, Mexico, at elevations from 1000 to 1300 m.

Geographic Variation: Morafka (in Bury, 1982) reported that individuals from Durango have much more yellow on the carapace and much less sexual dimorphism in plastral concavity than tortoises from the Sierra del Diablo, Chihuahua.

Habitat: The area is a rather xeric steppe to desert with low humidity and rainfall, and wide temperature fluctuations; vegetation includes grass and dry shrubs, and the soil is usually sandy. Bolson tortoises, like their relative, *Gopherus polyphemus*, are burrowers, digging a single-entranced hole which slopes steeply downward to a depth of 1.5 to 2.5 m and a length of up to 10 m (Morafka; in Bury, 1982). Here they may estivate for periods in summer, and hibernate during cold winters.

Natural History: Courtship and mating take place from May to August. Species and sex recognition is by head bobbing. Males trail the females and mount immediately when the females are caught. Females may also initiate courtship and pursue males. Nesting occurs from May to early July. Egg deposition in captives has been in a hole about 20 cm deep, outside but near the entrance of the burrow, and captive females have become aggressive toward humans at these nest sites. Up to three clutches (usually two) of three to nine white, brittle, almost spherical eggs are laid each year. Hatchlings emerge from late July to October in 75–100 days.

Predominant natural foods include tobossa grass, other grasses, *Opuntia* fruits, herbaceous annuals, and occasionally small insects.

Since populations are relatively small and mostly adult, and their habitat is one which has been slowly drying, *G. flavomarginatus* is considered endangered.

For more information on the species, see Legler and Webb (1961), Morafka (in Bury, 1982), and recent *Proceedings of the Desert Tortoise Council*.

Gopherus polyphemus (Daudin, 1802)

Gopher tortoise

Recognition: The oblong carapace (to 27 cm) is flat topped with abruptly descending sides, and lacks a cervical indentation. It is usually widest in front of the bridge, highest in the sacral region, and may be somewhat constricted behind the bridge. It drops off abruptly to the rear, and the posterior marginals are only slightly serrated. The cervical scute is about as broad as long. Vertebrals are broader than long; the 1st is usually the narrowest, the 3d broadest, and the 5th laterally expanded. Well-defined growth annuli occur on young tortoises, but the carapace becomes smoother with age. There are usually 11 marginals on each side, and the single supracaudal scute is undivided and downturned. The carapace is dark brown to grayish black; some individuals have lighter vertebral and pleural areolae. The yellow to gray plas-

Gopherus polyphemus

tron is well developed and covers much of the carapacial opening. Its forelobe is longer and slightly narrower than the hind lobe, which bears a well-defined posterior notch. The plastral formula is: abd > hum > gul >< fem > pect >< an. The paired gular scutes are thickened, project beyond the carapacial rim (especially in males), and may be upturned. The head is broad to moderate with a nonprojecting snout, and the upper jaws are neither notched nor hooked. The scales on the top of the head are small and irregular; the prefrontal scale is divided longitudinally, and there is a small single frontal. The iris is dark brown. Head, limbs, and tail are grayish black; limb sockets are yellow. The foreleg's anterior surface is covered with large, non- or slightly overlapping scales in seven or eight longitudinal rows. The hind limbs appear small.

Males have more concave plastra and larger integumentary glands under the chin than females have.

Distribution: *Gopherus polyphemus* ranges from South Carolina south along the Atlantic Coastal Plain through Florida, and west along the Gulf Coastal Plain to extreme eastern Louisiana.

Habitat: Gopher tortoises usually live on well-drained, sandy soils in ecotones between broad-leafed woodland and grasslands. Where the soil is deep enough, *G. polyphemus* digs a long, straight burrow of 2.5–6.2 m (Hansen, 1963). There usually is an enlarged chamber at the end of the burrow, where the tortoise sleeps, turns around, estivates during hot dry summers, and hibernates in the winter. Even juveniles dig burrows.

Natural History: Mating occurs in April and May. Courtship apparently consists of several basic sequential activities. The first of these Auffenberg (1966b) termed the "male orientation circle": the male walks in a circle and periodically stops and bobs his head, perhaps to attract the attention of a sexually responsive female. When a female approaches he bobs his head violently, then bites her on the forelegs, head, anterior edge of the carapace, and especially the gular projection. (This is probably the method of sex identification; females do not bite one another or males.) The female then backs in a semicircle, stops, and stretches her hind legs. Later she pivots 180°, so that the posterior part of her shell is nearest the male's head; this may be a primitive form of presentation. The male's attempts to mount—usually unsuccessful—are followed by more biting, which in turn is followed by a successful mounting and coition.

Auffenberg (1969) found that in some instances, just before the male's behavior changed from head bobbing to biting, the female rubbed the side of her face and chin across her outstretched front legs. Both adult males and females of *Gopherus* have a pair of glands on the chin; although small most of the year, these glands usually become swollen in the breeding season. Both sexes have an enlarged scale, more prominent than the rest, on each front leg near the elbow. Scent probably is transferred by the female from her chin to her front legs by means of this scale. Her chin rubbing seems to be a response to the male's head bobbing, and Auffenberg speculated that this action causes the male to change his behavior from bobbing to biting. His bobbing may provide the female with more than just a visual cue: he has glands beneath the lower jawbone, and the bobbing may also serve to spread a scent to signal the female.

Nesting occurs from late April to mid-July, and most nests are located at a considerable distance from the burrow, but some are dug in the burrow entrances. Nests extend to a depth of about 15 cm, and one to nine (usually five) eggs are laid in the single, annual clutch (Iverson, 1980). The white eggs are brittle shelled and almost spherical (38–46 mm, $\bar{\chi}$ = 43, in diameter). Hatching occurs from mid-August through September, and hatchlings have rounded (about 43 mm) carapaces.

Natural foods include large amounts of grasses and leaves, with occasional bits of hard fruits, bones, charcoal, and insects (Carr, 1952). Captives eat lettuce, grasses, dandelions, various fruits, and occasionally hamburger and canned dog food.

Unfortunately, good natural habitats for gopher tortoises are rapidly disappearing throughout its range, and this species should be considered at least threatened. Each state in which *G. polyphemus* occurs now protects it, and this has virtually eliminated it from the pet trade, but it is still eaten by some local people.

Gopherus agassizii (Cooper, 1863)

Desert tortoise

Recognition: The oblong carapace (to 49 cm) is rather flat topped, highest behind the center, has the marginals above the hind limbs flared, and the rear margin serrated. Its cervical scute is about as broad as long. Vertebrals are broader than long; the 1st is narrowest, and the 5th broadest and laterally expanded. Raised growth annuli surround the slightly raised vertebral and pleural areolae. Eleven marginals lie on each side, and the single supracaudal scute is undivided and downturned. The carapace is black to tan, and often the areolae are yellow or orange. The plastron is large and well developed; the paired elongated gulars are forked, upturned, and project anteriorly past the carapacial rim. Its forelobe is longer but slightly narrower than the hind lobe, which bears a deep posterior notch. The plastral formula is: abd > hum > fem >< gul > an >< pect. Only a single axillary scute occurs on the broad bridge. The plastral scutes are black to tan; some may have yellow centers. The head is somewhat rounded with a nonprojecting snout and a slightly hooked upper jaw. The angle between the upper triturating ridges is less than 65°. Dorsal head scales are small and irregularly shaped. Well-developed integumentary glands lie beneath the chin; these become prominent in males during the mating season. The head usually is tan but may be reddish brown; the iris is greenish yellow. Skin of the limbs is brown, that of the sockets and neck yellow. The anterior foreleg surfaces are covered with large, slightly overlapping scales in eight or more rows. The thighs contain numerous conical scales.

Males are larger than females and have longer, thicker tails, longer gular projections, concave plastrons, better developed chin glands, and more massive claws.

Distribution: In the United States, *Gopherus agassizii* ranges through southern Nevada, extreme southwestern Utah, southwestern California, and western Arizona. In Mexico it is found in northern Baja California, western Sonora (including Tiburón Island in the Gulf of California), and northwestern Sinaloa.

Geographic Variation: No subspecies have been described, but tortoises from Sonora may eventually be shown sufficiently different to warrant trinomial distinction.

Habitat: This tortoise inhabits desert oases, washes, canyon bottoms, and rocky hillsides having sandy or gravelly soil; it occurs to an altitude of at least 700 m. Thorn scrub, creosote bushes, and cacti are often present in the habitat.

Gopherus agassizii constructs a burrow, which is its home. The turtle digs by scraping alternately with the forelimbs, and when the hole becomes deep enough the turtle turns around and pushes the dirt out with its shoulders. The burrows are dug in dry, gravelly soil and are located under a bush either in an arroyo bank or at the base of a cliff. In cross section they are somewhat oval; they may be straight, curved, or forked; and many have enlarged chambers. Although sometimes just long enough to admit the tortoise, they are occasionally over 10 m in length. Woodbury and Hardy (1948) reported that two kinds of burrows—dens and summer holes—are used in southwestern Utah. Dens are horizontal tunnels dug in banks of washes, usually for distances of 2.4–4.6 m but occasionally for 6–9 m. Summer holes are scattered over the flats and benches and are dug downward at angles of about 20–40° for a distance of about 1–1.5 m. In Sonora, Mexico, Auffenberg (1969) found that the summer retreat was most commonly a shallow hollow dug into the base of an arroyo wall. Several tortoises may use the same shelter during a single season.

Natural History: Mating begins in the spring and continues into the next fall. Black (1976) reported the following sequence of courtship. The male approaches the indifferent female who moves away. He trails her (if she remains stationary, this trailing stage is unnecessary) and bobs his head at her, increasing the frequency of head bobs with duration. When she is caught, the male continues high-intensity head bobbing while circling her, usually in a counterclockwise direction. The female may still try to move away, or avoid the male by circling around him, but once she stops, the male starts biting her head and forelimbs, and occasionally her carapace. He continues to circle to keep in front of her, and sometimes rams her gular projections with his while giving an open mouth threat. Finally she withdraws into her shell and remains in place. The male then moves behind, always ready to bite at her head, and mounts. Hissing and grunting sounds are produced, but Weaver (1970) thought these are a product of copulatory effort, not auditory signals. Intromission soon follows. Nesting occurs mainly from May through July. Funnel-shaped nests are dug in sandy or friable soils to depths of 15 to 20 cm, and two or three clutches of 2 to 14, normally 5 or 6, eggs may be laid each year. Eggs are elliptical to nearly spherical (41.6–48.7 mm × 34.9–39.6 mm) with rough, brittle shells. Hatching occurs from mid-August to October, with the peak in September and early October; natural incubations vary from 90 to 120 days; hatchlings from late clutches may overwinter in the nest. Hatchlings are pale yellow to brown with rounded carapaces (36–48 mm).

Daily activity is determined by ambient temperature, and in the summer they avoid the hottest period of the day by remaining within their burrows, the only place with temperatures at ground level below the tortoise's lethal range, emerging only in the early morning and at dusk. In the winter, they hibernate in deeper burrows, to which some tortoises make lengthy migrations from their feeding areas.

The serrated jaws are well suited for shredding vegetation. The species is mostly herbivorous, subsisting on various grasses, cacti, and the blossoms of *Encelia canescens* and other desert Compositae. Captives eat radishes, dandelions, watermelon, cantaloupe, lettuce, cabbage, spinach, bananas, figs, peaches, clover, grasses, grapes, apples, bread, cheese, snails, insects, and bird eggshells.

For additional data on this tortoise, the reader is referred to Woodbury and Hardy (1948), Ernst and Barbour (1972), Bury (1982) and the *Proceedings of the Desert Tortoise Council.*

Gopherus berlandieri (Agassiz, 1857)

Texas tortoise

Recognition: The oblong carapace (to 22 cm) is rather flat topped, highest behind the center, and drops off abruptly to a serrated rear margin. The cervical scute is about as broad as long, but may be absent. Vertebrals are broader than long; the 1st is usually narrowest, and the 3d broadest. The carapace has a somewhat rough, ridged appearance, caused by well-marked growth annuli surrounding the slightly raised vertebral and pleural areolae. There are usually 11 marginals on each side, and the single supracaudal scute is undivided and downturned. The carapace is brown, and the areolae may be yellow. The yellow plastron is large and well developed with the paired gular scutes elongated, somewhat forked, and extending anteriorly (and sometimes upward) beyond the carapacial rim. Its forelobe is longer than and as broad as, or slightly narrower than, the hind lobe, which has a deep posterior notch. The plastral formula is: abd > hum > gul > fem > an >< pect. The bridge is broad, usually with two axillary scutes. The head is wedge shaped with a somewhat pointed, nonprojecting snout and a slightly hooked upper jaw. The angle between the upper triturating ridges is usually more than 65° but less than 70°. Dorsal head scales are large and irregularly shaped. Well-developed glands beneath the rami of the lower jaw become prominent in males during the mating season. Head, limbs, and tail are yellowish brown. Anterior foreleg surfaces are covered with large, slightly overlapping scales in seven or eight longitudinal rows. The thighs contain a series of small conical tubercles.

The males have slightly longer and narrower carapaces, a longer and more deeply forked gular projection, a concave plastron, and better developed chin glands.

Distribution: *Gopherus berlandieri* occurs from Texas south of Del Rio and San Antonio through eastern Coahuila and Nuevo León to southern Tamaulipas, Mexico.

Geographic Variation: No subspecies have been described, but populations differ in carapace length.

Habitat: This species is found in habitats ranging from near-desert in Mexico to scrub forests in humid and subtropical parts of southern Texas. Sandy, well-drained soils are preferred, and open scrub woods seem to be specially favored. In southern Texas the species occurs from sea level to 100 or 200 m; in Tamaulipas it ranges up to 884 m. Unlike the other species of *Gopherus* in the United States, *G. berlandieri* usually does not dig an extensive burrow. Instead, it uses its gular projection, forelimbs, and the lateral edges of the shell to push away the surface debris and soil to create a resting place, to which it often returns. This so-called pallet usually is located under a bush or the edge of a clump of cactus and is simply a ramp sloping just steeply enough to accomodate the anterior edge of the shell below the surface. There are usually several pallets within a tortoise's home range. Continued use of a pallet, together with the clearing away of accumulated soil and debris, tends to deepen it; the deepest one found by Auffenberg and Weaver (1969) was 1.5 m deep at its anterior end and 5 m long. They also found that pallet use is seasonal. Pallets in thick brush near Brownsville, Texas, are used throughout the year, but in summer proportionately more tortoises are found in pallets in such places. In winter a greater proportion of tortoises are found in pallets in open brush or in grassland. Texas tortoises also sometimes occupy empty mammal burrows of suitable size, and on occasion further excavate these.

Natural History: The mating season in Texas extends from June to September. The male trails the female, bobbing his head in her direction. When she is caught, he attempts to confront her, face to face. He stops her by biting her head, forefeet, and anterior carapace, and ramming her with his gular projection. She usually pivots to avoid this, but he moves with her. When she stops pivoting and withdraws her head, he continues to push her around. Finally he works around to the rear of her carapace and mounts, with his forefeet on the dorsal surface of her carapace and his hind feet firmly planted on the ground. Intromission follows. Nesting is known to occur in June and July, but the season may extend longer; females have contained shelled eggs in November. One to four eggs are laid in a clutch. Eggs are elongated (40.0–53.7 × 29.0–34.0 mm), white, and brittle shelled. Incubation lasts about 90 days (Judd and McQueen, 1980). Hatchlings are about 40–50 mm in both length and width; their shells are brown and cream in color.

Natural foods include stems, fruits, and flowers of *Opuntia* cacti, grasses, violets, asters, and other plants. Insects, snails, and feces are also eaten. For further data, see Auffenberg and Weaver (1969), Weaver (1970), Ernst and Barbour (1972), and Rose and Judd (in Bury, 1982).

Malacochersus Lindholm, 1929

The single species in this genus, *Malacochersus tornieri*, is the most divergent member of the Testudinidae.

Malacochersus tornieri (Siebenrock, 1903b)

African pancake tortoise

Recognition: The keelless, hingeless carapace (to 17.7 cm) is extremely flattened (some individuals are even swaybacked) and has nearly parallel sides. Lateral marginals are sometimes slightly upturned; posterior marginals are usually smooth in larger adults but may be slightly serrated in juveniles. There are 11 or 12 marginals on each side, and two supracaudals are present. Submarginal scutes are absent. The cervical scute is long and narrow in adults, and the vertebrals are broader than long with the middle three smallest. The 1st vertebral may be slightly pointed anteriorly, and the 5th is flared. There is usually an indentation in the cervical region and, posteriorly, only a single suprapygal bone is present. The most unique feature of the carapace is the retention of the juvenile fenestra (windowlike openings between the carapacial bones) into adulthood (see figure). This condition is found in the hatchling stage of all tortoises, but these openings, at least in the dermal shell, usually close by adulthood. Some resorption of the endochondral ribs occurs in *Malacochersus*, which also occurs in other tortoises, the costal and peripheral bones are thin, and the neural bones are also shortened and weakened. These adaptations allow much flexibility in the carapace and permit the pancake tortoise to crawl

Malacochersus tornieri

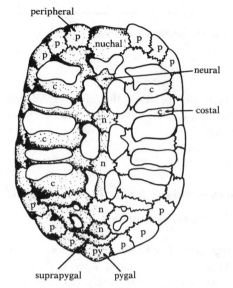

peripheral
nuchal
neural
costal
suprapygal
pygal

Malacochersus tornieri, carapacial bones

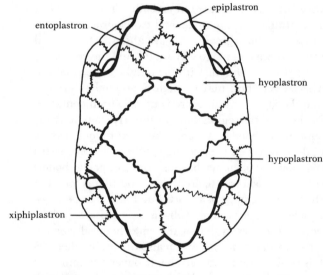

epiplastron
entoplastron
hyoplastron
hypoplastron
xiphiplastron

Malacochersus tornieri, plastral bones

into narrow crevices or under rocks to avoid predators and the heat of the sun. Apart from the fenestration, Loveridge and Williams (1957) felt the shell to be emydinelike and rather primitive. The entire morphology and development of the shell in *Malacochersus* has been well studied by Procter (1922). The carapace is yellow to tan with light areolae on the vertebrals and pleurals, and darker scute borders. Width of these dark scute borders is extremely variable, even within a single clutch, no two individuals look alike, and many have radiating yellow stripes crossing the dark borders. These variable patterns tend to camouflage the turtle in its natural dry habitat. The hingeless plastron is also persistently fenestrated (see figure). Its forelobe is only slightly anteriorly projecting, and the paired gular scutes are much broader than long, giving the appearance of lying transversely rather than longitudinally. There is a slight anterior notch between the gulars, and occasional individuals may also have an additional intergular scute. The hind lobe is posteriorly notched. Two, (rarely three) small axillary scutes and two to four inguinals are present on the bridge. The plastral formula is: abd > hum > pect >< fem > gul > an. The plastron is yellow with solid brown blotches or jagged brown radiations. The yellow-brown head is moderate in size with a nonprotruding snout and a hooked upper jaw which may be bi- or tricuspid. The premaxillae are ridgeless, but the triturating surfaces of the maxillae contain strong ridges. The edges of the jaws are slightly serrated. The prootic is well exposed dorsally and the quadrate encloses the stapes. The supranasal scales are large and usually in contact. There are also a single or double prefrontal scale and a large, sometimes subdivided, frontal scale. The other head scales are small. Limbs and tail are yellow brown. The toes lack webs but bear strong digging claws. The anterior surface of the forelimb is covered with large overlapping scales.

The karyotype is 2n = 52 (Dowler and Bickham, 1982).

Males have longer, thicker tails than do the slightly larger females.

Distribution: *Malacochersus tornieri* occurs in East Africa from Njoro east to Malindi in Kenya, southward in Tanzania to Busisi, Smith Sound, Lake Victoria, and southeastward through Ugogo to Lindi on the Indian Ocean (Loveridge and Williams, 1957).

Geographic Variation: No subspecies are currently recognized. However, Loveridge (1923) described a separate form, *procterae*, from within the range of *M. tornieri*, which has a distinct shell pattern, very elongated and large prefrontal scales, but no supranasals. It is possible this is a valid species, but unfortunately it is represented by only the holotype in the British Museum of Natural History.

Habitat: *Malacochersus tornieri* occupies a small home range in arid scrub and thornbush thickets with abundant boulders, flat rocks, and rocky outcrops on low hills and savannas at altitudes to over 1800 m. It is quite agile and a good climber. When disturbed, the tortoise wedges itself under rocks or within crevices between rocks to such an extent as to make its extraction extremely difficult. The pliability of its shell aids in this, and it was previously thought that the tortoises inflated their lungs to help expand the shell to lock them into place. However, Ireland and Gans (1972) have shown that no significant or sustained increase in intrapulmonary pressure is associated with this wedging action. Instead, the tortoises dig in their foreclaws and rotate the forelegs outward to maintain their position in the rocks.

Natural History: Loveridge and Williams (1957) reported courtship consisted of the male biting the limbs of the female as he trailed her, at times climbing on her back and biting at her head. Males may also crawl beneath females and try to overturn them. Mating was observed by them in January and February. Nesting in Africa seems to occur in July or August, but in captivity this tortoise has produced clutches over most of the year. Hatching in the wild may occur in December. A single elongated (44–48 × 26–28 mm), hard-shelled egg is laid in each clutch, and several clutches may be laid each year. Hatchlings have rounded carapaces about 40 mm long.

Feeding occurs early in the morning and grasses are predominantly eaten. In captivity, *M. tornieri* readily feeds on a variety of vegetables, grasses, and some fruits. It is quite hardy and does well if properly cared for.

Wild and captive specimens often bask, and Loveridge and Williams (1957) reported that they may estivate beneath flat rocks during the hottest months.

Bibliography

Agassiz, L. 1857. Contributions to the natural history of the United States of America, vol. 1, 2. Little, Brown and Co., Boston. 452 p.

Ahl, E. 1932. Beschreibung einer neuen Schildkröte aus Australien. Sitz. Ber. Gesinaturf. Freunde Berlin 1932:127–129.

Ahmad, N. 1955. On edible tortoises and turtles of East Pakistan. E. Bengal Dir. Fish. 18 p.

Alho, C. J. R., and L. F. M. Padua. 1982. Reproductive parameters and nesting behavior of the Amazon turtle *Podocnemis expansa* (Testudinata: Pelomedusidae) in Brazil. Canadian J. Zool. 60:97–103.

Allen, E. R., and W. T. Neill. 1950. The alligator snapping turtle, *Macrochelys temminckii*, in Florida. Spl. Publ. Ross Allen's Rept. Inst. 4:1–15.

Alvarez del Toro, M. 1960. Los reptiles de Chiapas. Inst. Zool. Estado Chiapas. 204 p.

Alvarez del Toro, M., R. A. Mittermeier, and J. B. Iverson. 1979. River turtle in danger. Oryx 15:170–173.

Anderson, J. 1875. Description of some new Asiatic mammals and Chelonia. Ann. Mag. Natur. Hist. (London) (4)16:282–285.

———. 1878 [1879]. Anatomical and zoological researches, comprising an account of the zoological results of the two expeditions to Western Yunnan in 1868 and 1875, &c. 1st vol. text, 2d vol. plates. Reptilia: Chelonia, pp. 705–794. London.

Anderson, S. C. 1979. Synopsis of the turtles, crocodiles, and amphisbaenians of Iran. Proc. California Acad. Sci. 41:501–528.

Andreu, G. C. 1965. Estudio preliminar sobre los tortugas de aqua dulce en Mexico. An. Inst. Nac. Invest. Biol.-Pesq. 1:361–401.

Annandale, N. 1912. The Indian mudturtles (Trionychidae). Rec. Indian Mus. (Calcutta) 7:151–179.

———. 1913. The tortoises of Chota Nagpur. Rec. Indian Mus. (Calcutta) 9:63–78.

———. 1915. Herpetological notes and descriptions. Rec. Indian Mus. (Calcutta) 11:341–347.

Arndt, R. G. 1977. Notes on the natural history of the bog turtle, *Clemmys muhlenbergi* (Schoepff), in Delaware. Chesapeake Sci. 18:67–76.

Arnold, E. N. 1976. Fossil reptiles of Aldabra Atoll, Indian Ocean. Bull. British Mus. Natur. Hist. (Zool.) 29:83–116.

———. 1979. Indian Ocean giant tortoises: their systematics and island adaptations. Phil. Trans. Royal Soc. London, B 286:127–145.

Arnold, E. N., and J. A. Burton. 1978. A field guide to the reptiles and amphibians of Britain and Europe. Collins, London. 272 p.

Ashley, L. M. 1962. Laboratory anatomy of the turtle. Wm. C. Brown Co. Publ., Dubuque, Iowa. 48 p.

Ashton, R. E., Jr., and P. S. Ashton. 1985. Handbook of reptiles and amphibians of Florida. Part 2. Lizards, turtles, and crocodilians. Windward Publ., Inc., Miami, Florida. 191 p.

Atkin, N. B., G. Mattinson, W. Becak, and S. Ohno. 1965. The comparative DNA content of 19 species of placental mammals, reptiles, and birds. Chromosoma 17:1–10.

Auffenberg, W. 1964. Notes on the courtship of the land tortoise *Geochelone travancorica* (Boulenger). J. Bombay Natur. Hist. Soc. 61:247–253.

———. 1965. Sex and species discrimination in two sympatric South American tortoises. Copeia 1965:335–342.

———. 1966a. The carpus of land tortoises (Testudininae). Bull. Florida St. Mus. Biol. Sci. 10:159–192.

———. 1966b. On the courtship of *Gopherus polyphemus*. Herpetologica 22:113–117.

———. 1969. Tortoise behavior and survival. Rand McNally and Co., Chicago. 38 p.

———. 1971. A new fossil tortoise, with remarks on the origin of South American testudines. Copeia 1971:106–117.

———. 1974. Checklist of fossil land tortoises (Testudinidae). Bull. Florida St. Mus. Biol. Sci. 18:121–251.

———. 1976. The genus *Gopherus* (Testudinidae): Pt. I. Osteology and relationships of extant species. Bull. Florida St. Mus. Biol. Sci. 20:47–110.

———. 1977. Display behavior in tortoises. Amer. Zool. 17:241–250.

———. 1978. Courtship and breeding behavior in *Geochelone radiata* (Testudines: Testudinidae). Herpetologica 34:277–287.

———. 1981. The fossil turtles of Olduvai Gorge, Tanzania, Africa. Copeia 1981:509–522.

Auffenberg, W., and R. Franz. 1978a. *Gopherus*. Catalog. Amer. Amphib. Rept. 211:1–2.

———. 1978b. *Gopherus agassizii*. Catalog. Amer. Amphib. Rept. 212:1–2.

———. 1978c. *Gopherus berlandieri*. Catalog. Amer. Amphib. Rept. 213:1–2.

———. 1978d. *Gopherus flavomarginatus*. Catalog. Amer. Amphib. Rept. 214:1–2.

———. 1978e. *Gopherus polyphemus*. Catalog. Amer. Amphib. Rept. 215:1–2.

Auffenberg, W., and J. B. Iverson. 1979. Demography of terrestrial turtles. *In* M. Harless and H. Morlock (eds.), Turtles: perspectives and research, pp. 541–569. Wiley-Interscience, New York.

Auffenberg, W., and W. G. Weaver, Jr. 1969. *Gopherus berlandieri* in southeastern Texas. Bull. Florida St. Mus. Biol. Sci. 13:141–203.

Ayres, M., M. M. Sampaio, R. M. S. Barros, L. B. Dias, and O. R. Cunha. 1969. A karyological study of the turtles from the Brazilian Amazon Region. Cytogenetics 8:401–409.

Babcock, H. L. 1937. A new subspecies of the red-bellied terrapin *Pseudemys rubriventris* (LeConte). Occ. Pap. Boston Soc. Natur. Hist. 8:293–294.

————. 1971. Turtles of the northeastern United States. Dover Publ., Inc., New York. 105 p.

Bachere, E. 1981. Le caryotype de la Tortue verte, *Chelonia mydas* L. dans l' Ocean Indien (Canal du Mozambique). C. R. Acad. Sci., Paris 292:1129–1131.

Baird, S. F., and C. Girard. 1852. Descriptions of new species of reptiles collected by the U. S. Exploring Expedition under the command of Capt. Charles Wilkes, U. S. N. Proc. Acad. Natur. Sci. Philadelphia 6:174–177.

Barbour, T. 1935. A new *Pseudemys* from Cat Island, Bahamas. Occ. Pap. Boston Soc. Natur. Hist. 8:205–206.

Barbour, T., and A. F. Carr. 1938. Another Bahamian freshwater tortoise. Proc. New England Zool. Club. 17:75–76.

————. 1940. Antillean terrapins. Mem. Mus. Comp. Zool. Harvard 54:381–415.

————. 1941. Terrapin from Grand Cayman. Proc. New England Zool. Club. 18:57–60.

Barros, R. M., M. Ayres, M. M. Sampaio, O. Cunha, and F. Assis. 1972. Karyotypes of two subspecies of turtles from the Amazon region of Brazil. Caryologia 25:463–469.

Barros, R. B., M. M. Sampaio, M. F. Assis, M. Ayres, and O. Cunha. 1975. A karyological study of *Geoemyda punctularia punctularia*. Acta Amazonica 5:95–96.

————. 1976. General consideration on the karyotype evolution of Chelonia from the Amazon Region of Brazil. Cytologia 41:559–565.

Barton, A. J., and J. W. Price. 1955. Our knowledge of the bog turtle, *Clemmys muhlenbergii*, surveyed and augmented. Copeia 1955:159–165.

Baur, G. 1888. Osteologische Notizen über Reptilien. Zool. Anz. (296):1–5.

————. 1889a. Die systematische Stellung von *Dermochelys* Blainv. Biol. Centralbl. 9:149–191.

————. 1889b. The relationship of the genus *Deirochelys*. Amer. Natur. 23:1099–1100.

————. 1890a. Two new species of tortoises from the South. Science 16:262–263.

————. 1890b. The gigantic land tortoises of the Galapagos Islands. Amer. Natur. 23:1039–1057.

————. 1891. Notes on the trionychian genus *Pelochelys*. Ann. Mag. Natur. Hist. (London) (6)7:445–446.

————. 1893a. Notes on the classification and taxonomy of the Testudinata. Proc. Amer. Phil. Soc. 31:210–225.

————. 1893b. Two new species of North American Testudinata. Amer. Natur. 27:675–676.

Belkin, D. A., and C. Gans. 1968. An unusual chelonian feeding niche. Ecology 49:768–769.

Bell, T. 1825. A monograph of the tortoises having a moveable sternum, with remarks on their arrangement and affinities. Zool. J., London. 2:299–310.

————. 1827. On two new genera of land tortoises. Trans. Linn. Soc. London 15:392–401.

————. 1828. Descriptions of three new species of land tortoises. Zool. Jour. 3:419–421.

————. 1834. Characters of a new genus of freshwater tortoise. Proc. Zool. Soc. London 1834:17.

Bergoinioux, F. M. 1955. Chelonia. Traite de Paleontologie 5:487–544.

Berry, J. F. 1978. Variation and systematics in the *Kinosternon scorpioides* and *K. leucostomum* complexes (Reptilia: Testudines: Kinosternidae) of Mexico and Central America. Ph.D. Diss., Univ. Utah, Salt Lake City, Utah.

Berry, J. F., and J. M. Legler. 1980. A new turtle (genus *Kinosternon*) from Sonora, Mexico. Contrib. Sci. Natur. Hist. Mus. Los Angeles Co. (325):1–12.

Berry, J. F., and J. B. Iverson. 1980a. A new species of mud turtle, genus *Kinosternon*, from Oaxaca, Mexico. J. Herpetol. 14:313–320.

————. 1980b. *Kinosternon herrerai*. Catalog. Amer. Amphib. Rept. 239:1–2.

Bertl, J., and F. C. Killebrew. 1983. An osteological comparison of *Graptemys caglei* Haynes and McKown and *Graptemys versa* Stejneger (Testudines: Emydidae). Herpetologica 39:375–382.

Bickham, J. W. 1975. A cytosystematic study of turtles in the genera *Clemmys*, *Mauremys* and *Sacalia*. Herpetologica 31:198–204.

————. 1980. *Chrysemys decorata*. Catalog. Amer. Amphib. Rept. 235:1–2.

————. 1981. Two-hundred-million-year-old chromosomes: deceleration of the rate of karyotypic evolution in turtles. Science 212:1291–1293.

Bickham, J. W., and R. J. Baker. 1976a. Chromosome homology and evolution of emydid turtles. Chromosoma 54:201–219.

————. 1976b. Karyotypes of some neotropical turtles. Copeia 1976:703–708.

————. 1979. Canalization model of chromosomal evolution. Bull. Carnegie Mus. Natur. Hist. (13):70–84.

Bickham, J. W., K. A. Bjorndal, M. W. Haiduk, and W. E. Rainey. 1980. The karyotype and chromosomal banding patterns of the green turtle *(Chelonia mydas)*. Copeia 1980: 540–543.

Bickham, J. W., and J. L. Carr. 1983. Taxonomy and phylogeny of the higher catagories of cryptodiran turtles based on a cladistic analysis of chromosomal data. Copeia 1983:918–932.

Bishop, S. C., and F. J. W. Schmidt. 1931. The painted turtles of the genus *Chrysemys*. Zool. Ser. Field Mus. Natur. Hist. (293)18:123–139.

Biswas, S., L. N. Acharjyo, and S. Mohapatra. 1978. Notes on distribution, sexual dimorphism and growth in captivity of *Geochelone elongata* (Blyth). J. Bombay Natur. Hist. Soc. 75:928–930.

Bjorndal, K. A. (ed.). 1981. Biology and conservation of sea turtles. Smithsonian Inst. Press and World Wildlife Fund, Washington, D. C. 583 p.

Black, J. H. 1976. Observations on courtship behavior in the desert tortoise. Great Basin Natur. 36:467–471.

Blackwell, K. 1966. Coital behaviour of the African tortoise *Kinixys erosa*. British J. Herpetol. 3:289.

————. 1968. Some observations on the hatching and growth of the African tortoise *Kinixys homeana*. British J. Herpetol. 4:40.

Blainville, H. M. de. 1816. Prodrome d'une nouvelle distribution systematique de regne animal. Reptiles. Bull. Sci. Soc. Philom. Paris (111):119.

Bleakney, S. 1958. Postglacial dispersal of the turtle *Chrysemys picta*. Herpetologica 14:101–104.

————. 1963. Notes on the distribution and life histories of turtles in Nova Scotia. Canadian Field-Natur. 77:67–76.

Blyth, E. 1853. Notices and descriptions of various reptiles, new or little known. J. Asiatic Soc. Bengal, Calcutta 22:639–655.

————. 1856. Notabilia contained in the collection presented by Capt. Berdmore and Mr. Theobald. Proc. Asiatic Soc., Bengal 24:713–723.

————. 1863. Report on the collections presented by Capt. Berdmore and Mr. Theobald. J. Asiatic Soc. Bengal, Calcutta 32:80–86.

Bocourt, F. 1868. Description de quelques cheloniens

nouveaux appartenant à la fauna Mexicaine. Ann. Sci. Nat. Zool., Paris (5)10:121–122.

Boddaert, P. 1770. Over de Kraakbeenige Schilpad. De Testudine cartilagainea. Amsterdam. 40 p.

Bogert, C. M. 1943. A new box turtle from southeastern Sonora, Mexico. Amer. Mus. Novitates (1226):1–7.

Bojanus, L. H. 1970. Anatome Testudinis Europaeae. Soc. Stud. Amphib. Rept. 178 p.

Bonaparte, C. L. J. L. 1836. Cheloniorum tabula analytica. Rome.

Booth, J., and J. A. Peters. 1972. Behavioural studies on the green turtle (*Chelonia mydas*) in the sea. Animal Behav. 20:808–812.

Boulenger, G. A. 1886. On some South-African tortoises allied to *Testudo geometrica*. Proc. Zool. Soc. London 1886:540–542.

———. 1888a. Description of a new land-tortoise from South Africa, from a specimen living in the Society's Gardens. Proc. Zool. Soc. London 1888:251.

———. 1888b. On the chelydoid chelonians of New Guinea. Ann. Mus. Civicico Storia Natur. Genova. (2)6:449–452.

———. 1889. Catalogue of the chelonians, rhynchocephalians, and crocodiles in the British Museum (Natural History). Taylor and Francis, London. 311 p.

———. 1895. On American box-tortoises. Ann. Mag. Natur. Hist. (London) (6)15:330–331.

———. 1902. Descriptions of new batrachians and reptiles from northwestern Ecuador. Ann. Mag. Natur. Hist. (London) (7)9:51–57.

———. 1903. Report on the batrachians and reptiles. *In* N. Annandale and H. C. Robinson (eds.), Fasciculi Malayensis, anthropological and zoological results of an expedition to Perak and the Siamese Malay States, 1901–1902, Zoology, pp. 131–170. Univ. Press, Liverpool.

———. 1906. Descriptions of new reptiles from Yunnan. Ann. Mag. Natur. Hist. (London) (7)17:567–568.

———. 1907. A new tortoise from Travancore. J. Bombay Natur. Hist. Soc. 17:560–564.

Bour, R. 1973. Contribution à la connaissance de *Phrynops nasutus* (Schweigger: 1812) et *Phrynops tuberculatus* (Luederwaldt: 1926). Description d'une nouvelle sous-espèce originaire du Paraguay, *Phrynops tuberculatus vanderhaegei*. (Testudinata-Pleurodira-Chelidae). Bull. Soc. Zool. France 98:175–190.

———. 1978. Les tortues actuelles de Madagascar (République malgache): liste systématique et description de deux sous-espèces nouvelles (Reptilia-Testudines). Bull. Soc. Et. sci. Anjou N. S. 10:141–154.

———. 1980a. Essai sur la taxinomie des Testudinidae actuels (Reptilia, Chelonii). Bull. Mus. Natn. Hist. Nat., Paris (4)2:541–546.

———. 1980b. Position systématique de *Geoclemys palaeannamitica* Bourret, 1941 (Reptilia-Testudines-Emydidae). Amphibia-Reptilia 1:149–159.

———. 1981. Etude systématique du genre emdémique Malgache *Pyxis* Bell, 1827. (Reptilia, Chelonii). Bull. Mens. Soc. Linn. Lyon. 50(4, 5):132–144, 154–176.

———. 1982. Contribution à la connaissance des Tortues terrestres des Seychelles: définition du genre emdémique et description d'une espèce nouvelle probablement originaire des îles granitiques et au bord de l'extinction. C. R. Acad. Sci. Paris 295:117–122.

———. 1983. Trois populations endémiques du genre *Pelusios* (Reptilia, Chelonii, Pelomedusidae) aux îles Seychelles; relations avec les espèces africaines et malgaches. Bull. Mus. Natn. Hist. Nat., Paris (4)5:343–382.

———. 1984. Note sur *Pelusios williamsi* Laurent, 1965 (Chelonii, Pelomedusinae). Rev. fr. Aquariol. 11:27–32.

———. 1986. Note sur *Pelusios adansonii* (Schweigger, 1812) et sur une nouvelle espèce affine du Kenya (Chelonii, Pelomedusidae). Stud. Palaeocheloniol. 2:23–54.

Bourn, D. 1977. Reproductive study of giant tortoises on Aldabra. J. Zool. (London) 182:27–38.

Bourn, D., and M. Coe. 1978. The size, structure and distribution of the giant tortoise population of Aldabra. Phil. Trans. Royal Soc. London, B 282:139–175.

Bourret, R. 1939. Notes herpetologiques sur l'Indochine Française, XVIII. Reptiles et batraciens reçus au Laboratoire des Science Naturelles de l'Université au cours de l'année 1939. Descriptions de quatre espèces et d'une variété nouvelles. Bull. Gen. Instr. Publ., Hanoi 1939(4):5–39.

———. 1941. Les tortues de l'Indochine. Inst. Oceanograph. l'Indochine. 235 p.

Boycott, R. 1986. A review of *Homopus signatus* (Schoepff) with notes on related species (Cryptodira: Testudinidae). J. Herpetol. Assoc. Africa 32:10–16.

Bramble, D. M. 1974. Emydid shell kinesis: biomechanics and evolution. Copeia 1974:707–727.

———. 1982. *Scaptochelys*: generic revision and evolution of gopher tortoises. Copeia 1982:852–867.

Bramble, D. M., and J. H. Hutchison. 1981. A reevaluation of plastral kinesis in African turtles of the genus *Pelusios*. Herpetologica 37:205–212.

Bramble, D. M., J. H. Hutchison, and J. M. Legler. 1984. Kinosternid shell kinesis: structure, function and evolution. Copeia 1984:456–475.

Broadley, D. G. 1981a. A review of the genus *Pelusios* Wagler in southern Africa (Pleurodira: Pelomedusidae). Occ. Pap. Natl. Mus. Rhodesia, B, Nat. Sci. 6:633–686.

———. 1981b. A review of the populations of *Kinixys* (Testudinidae) occurring in south-eastern Africa. Ann. Cape Prov. Mus. (Natur. Hist.) 13:195–216.

———. 1983. Neural pattern—a neglected taxonomic character in the genus *Pelusios* Wagler (Pleurodira: Pelomedusidae). *In* A.G.J. Rhodin and K. Miyata (eds.). Advances in herpetology and evolutionary biology, pp. 159–168. Mus. Comp. Zool. Harvard, Cambridge, Mass.

Brock, V. E. 1947. The establishment of *Trionyx sinensis* in Hawaii. Copeia 1947:142.

Brongersma, L. C. 1968. Miscellaneous notes on turtles, IIA. *Dermochelys coriacea*. Proc. Kon. Nederl. Acad. Wetensch. C, 72:76–102.

———. 1972. European Atlantic turtles. E. J. Brill, Leiden. 318 p.

Brown, L. E., and D. Moll. 1979. The status of the nearly extinct Illinois mud turtle with recommendations for its conservation. Milwaukee Publ. Mus. Spl. Publ. Biol. Geol. (3):1–49.

Brown, W. S. 1974. Ecology of the aquatic box turtle, *Terrapene coahuila* (Chelonia, Emydidae) in northern Mexico. Bull. Florida St. Mus. Biol. Sci. 19:1–67.

Brumwell, M. J. 1940. Notes on the courtship of the turtle, *Terrapene ornata*. Trans. Kansas Acad. Sci. 43:391.

Bull, J. J., and J. M. Legler. 1980. Karyotypes of side-necked turtles (Testudines: Pleurodira). Canadian J. Zool. 58:828–841.

Bull, J. J., R. G. Moon, and J. M. Legler. 1974. Male heterogamety in kinosternid turtles (genus *Staurotypus*). Cytogenet. Cell. Genet. 13:419–425.

Bull, J. J., and R. C. Vogt. 1979. Temperature-dependent sex determination in turtles. Science 206:1186–1188.

Burger, W. L. 1952. A neglected subspecies of the turtle *Pseudemys scripta*. Report Reelfoot Lake Biol. St. 16:75–79.

Burbidge, A. A., J. A. W. Kirsch, and A. R. Main. 1974. Relationships within the Chelidae (Testudines: Pleurodira) of Australia and New Guinea. Copeia 1974:392–409.

Bury, R. B. 1970. *Clemmys marmorata*. Catalog. Amer. Amphib. Rept. 100:1–3.

———. (ed.). 1982. North American tortoises: conservation and ecology. U. S. Dept. Interior, Fish, Wildl. Serv., Wildlife Res. Rept. (12):1–126.

Bury, R. B., and C. H. Ernst. 1977. *Clemmys*. Catalog. Amer. Amphib. Rept. 203:1–2.

Busack, S. D., and C. H. Ernst. 1980. Variation in Mediterranean populations of *Mauremys* Gray 1869 (Reptilia, Testudines, Emydidae). Ann. Carnegie Mus. Natur. Hist. 49:251–264.

Bustard, H. R. 1972. Sea turtles, their natural history and conservation. Taplinger Publ. Co., New York. 220 p.

Bustard, H. R., J. P. Greenham, and C. Limpus. 1975. Nesting behavior of loggerhead and flatback turtles in Queensland, Australia. Proc. Kon. Nederl. Akad. Wetensch. C, 78:111–122.

Cagle, F. R. 1950. The life history of the slider turtle, *Pseudemys scripta troostii* (Holbrook). Ecol. Monogr. 20:31–54.

———. 1953a. Two new subspecies of *Graptemys pseudogeographica*. Occ. Pap. Mus. Zool. Univ. Michigan (546):1–17.

———. 1953b. The status of the turtle *Graptemys oculifera* (Baur). Zoologica 38:137–144.

———. 1954a. Two new species of the genus *Graptemys*. Tulane Stud. Zool. 1:167–186.

———. 1954b. Observations on the life cycles of painted turtles (genus *Chrysemys*). Amer. Midl. Natur. 52:225–235.

Cahn, A. R. 1937. The turtles of Illinois. Illinois Biol. Monogr. 35:1–218.

Caldwell, D. K. 1962a. Sea turtles in Baja California waters (with special reference to those of the Gulf of California), and the description of a new subspecies of north-eastern Pacific green turtle. Los Angeles Co. Mus. Contrib. Sci. 61:1–31.

———. 1962b. Comments on the nesting behavior of Atlantic loggerhead sea turtles, based primarily on tagging returns. Quart. J. Florida Acad. Sci. 25:287–302.

Caldwell, D. K., A. F. Carr, and L. H. Ogren. 1959. Nesting and migration of the Atlantic loggerhead turtle. Bull. Florida St. Mus. Biol. Sci. 4:295–308.

Caldwell, J. P., and J. T. Collins. 1981. Turtles in Kansas. AMS Publ., Lawrence, Kansas. 67 p.

Cann, J. 1978. Tortoises of Australia. Angus and Robertson, Publ., Sydney. 79 p.

Cansdale, G. 1955. Reptiles of West Africa. Penguin Books, London. 104 p.

Cantor, T. 1842. General features of Chusan, with remarks on the flora and fauna of that island. Ann. Mag. Natur. Hist. (London) 11:481–493.

Carpenter, C. C. 1966. Notes on the behavior and ecology of the Galapagos tortoise on Santa Cruz Island. Proc. Oklahoma Acad. Sci. (1965) 46:28–32.

Carpenter, C. C., and G. W. Ferguson. 1977. Variation and evolution of stereotyped behavior in reptiles. *In* C. Gans and D. W. Tinkle (eds.), Biology of the Reptilia, vol. 7, pp. 335–554. Academic Press, London.

Carr, A. F., Jr. 1937a. A new turtle from Florida, with notes on *Pseudemys floridana mobiliensis* (Holbrook). Occ. Pap. Mus. Zool. Univ. Michigan (348):1–7.

———. 1937b. The status of *Pseudemys scripta* and *Pseudemys troostii*. Herpetologica 1:75–77.

———. 1938a. A new subspecies of *Pseudemys floridana*, with notes on the *floridana* complex. Copeia 1938:105–109.

———. 1938b. Notes on the *Pseudemys scripta* complex. Herpetologica 1:131–135.

———. 1938c. *Pseudemys nelsoni*, a new turtle from Florida. Occ. Pap. Boston Soc. Natur. Hist. 8:305–310.

———. 1942a. A new *Pseudemys* from Sonora, Mexico. Amer. Mus. Novitates (1181):1–4.

———. 1942b. The status of *Pseudemys floridana texana*, with notes on parallelism in *Pseudemys*. Proc. New England Zool. Club 21:69–76.

———. 1952. Handbook of turtles. The turtles of the United States, Canada, and Baja California. Cornell Univ. Press, Ithaca, New York. 542 p.

———. 1967. So excellent a fishe. A natural history of sea turtles. Natur. Hist. Press, Garden City, New York. 248 p.

Carr, A. F., Jr., M. H. Carr, and A. B. Meylan. 1978. The ecology and migrations of sea turtles, 7. The West Caribbean green turtle colony. Bull. Amer. Mus. Natur. Hist. 162:1–46.

Carr, A. F., and P. J. Coleman. 1974. Seafloor spreading theory and the odyssey of the green turtle. Nature 249:128–130.

Carr, A. F., and J. W. Crenshaw, Jr. 1957. A taxonomic reappraisal of the turtle *Pseudemys alabamensis* Baur. Bull. Florida St. Mus. Biol. Sci. 2:25–42.

Carr, A. F., and L. Giovannoli. 1957. The ecology and migrations of sea turtles, 2. Results of field work in Costa Rica, 1955. Amer. Mus. Novitates (1835):1–32.

Carr, A. F., H. Hirth, and L. Ogren. 1966. The ecology and migrations of sea turtles, 6. The hawksbill turtle in the Caribbean Sea. Amer. Mus. Novitates (2248):1–29.

Carr, A. F., and L. T. Marchand. 1942. A new turtle from the Chipola River, Florida. Proc. New England Zool. Club. 20:95–100.

Carr, A. F., A. Meylan, J. Mortimer, K. Bjorndal, and T. Carr. 1982. Surveys of sea turtle populations and habitats in the Western Atlantic. U. S. Dept. Comm., Nova Tech. Mem. NMFS-SEFC-91. 82 p.

Carr, A. F., and L. Ogren. 1959. The ecology and migrations of sea turtles, 3: *Dermochelys* in Costa Rica. Amer. Mus. Novitates (1958):1–29.

———. 1960. The ecology and migrations of sea turtles, 4. The green turtle in the Caribbean Sea. Bull. Amer. Mus. Natur. Hist. 121:1–48.

Carr, A. F., L. Ogren, and C. McVea. 1980. Apparent hibernation by the Atlantic loggerhead turtle *Caretta caretta* off Cape Canaveral, Florida. Biol. Conserv. 19:7–14.

Carr, J. L., and J. W. Bickham. 1981. Sex chromosomes of the Asian black pond turtle, *Siebenrockiella crassicollis* (Testudines: Emydidae). Cytogenet. Cell. Genet. 31:178–183.

———. 1986. Phylogenetic implications of karyotype variation in the Batagurinae (Testudines: Emydidae). Genetica 70:89–106.

Carr, J. L., J. W. Bickham, and R. H. Dean. 1981. The karyotype and chromosomal banding patterns of the Central American river turtle *Dermatemys mawii*. Herpetologica 37:92–95.

Castaño-Mora, O. V., and F. Medem. 1983. Datos preliminares sobre la reproduccion de *Rhinoclemmys melanosterna* Gray (Reptilia: Quelonia: Emydidae). Lozania (Acta Zool. Colombiana) (47):1–5.

Chaudhuri, B. L. 1912. Aquatic tortoises of the Middle Ganges and Brahmaputra. Rec. Indian Mus. (Calcutta) 7:212–214.

Chavez, H., M. Contreras G., and T. P. E. Hernandez D. 1968. On the coast of Tamaulipas. II. Int. Turtle and Tortoise Soc. J. 2(5):16–19, 27–34.

Chessman, B. C. 1986. Diet of the Murray turtle, *Emydura macquarii* (Gray) (Testudines: Chelidae). Australian Wildl. Res. 13:65–69.

Christiansen, J. L., and A. E. Dunham. 1972. Reproduction of the yellow mud turtle *(Kinosternon flavescens flavescens)* in New Mexico. Herpetologica 28:130–137.

Christiansen, J. L., and E. O. Moll. 1973. Latitudinal reproductive variation within a single subspecies of painted turtle, *Chrysemys picta bellii.* Herpetologica 29:152–163.

Clark, R. J. 1963. On the possibility of an autumnal mating in the tortoise *(Testudo graeca ibera).* British J. Herpetol. 3:85–86.

Clay, B. T. 1981. Observations on the breeding biology and behaviour of the long-necked tortoise, *Chelodina oblonga.* J. Roy. Soc. West. Australia 4:27–32.

Cloudsley-Thompson, J. L. 1970. On the biology of the desert tortoise *Testudo sulcata* in Sudan. J. Zoology (London) 160:17–33.

Cogger, H. G. 1975. Reptiles and Amphibians of Australia. A. H. and A. W. Reed, Sydney. 584 p.

Coker, R. E. 1906. The cultivation of the diamond-back terrapin. North Carolina Geol. Surv. Bull. 14:1–69.

Conant, R. 1951a. The reptiles of Ohio. Univ. Notre Dame Press, Notre Dame, Indiana. 284 p.

———. 1951b. The red-bellied terrapin, *Pseudemys rubriventris* (LeConte), in Pennsylvania. Ann. Carnegie Mus. 32:281–290.

———. 1975. A field guide to reptiles and amphibians of eastern and central North America. Houghton Mifflin, Boston. 429 p.

Conant, R., and J. F. Berry. 1978. Turtles of the family Kinosternidae in the southwestern United States and adjacent Mexico: identification and distribution. Amer. Mus. Novitates (2642):1–18.

Conant, R., and C. J. Goin. 1948. A new subspecies of soft-shelled turtle from the Central United States, with comments on the application of the name *Amyda.* Occ. Pap. Mus. Zool. Univ. Michigan (510):1–19.

Congdon, J. D., J. W. Gibbons, and J. L. Greene. 1983. Parental investment in the chicken turtle *(Deirochelys reticularia).* Ecology 64:419–425.

Congdon, J. D., and D. W. Tinkle. 1982. Reproductive energetics of the painted turtle *(Chrysemys picta).* Herpetologica 38:228–237.

Cope, E. D. 1865. Third contribution to the herpetology of tropical America. Proc. Acad. Natur. Sci. Philadelphia 17:185–198.

———. 1870. Seventh contribution to the herpetology of tropical America. Proc. Amer. Philos. Soc. 1869. 11:147–169.

———. 1876. On the Batrachia and Reptilia of Costa Rica with notes on the herpetology and ichthyology of Nicaragua and Peru. J. Acad. Natur. Sci. Philadelphia 2:93–154.

———. 1887. Catalogue of batrachians and reptiles of Central America and Mexico. Bull. U. S. Natl. Mus. 32:1–98.

Cooper, J. G. 1863. Description of *Xerobates agassizii.* Proc. California Acad. Sci. 2:118–123.

Cornalia, E. 1849. Vertebratorum synopsis in Museo Mediolanense Extantium: 301–315.

Cox, W. A., and K. R. Marion. 1978. Observations on the female reproductive cycle and associated phenomena in spring-dwelling populations of *Sternotherus minor* in North Florida (Reptilia:Testudines). Herpetologica 34:20–33.

Cox, W. A., M. C. Nowak, and K. R. Marion. 1980. Observations on courtship and mating behavior in the musk turtle, *Sternotherus minor.* J. Herpetol. 14:200–204.

Crenshaw, J. W. 1965. Serum protein variation in an interspecies hybrid swarm of turtles of the genus *Pseudemys.* Evolution 19:1–15.

Crumly, C. R. 1982a. A cladistic analysis of *Geochelone* using cranial osteology. J. Herpetol. 16:215–234.

———. 1982b. The "parietal" foramen in turtles. J. Herpetol. 16:317–320.

———. 1984. The cranial morphometry of Galapagos tortoises. Proc. California Acad. Sci. 43:111–121.

Cunha, O. R. da. 1970. Una nova subespécie de qualônio, *Kinosternon scorpioides carajasensis* da Serra dos Carajás, Pará. Bol. Mus. Paraense Emilio Geoldi (73):1–11.

Curl, D. A., I. C. Scoones, and M. K. Guy. 1985. The Madagascan tortoise *Geochelone yniphora:* current status and distribution. Biol. Conserv. 34:35–54.

Cuvier, G. 1825. Recherches sur les ossemens fossiles de quadropedes, 3rd. ed. Paris.

Daudin, F. M. 1802. Histoire naturelle, générale et particulière des reptiles. 2. F. Dufart, Paris. 432 p.

Davis, J. D., and C. G. Jackson, Jr. 1973. Notes on the courtship of a captive male *Chrysemys scripta taylori.* Herpetologica 29:62–64.

Dean, R. H., and J. W. Bickham. 1983. *Staurotypus salvini.* Catalog. Amer. Amphib. Rept. 327:1–2.

Degenhardt, W. G., and J. L. Christiansen. 1974. Distribution and habitats of turtles in New Mexico. Southwestern Natur. 19:21–46.

Deraniyagala, P. E. P. 1930. Testudinate evolution. Proc. Zool. Soc. London 48:1057–1070.

———. 1933. The loggerhead turtles (Carettidae) of Ceylon. Ceylon J. Sci. (B) 18:61–72.

———. 1939. The tetrapod reptiles of Ceylon. Vol. 1. Testudinates and crocodilians. Dulau and Co., Ltd., London. 412 p.

de Rooij, N. 1915. The reptiles of the Indo-Australian Archipelago. Vol. 1. Lacertilia, Chelonia, Emydosauria. E. J. Brill, Leiden. 384 p.

Derr, J. N., J. W. Bickham, I. F. Greenbaum, A. G. J. Rhodin, and R. A. Mittermeier. 1987. Biochemical systematies and evolution in the South American turtle genus *Platemys* (Pleurodira: Chelidae). Copeia 1987:370–375.

DeSmet, W. H. O. 1978. The chromosomes of 11 species of Chelonia (Reptilia). Acta Zool. Pathol. Antverp. 70:15–34.

DeSola, R. 1930. The liebespiel of *Testudo vandenburghii,* a new name for the mid-Albemarle Island Galapagos tortoise. Copeia 1930:79–80.

DeWaal, S. W. P. 1980. The Testudines (Reptilia) of the Orange Free State, South Africa. Nav. Nas. Mus. Rep. Suid-Afrika 4:85–91.

Dobie, J. L. 1981. The taxonomic relationship between *Malaclemys* Gray, 1844 and *Graptemys* Agassiz, 1857 (Testudines: Emydidae). Tulane Stud. Zool. Bot. 23:85–102.

Dobie, J. L., and D. R. Jackson. 1979. First fossil record for the diamondback terrapin, *Malaclemys terrapin* (Emydidae), and comments on the fossil record of *Chrysemys nelsoni* (Emydidae). Herpetologica 35:139–145.

Dodd, C. K., Jr. 1982. A controversy surrounding an endangered species listing: the case of the Illinois mud turtle. Smithsonian Herpetol. Inform. Serv. (55):1–22.

Douglass, J. F. 1975. Bibliography of the North American land tortoises (Genus *Gopherus*). U. S. Fish and Wildlife Serv. Spl. Sci. Rept. Wildlife (190):1–60.

———. 1977. Supplement to the bibliography of the North American land tortoises (Genus *Gopherus*). Smithsonian Herpetol. Inform. Serv. (39):1–18.

Dowler, R. C., and J. W. Bickham. 1982. Chromosomal relationships of the tortoises (family Testudinidae). Genetica 58:189–197.

Duda, P. L., and V. K. Gupta. 1981. Courtship and mating behaviour of the Indian soft shell turtle, *Lissemys punctata punctata*. Proc. Indian Acad. Sci. (Anim. Sci.) 90:453–461.

Duellman, W. E. 1963. Amphibians and reptiles of the rainforests of southern El Petén, Guatemala. Univ. Kansas Publ. Mus. Natur. Hist. 15:205–249.

———. 1965. Amphibians and reptiles from the Yucatan Peninsula, Mexico. Univ. Kansas Publ. Mus. Natur. Hist. 15:577–614.

Duerden, J. E. 1906. South African tortoises of the genus *Homopus*, with description of a new species. Rec. Albany Mus. 1:405–411.

Duméril, Andre M. C. 1806. Zoologie analytique, ou methode naturelle de classification des animaux, rendue plus facile a l'aide de tableaux synoptiques. Allais, Paris. 344 p.

Duméril, Andre M. C., and G. Bibron. 1835. Erpétologie générale ou histoire naturelle complète des reptiles. Vol. 2. Libraire Encyclopedique de Roret, Paris. 680 p.

Duméril, Andre M. C., G. Bibron, and Auguste Duméril. 1851. Catalogue méthodique de la collection des reptiles du Muséum d'Histoire Naturelle. Gide and Boudry, Paris. 224 p.

Duméril, Auguste. 1852. Description des reptiles nouveaux ou imparfaitement connus de la collection du Muséum, et remarques sur la classification et les charactéres des reptiles. Arch. Mus. Hist. Nat., Paris 6:209–264.

———. 1856. Note sur les reptiles du Gabon. Rev. Mag. Zool., Paris 2, 8:369–377.

Duméril, Auguste, and M. F. Bocourt. 1870. Etudies sur les reptiles. Mission scientifique au Mexique et dans l'Amerique Centrale . . . recherches zoologiques. Part 3. Imprimerie Imperiale, Paris. 1012 p.

Dunn, E. R. 1930. A new *Geoemyda* from Costa Rica. Proc. New England Zool. Club 12:31–34.

Dunson, W. A., and E. O. Moll. 1980. Osmoregulation in sea water of hatchling emydid turtles, *Callagur borneoensis*, from a Malaysian sea beach. J. Herpetol. 14:31–36.

Eglis, A. 1962. Tortoise behavior: a taxonomic adjunct. Herpetologica 18:1–8.

Engberg, N. J. 1978. The seductive snakeneck *(Chelodina novaeguineas)*. Turtles 1(1):4–7.

Ernst, C. H. 1967a. Intergradation between the painted turtles *Chrysemys picta picta* and *Chrysemys picta dorsalis*. Copeia 1967:131–136.

———. 1967b. Serum protein analysis: a taxonomic tool. Int. Turtle and Tort. Soc. J. 1(3):34–36.

———. 1970a. The status of the painted turtle, *Chrysemys picta*, in Tennessee and Kentucky. J. Herpetol. 4:39–45.

———. 1970b. Reproduction in *Clemmys guttata*. Herpetologica 26:228–232.

———. 1971a. *Chrysemys picta*. Catalog. Amer. Amphib. Rept. 106:1–4.

———. 1971b. Population dynamics and activity cycles of *Chrysemys picta* in southeastern Pennsylvania. J. Herpetol. 5:151–160.

———. 1971c. Observations of the painted turtle, *Chrysemys picta*. J. Herpetol. 5:216–220.

———. 1971d. Observations on the egg and hatchling of the American turtle, *Chrysemys picta*. British J. Herpetol. 4:224–227.

———. 1972a. *Clemmys guttata*. Catalog. Amer. Amphib. Rept. 124:1–2.

———. 1972b. *Clemmys insculpta*. Catalog. Amer. Amphib. Rept. 125:1–2.

———. 1973. The distribution of the turtles of Minnesota. J. Herpetol. 7:42–47.

———. 1974a. *Kinosternon baurii*. Catalog. Amer. Amph. Rept. 161:1–2.

———. 1974b. Observations on the courtship of male *Graptemys pseudogeographica*. J. Herpetol. 8:377–378.

———. 1976. Ecology of the spotted turtle, *Clemmys guttata* (Reptilia, Testudines, Testudinidae), in southeastern Pennsylvania. J. Herpetol. 10:25–33.

———. 1977a. *Clemmys muhlenbergii*. Catalog. Amer. Amphib. Rept. 204:1–2.

———. 1977b. Biological notes on the bog turtle, *Clemmys muhlenbergii*. Herpetologica 33:241–246.

———. 1978. A revision of the neotropical turtle genus *Callopsis* (Testudines: Emydidae: Batagurinae). Herpetologica 34:113–134.

———. 1980a. *Rhinoclemmys annulata*. Catalog. Amer. Amphib. Rept. 250:1–2.

———. 1980b. *Rhinoclemmys areolata*. Catalog. Amer. Amphib. Rept. 251:1–2.

———. 1980c. *Rhinoclemmys funerea*. Catalog. Amer. Amphib. Rept. 263:1–2.

———. 1980d. *Rhinoclemmys nasuta*. Catalog. Amer. Amphib. Rept. 264:1.

———. 1981a. Courtship of the African helmeted turtle *(Pelomedusa subrufa)*. British J. Herpetol. 6:141–142.

———. 1981b. *Rhinoclemmys*. Catalog. Amer. Amphib. Rept. 274:1–2.

———. 1981c. *Rhinoclemmys pulcherrima*. Catalog. Amer. Amphib. Rept. 275:1–2.

———. 1981d. *Rhinoclemmys punctularia*. Catalog. Amer. Amphib. Rept. 276:1–2.

———. 1981e. *Rhinoclemmys rubida*. Catalog. Amer. Amphib. Rept. 277:1–2.

———. 1981f. *Phrynops gibbus*. Catalog. Amer. Amphib. Rept. 279:1–2.

———. 1981g. Courtship behavior of male *Terrapene carolina major* (Reptilia, Testudines, Emydidae). Herpetol. Rev. 12(1):7–8.

———. 1983a. *Platemys pallidipectoris*. Catalog. Amer. Amphib. Rept. 325:1–2.

———. 1983b. *Platemys spixii*. Catalog. Amer. Amphib. Rept. 326:1–2.

———. 1983c. *Platemys radiolata*. Catalog. Amer. Amphib. Rept. 339:1–2.

———. 1984. Geographic variation in the Neotropical turtle, *Platemys platycephala*. J. Herpetol. (1983) 17:345–355.

———. 1987a. *Platemys, Platemys platycephala*. Catalog. Amer. Amphib. Rept. 405:1–4.

———. 1988a. Redescriptions of two Chinese *Cuora* (Testudines: Emydidae). Proc. Biol. Soc. Washington 101:155–161.

———. 1988b. *Cuora mccordi*, a new Chinese box turtle from Guangxi Province. Proc. Biol. Soc. Washington 101:466–470.

Ernst, C. H., and R. W. Barbour. 1972. Turtles of the United States. Univ. Press Kentucky, Lexington. 347 p.

Ernst, C. H., R. W. Barbour, E. M. Ernst, and J. R. Butler. 1974. Subspecific variation and intergradation in Florida *Kinosternon subrubrum*. Herpetologica 30:317–320.

Ernst, C. H., and R. Bruce Bury. 1977. *Clemmys muhlenbergii*. Catalog. Amer. Amphib. Rept. 204:1–2.

——. 1982. *Malaclemys, M. terrapin*. Catalog. Amer. Amphib. Rept. 299:1–4.

Ernst, C. H., and E. M. Ernst. 1969. The turtles of Kentucky. Int. Turtle and Tort. Soc. J. 3(5):13–15.

——. 1971. The taxonomic status and zoogeography of the painted turtle, *Chrysemys picta*, in Pennsylvania. Herpetologica 27:390–396.

——. 1973. Biology of *Chrysemys picta bellii* in southwestern Minnesota. J. Minnesota Acad. Sci. 38:77–80.

——. 1980. Relationships between North American turtles of the *Chrysemys* complex as indicated by their endoparasitic helminths. Proc. Biol. Soc. Washington 93:339–345.

Ernst, C. H., and J. A. Fowler. 1977. Taxonomic status of the turtle, *Chrysemys picta*, in the northern peninsula of Michigan. Proc. Biol. Soc. Washington 90:685–689.

Ernst, C. H., and B. G. Jett. 1969. An intergrade population of *Pseudemys scripta elegans × Pseudemys scripta troosti* in Kentucky. J. Herpetol. 3:103.

Ernst, C. H., and W. P. McCord. 1987. Two new turtles from Southeast Asia. Proc. Biol. Soc. Washington 100:624–628.

Eschscholtz, J. F. 1829. Zoologischer Atlas, enthaltend Abbildungen und Beschreibungen neuer Thierarten, während des Flottcapitains von Kotzebue zweiter Reise um die Welt, auf der Russisch-Kaiserlichen Kriegsschlupp Predpriaetië in den Jahren 1823–1826, part 1, iv+17+15. G. Reimer, Berlin.

Estes, R., and J. H. Hutchison. 1980. Eocene lower vertebrates from Ellesmere Island, Canadian Arctic Archipelago. Palaeogeogr., Palaeoclimatol., Palaeoecol. 30:325–347.

Evans, L. T. 1953. The courtship pattern of the box turtle, *Terrapene c. carolina*. Herpetologica 9:189–192.

Ewert, M. A. 1979. The embryo and its egg: development and natural history. *In* M. Harless and H. Morlock (eds.), Turtles: perspectives and research, pp. 333–413. Wiley-Interscience, New York.

Ewing, H. E. 1943. Continued fertility in female box turtles following mating. Copeia 1943:112–114.

Fahey, K. M. 1980. A taxonomic study of the cooter turtles, *Pseudemys floridana* (LeConte) and *Pseudemys concinna* (LeConte), in the Lower Red River, Atchafalaya River, and Mississippi River basins. Tulane Stud. Zool. Bot. 22:49–66.

Falcoff, E., and B. Fanconnier. 1965. In vitro production of an interferon-like inhibitor of viral multiplication by a poikilothermic animal cell, the tortoise *(Testudo graeca)*. Proc. Soc. Exp. Biol. Med. 118:609–612.

Fan, T. 1931. Preliminary report of reptiles from Yaoshan, Kwangsi, China. Bull. Dept. Biol. Col. Sci. Sun Yatsen Univ. (11):1–154.

Fang, P. W. 1930. Notes on chelonians of Kwangsi, China. Sinensia 1:95–135.

——. 1934. Notes on some chelonians of China. Sinensia 4:145–199.

Feldman, F. 1979. Mating observed in captive New Guinea snake-necked turtles. Notes from NOAH 7(3):9.

Felger, R. S., K. Cliffton, and P. J. Regal. 1976. Winter dormancy in sea turtles: independent discovery and exploitation in the Gulf of California by two local cultures. Science 191:283–285.

Feuer, R. C. 1970. Key to the skulls of Recent adult North and Central American turtles. J. Herpetol. 4:69–75.

Fitch, H. S., and M. V. Plummer. 1975. A preliminary ecological study of the soft-shelled turtle *Trionyx muticus* in the Kansas River. Israel J. Zool. 24:28–42.

Fitzsimons, V. F. M. 1932. Preliminary descriptions of new forms of South African Reptilia and Amphibia, from the Vernay-Lang Kalahari Expedition, 1930. Ann. Transv. Mus. 15:35–40.

Fitzinger, L. 1826. Neue Classification der Reptilien nach ihren naturlichen Verwandschaften. . . . Wien. 66 p.

——. 1835. Entwurf einer systematischen Anordnung der Schildkröten nach den Grundsatzen der naturlichen Methode. Ann. Mus. Wien 1:103–128.

——. 1843. Systema reptilium. Fasciculus primus, Amblyglossae. Vindobonae, Braumuller et Seidel, Vienna. 106 p.

Flower, S. S. 1899. Notes on a second collection of reptiles made in the Malay Peninsula and Siam, from November 1896 to September 1898, with a list of species recorded from these countries. Proc. Zool. Soc. London 1899:600–696.

Folkerts, G. W., and R. H. Mount. 1969. A new subspecies of the turtle *Graptemys nigrinoda* Cagle. Copeia 1969:677–682.

Forbes, W. C., Jr. 1967. A cytological study of the Chelonia. Diss. Abstr. B 27:4169–4170.

Forskål, P. 1775. Descriptiones Animalium . . . Orienti. Haonioe. 164 p.

Fowler, H. W. 1906. Some cold blooded vertebrates from the Florida Keys. Proc. Acad. Natur. Sci. Philadelphia 58:77–113.

Fowler de Neira, L. E., and M. K. Johnson. 1985. Diets of giant tortoises and feral burros on Volcán Alcedo, Galapagos. J. Wildl. Manage. 49:165–169.

Frair, W. 1963. Blood group studies with turtles. Science 140:1412–1414.

——. 1964. Turtle family relationships as determined by serological tests. *In* C. A. Leone (ed.), Taxonomic biochemistry and seriology, pp. 535–544. Ronald Press, New York.

——. 1969. Aging of serum proteins and serology of marine turtles. Serological Mus. Bull. (42):1–10.

——. 1972. Taxonomic relations among chelydrid and kinosternid turtles elucidated by serological tests. Copeia 1972:97–108.

——. 1979. Taxonomic relations among sea turtles elucidated by serological tests. Herpetologica 35:239–244.

——. 1980. Serological survey of pleurodiran turtles. Comp. Biochem. Physiol. 65B:505–511.

——. 1982a. Serological studies of the red turtle, *Phrynops rufipes*. HERP, Bull. New York Herpetol. Soc. 17:4–9.

——. 1982b. Serum electrophoresis and sea turtle classification. Comp. Biochem. Physiol. 72B:1–4.

——. 1982c. Serological studies of *Emys, Emydoidea* and some other testudinid turtles. Copeia 1982:976–978.

Frair, W., R. G. Ackman, and N. Mrosovsky. 1972. Body temperature of *Dermochelys coriacea*: warm turtle from cold water. Science 177:791–793.

Frair, W., R. A. Mittermeier, and A. G. J. Rhodin. 1978. Blood chemistry and relations among *Podocnemis* turtles (Pleurodira, Pelomedusidae). Comp. Biochem. Physiol. 61B:139–143.

Frair, W., and B. K. Shah. 1982. Sea turtle blood serum protein concentrations correlated with carapace length. Comp. Biochem. Physiol. 73A:337–339.

Franz, R., and R. J. Bryant (eds.). 1980. The dilemma of the gopher tortoise—is there a solution. Proc. 1st Ann. Meet. Gopher Tortoise Council. 80 p.

Frazier, J. 1980. Exploitation of marine turtles in the Indian Ocean. Human Ecology 8:329–370.

Freiberg, M. A. 1936. Una nueva tortuga del norte Argentino. Physis; Rev. Soc. Arg. C. Nat. 10:169.

———. 1945. Tortuga del genero *Platemys* Wagler. Physis 20:19–23.

———. 1967. Tortugas de la Argentina. Cienc. Invest. 23:351–363.

———. 1969. Una nueva subspecie de *Pseudemys dorbignyi* (Duméril et Bibron) (Reptilia, Chelonia, Emydidae). Physis 28:299–314.

———. 1973. Dos nuevas tortugas terrestres de Argentina. Bol. Soc. Biol. Concepción 46:81–93.

———. 1981. Turtles of South America. T. F. H. Publ., Inc., Neptune, New Jersey. 125 p.

Fretey, J. 1976. Reproduction de *Kinosternon scorpioides scorpioides* (Linné) (Testudinata, Kinosternidae). Bull. Soc. Zool. France 101:732–733.

———. 1981. Tortues marines de Guyane. World Wildlife Fund, Paris. 136 p.

Fritts, T. H. 1983. Morphometrics of Galapagos tortoises: evolutionary implications. *In* R. I. Bowman, M. Berson, and A. Leviton (eds.), Patterns of evolution in Galapagos organisms, pp. 107–122. Pacific Div. Amer. Assoc. Adv. Sci., San Francisco, California.

———. 1984. Evolutionary divergence of giant tortoises in Galapagos. Biol. J. Linnean Soc. 21:165–176.

Fukada, H. 1965. Breeding habits of some Japanese reptiles (critical review). Bull. Kyoto Gak. Univ. Ser. B 27:65–82.

Gaffney, E. S. 1972. An illustrated glossary of turtle skull momenclature. Amer. Mus. Novitates (2486):1–33.

———. 1975a. Phylogeny of the chleydrid turtles: a study of shared derived characters in the skull. Fieldiana: Geol. 33:157–178.

———. 1975b. A phylogeny and classification of the higher categories of turtles. Bull. Amer. Mus. Natur. Hist. 155:387–436.

———. 1977. The sidenecked turtle family Chelidae: a theory of relationships using shared derived characters. Amer. Mus. Novitates (2620):1–28.

———. 1979a. Fossil chelid turtles of Australia. Amer. Mus. Novitates (2681):1–23.

———. 1979b. Comparative cranial morphology of Recent and fossil turtles. Bull. Amer. Mus. Natur. Hist. 164:65–376.

———. 1984. Historical analysis of theories of chelonian relationship. System. Zool. 33:282–301.

Gaffney, E. S., and R. Zangerl. 1968. A revision of the chelonian genus *Bothremys* (Pleurodira: Pelomedusidae). Fieldiana: Geol. 16:193–239.

Garman, S. 1880. On certain species of Chelonioidae. Bull. Mus. Comp. Zool. 6:123–126.

———. 1884. The reptiles of Burmuda. Bull. U. S. Natl. Mus. (25):285–303.

———. 1891. On a tortoise found in Florida and Cuba, *Cinosternum baurii*. Bull. Essex Inst. 23:141–144.

———. 1917. The Galapagos tortoises. Mem. Mus. Comp. Zool. Harvard 30:261–296.

Gaymer, R. 1968. The Indian Ocean giant tortoise *Testudo gigantea* on Aldabra. J. Zool. (London) 154:341–363.

Geoffroy St.-Hilaire, E. F. 1809. Sur les tortues molles, nouveau genre sous le nom de *Trionyx* et sur la formation des carapaces. Ann. Mus. Hist. Nat. (Paris) 14:1–20.

Gibbons, J. W. 1968. Population structure and survivorship in the painted turtle, *Chrysemys picta*. Copeia 1968:260–268.

———. 1982. Reproductive patterns in freshwater turtles. Herpetologica 38:222–227.

———. 1983. Reproductive characteristics and ecology of the mud turtle, *Kinosternon subrubrum* (Lacépède). Herpetologica 39:254–271.

Gibbons, J. W., and J. W. Coker. 1977. Ecological and life history aspects of the cooter, *Chrysemys floridana* (LeConte). Herpetologica 33:29–33.

Gilmore, C. W. 1922. A new fossil turtle, *Kinosternon arizonense*, from Arizona. Proc. U. S. Natl. Mus. 62(2451):1–8.

Girgis, S. 1961. Aquatic respiration in the common Nile turtle, *Trionyx triunguis* (Forskål). Comp. Biochem. Physiol. 3:206–217.

Glass, B. P., and N. Hartweg. 1951. *Kinosternon murrayi*, a new muskturtle of the *hirtipes* group from Texas. Copeia 1951:50–52.

Glass, H. and W. Meusel. 1969. Die Süsswasserschildkröten Europas. A. Ziemsen Verlag, Wittenberg Lutherstadt. 77 p.

Gmelin, S. G. 1774. Reise durch Russland zur Untersuchung der drey Natur-Reich. Vol. 3. St. Petersburg, Russia.

———. 1789. Caroli a Linné, . . . Systema Naturae . . . . 13th ed., vol. 1.

Goldstein, S., and C. C. Lin. 1972. Somatic cell hybrids between cultured fibroblasts from the Galapagos tortoise and the golden hamster. Exp. Cell. Res. 73:266–269.

Goode, J. 1967. Freshwater tortoises of Australia and New Guinea (in the family Chelidae). Lansdowne Press, Melbourne. 154 p.

Goode, J., and J. Russell. 1968. Incubation of eggs of three species of chelid tortoises, and notes on their embryological development. Australian J. Zool. 16:749–761.

Gorman, G. C. 1973. The chromosomes of the Reptilia, a cytotaxonomic interpretation. *In* A. B. Chiarelli and E. Capanna (eds.), Cytotaxonomy and vertebrate evolution, pp. 349–424. Academic Press, New York.

Grandidier, A. 1867. Liste des reptiles nouveaux découverts, en 1866, sur la côte sud-ouest de Madagascar. Rev. Mag. Zool., Paris (2)19:232–234.

Gray, J. E. 1825. A synopsis of the genera of reptiles and Amphibia, with a description of some new species. Ann. Phil. (new ser.) 10:193–217.

———. 1830–1835. Illustrations of Indian Zoology. Vols. 1, 2. London.

———. 1831a. Synopsis of the species of the Class Reptilia. *In* E. Griffith, The Animal Kingdom . . . by the Baron Cuvier . . . . 110 p.

———. 1831b. Synopsis Reptilium or short descriptions of the species of reptiles. Part 1. Cataphracta, tortoises, crocodiles, and enaliosaurians. London. 85 p.

———. 1831c. Characters of a new genus of freshwater tortoise from China. Proc. Zool. Soc. London 1831:106–107.

———. 1834a. Characters of several new species of freshwater tortoises *(Emys)* from India and China. Proc. Zool. Soc. London 1834:53–54.

———. 1834b. Characters of two new genera of reptiles *(Geoemyda* and *Gehyra)*. Proc. Zool. Soc. London. 1834:99–100.

———. 1841. Appendix. *In* Capt. George Gray's "Journals of an expedition to Northern and Western Australia," pp. 445–446. London.

———. 1842. The Zoological Miscellany. Part 2. London.

———. 1844. Catalogue of the tortoises, crocodiles, and

amphisbaenians in the Collection of the British Museum. London, 80 p.

———. 1847. Description of a new genus of Emydidae. Proc. Zool. Soc. London 1847:55–56.

———. 1849. Description of a new species of box tortoise from Mexico. Proc. Zool. Soc. London (1848) 17:16–17.

———. 1852. Descriptions of a new genus and some new species of tortoises. Proc. Zool. Soc. London 1852:133–135.

———. 1855. Catalogue of shield reptiles in the collection of the British Museum. Part I. Testudinata (tortoises). Taylor and Francis, London. 79 p.

———. 1856. Description of a new species of *Chelodina* from Australia. Proc. Zool. Soc. London 1856:369–371.

———. 1860a. On some new species of Mammalia and tortoises from Cambojia. Ann. Mag. Natur. Hist. (London) (3)6:217–218.

———. 1860b. Description of a new species of *Geoclemmys* from Ecuador. Proc. Zool. Soc. London. 1860:231–232.

———. 1861. On a new species of water-tortoise *(Geoclemmys melanosterna)* from Darien. Proc. Zool. Soc. London 1861:204–205.

———. 1862. Notice of a new species of *Cyclemys* from the Lao Mountains, in Siam. Ann. Mag. Natur. Hist. (London) (3)10:157.

———. 1863a. Observations on the box tortoises, with the descriptions of three new Asiatic species. Proc. Zool. Soc. London 1863:173–177.

———. 1863b. On the species of *Chelymys* from Australia, with the description of a new species. Ann. Mag. Natur. Hist. (London) (3)12:98–99.

———. 1863c. Notice of a new species of *Pelomedusa* from Natal. Ann. Mag. Natur. Hist. (London) (3)12:99–100.

———. 1863d. Notice of a new species of *Batagur* from northwestern India. Proc. Zool. Soc. London 1863:253.

———. 1863e. Notice of a new species of *Kinixys* and other tortoises from Central Africa. Ann. Mag. Natur. Hist. (London) (3)12:381–382.

———. 1864a. Revision of the species of Trionychidae found in Asia and Africa, with the descriptions of some new species. Proc. Zool. Soc. London 1864:76–98.

———. 1864b. Description of a new species of *Staurotypus (S. salvinii)* from Guatemala. Proc. Zool. Soc. London 1864:127–128.

———. 1867. Description of a new Australian tortoise *(Elseya latisternum)*. Ann. Mag. Natur. Hist. (London) (3)20:43–45.

———. 1869a. Notes on the families and genera of tortoises (Testudinata), and on the characters afforded by the study of their skulls. Proc. Zool. Soc. London 1869:165–225.

———. 1869b. Description of *Mauremys laniaria*, a new freshwater tortoise. Proc. Zool. Soc. London 1869:499–500.

———. 1870a. Notice of a new Chilian tortoise *(Testudo chilensis)*. Ann. Mag. Natur. Hist. (London) (4)6:190–191.

———. 1870b. Supplement to the Catalogue of Shield Reptiles in the Collection of the British Museum. Pt. 1, Testudinata (tortoises). Taylor and Francis, London. 120 p.

———. 1871. Notes on Australian freshwater tortoises. Ann Mag. Natur. Hist. (London) (4)8:366.

———. 1872. Notes on the mud-tortoises of India (*Trionyx* Geoffroy). Ann. Mag. Natur. Hist. (London) (4)10:326–340.

———. 1873a. Handlist of the specimens of shield reptiles in the British Museum. London. 124 p.

———. 1873b. On a new freshwater tortoise from Borneo

(Orlitia borneensis). Ann. Mag. Natur. Hist. (London) (4)11:156–157.

———. 1873c. Observations on chelonians, with descriptions of new genera and species. Ann. Mag. Natur. Hist. (London) (4)11:289–308.

———. 1873d. Notes on Chinese mud-tortoises (Trionychidae), with the description of a new species sent to the British Museum by Mr. Swinhoe, and observations on the male organ of this family. Ann. Mag. Natur. Hist. (London) (4)12:156–161.

Greer, A. E., J. D. Lazell, and R. M. Wright. 1973. Anatomical evidence for a countercurrent heat exchange in the leatherback turtle *(Dermochelys coriacea).* Nature 244:181.

Greig, J. C., and P. D. Burdett. 1976. Patterns in the distribution of Southern African terrestrial tortoises (Cryptodira: Testudinidae). Zool. Africana 11:249–273.

Groombridge, B., E. O. Moll, and J. Vijaya. 1983. Rediscovery of a rare Indian turtle. Oryx 17:130–134.

Groves, J. D. 1983. Taxonomic status and zoogeography of the painted turtle *Chrysemys picta* (Testudines: Emydidae), in Maryland. Amer. Midl. Natur. 109:274–279.

Grubb, P. 1971. The growth, ecology, and population structure of giant tortoises on Aldabra. Phil. Trans. Royal Soc. London, B 260:327–372.

———. 1971. Comparative notes on the behavior of *Geochelone sulcata.* Herpetologica 27:328–332.

Günther, A. 1864. Reptiles of British India. Robert Hardwicks, London. 452 p.

———. 1875. Description of the living and extinct races of gigantic land-tortoises. Parts I, II. Introduction, and the tortoises of the Galapagos Islands. Phil. Trans. Royal Soc. London 165:251–284.

———. 1877. The gigantic land-tortoises (living and extinct) in the collection of the British Museum. London. 96 p.

———. 1882. Description of a new species of tortoise *(Geoemyda impressa)* from Siam. Proc. Zool. Soc. London 1882:343–346.

———. 1885. Reptilia and Batrachia, xx + 326 pp. *In* F. D. Godman and O. Salvin, Biologia Centrali-Americana. Dulau and Co., London.

Haiduk, M. W., and J. W. Bickham. 1982. Chromosomal homologies and evolution of testudinoid turtles with emphasis on the systematic placement of *Platysternon.* Copeia 1982:60–66.

Hansen, K. L. 1963. The burrow of the gopher tortoise. Quart. J. Florida Acad. Sci. 26:353–360.

Harding, J. H. 1983. *Platemys platycephala* (twistneck turtle). Reproduction. Herp. Review 14:22.

Harding, J. H., and T. J. Bloomer. 1979. The wood turtle, *Clemmys insculpta* . . . a natural history. HERP, Bull. New York Herpetol. Soc. 15:9–26.

Hardy, L. M., and R. W. McDiarmid. 1969. The amphibians and reptiles of Sinaloa, Mexico. Univ. Kansas Publ. Mus. Natur. Hist. 18:39–252.

Harlan, R. 1827. Description of a land tortoise, from the Galapagos Islands, commonly known as the "Elephant Tortoise." J. Acad. Natur. Sci. Philadelphia 5:284–292.

Harless, M., and H. Morlock. (eds.). 1979. Turtles: perspectives and research. John Wiley & Sons, New York. 695 p.

Hartman, W. L. 1958. Intergradation between two subspecies of painted turtle, genus *Chrysemys.* Copeia 1958:261–265.

Hartweg, N. 1934. Description of a new kinosternid from Yucatan. Occ. Pap. Mus. Zool., Univ. Michigan (277):1–2.

————. 1939a. A new American *Pseudemys.* Occ. Pap. Mus. Zool., Univ. Michigan (397):1–4.

————. 1939b. Further notes on the *Pseudemys scripta* complex. Copeia 1939:55.

Hattan, L. R., and D. H. Gist. 1975. Seminal receptacles in the eastern box turtle, *Terrapene carolina.* Copeia 1975:505–510.

Hay, O. P. 1898. On *Protostega,* the systematic position of *Dermochelys,* and the morphogeny of the chelonian carapace and plastron. Amer. Natur. 32:929–948.

————. 1908. The fossil turtles of North America. Carnegie Inst. Washington Publ. 75:1–568.

Hay, W. P. 1904a. A revision of *Malaclemmys,* a genus of turtles. Bull. U. S. Bur. Fish. 24:1–20.

————. 1904b. On the existing genera of the Trionychidae. Proc. Amer. Phil. Soc. 42:268–274.

Haynes, D. 1976. *Graptemys caglei.* Catalog. Amer. Amphib. Rept. 184:1–2.

Haynes, D., and R. R. McKown. 1974. A new species of map turtle (genus *Graptemys*) from the Guadalupe River system in Texas. Tulane Stud. Zool. Bot. 18:143–152.

Henderson, G. E. (ed.). 1978. Proceedings of the Florida and interregional conference on sea turtles, 24–25 July 1976, Jensen Beach, Florida. Florida Mar. Res. Publ. (33):1–66.

Henderson, J. R. 1912. Preliminary note on a new tortoise from South India. Rec. Indian Mus. (Calcutta) 7:217–218.

Hendrickson, John R. 1958. The green sea turtle, *Chelonia mydas* (Linn.) in Malaya and Sarawak. Proc. Zool. Soc. London 130:455–535.

————. 1965. Reptiles of the Galapagos. Pacific Discovery 18(5):28–36.

————. 1966. The Galapagos tortoises, *Geochelone* Fitzinger 1835 (*Testudo* Linnaeus 1758 in part). *In* R. I. Bowman (ed.), The Galapagos, pp. 252–257. Proc. Symp. Galapagos Int. Sci. Proj. Univ. California Press, Berkeley.

Hewitt, J. 1927. Further descriptions of reptiles and batrachians from South Africa. Rec. Albany Mus. 3:371–415.

————. 1931. Descriptions of some African tortoises. Ann. Natal Mus. 6:461–506.

————. 1933. On the Cape species and subspecies of the genus *Chersinella* Gray. Part I. Ann. Natal Mus. 7:255–293.

————. 1934. On the Cape species and subspecies of the genus *Chersinella* Gray. Part II. Ann. Natal. Mus. 7:303–349.

————. 1935. Some new forms of batrachians and reptiles from South Africa. Rec. Albany Mus. 4:283–357.

Hidalgo, H. 1982. Courtship and mating behavior in *Rhinoclemmys pulcherrima incisa* (Testudines: Emydidae: Batagurinae). Trans. Kansas Acad. Sci. 85:82–95.

Hine, M. L. 1982. Notes on the marginated tortoise (*Testudo marginata*) in Greece and in captivity. British Herpetol. Soc. Bull. (5):35–38.

Hirayama, R. 1984. Cladistic analysis of batagurine turtles (Batagurinae: Emydidae: Testudinoidea); a preliminary result. Stud. Geol. Salmanticensia. Spl. vol. 1:141–157.

Hirth, H. F. 1971. Synopsis of biological data on the green turtle, *Chelonia mydas* (Linnaeus) 1758. FAD Fish. Synopsis (85), page var.

————. 1980a. *Chelonia.* Catalog. Amer. Amphib. Rept. 248:1–2.

————. 1980b. *Chelonia mydas.* Catalog. Amer. Amphib. Rept. 249:1–4.

Hirth, H. F., and A. Carr. 1970. The green turtle in the Gulf of Aden and Seychelles Islands. Ver. Kon. Nederl. Akad. Wetenschap. Nat. 58:1–44.

Hodge, R. P. 1981. *Chelonia mydas agassizi* (Pacific Green Turtle). Herp. Review 12(3):83–84.

Hodsdon, L. A., and J. F. W. Pearson. 1943. Notes on the discovery and biology of two Bahaman fresh-water turtles of the genus *Pseudemys.* Quart. J. Florida Acad. Sci. 6:17–23.

Holbrook, J. E. 1836. North American herpetology; or a description of the reptiles inhabiting the United States. Vol. 1. J. Dobson and Son, Philadelphia. 55 p.

————. 1838. North American herpetology. Vol. 3. J. Dobson and Son, Philadelphia. 122 pp.

Holman, J. A. 1964. Observations on dermatemyid and staurotypine turtles from Veracruz, Mexico. Herpetologica 19:277–279.

————. 1977. Comments on turtles of the genus *Chrysemys* Gray. Herpetologica 33:274–276.

Holman, J. A., and J. H. Harding. 1977. Michigan's turtles. Publ. Mus., Michigan St. Univ., Ed. Bull. (3):1–40.

Holmstrom, W. F., Jr. 1978. Preliminary observations on prey herding in the matamata turtle, *Chelus fimbriatus* (Reptilia, Testudines, Chelidae). J. Herpetol. 12:573–574.

Honegger, R. C. 1972. Die Reptilien-Bestände auf den Galápagos. Inseln 1972. Natur und Museum 102:437–454.

Hoogmoed, M. S., and C. R. Crumly. 1984. Land tortoise types in the Rijksmuseum van Natuurlijke Historie with comments on nomenclature and systematics (Reptilia: Testudines: Testudinidae). Zool. Med. Leiden 58:241–259.

Hornell, J. 1927. The turtle fisheries of the Seychelles Islands. H. M. Stationery Office, London, 55 p.

Houseal, T. W., J. W. Bickham, and M. D. Springer. 1982. Geographic variation in the yellow mud turtle, *Kinosternon flavescens.* Copeia 1982:567–580.

Hsu, H. E. 1930. Preliminary note on a new variety of *Cyclemys flavomarginata* from China. Contr. Biol. Lab. Sci. Soc. China, Zool. Ser. 6(1):1–7.

Huang, C. C., and H. F. Clark. 1969. Chromosome studies of the cultured cells of two species of side-necked turtles (*Podocnemis unifilis* and *P. expansa*). Chromosoma (Berlin) 26:245–253.

Hughes, G. R. 1974. The sea turtles of south-east Africa. I, II. S. African Assoc. Mar. Biol. Res., Ocean. Res. Inst. Invest. Rep. (35, 36).

Hulse, A. C. 1982. Reproduction and population structure in the turtle, *Kinosternon sonoriense.* Southwest. Natur. 27:447–456.

Hunt, T. J. 1958. The ordinal name for tortoises, terrapins and turtles. Herpetologica 14:148–150.

Hutchison, J. H., and D. M. Bramble. 1981. Homology of the plastral scales of the Kinosternidae and related turtles. Herpetologica 37:73–85.

Inchaustegui Miranda, S. J. 1973. Las tortugas Dominicanas de agua dulce *Chrysemys decussata vicina* y *Chrysemys decorata* (Testudinata, Emydidae). Thesis, Univ. Auton. Santo Domingo.

Ireland, L. C., and C. Gans. 1972. The adaptive significance of the flexible shell of the tortoise *Malacochersus tornieri.* Anim. Behav. 20:778–781.

Iverson, J. B. 1976a. The genus *Kinosternon* in Belize (Testudines: Kinosternidae). Herpetologica 32:258–262.

————. 1976b. *Kinosternon sonoriense.* Catalog. Amer. Amphib. Rept. 176:1–2.

————. 1977a. *Kinosternon subrubrum.* Catalog. Amer. Amphib. Rept. 193:1–4.

———. 1977b. *Sternotherus depressus*. Catalog. Amer. Amphib. Rept. 194:1–2.

———. 1977c. Geographic variation in the musk turtle, *Sternotherus minor*. Copeia 1977:502–517.

———. 1977d. Reproduction in freshwater and terrestrial turtles in North Florida. Herpetologica 33:205–212.

———. 1977e. *Sternotherus minor*. Catalog. Amer. Amphib. Rept. 195:1–2.

———. 1978a. Reproductive cycle of female loggerhead musk turtles *(Sternotherus minor minor)* in Florida. Herpetologica 34:33–39.

———. 1978b. Variation in striped mud turtles, *Kinosternon baurii* (Reptilia, Testudines, Kinosternidae). J. Herpetol. 12:135–142.

———. 1979a. A taxonomic reappraisal of the yellow mud turtles, *Kinosternon flavescens* (Testudines: Kinosternidae). Copeia 1979:212–225.

———. 1979b. *Sternotherus carinatus*. Catalog. Amer. Amphib. Rept. 226:1–2.

———. 1980a. The reproductive biology of *Gopherus polyphemus* (Chelonia: Testudinidae). Amer. Midl. Natur. 103:353–359.

———. 1980b. *Kinosternon acutum*. Catalog. Amer. Amphib. Rept. 261:1–2.

———. 1980c. *Kinosternon angustipons*. Catalog. Amer. Amphib. Rept. 262:1–2.

———. 1981a. Biosystematics of the *Kinosternon hirtipes* species group (Testudines: Kinosternidae). Tulane Stud. Zool. Bot. 23:1–74.

———. 1981b. *Kinosternon dunni*. Catalog. Amer. Amphib. Rept. 278:1–2.

———. 1982a. *Terrapene coahuila*. Catalog. Amer. Amphib. Rept. 288:1–2.

———. 1982b. *Terrapene nelsoni*. Catalog. Amer. Amphib. Rept. 289:1–2.

———. 1983a. *Kinosternon creaseri*. Catalog. Amer. Amphib. Rept. 312:1–2.

———. 1983b. *Staurotypus triporcatus*. Catalog. Amer. Amphib. Rept. 328:1–2.

———. 1983c. *Kinosternon oaxacae*. Catalog. Amer. Amphib. Rept. 338:1–2.

———. 1985a. *Kinosternon hirtipes*. Catalog. Amer. Amphib. Rept. 361:1–4.

———. 1985b. *Staurotypus*. Catalog. Amer. Amphib. Rept. 362:1–2.

———. 1986. A checklist with distribution maps of the turtles of the world. Privately printed. Paust Printing, Richmond, Indiana. 283 p.

Iverson, J. B., and J. F. Berry. 1979. The mud turtle genus *Kinosternon* in northeastern Mexico. Herpetologica 35:318–324.

———. 1980. *Claudius, C. angustatus*. Catalog. Amer. Amphib. Rept. 236:1–2.

Iverson, J. B., and R. A. Mittermeier. 1980. Dermatemydidae, *Dermatemys, D. mawii*. Catalog. Amer. Amphib. Rept. 237:1–4.

Jackson, C. G., Jr., and J. D. Davis. 1972. A quantitative study of the courtship display of the red-eared turtle, *Chrysemys scripta elegans*. Herpetologica 28:58–64.

Jackson, D. R. 1978a. *Chrysemys nelsoni*. Catalog. Amer. Amphib. Rept. 210:1–2.

———. 1978b. Evolution and fossil record of the chicken turtle *Deirochelys*, with a re-evaluation of the genus. Tulane Stud. Zool. Bot. 20:35–55.

Janzen, D. H. 1980. Two potential coral snake mimics in a tropical deciduous forest. Biotropica 12:77–78.

Jaques, J. 1966. Some observations on the Cape terrapin. African Wild Life 20:137–150.

Jerdon, T. C. 1870. Notes on Indian herpetology. Proc. Asiatic Soc. Bengal 1870:66–85.

Johnson, R. M. 1954. The painted turtle, *Chrysemys picta picta*, in eastern Tennessee. Copeia 1954:298–299.

Judd, F. W., and J. C. McQueen. 1980. Incubation, hatching, and growth of the tortoise, *Gopherus berlandieri*. J. Herpetol. 14:377–380.

Juvik, J. O., A. J. Andrianarivo, and C. P. Blanc. 1981. The ecology and status of *Geochelone yniphora;* a critically endangered tortoise in Northwestern Madagascar. Biol. Conserv. 19:297–316.

Kaudern, W. 1922. Sauropsidien aus Madagascar. Zool. Jahrb. Syst. 45:416–458.

Khan, M. A. R. 1982. Chelonians of Bangladesh and their conservation. J. Bombay Natur. Hist. Soc. 79:110–116.

Khozatsky, L. I., and M. Mlynarski. 1966. *Agrionemys* - nouveau genre de tortue terrestris (Testudinidae). Bull. Acad. Polonaise Sci. 14:123–125.

Killebrew, F. 1975a. Mitotic chromosomes of turtles: I. The Pelomedusidae. J. Herpetol. 9:281–285.

———. 1975b. Mitotic chromosomes of turtles. III. The Kinosternidae. Herpetologica 31:398–403.

———. 1976. Mitotic chromosomes of turtles. II. The Chelidae. Texas J. Sci. 27:149–154.

———. 1977a. Mitotic chromosomes of turtles. IV. The Emydidae. Texas J. Sci. 29:245–253.

———. 1977b. Mitotic chromosomes of turtles. V. The Chelydridae. Southwest. Natur. 21:547–548.

Killebrew, F., and R. R. McKown. 1978. Mitotic chromosomes of *Gopherus berlandieri* and *Kinixys belliana belliana* (Testudines, Testudinidae). Southwest. Natur. 23:162–164.

Koshikawa, A. 1982. Three new species of reptiles from Hainan Island, Guangdong Province. Translation and Introduction. Smithsonian Herpetol. Inform. Serv. (53): 1–9.

Krefft, G. 1876. Notes on Australian animals in New Guinea with description of a new species of fresh water tortoise belonging to the genus *Euchelymys* (Gray). Ann. Mus. Civ. Stor. Nat. Genova 8:390–394.

Kuhl, H. 1820. Beiträge zur Zoologie und vergleichanden Anatomie . . . Frankfurt a. M. 2 vols.

Lacépède, B. G. E. 1788. Histoire naturelle des quadrupèdes ovipares et des serpens. Vol. 1 [Ovipares]. 18+651 p. Paris.

Lahanas, P. N. 1986. *Graptemys nigrinoda*. Catalog. Amer. Amphib. Rept. 396:1–2.

Lamar, W. W. 1986. The turtles of Venezuela (a review). Herpetologica 42:139–144.

Lamar, W. W., and F. Medem. 1982. Notes on the chelid turtle *Phrynops rufipes* in Colombia (Reptilia: Testudines: Chelidae). Salamandra 18:305–321.

Lamb, T. 1983. The striped mud turtle *(Kinosternon bauri)* in South Carolina, a confirmation through multivariate character analysis. Herpetologica 39:383–390.

Landers, J. L., W. A. McRae, and J. A. Garner. 1982. Growth and maturity of the gopher tortoise in southwestern Georgia. Bull. Florida St. Mus. Biol. Sci. 27:81–110.

Lataste, F. 1886. Description d'une tortue nouvelle du Haut-Sénégal *(Homopus nogueyi)*. Le Naturaliste (2)8:286–287.

Laurent, R. F. 1956. Contribution à l'herpétologie de la Région des Grandes Lacs de l'Afrique centrale. I. Généralités. II. Chéloniens. III. Ophidens. Ann. Mus. Roy. Congo Belge, Zool. 48:1–390.

———. 1962. On the races of *Kinixys belliana* Gray. Breviora (176):1–6.

———. 1965. A contribution to the knowledge of the genus *Pelusios* (Wagler). Ann. Mus. r. Afr. Cent. 135:1–33.

LeConte, J. 1830. Description of the species of North American tortoises. Ann. Lyceum Natur. Hist. New York 3:91–131.

———. 1854. Description of four new species of *Kinosternum*. Proc. Acad. Natur. Sci. Philadelphia 7:180–190.

Lee, R. C. 1969. Observing the tortuga blanca *(Dermatemys mawi)*. Int. Turtle Tortoise Soc. J. 3(3):32–34.

Legler, J. M. 1955. Observations on the sexual behavior of captive turtles. Lloydia 18:95–99.

———. 1959. A new tortoise, genus *Gopherus*, from Northcentral Mexico. Univ. Kansas Publ. Mus. Natur. Hist. 11:335–343.

———. 1960a. A new subspecies of slider turtle *(Pseudemys scripta)* from Coahuila, Mexico. Univ. Kansas Publ. Mus. Natur. Hist. 13:73–84.

———. 1960b. Natural history of the ornate box turtle, *Terrapene ornata ornata* Agassiz. Univ. Kansas Publ. Mus. Natur. Hist. 11:527–669.

———. 1963. Tortoises *(Geochelone carbonaria)* in Panama: distribution and variation. Amer. Midl. Natur. 70:490–503.

———. 1965. A new species of turtle, genus *Kinosternon*, from Central America. Univ. Kansas Publ. Mus. Natur. Hist. 15:615–625.

———. 1966. Notes on the natural history of a rare Central American turtle, *Kinosternon angustipons* Legler. Herpetologica 22:118–122.

———. 1978. Observations on behavior and ecology in an Australian turtle, *Chelodina expansa* (Testudines: Chelidae). Canadian J. Zool. 56:2449–2453.

Legler, J. M., and J. Cann. 1980. A new genus and species of chelid turtle from Queensland, Australia. Los Angeles Co. Mus. Natur. Sci. Contrib. Sci. (324):1–18.

Legler, J. M., and R. G. Webb. 1961. Remarks on a collection of Bolson tortoises, *Gopherus flavomarginatus*. Herpetologica 17:26–37.

———. 1970. A new slider turtle *(Pseudemys scripta)* from Sonora, Mexico. Herpetologica 26:157–168.

Lenarz, M. S., N. B. Frazer, M. S. Ralston, and R. C. Most. 1981. Seven nests recorded for loggerhead turtle *(Caretta caretta)* in one season. Herp. Review 12:9.

Lescure, J., and J. Fretey. 1975. Etude taxinomique de *Phrynops (Batrachemys) nasutus* (Schweigger) (Testudinata, Chelidae). Bull. Mus. Nat. Hist. Natur., Paris (337):1317–1328.

Lesson, R. P. 1830. Centurie zoologique. F. G. Levrault, Paris. 235 p.

Le Sueur, C. A. 1817. An account of the American species of tortoise, not noticed in the systems. J. Acad. Natur. Sci. Philadelphia 1:86–88.

———. 1827. Note sur deux espèces de tortues du genre *Trionyx* Gffr. St. H. Mém. Mus. Hist. Nat. Paris 15:257–268.

Leuck, B. E., and C. C. Carpenter. 1981. Shell variation in a population of three-toed box turtles *(Terrapene carolina triunguis)*. J. Herpetol. 15:53–58.

Li, Z. Y. 1958. Report on the investigation of reptiles of Hainan Island. Chinese J. Zool. 2:234–239.

Limpus, C. J. 1971. The flatback turtle, *Celonia depressa*

Garman in Southeast Queensland, Australia. Herpetologica 27:431–446.

Lindholm, W. A. 1929. Revidiertes Verzeichnis der Gattungen der rezenten Schildkröten nebst Notizen zur Nomenklatur einiger Arten. Zool. Anz. (Liepzig) 81:275–295.

———. 1931. Über eine angebliche *Testudo*-Art aus Südchina. Zool. Anz. (Liepzig) 97:27–30.

Linnaeus, C. 1758. Systema Naturae, 10th ed. Holmiae. 1:1–824.

———. 1766. Systema Naturae, 12th ed. Halae Magdeborgicae. 1:1–532.

Lortet, L. 1883. Etudes zoologiques sur la faune du Lac de Tiberiade. Arch. Mus. Hist. Nat., Lyon 3:99–189.

Loveridge, A. 1923. Notes on East African tortoises collected 1921–1923, with the description of a new species of soft land tortoise. Proc. Zool. Soc. London 1923:923–933.

———. 1935. Scientific results of an expedition to rain forest regions in eastern Africa. I. New reptiles and amphibians from East Africa. Bull. Mus. Comp. Zool. Harvard 79:1–19.

———. 1941. Revision of the African terrapin of the family Pelomedusidae. Bull. Mus. Comp. Zool. Harvard 88:467–524.

Loveridge, A., and E. E. Williams. 1957. Revision of the African tortoises and turtles of the suborder Cryptodira. Bull. Mus. Comp. Zool. Harvard 115:163–557.

Lovich, J. E. 1985. *Graptemys pulchra*. Catalog. Amer. Amphib. Rept. 360:1–2.

Lovich, J. E., C. H. Ernst, and S. W. Gotte. 1985. Geographic variation in the Asiatic turtle *Chinemys reevesii* (Gray), and the status of *Geoclemys grangeri* Schmidt. J. Herpetol. 19:238–245.

Luederwaldt, H. 1926. Os chelonios Brasileiros com a lista dos especies do Museu Paulista. Rev. Mus. Paulista 14:405–468.

Lynn, W. G., and C. Grant. 1940. The herpetology of Jamaica. Bull. Inst. Jamaica Sci. Ser. (1):1–148.

MacCulloch, R. D. 1981. Variation in the shell of *Chrysemys picta belli* from southern Saskatchewan. J. Herpetol. 15:181–185.

MacFarland, C. G., and W. G. Reeder. 1974. Cleaning symbiosis involving Galapagos tortoises and two species of Darwin's finches. Zeit. Tierpsychol. 34:464–483.

MacFarland, C. G., J. Villa, and B. Toro. 1974a. The Galapagos giant tortoises *(Geochelone elephantopus)*. Part I: Status of the surviving populations. Biol. Conserv. 6:118–133.

———. 1974b. The Galapagos giant tortoises *(Geochelone elephantopus)*. Part II: Conservation methods. Biol. Conserv. 6:198–212.

Mahmoud, I. Y. 1967. Courtship behavior and sexual maturity in four species of kinosternid turtles. Copeia 1967:314–319.

———. 1968. Feeding behavior in kinosternid turtles. Herpetologica 24:300–305.

———. 1969. Comparative ecology of the kinosternid turtles of Oklahoma. Southwest. Natur. 14:31–66.

Mao, S. H. 1971. Turtles of Taiwan. Comm. Press, Ltd. Taipei. 128 p.

Marlow, R. W., and J. L. Patton. 1981. Biochemical relationships of the Galápagos giant tortoises *(Geochelone elephantopus)*. J. Zool. (London) 195:413–422.

Matthey, R. 1931. Chromosomes de reptiles sauriens, ophidiens, chéloniens. L'évolution de la formule chromosomiale chez les sauriens. Rev. Suisse Zool. 38:117–186.

McCoy, C. J. 1973. *Emydoidea, E. blandingii.* Catalog. Amer. Amphib. Rept. 136:1–4.

McCoy, C. J., and R. C. Vogt. 1985. *Pseudemys alabamensis.* Catalog. Amer. Amphib. Rept. 371:1–2.

———. 1987. *Graptemys flavimaculata.* Catalog. Amer. Amphib. Rept. 403:1–2.

McDowell, S. B. 1964. Partition of the genus *Clemmys* and related problems in the taxonomy of the aquatic Testudinidae. Proc. Zool. Soc. London 143:239–279.

———. 1983. The genus *Emydura* (Testudines: Chelidae) in New Guinea with notes on the penial morphology of Pleurodira. *In* A. G. J. Rhodin and K. Miyata (eds.), Advances in herpetology and evolutionary biology, pp. 169–189. Mus. Comp. Zool. Harvard, Cambridge, Mass.

McKeown, S., J. O. Juvik, and D. E. Meier. 1982. Observations on the reproductive biology of the land tortoises *Geochelone emys* and *yniphora* in the Honolulu Zoo. Zoo Biology 1:223–235.

McKeown, S., and R. G. Webb. 1982. Softshell turtles in Hawaii. J. Herpetol. 16:107–111.

McKown, R. R. 1972. Phylogenetic relationships within the turtle genera *Graptemys* and *Malaclemys.* Ph.D. Diss., Univ. Texas, Austin, Texas.

McMorris, J. R. 1975. Notes on *Geochelone impressa.* Chelonia 2(2):5–7.

———. 1976. The generic reassignment of *Geomyda tcheponensis* Bourret. Chelonia 3(2):10–11.

Meany, D. B. 1979. Nesting habits of the Alabama red-bellied turtle, *Pseudemys alabamensis.* J. Alabama Acad. Sci. 50:113.

Medem, F. 1960. Datos zoo- geográficos y ecologícos sobre los Crocodylia y Testudinata de los rios Amazonas, Putumayo y Caqueta. Caldasia 8:341–351.

———. 1961. Contribuciónes al conocimiento sobre la morfología, ecología y distribución geográfica de la tortuga *Kinosternon dunni* K. P. Schmidt. Novedades Colombianas 1:446–476.

———. 1962. La distribución geográfica y ecología de los Crocodylia y Testudinata en el Departamento del Choco. Rev. Acad. Colomb. Cien. Ex. Fis. Nat. 11:279–303.

———. 1964. Morphologie, Ökologie und Verbreitung der Schildkröte *Podocnemis unifilis* in Kolumbien. Senck. Biol. 45:353–368.

———. 1966. Contribuciónes al conocimiento sobre la ecología y distribución geográfica de *Phrynops (Batrachemys) dahli*; (Testudinata, Pleurodira, Chelidae) Caldasia 9:467–489.

———. 1975. La reproducción de la "Icotea." Caldasia 11:83–106.

———. 1977. Contribución al conocimiento sobre la taxonomía, distribución geográfica y ecología de la tortuga "Bache" *(Chelydra serpentina acutirostris).* Caldasia 12:41–101.

———. 1983a. Reproductive data on *Platemys platycephala* (Testudines: Chelidae) in Colombia. *In* A. G. J. Rhodin and K. Miyata (eds.), Advances in herpetology and evolutionary biology, pp. 429–434. Mus. Comp. Zool. Harvard, Cambridge, Mass.

———. 1983b. La reproducción de la tortuga "cabezon" *Peltocephalus tracaxa* (Spix), 1824, (Testudines. Pelomedusidae), en Colombia. Lozania (Acta Zool. Columbiana) (41):1–12.

Meek, R. and R. Inskeep. 1981. Aspects of the field biology of a population of Hermann's tortoise *(Testudo hermanni)* in southern Yugoslavia. British J. Herpetol. 6:159–164.

Mehrtens, J. M. 1970. *Orlitia* the Bornean terrapin. Int. Turtle Tort. Soc. J. 5(1):6–7, 33.

Mell, R. 1922. Beiträge zur Fauna sinica. I. Die Vertebraten Südchinas; Feldlisten und Feldnoten der Säuger, Vögel, Reptilien, Batrachier. Arch. Naturg. 88:1–134.

———. 1929. Beiträge zur Fauna sinica. IV. Gründzuge einer Ökologie der chinesischen Reptilien und einer herpetologischen Tiergeographie Chinas. Berlin. 282 p.

———. 1938. Beiträge zur Fauna sinica. VI. Aus der Biologie chiesischer Schildkröten. Arch. Naturg. n.s. 7:390–475.

Merkle, D. A. 1975. A taxonomic analysis of the *Clemmys* complex (Reptilia: Testudines) using starch gel electrophoresis. Herpetologica 31:162–166.

Merrem, B. 1820. Tentamen Systematic Amphiborum. Krieger, Marburg. 191 p.

Mertens, R. 1954a. Zur Kenntnis der Schildkrötenfauna Venezuelas. Senck. Biol. 35:3–7.

———. 1954b. Bemerkenswerte Schildkröten aus Süd- und Zentralamerika. Die Aquar. Terr. Zeit. 7:239–242.

———. 1967. Bemerkenswerte Susswasserschildkröten aus Brasilien. Senck. Biol. 48:71–82.

———. 1969a. Eine neue Rasse der Dachschildkröte, *Kachuga tecta.* Senck. Biol. 50:23–30.

———. 1969b. Eine neue Halswender-Schildkröte aus Peru. Senck. Biol. 50:132.

———. 1971. Die Stachelschildkröte *(Heosemys spinosa)* und ihre Verwandten. Salamandra 7:49–54.

Metzger, S., and R. W. Marlow. 1986. The status of the Pinzon Island giant tortoise. Noticias de Galápagos 1986:18–20.

Meyer, A. B. 1874. *Platemys novaequineas* n. sp. Dr. W. H. Peters legte vor: Eine mittheilung von Hrn. Adolf Bernhard Meyer über die vonihm auf Neu-Guinea under den inseln Jobi, Mysore und Mafoor im Jahr 1873 gesammelten amphibien. Monatsber. dt. Akad. Wiss. Berlin 39:128–140.

Meylan, P. 1984. Evolutionary relationships of Recent trionychid turtles: evidence from shell morphology. Stud-Geol. Salmanticensia spl. vol. 1:169–188.

———. 1987. The phylogenetic relationships of soft-shelled turtles (Family Trionychidae). Bull. Amer. Mus. Natur. Hist. 186:1–101.

Meylan, P., and R. G. Webb. 1988. *Trionyx swinhoei* (Gray) 1873, a valid species of living soft-shelled turtle (Family Trionychidae) from China. J. Herpetol. 22:118–119.

Mikan, J. C. 1820. Delectus florae et faunae brasiliensis. Antoinii Strauss, Wien. 47 p.

Miller, J. F. 1776–1782. Various subjects of natural history, wherein are delineated birds, animals, and many curious plants, etc. Letterpress, London.

Milstead, W. W. 1967. Fossil box turtles *(Terrapene)* from central North America, and box turtles of eastern Mexico. Copeia 1967:168–179.

———. 1969. Studies on the evolution of box turtles (genus *Terrapene*). Bull. Florida St. Mus. Biol. Sci. 14:1–108.

Milstead, W. W., and D. W. Tinkle. 1967. *Terrapene* of western Mexico, with comments on the species groups in the genus. Copeia 1967:180–187.

Milstein, P. Le. S. 1968. Hibernation of the Kalahari geometrid tortoise and other species in the Transvaal. Fauna & Flora 19:42–44.

Minton, S. A., Jr. 1966. A contribution to the herpetology of West Pakistan. Bull. Amer. Mus. Natur. Hist. 134:27–184.

Mitsukuri, K. 1905. The cultivation of marine and fresh-water animals in Japan. The snapping turtle, or softshell tortoise, "suppon." Bull. U. S. Bur. Fish. 24:260–266.

Mittermeier, R. A., A. G. J. Rhodin, R. da Rocha e Silva, and N. Araujo de Oliveira. 1980. Rare Brazilian sideneck turtle. Oryx 15:473–475.

Mittermeier, R. A., and R. A. Wilson. 1974. Redescription of *Podocnemis erythrocephala* (Spix, 1824), an Amazonian pelomedusid turtle. Pap. Avul. Zool., S. Paulo 28:147–162.

Mlynarski, M. 1969. Fossile Schildkröten. A. Ziemsen Verlag, Wittenberg Lutherstadt. 128 p.

Moll, D. 1986. The distribution, status, and level of exploitation of the freshwater turtle *Dermatemys mawei* in Belize, Central America. Biol. Conserv. 35:87–96.

Moll, E. O. 1973. Latitudinal and intersubspecific variation in reproduction of the painted turtle, *Chrysemys picta*. Herpetologica 29:307–318.

———. 1978. Drumming along the Perak. Natur. Hist. 87:36–43.

———. 1979. Reproductive cycles and adaptations. *In* M. Harless and H. Morlock (eds.), Turtles: perspectives and research, pp. 305–331. Wiley-Interscience, New York.

———. 1985. Comment: Sexually dimorphic plastral kinesis—the forgotten papers. Herp. Review 16:16.

———. 1987. Survey of the freshwater turtles of India. Part II: The genus *Kachuga*. J. Bombay Natur. Hist. Soc. 84:7–25.

Moll, E. O., B. Groombridge, and J. Vijaya. 1986. Redescription of the cane turtle with notes on its natural history and classification. J. Bombay Natur. Hist. Soc., suppl. 83:112–126.

Moll, E. O., and J. M. Legler. 1971. The life history of a neotropical slider turtle, *Pseudemys scripta* (Schoepff), in Panama. Bull. Los Angeles Co. Mus. Natur. Hist. Sci. (11):1–102.

Moll, E. O., K. E. Matson, and E. B. Krehbiel. 1981. Sexual and seasonal dichromatism in the Asian river turtle *Callagur borneoensis*. Herpetologica 37:181–194.

Moll, E. O., and J. Vijaya. 1986. Distributional records for some Indian turtles. J. Bombay Natur. Hist. Soc. 83:57–62.

Moon, R. G. 1974. Heteromorphism in a kinosternid turtle. Mammalian Chromosome Newsletter 15:10–11.

Morreale, S. J., G. J. Ruiz, J. R. Spotila, and E. A. Standora. 1982. Temperature-dependent sex determination: current practices threaten conservation of sea turtles. Science 216:1245–1247.

Mosauer, W. 1934. The reptiles and amphibians of Tunisia. Univ. California Los Angeles, Publ. Biol. Sci. 1:46–63.

Mosimann, J. E., and G. B. Rabb. 1953. A new subspecies of the turtle *Geoemyda rubida* (Cope) from western Mexico. Occ. Pap. Mus. Zool., Univ. Michigan (548):1–7.

Mount, R. H. 1975. The reptiles and amphibians of Alabama. Auburn Univ. Agric. Exp. Stat., Auburn, Alabama. 347 p.

Mrosovsky, N. 1982. Sex ratio bias in hatchling sea turtles from artificially incubated eggs. Biol. Conserv. 23:309–314.

———. 1983. Conserving sea turtles. British Herpetol. Soc., London. 176 p.

Müller, L. 1935. Über eine neue *Podocnemis*-Art *(Podocnemis vogli)* aus Venezuela nebst ergänzenden Bemerkungen über die systematischen Merkmale der ihr nächstverwardten Arten. Zool. Anz. (Liepzig) 110:97–109.

Müller, L., and W. Hellmich. 1936. Wissenschaftliche Ergebnisse der Deutschen Gran Chaco-Expedition (Leiter: Professor Dr. Hans Krieg, Munchen). Amphibien und Reptilien. I. Teil: Amphibia, Chelonia, Loricata. Strecker und Schröder, Stuttgart. 120 p.

Murphy, J. B., and W. E. Lamoreaux. 1978. Mating behavior in three Australian chelid turtles (Testudines: Pleurodira: Chelidae). Herpetologica 34:398–405.

Nakamura, K. 1935. Studies on reptilian chromosomes. VII. Chromosomes of a turtle *Clemmys japonica* (Temm. and Schl.). Mem. Col. Sci. Kyoto Imp. Univ. Ser. B, 10:404–406.

———. 1937. On the chromosomes of some chelonians. (A preliminary note). Japanese J. Genetics 13:240.

———. 1949. A study of chromosomes in some chelonians with notes on chromosomal formula in the Chelonia. Kromosomo 5:205–213.

Neill, W. T. 1951. The taxonomy of North American soft-shelled turtles, genus *Amyda*. Publ. Res. Div. Ross Allen's Rept. Inst. 1:7–24.

Nikolskii, A. M. 1896. Diagnosis Reptilium et Amphibiorum novorum in Persia orientali a N. Zarudny Collectorum. Ann. Mus. Zool. Acad. Imp. Sci. St. Pétersbourg 4:369–372.

———. 1915. Fauna of Russia and adjacent countries. Reptiles. 1. Chelonia and Sauria. Petrograd. 352 p.

Obst, F. J. 1978. Beiträge zur Kenntnis der Testudiniden Madagaskars (Reptilia, Chelonia, Testudinidae). Zool. Abh. Mus. Tierk. Dresden 35:31–54.

———. 1980. Ergänzende Bemerkungen zu den Testudiniden Madagaskars (Reptilia, Chelonia, Testudinidae) Zool. Abh. Mus. Tierk. Dresden. 36:229–232.

———. 1983. Beiträge zur Kenntnis der Landschildkröten-Gattung *Manouria* Gray, 1852. (Reptilia, Testudines, Testudinidae). Zool. Abh. Mus. Tierk. Dresden 38:247–256.

———. 1985. Schmuckschildkröten. Die Gattung *Chrysemys*. A. Ziemsen Verlag, Wittenberg Lutherstadt. 127 p.

———. 1986. Turtles, tortoises and terrapins. St. Martin's Press, New York. 231 p.

Ogilby, J. D. 1890. Description of a new Australian tortoise. Rec. Australian Mus. 1:56–59.

Oguma, K. 1937. Studies on sauropsid chromosomes III. The chromosomes of the soft-shelled turtle, *Amyda japonica*. (Temminck and Schleg.), as additional proof of female heterogamety in the Reptilia. Japanese J. Genetics. 24:247–264.

Owen, R. 1853. Descriptive Catalogue of the Osteological series contained in the museum of the Royal College of Surgeons. London.

Owens, D. W. (ed.). 1980. Behavioral and reproductive biology of sea turtles. Amer. Zool. 20:485–617.

Pallas, P. S. 1814. Animalia monocardia seu frigidi sanguines. Imperii Russo-Asiatici 3:1–428.

Paolillo O., A. 1985. Description of a new subspecies of the turtle *Rhinoclemmys punctularia* (Daudin) (Testudines: Emydidae) from southern Venezuela. Amphibia-Reptilia 6:293–305.

Parsons, J. J. 1962. The green turtle and man. Univ. Florida Press, Gainesville. 126 p.

Parsons, T. S. 1968. Variation in the choanal structure of Recent turtles. Canadian J. Zool. 46:1235–1263.

Parsons, T. S., and E. E. Williams. 1961. Two Jurassic turtle skulls: a morphological study. Bull. Mus. Comp. Zool. Harvard 125:43–107.

Pawley, R. 1975. Man and tortoise. Field Mus. Natur. Hist. Bull. 46(10):13–18.

Peters, W. K. H. 1854. Übersicht der auf seiner Reise nach

Mossambique beobachteten Schildkröten. Monatsber. Acad. Wiss. Berlin 1854:215–216.

———. 1862. [No title.] Monatsber. Acad. Wiss. Berlin 1862:626–627.

———. 1868. Eine Mittheilung über eine neue Nagergattung, *Chiropodomys pencillatus*, so wie über einige neue oder weniger bekannte Amphibien und Fische. Amphibien. Monatsber. Akad. Wiss. Berlin. 1868:448–453.

———. 1870. Über *Platemys tuberosa*, eine neue Art von Schildkröten aus British-Guiana. Sitz. Akad. Wiss. Wien. 187:311–313.

Petzold, H. G. 1963. Über einige Schildkröten aus Nord-Vietnam im Tierpark, Berlin. Senck. Biol. 44:1–20.

———. 1965. *Cuora galbinifrons* und andere südostasiatische Schildkröten im Tierpark Berlin. Aquar. Terr.-Zeit. 18(3/4):87–91, 119–121.

Philippi, R. A. 1899. Los tortugas chilenas. Anal. Univ. Chile (Santiago). 104:727–736.

Pieau, C. 1971. Sur la proportion sexuelle chez les embryons de deau Chéloniens (*Testudo graeca* L. et *Emys orbicularis* L.) issus d'oeufs incubés artificiellement. C. R. Acad. Sci. (Paris) 272:3071–3074.

———. 1972. Effects de la température sur le développement des glandes génitales chez les embryons de deux Chéloniens, *Emys orbicularis* L. et *Testudo graeca* L. C. R. Acad. Sci. (Paris) 274:719–722.

———. 1973. Nouvelles données expérimentales concernant les effects de la température sur la différenciation sexuelle chez les embryons de Chéloniens. C. R. Acad. Sci. (Paris) 277:2789–2792.

Plummer, M. V. 1977a. Notes on the courtship and mating behavior of the softshell turtle, *Trionyx muticus* (Reptilia, Testudines, Trionychidae). J. Herpetol. 11:90–92.

———. 1977b. Reproduction and growth in the turtle *Trionyx muticus*. Copeia 1977:440–447.

Poglayen, I. 1965. Observations on Herrera's mud turtle *Kinosternon herrerai* Stejneger. Intern. Zoo Yearb. 5:171–173.

Pope, C. H. 1929. Notes on reptiles from Fukien and other Chinese provinces. Bull. Amer. Mus. Natur. Hist. 58:335–487.

———. 1934. A new emydid turtle of the genus *Geoclemys* from Kwangtung Province, China. Amer. Mus. Novitates (691):1–2.

———. 1935. Natural history of Central Asia. Vol. 10. The reptiles of China. Amer. Mus. Natur. Hist. 604 p.

———. 1939. Turtles of the United States and Canada. Alfred A. Knopf, New York. 343 p.

Pough, F. H., and M. B. Pough. 1968. The systematic status of painted turtles (*Chrysemys*) in the northeastern United States. Copeia 1968:612–618.

Pritchard, P. C. H. 1964. Turtles of British Guiana. Proc. British Guiana Mus. Zoo. (39):19–45.

———. 1967. Living turtles of the world. T. F. H. Publ., Inc., Jersey City, New Jersey. 288 p.

———. 1969. Sea turtles of the Guianas. Bull. Florida St. Mus. Biol. Sci. 13:85–140.

———. 1971. The leatherback or leathery luth. IUCN Monogr. (1):1–39.

———. 1973. International migrations of South American sea turtles (Cheloniidae and Dermochelidae). Animal Behav. 21:18–27.

———. 1975. Distribution of tortoises in tropical South America. Chelonia 2(1):3–10.

———. 1979. Encyclopedia of turtles. T. F. H. Publ., Inc., Neptune, New Jersey. 895 p.

———. 1980. *Dermochelys, D. coriacea*. Catalog. Amer. Amphib. Rept. 238:1–4.

———. 1982. Nesting of the leatherback turtle, *Dermochelys coriacea* in Pacific Mexico, with a new estimate of the World population status. Copeia 1982:741–747.

———. 1987. *Phrynops zulie*. Catalog. Amer. Amphib. Rept. 404:1–2.

Pritchard, P. C. H., and R. Marquez M. 1973. Kemp's ridley turtle or Atlantic ridley *Lepidochelys kempi*. IUCN Monogr. (2):1–30.

Pritchard, P. C. H., and P. Trebbau. 1984. Turtles of Venezuela. Soc. Stud. Amphib. Rept. 403 p.

Proceedings of the Annual Symposia of the Desert Tortoise Council. 1976–. Desert Tortoise Council, Inc., Long Beach, California.

Procter, J. B. 1922. A study of the remarkable tortoise, *Testudo loveridgii* Blgr., and the morphogeny of the chelonian carapace. Proc. Zool. Soc. London 1922:483–526.

Rafinesque, C. S. 1814. Specchio delle Scienze o giornale Enciclopedico di Sicilia. 2 vols. Palermo.

———. 1832. Descriptions of two new genera of soft shell turtles of North America. Atlantic J. Friend. Knowledge, Philadelphia 1:64–65.

Ramsey, E. P. 1887. On a new genus and species of fresh water tortoise from the Fly River, New Guinea. Proc. Linnean Soc. New South Wales 1:158–162.

Rau, R. 1969. Über die Geometrische Landschildkröte (*Testudo geometrica*). Salamandra 5:36–45.

———. 1976. Weitere Angaben über die Geometrische Landschildkröte, *Testudo geometrica*, 2. Salamandra 12:165–175.

Rebel, T. P. (ed.). 1974. Sea turtles and the turtle industry of the West Indies, Florida, and the Gulf of Mexico. Univ. Miami Press. 250 p.

Reed, C. A. 1957. Non-swimming water turtles in Iraq. Copeia 1957:51.

Reynolds, S. L., and M. E. Seidel. 1982. *Sternotherus odoratus*. Catalog. Amer. Amphib. Rept. 287:1–4.

Rhodin, A. G. J., F. Medem, and R. A. Mittermeier. 1981. The occurrence of neustophagia among podocnemine turtles. British J. Herpetol. 6:175–176.

Rhodin, A. G. J., and R. A. Mittermeier. 1976. *Chelodina parkeri*, a new species of chelid turtle from New Guinea, with a discussion of *Chelodina siebenrocki* Werner, 1901. Bull. Mus. Comp. Zool. Harvard 147:465–488.

———. 1977. Neural bones in chelid turtles from Australia and New Guinea. Copeia 1977:370–372.

———. 1983. Description of *Phrynops williamsi*, a new species of chelid turtle of the South American *P. geoffroanus* complex. *In* A. G. J. Rhodin and K. Miyata (eds.), Advances in herpetology and evolutionary biology, pp. 58–73. Mus. Comp. Zool. Harvard. Cambridge, Mass.

Rhodin, A. G. J., R. A. Mittermeier, A. L. Gardner, and F. Medem. 1978. Karyotypic analysis of the *Podocnemis* turtles. Copeia 1978:723–728.

Rhodin, A. G. J., R. A. Mittermeier, and J. R. McMorris. 1984. *Platemys macrocephala*, a new species of chelid turtle from central Bolivia and the Pantanal Region of Brazil. Herpetologica 40:38–46.

Rhodin, A. G. J., R. A. Mittermeier, and R. da Rocha e Silva. 1982. Distribution and taxonomic status of *Phrynops hogei*, a rare chelid turtle from southeastern Brazil. Copeia 1982:179–181.

Richmond, N. D. 1964. The mechanical functions of the testudinate plastron. Amer. Midl. Natur. 72:50–56.

Risley, P. L. 1933. Observations on the natural history of

the common musk turtle, *Sternotherus odoratus* (Latreille). Pap. Michigan Acad. Sci. Arts.-Lett. 17:685–711.

Ritgen, F. A. 1828. Versuch einer natürlichen Eintheilung der Amphibien. Nova Acta Physico-Medica Acad. Caes. Leopold.-Carol. Natur. Curio. 14:246–284.

Rochat, K. C., N. N. Deane, and A. M. Erasmus. 1962. Foods and feeding. Water tortoise *Pelomedusa subrufa*. Lammergeyer 2:69.

Rödel, M. O. 1985. Zum Verhalten von *Scalia bealei* (Gray, 1831). Salamandra 21:123–131.

Rodhouse, P., R. W. A. Barling, W. I. C. Clark, A.-L. Kinmonth, E. M. Mark, D. Roberts, L. E. Armitage, P. R. Austin, S. P. Baldwin, A. d'A. Bellairs, and P. J. Nightingale. 1975. The feeding and ranging behaviour of Galapagos giant tortoises *(Geochelone elephantopus)*. J. Zoology (London) 176:297–310.

Romer, A. S. 1956. Osteology of the reptiles. Univ. Chicago Press, Chicago, Illinois. 772 p.

———. 1966. Vertebrate paleontology. Univ. Chicago Press, Chicago, Illinois. 468 p.

———. 1968. Notes and comments on vertebrate paleontology. Univ. Chicago Press, Chicago, Illinois. 304 p.

Romer, J. D. 1978. Annotated checklist with keys to the chelonians of Hong Kong. Mem. Hong Kong Natur. Hist. Soc. (12):1–10.

Rose, F. L., R. Drotman, and W. G. Weaver. 1969. Electrophoresis of chin gland extracts of *Gopherus* (tortoises). Comp. Biochem. Physiol. 29:847–851.

Rothschild, W. 1901. On a new land-tortoise from the Galapagos Islands. Novitates Zool. 8:372.

———. 1902. Description of a new species of gigantic land tortoise from the Galapagos Islands. Novitates Zool. 9:619.

———. 1903. Description of a new species of gigantic land tortoise from Indefatigable Island. Novitates Zool. 10:119.

———. 1915a. The giant land tortoises of the Galapagos Islands in the Tring Museum. Novitates Zool. 22:403–417.

———. 1915b. On the gigantic land tortoises of the Seychelles and Aldabra-Madagascar Group with some notes on certain forms of the Mascarene Group. Novitates Zool. 22:418–442.

Roze, J. A. 1964. Pilgrim of the river. Natural History 73:34–41.

Rüppell, E. 1835. Neue wilbelthiere zu der Fauna von Abyssinian gehörig. III. Amphibien. Frankfurt am. M. 18 p.

Sachsse, W. 1973. *Pyxidea mouhotii*, eine landbewohende Emydide Südoasiens (Testudines). Salamandra 9:49–53.

Sachsse, W. 1975. *Chinemys reevesii* var. *unicolor* und *Clemmys bealei* var. *quadriocellata*—Ausprägungen von Sexualdimorphismus der beiden "Nominatformen." Salamandra 11:20–26.

Sachsse, W., and A. A. Schmidt. 1976. Nachzucht in der zweiten Generation von *Staurotypus salvinii* mit weiteren Beobachtungen zum Fortpflanzungsverhalten (Testudines, Kinosternidae). Salamandra 12:5–16.

Sampaio, M. M., R. M. Barros, M. Ayres, and O. R. Cunha. 1971. A karyological study of two species of tortoises from the Amazon Region of Brazil. Cytologia (Tokyo) 36:199–204.

Sasaki, M., and M. Itoh. 1967. Preliminary notes on the karyotype of two species of turtles, *Clemmys japonica* and *Geoclemys reevesii*. Chromosome Infor. Serv. (8):21–22.

Satyamurti, S. T. 1962. Guide to the lizards, crocodiles, turtles and tortoises in the reptile gallery. Madras Gov't. Mus. Madras, India. 45 p.

Savage, J. M. 1953. Remarks on the Indo-chinese turtle *Annamemys merkleni*. Ann. Mag. Natur. Hist. (London) (12)6:468–472.

Schlegel, H., and S. Müller. 1844 Over de Schildpadden van den Indischen Archipal., beschrijving einer nieuwe soort van Sumatra. *In* Temminck, C. J., Verh-Nat. Gesch. Nederland., Oost-Indie, 1839–44. Zool., pt. 3, Reptilia, pp. 29–36.

Schleich, H.-H., and U. Gruber. 1984. Eine neue Grosskopfschildkröte, *Platysternon megacephalum tristernalis* nov. ssp., aus Yunnan, China. Spixiana 7:67–73.

Schmidt, A. A. 1966. Morphologische Unterschiede bei *Chelus fimbriatus* verschiedener Herkunft. Salamandra 2:74–78.

———. 1970. Zur Fortpflanzung der Kreuzbrustschildkröte *(Staurotypus salvinii)* in Gefangenschaft. Salamandra 6:3–10.

Schmidt, K. P. 1919. Contributions to the herpetology of the Belgian Congo based on the collection of the American Congo Expedition, 1909–1915. Part I. Turtles, crocodiles, lizards, and chameleons. Bull. Amer. Mus. Natur. Hist. 39:385–684.

———. 1925. New reptiles and a new salamander from China. Amer. Mus. Novitates (157):1–5.

———. 1927a. Notes on Chinese reptiles. Bull. Amer. Mus. Natur. Hist. 54:467–551.

———. 1927b. The reptiles of Hainan. Bull. Amer. Mus. Natur. Hist. 54:395–465.

———. 1928. Scientific survey of Porto Rico and the Virgin Islands. New York Acad. Sci. 160 p.

———. 1947. A new kinosternid turtle from Colombia. Fieldiana: Zool. 31:109–112.

Schmidt, K. P., and D. W. Owens. 1944. Amphibians and reptiles of northern Coahuila, Mexico. Zool. Ser. Field Mus. Natur. Hist. 29:97–115.

Schneider, J. G. 1783. Allgemeine Naturgeschichte der Schildkröten, nebst einem System. Verseichnisse der einzelnen Arten. Leipzig. 364 p.

———. 1792. Beschreibung und Abbildung einer neun Wasserschildkröte. Schr. Ges. Naturf. Fruende Berlin 10:259–283.

Schoepff, J. D. 1792–1801. Historia Testudinum iconibus illustrata. Palm, Erlangae. 136 p. [Latin version of Naturgeschichte der Schildkröten mit Abbildungen erlautert. Palm, Erlangen. 160 p. 1792:1–32; 1793:33–88; 1795:89–136; 1801:137–160.]

Schomburgk, M. R. 1848. Reisen in British Guiana in . . . 1840–1844. I. Fauna. Amphibien, pp. 645–661. Leipzig.

Schwartz, A. 1955. The diamondback terrapins *(Malaclemys terrapin)* of peninsular Florida. Proc. Biol. Soc. Washington 68:157–164.

———. 1956a. Geographic variation in the chicken turtle. Fieldiana: Zool. 34:461–503.

———. 1956b. The relationships and nomenclature of the soft-shelled turtles (genus *Trionyx*) of the southeastern United States. Charleston Mus. Leaflet 26:1–21.

Schwartz, A., and R. Thomas. 1975. A check-list of West Indian amphibians and reptiles. Carnegie Mus. Natur. Hist. Spl. Publ. (1):1–216.

Schweigger, A. F. 1812. Monographiae Cheloniorum. Königsberg. Arch. Naturwiss. Math. 1:271–368, 406–458.

———. 1814. Prodromi monographie cheloniorum. Regiomonti. 58 p.

Seeliger, L. M. 1945. Variation in the Pacific mud turtle. Copeia 1945:150–159.

Seidel, M. E. 1978. *Kinosternon flavescens*. Catalog. Amer. Amphib. Rept. 216:1–4.

——. 1981. A taxonomic analysis of pseudemyd turtles (Testudines: Emydidae) from the New River and phenetic relationships in the subgenus *Pseudemys*. Brimleyana (6):25–44.

——. 1988. Revision of the West Indian emydid turtles (Testudines). Amer. Mus. Novitates (2918):1–41.

Seidel, M. E., and M. D. Adkins. 1987. Biochemical comparisons among West Indian *Trachemys* (Emydidae: Testudines). Copeia 1987:485–489.

Seidel, M. E., and S. J. Inchaustegui Miranda. 1984. Status of the trachemyd turtles (Testudines: Emydidae) on Hispaniola. J. Herpetol. 18:468–479.

Seidel, M. E., J. B. Iverson, and M. D. Adkins. 1986. Biochemical comparisons and phylogenetic relationships in the family Kinosternidae (Testudines). Copeia 1986:285–294.

Seidel, M. E., and R. V. Lucchino. 1981. Allozymic and morphological variation among the musk turtles *Sternotherus carinatus*, *S. depressus*, and *S. minor* (Kinosternidae). Copeia 1981:119–128.

Seidel, M. E., and S. L. Reynolds. 1980. Aspects of evaporative water loss in the mud turtles *Kinosternon hirtipes* and *Kinosternon flavescens*. Comp. Biochem. Physiol. 67A:593–598.

Seidel, M. E., S. L. Reynolds, and R. V. Lucchino. 1981. Phylogenetic relationships among musk turtles (Genus *Sternotherus*) and genic variation in *Sternotherus odoratus*. Herpetologica 37:161–165.

Seidel, M. E., and H. M. Smith. 1986. *Chrysemys, Pseudemys, Trachemys* (Testudines: Emydidae): Did Agassiz have it right? Herpetologica 42:242–248.

Seigel, R. A. 1980. Courtship and mating behavior of the diamondback terrapin *Malaclemys terrapin tequesta*. J. Herpetol. 14:420–421.

Sexton, O. J. 1959. Spatial and temporal movements of a population of the painted turtle, *Chrysemys picta marginata* (Agassiz). Ecol. Monogr. 29:113–140.

——. 1960. Notas sobre la reproducción de una tortuga Venezolana, la *Kinosternon scorpioides*. Soc. Cienc. Natur. LaSalle Mem. 20:189–197.

Shaffer, J. C., and C. H. Ernst. 1979. The giant land tortoise of Aldabra, *Geochelone gigantea*. Bull. Maryland Herpetol. Soc. 15:46–55.

Shaw, G. 1802. General zoology, or systematic zoology. G. Kearsley, London. 312 p.

Shaw, G., and J. E. Smith. 1793. Zoology and Botany of New Holland and the Isles Adjacent. London

Shealy, R. M. 1976. The natural history of the Alabama map turtle, *Graptemys pulchra* Baur, in Alabama. Bull. Florida St. Mus. Biol. Sci. 21:47–111.

Siebenrock, F. 1901. Beschreibung einer neuen Schildkrötengattung aus der Familie Chelydidae von Australien: *Pseudemydura*. Anz. Akad. Wiss. Wien. 22:248–250.

——. 1903a. Schildkröten des ostlichen Hinterindien. Sitz. Acad. Wiss. Wien. Math.-natur. Kl. 112:333–353.

——. 1903b. Über zwei seltene und eine neue Schildkröte des Berliner Museum. Sitz. Akad. Wiss. Wien. Math-natur. Kl. 112:439–445.

——. 1906a. Schildkröten von Ostafrika and Madagaskar. *In* A. Voeltzkow, Reise in Ostafrika in den Jahren 1903–1905, 2:1–40. Stuttgart.

——. 1906b. Zur Kenntnis der Schildkrötenfauna der Insel Hainan. Zool. Anz. 30:578–586.

——. 1906c. Eine neue *Cinosternum*-Art aus Florida. Zool. Anz. 30:727–728.

——. 1907. Über einige, zum Teil seltene Schildkröten aus Südchina. Sitz. Acad. Wiss. Wien. Math.-natur. K1. 116:1741–1776.

——. 1909. Synopsis der rezenten Schildkröten mit Berücksichtigung der in historischer Zeit ausgestorbenen Arten. Zool. Jahrb., suppl. 10:427–618.

——. 1914. Eine neue *Chelodina*-Art aus Westaustralien. Anz. Akad. Wiss. Wien 17:386–387.

Sites, J. W., Jr., J. W. Bickham, and M. W. Haiduk. 1979. Derived X chromosome in the turtle genus *Staurotypus*. Science 206:1410–1412.

Sites, J. W., Jr., J. W. Bickham, M. W. Haiduk, and J. B. Iverson. 1979. Banded karyotypes of six taxa of kinosternid turtles. Copeia 1979:692–698.

Sites, J. W., Jr., J. W. Bickham, B. A. Pytel, I. F. Greenbaum, and B. B. Bates. 1984. Biochemical characters and the reconstruction of turtle phylogenies: Relationships among batagurine genera. Syst. Zool. 33:137–158.

Smith, A. 1838. Illustrations of the zoology of South Africa. Reptilia. London. 28 p.

Smith, H. M., and B. P. Glass. 1947. A new musk turtle from southeastern United States. J. Washington Acad. Sci. 37:22–24.

Smith, H. M., and L. W. Ramsey. 1952. A new turtle from Texas. Wasmann J. Biol. 10:45–54.

Smith, H. M., and R. B. Smith. 1979. Synopsis of the herpetofauna of Mexico. Vol. 6. Guide to Mexican turtles. John Johnson, North Bennington, Vermont. 1044 p.

Smith, M. A. 1916. A list of the crocodiles, tortoises, turtles and lizards at present known to inhabit Siam. J. Natur. Hist. Soc. Siam 2:48–57.

——. 1923. On a collection of reptiles and batrachians from the Island of Hainan. J. Natur. Hist. Soc. Siam 6:195–212.

——. 1931. The fauna of British India, including Ceylon and Burma. Reptilia and Amphibia, vol. 1, Loricata and Testudines. Taylor and Francis, London. 185 p.

Smith, M. H., H. O. Hillestad, M. N. Manlove, P. O. Straney, and J. M. Dean. 1977. Management implications of genetic variability in loggerhead and green sea turtles. 13th Congr. Game Biol., pp. 302–312.

Smith, P. W. 1951. A new frog and a new turtle from the western Illinois sand prairies. Bull. Chicago Acad. Sci. 9:189–199.

——. 1961. The amphibians and reptiles of Illinois. Illinois Natur. Hist. Surv. Bull. (28):1–298.

Snow, D. W. 1964. The giant tortoises of the Galapagos Islands. Their present status and future chances. Oryx 7:277–290.

Song, M. 1984. A new species of the turtle genus *Cuora* (Testudoformes: Testudinidae). Acta Zootaxonomica Sinica 9:330–332.

Sonnini, C. S., and P. A. Latreille. 1802. Histoire naturelle des reptiles avec figures dessinées d'apres nature. I. Deterville, Paris. 280 p.

Spence, T., R. Fairfax, and I. Loach. 1979. The western Australian swamp tortoise *Pseudemydura umbrina*. Int. Zoo. Yearb. 19:58–60.

Spix, J. B. de. 1824. Animalia nova, species novae Testudinum et ranarum quas in itinere per Brasiliam annis 1817–1820 jussu et auspiciis Maximiliana Josephi 2. Leipzig. 24 p.

Stebbins, R. 1985. A field guide to western reptiles and amphibians. 2nd ed. Houghton Mifflin Co., Boston. 336 p.

Stejneger, L. 1902. Some generic names of turtles. Proc. Biol. Soc. Washington 15:235–238.

———. 1909. Generic names of some chelyid turtles. Proc. Biol. Soc. Washington 22:125–128.

———. 1918. Description of a new snapping turtle and a new lizard from Florida. Proc. Biol. Soc. Washington 31:89–92.

———. 1925a. Chinese amphibians and reptiles in the United States National Museum. Proc. U. S. Natl. Mus. 66:1–115.

———. 1925b. New species and subspecies of American turtles. J. Washington Acad. Sci. 15:462–463.

———. 1944. Notes on the American soft-shell turtles with special reference to *Amyda agassizii*. Bull. Mus. Comp. Zool. Harvard 94:1–75.

Sternberg, J. 1981. The worldwide distribution of sea turtle nesting beaches. Sea Turtle Rescue Fund, Center Environ. Ed., Washington, D.C.

Stickel, L. F. 1950. Population and home range relationships of the box turtle, *Terrapene c. carolina* (Linnaeus). Ecol. Monogr. 20:351–378.

Stimson, A. F. 1986. On the correct zoological name of the Australasian turtle genus *Emydura* Bonaparte. J. Herpetol. 20:279–280.

Stock, A. D. 1972. Karyological relationships in turtles (Reptilia: Chelonia). Canadian J. Genet. Cytol. 14:859–868.

Stoddart, D. R., D. Cowx, C. Peet, and J. R. Wilson. 1982. Tortoises and tourists in the western Indian Ocean: The Curieuse Experiment. Biol. Conserv. 24:67–80.

Stoneburner, D. L. 1980. Body depth: an indicator of morphological variation among nesting groups of adult loggerhead sea turtles *(Caretta caretta)*. J. Herpetol. 14:205–206.

Storer, T. I. 1930. Notes on the range and life-history of the Pacific fresh-water turtle, *Clemmys marmorata*. Univ. California Publ. Zool. 32:429–441.

Street, D. 1979. The reptiles of northern and central Europe. B. T. Batsford Ltd., London. 268 p.

Surface, H. A. 1908. First report on the economic features of the turtles of Pennsylvania. Zool. Bull. Div. Zool. Pennsylvania Dept. Agric. 6:105–196.

Swindells, R. J., and F. C. Brown. 1964. Ability of *Testudo elongata* to withstand excessive heat. British J. Herpetol. 3:166.

Swingland, I. R., and M. Coe. 1978. The natural regulation of giant tortoise populations on Aldabra Atoll. Reproduction. J. Zool. (London) 186:285–309.

———. 1979. The natural regulation of giant tortoise populations on Aldabra Atoll: recruitment. Phil. Trans. Royal Soc. London, B 286:177–188.

Taylor, E. H. 1920. Philippine turtles. Philippine J. Sci. 16:111–144.

———. 1933. Observations on the courtship of turtles. Univ. Kansas Sci. Bull. 21:269–271.

———. 1970. The turtles and crocodiles of Thailand and adjacent waters. Univ. Kansas Sci. Bull. 49:87–179.

Taylor, W. E. 1895. The box turtles of North America. Proc. U. S. Natl. Mus. 17:573–588.

Temminck, C. J., and H. Schlegel. 1835. Reptilia. Chelonii. *In* Ph. Fr. de Siebold, 1833–1838, Fauna Japonica. Leyden. 144 p.

Theobald, W. 1876. Descriptive catalogue of the reptiles of British India. Thacher, Spink and Co., Calcutta. 238 p.

Throp, J. L. 1976. Honolulu honeymoon. Animal Kingdom. Aug.–Sept. 1976:25–29.

Thunberg, C. P. 1785–1788. [No title]. Museum naturalium Academiae Upsaliensis . . . . Kongliga Svenska Vetenskapsakademien, Stockholm. Handlinger 8.

Timmerman, W. W., and D. L. Auth. 1988. *Heosemys leytensis* (Leyte pond turtle). Herp. Review 19:21

Tinkle, D. W. 1958. The systematics and ecology of the *Sternotherus carinatus* complex (Testudinata, Chelydridae). Tulane Stud. Zool. 6:1–56.

———. 1959. The relation of the fall line to the distribution and abundance of turtles. Copeia 1959:167–170.

———. 1961. Geographic variation in reproduction, size, sex ratio and maturity of *Sternothaerus odoratus* (Testudinata: Chelydridae). Ecology 42:68–76.

Tinkle, D. W., J. D. Congdon, and P. C. Rosen. 1981. Nesting frequency and success: implications for the demography of painted turtles. Ecology 62:1426–1432.

Tinkle, D. W., and R. G. Webb. 1955. A new species of *Sternotherus* with a discussion of the *Sternotherus carinatus* complex (Chelonia, Kinosternidae). Tulane Stud. Zool. 3:53–67.

Vaillant, L. 1885. Sur une tortue terrestre d'espèce nouvelle, rapportée par M. Humbolt au Museum d'Histoire naturelle. Bull. C. R. Acad. Sci., Paris 101:440–441.

Valenciennes, A. 1833. *In* G. Bibron and J. B. G. M. Bory de Saint-Vincent. Reptiles et Poissons (of the Morea). Fr. Expéd. Sci. Morée. 3(1):1–209.

Van Denburgh, J. 1895. A review of the herpetology of Lower California. Part 1. Reptiles. Proc. California Acad. Sci. (2)5:77–162.

———. 1907. Expedition of the California Academy of Sciences to the Galapagos Islands, 1905–1906. I. Preliminary descriptions of four new races of gigantic land tortoises from the Galapagos Islands. Proc. California Acad. Sci. 1:1–6.

———. 1914. Expedition of the California Academy of Sciences to the Galapagos Islands, 1905–1906. X. The gigantic land tortoises of the Galapagos Archipelago. Proc. California Acad. Sci. (4)2:203–374.

Vanzolini, P. E. 1977. A brief biometrical note on the reproductive biology of some South American *Podocnemis* (Testudines: Pelomedusiidae). Pap. Avul. Zool., São Paulo 31:79–102.

Vanzolini, P. E., A. M. M. Ramos-costa, and L. J. Vitt. 1980. Repteis das caatingas. Acad. Brasil. Cienc., Rio de Janeiro. 161 p.

Vestjens, W. J. M. 1969. Nesting, egg-laying and hatching of the snake-necked tortoise at Canberra, A. C. T. Australian Zool. 15:141–149.

Vijaya, J. 1982a. Rediscovery of the forest cane turtle *(Heosemys silvatica)* of Kerala. Hamadryad 7(3):2–3.

———. 1982b. *Kachaga tecta* hatching at the Snake Park. Hamadryad 7(3):14–15.

Villiers, A. 1958. Tortues et crocodiles de l'Afrique Noire Française. Inst. Franç. D'Afrique Noire Initiat. Afriq. 15:1–354.

Vogt, R. C. 1980. Natural history of the map turtles *Graptemys pseudogeographica* and *G. ouachitensis* in Wisconsin. Tulane Stud. Zool. Bot. 22:17–48.

———. 1981. *Graptemys versa*. Catalog. Amer. Amphib. Rept. 280:1–2.

Vogt, R. C., and C. J. McCoy. 1980. Status of the emydine

turtle genera *Chrysemys* and *Pseudemys*. Ann. Carnegie Mus. 49:93–102.

Vuillemin, S. and C. Domergue. 1972. Contribution a l'étude de la faune de Madagascar: description de *Pyxoides brygooi* n. gen. n. sp. (Testudinidae). Ann. Univ. Madagascar. Ser. Sci. Natur. Math. (9):193–200.

Waagen, G. N. 1984. Sexually dimorphic plastral kinesis in *Heosemys spinosa*. Herp. Review 15:33–34.

Wagler, J. 1830. Natürliches System der Amphibien, mit Vorangehander Classification der Säugthiere und Vögel. J. G. Cotta'schen Buchhandlung, Munchen. 354 p.

Wahlquist, H. 1970. Sawbacks of the Gulf Coast. Int. Turtle & Tortoise Soc. J. 4(4):10–13, 28.

Wahlquist, H., and G. W. Folkerts. 1973. Eggs and hatchlings of Barbour's map turtle, *Graptemys barbouri* Carr and Marchand. Herpetologica 29:236–237.

Wallin, L. 1977. The Linnean type-specimen of *Testudo geometrica*. Zoon 5:77–78.

Ward, J. P. 1978. *Terrapene ornata*. Catalog. Amer. Amphib. Rept. 217:1–4.

———. 1984. Relationships of chrysemyd turtles of North America (Testudines: Emydidae). Spl. Publ. Mus. Texas Tech. Univ. (21):1–50.

Waters, J. H. 1964. Subspecific intergradation in the Nantucket Island, Massachusetts, population of the turtle *Chrysemys picta*. Copeia 1964:550–553.

———. 1969. Additional observations of southeastern Massachusetts insular and mainland populations of painted turtles, *Chrysemys picta*. Copeia 1969:179–182.

Weaver, W. G., Jr. 1970. Courtship and combat behavior in *Gopherus berlandieri*. Bull. Florida St. Mus. Biol. Sci. 15:1–43.

Weaver, W. G., Jr., and J. S. Robertson. 1967. A re-evaluation of fossil turtles of the *Chrysemys scripta* group. Tulane Stud. Geol. 5:53–66.

Weaver, W. G., Jr., and F. L. Rose. 1967. Systematics, fossil history, and evolution of the genus *Chrysemys*. Tulane Stud. Zool. 14: 63–73.

Webb, R. G., 1959. Description of a new softshell turtle from the southeastern United States. Univ. Kansas Publ. Mus. Natur. Hist. 11:517–525.

———. 1962. North American Recent soft-shelled turtles (Family Trionychidae). Univ. Kansas Publ. Mus. Natur. Hist. 13:429–611.

———. 1973a. *Trionyx ater*. Catalog. Amer. Amphib. Rept. 137:1.

———. 1973b. *Trionyx ferox*. Catalog. Amer. Amphib. Rept. 138: 1–3.

———. 1973c. *Trionyx muticus*. Catalog. Amer. Amphib. Rept. 139:1–2.

———. 1973d. *Trionyx spiniferus*. Catalog. Amer. Amphib. Rept. 140:1–4.

———. 1980a. The identity of *Testudo punctata* Lacépède, 1788 (Testudines, Trionychidae). Bull. Mus. Natn. Hist. Nat., Paris (4)2:547–557.

———. 1980b. The trionychid turtle *Trionyx steindachneri* introduced in Hawaii? J. Herpetol. 14:206–207.

———. 1982. Taxonomic notes concerning the trionychid turtle *Lissemys punctata* (Lacépède). Amphibia-Reptilia 3:179–184.

Webb, R. G., and J. M. Legler. 1960. A new softshell turtle (genus *Trionyx*) from Coahuila, Mexico. Univ. Kansas Sci. Bull. 40:21–30.

Wells, R. W., and C. R. Wellington. 1984. A synopsis of the class Reptilia in Australia. Australian J. Herpetol. 1:73–129.

Wermuth, H. 1952. *Testudo hermanni robertmertensi* n. sub-

sp. und ihr Vorkommen in Spanien. Senckenbergiana 33:157–164.

———. 1969. Eine neue Grosskopfschidkröte, *Platysternon megacephalum vogeli*, n. ssp. Aquar.-Terr. Zeitsch. 22:372–374.

Wermuth, H., and R. Mertens. 1961. Schildkröten, Krocodile, Brüchenechsen. G. Fischer, Jena. 422 p.

———. 1977. Liste der rezenten Amphibien und Reptilien. Testudines, Crocodylia, Rhynchocephalia. Das Tierreich, Berlin 28:1–174.

Werner, F. 1901. Über Reptilien und Batrachier aus Ecuador und Neu-Guinea. Verh. Zool. Bot. Ges. Wien 51:593–603.

Whelan, J. P., and J. Coakley. 1982. *Geochelone elegans* (Indian star tortoise) fecundity. Herp. Review 13:97.

Whetstone, K. N. 1978. Additional record of the fossil snapping turtle *Macroclemys schmidti* from the Marshland Formation (Miocene) of Nebraska with notes on interspecific skull variation within the genus *Macroclemys*. Copeia 1978:159–162.

Whitaker, R. 1983. World's rarest turtle (we think) lays eggs in captivity. Hamadryad 8(1):13.

Whittow, G. C., and G. H. Balazs. 1982. Basking behavior of the Hawaiian green turtle *(Chelonia mydas)*. Pacific Science 36:129–139.

Wied-Neuweid, M. A. P. 1839. Reise in das innere Nörd-America in den Jahren 1832 bis 1834. J. Hoelscher, Coblenz. 653 p.

———. 1865. Verzeichniss der Reptilien, welche auf einer Reise im nördlichen America beobachtet wurden. Nova Acta Acad. Leopold.-Carol. 32:1–146.

Wiegmann, A. F. A. 1828. Beyträge zur Amphibienkunde. Isis v. Oken, 21:364–383.

———. 1835. Beiträge zur Zoologie, gesammelt auf einer Reise um die Erde, von Dr. F. J. F. Meyen. Amphibien, Nova Acta Acad. Leopold.-Carol. 17:185–268.

Wilhoft, D. C., E. Hotaling, and P. Franks. 1983. Effects of temperature on sex determination in embryos of the snapping turtle, *Chelydra serpentina*. J. Herpetol. 17:38–42.

Williams, E. E. 1950a. Variation and selection of the cervical central articulations of living turtles. Bull. Amer. Mus. Natur. Hist. 94:505–562.

———. 1950b. *Testudo cubensis* and the evolution of Western Hemisphere tortoises. Bull. Amer. Mus. Natur. Hist. 95:1–36.

———. 1952. A new fossil tortoise from Mona Island, West Indies, and a tentative arrangement of the tortoises of the world. Bull. Amer. Mus. Natur. Hist. 99:541–560.

———. 1954a. A key and description of the living species of the genus *Podocnemis* (*sensu* Boulenger) (Testudines, Pelomedusidae). Bull. Mus. Comp. Zool. Harvard 111:279–295.

———. 1954b. A new Miocene species of *Pelusios* and the evolution of that genus. Breviora (25):1–7.

———. 1956. *Pseudemys scripta callirostris* from Venezuela with a general survey of the *scripta* series. Bull. Mus. Comp. Zool. Harvard 115:145–160.

———. 1960. Two species of tortoises in northern South America. Breviora (120):1–13.

Williams, E. E., and S. B. McDowell. 1952. The plastron of soft-shelled turtles (Testudinata, Trionychidae): a new interpretation. J. Morphol. 90:263–280.

Williams, K. L., and P. Han. 1964. A comparison of the density of *Terrapene coahuila* and *T. carolina*. J. Ohio Herpetol. Soc. 4:105.

Winokur, R. M., and J. M. Legler. 1975. Chelonian mental glands. J. Morphol. 147:275–291.

Wirot, N. 1979. The turtles of Thailand. Siamfarm Zool. Gard. 222 p.

Witzell, W. N. 1983. Synopsis of biological data on the hawksbill turtle, *Eretmochelys imbricata* (Linnaeus, 1766). FAO Fish. Synop. 137:1–78.

Wood, F. G. 1953. Mating behavior of captive loggerhead turtles, *Caretta caretta caretta*. Copeia 1953:184–186.

Wood, R. C. 1973. A possible correlation between the ecology of living African pelomedusid turtles and their relative abundance in the fossil record. Copeia 1973:627–629.

———. 1976. *Stupendemys geographicus*, the world's largest turtle. Breviora (436):1–31.

———. 1977. Evolution of the emydine turtles *Graptemys* and *Malaclemys* (Reptilia, Testudines, Emydidae). J. Herpetol. 11:415–421.

Wood, R. C., and R. T. J. Moody. 1976. Unique arrangement of carapace bones in the South American chelid turtle *Hydromedusa maximiliani* (Mikan). Zool. J. Linn. Soc. 59:69–78.

Woodbury, A. M., and R. Hardy. 1948. Studies of the desert tortoise, *Gopherus agassizii*. Ecol. Monogr. 18:145–200.

Woolley, P. 1957. Colour change in a chelonian. Nature (London) 179:1255–1256.

Worrell, E. 1964. Reptiles of Australia. Angus and Robertson Ltd., Sydney. 207 p.

Yntema, C. L. 1976. Effects of incubation temperature on sexual differentiation in the turtle, *Chelydra serpentina*. J. Morphol. 150:453–462.

Yntema, C. L., and N. Mrosovsky. 1982. Critical periods and pivotal temperatures for sexual differentiation in loggerhead sea turtles. Canadian J. Zool. 60:1012–1016.

Zangerl, R. 1969. The turtle shell. *In* C. Gans and T. S. Parsons (eds.), Biology of the Reptilia, I, pp. 311–339. Academic Press, New York.

Zangerl, R., and F. Medem. 1958. A new species of chelid turtle, *Phrynops (Batrachemys) dahli*, from Colombia. Bull. Mus. Comp. Zool. Harvard 119:375–390.

Zappalorti, R. T. 1976. The amateur zoologist's guide to turtles and crocodilians. Stackpole Books, Harrisburg, Penna. 208 p.

Zhao, X. 1986. *Geochelone impressa* (Guenther) discovered from Shaoyang, Hunan. Acta Herpetol. Sinica 5:313.

Zug, G. R. 1966. The penial morphology and the relationships of cryptodiran turtles. Occ. Pap. Mus. Zool. Univ. Michigan (647):1–24.

———. 1971a. Buoyancy, locomotion, morphology of the pelvic girdle and hindlimb, and systematics of cryptodiran turtles. Misc. Publ. Mus. Zool., Univ. Michigan (142):1–98.

———. 1971b. American musk turtles, *Sternothaerus* or *Sternotherus*? Herpetologica 27:446–449.

———. 1977. The matamata (Testudines: Chelidae) is *Chelus* not *Chelys!* Herpetologica 33:53–54.

———. 1986. *Sternotherus*. Catalog. Amer. Amphib. Rept. 397: 1–3.

Zug, G. R., and A. Schwartz. 1971. *Deirochelys, D. reticularia*. Catalog. Amer. Amphib. Rept. 107:1–3.

Glossary of
scientific names

abaxillare aberrant axillary scute
abingdonii from Abingdon Island, Galapagos
Acinixys fixed back
acutirostris sharp, pointed snout
acutum pointed, sharp
adansonii named for M. Adanson
agassizii named for L. Agassiz
Agrionemys field turtle
alabamensis from Alabama
alamosae from Alamos, Sonora, Mexico
albogulare white gular scute
amboinensis from Amboina Island, Indonesia
Amyda turtle
andersoni named for J. Anderson
angulata angled
angusta, angustatus narrow
angustipons narrow bridge
annamensis from Annam (Vietnam)
Annamemys Annam (Vietnam) turtle
annandalei named for N. Annandale
annulata ringed
Apalone soft
arachnoides spiderlike
areolata, areolatus small open space
arizonense from Arizona
asper rough
Aspideretes shield, oars
ater black
aubryi named for Mons. Aubry
australis from Australia
babcocki named for H. L. Babcock
bangsi named for O. Bangs
barbouri named for T. Barbour
baska bewitching
Batagur a word without meaning, coined by Gray
Batrachemys frog tortoise
bauri, baurii named for G. Baur
bealei named after "Beale Esq."
bechuanicus from Bechuanaland
becki named for R. H. Beck
belliana named for T. Bell
bellii named for T. Bell
berlandieri named for L. Berlandier

bibroni named for G. Bibron
bissa biting
blandingii named for W. Blanding
brygooi named for E.-R. Brygoo
borneensis, borneoensis from Borneo
boulengeri named for G. A. Boulenger
brasiliensis from Brazil
broadleyi named for D. G. Broadley
caglei named for F. R. Cagle
Callagur beautiful turtle
callirostris beautiful nose
calvata smooth, bald
carajasensis from the Serra dos Carajas Plateau, Para, Brazil
carbonaria blackish
Caretta a tortoise shell
Carettochelys tortoise shell
carinatum keeled
carolina named after Carolina
cartilaginea cartilage
caspica from the Caspian Sea
castaneus of chestnut color
castanoides like *castaneus*
cataspila downward spot
centrata centered
chapalaense from Lake Chapala, Jalisco, Mexico
chathamensis from Chatham Island, Galapagos
Chelodina tortoise
Chelonia tortoise
Chelus tortoise
Chelydra a water serpent
Chersina land tortoise
chichiriviche a coastal hill in Edo. Falcon, Venezuela
chilensis from Chile
Chinemys Chinese turtle
Chitra picturesque, beautiful
chriskarannarum named for Christine, Karen, and Anne McCord
chrysea golden one
Chrysemys golden turtle
circumdata around, about
Claudius defective
Clemmys tortoise
coahuila from Coahuila, Mexico
concinna elegant, neatly arranged
coriacea leathery
coronata crowned
crassicollis thick neck
creaseri named for E. P. Creaser
cruentatum stained with blood
Cuora a colloquial Indonesian name for *C. amboinensis*
Cyclanorbis circular orbit
Cyclemys circle turtle
Cycloderma circle of skin
dahli named for G. Dahl
darwini named for C. Darwin
decorata elegantly marked
decussata crossed
Deirochelys hill or humped turtle
delticola delta inhabiting
dentata toothed
denticulata toothed sides
depressa, depressum flattened
Dermatemys skin turtle
Dermochelys leather tortoise
dhongoka from the Indian colloquial name for the species, dhon

diademata cross band
Dogania presumably referring to the proper name Dogan
dorbigni named for A. d'Orbigny
dorsalis dorsal
dumeriliana named for Andre M. C. Duméril
dunni named for E. R. Dunn
durangoense from Durango, Mexico
edeniana from Eden
elegans elegant
elephantopus sounding like an elephant
elongata elongated
Elseya named for Dr. J. R. Elsey
emoryi named for W. H. Emory
Emydoidea *Emys*-like
Emydura freshwater turtle
Emys, emys tortoise
ephippium mounted as on a horse, saddlelike
Eretmochelys oar turtle
erosa eroded, gnawed off
Erymnochelys strong turtle
erythrocephala red head
euphraticus from the Euphrates River
expansa broadened, expanded
femoralis pertaining to the thigh
ferox wild, savage
fimbriatus fringed
flammigera to show fire
flavescens yellowish colored
flavimaculata yellow spotted
flavomarginata, flavomarginatus yellow border
floridana from Florida
formosa finely formed, beautiful
forsteni named for Dr. E. A. Forsten
frenatum bridled
funerea funereal, black
gabonensis from Gabon, Africa
gaigeae named for H. T. Gaige
galbinifrons greenish-yellow front
gangeticus from the Ganges River
Geochelone land turtle
Geoclemys earth tortoise
Geoemyda earth turtle
geoffroanus named for E. Geoffroy Saint-Hilaire
geographica geographic, maplike
geometricus a land-measurer
gibbus bent, hunched
gigantea, gigas giant, huge
Gopherus burrower
grandis huge, large, great
granosa granulated
Graptemys inscribed turtle
grayi named for J. E. Gray
graeca Greek
guadalupensis from the Guadalupe River, Texas
guntheri named for A. C. L. G. Günther
guttata spotted
hainanensis from Hainan Island, China
hamiltonii named for Dr. F. Buchanan-Hamilton
Hardella a proper name, presumably Hardy
hartwegi named for N. E. Hartweg
Heosemys eastern turtle
hermanni named for J. Hermann
herrerai named for A. L. Herrera
Hieremys sacred turtle
hieroglyphica mysterious markings
hilarii named for J. G. Saint-Hilarie
hiltoni named for J. W. Hilton

hippocrepis horseshoe
hirtipes rough foot
hogei named for A. R. Hoge
homeana named for Sir E. Home
Homopus similar feet
hoodensis from Hood Island, Galapagos
horsfieldi named for Dr. T. Horsfield
hoyi named for P. R. Hoy
hurum a word without meaning, coined by Gray
Hydromedusa water dweller
ibera from Georgia, Caucasus of Russia
imbricata overlapping
impressa pressed into, impressed
incisa cut into, indented
indica from India
indopeninsularis from the peninsula of India
Indotestudo Indian tortoise
insculpta engraved
integrum whole, unchanged
intergularis pertaining to the intergular scute
japonica from Japan
Kachuga, kachuga from the Hindi *Kachua*, or the Hindustani *Kachuva*, a turtle
kempii named for R. M. Kemp
Kinixys movable
Kinosternon movable breast
klauberi named for L. M. Klauber
kleinmanni named for Mons. Kleinmann
kohnii named for G. Kohn
kreffti named for G. Krefft
kwangtungensis from Kwangtung Province, China
latisternum broad breast
laurenti named for R. F. Laurent
leithii named for Dr. A. H. Leith
Lepidochelys scaly turtle
leprosa leprous
leucostomum white mouth
leukops white eyes
lewyana belonging to M. Lewy
leytensis from Leyte, Philippines
Lissemys smooth turtle
littoralis from the seashore
longicollis long neck
longifemorale long femoral scute
luteola yellowish color
lutescens yellow-colored
macquarrii from Macquarie River, New South Wales, Australia
macrocephala large head
Macroclemys large tortoise
macrospilota large-spotted
madagascariensis from Madagascar
magdalense from the Magdalena Valley, Michoacan, Mexico
major larger, major
Malaclemys gentle tortoise
Malacochersus soft, dry land
Malayemys Malayan tortoise
malonei named for J. V. Malone
manni named for W. Mann
Manouria rare, domed one
marginata marginated
marmorata marbled
Mauremys dark turtle
mawii named for Lt. Mawe, British Royal Navy
maximiliani named for Prince Maximilian Wied-Neuwied
mccordi named for W. P. McCord

megacephalum, megalocephala huge head
Melanochelys black turtle
melanonota black back
melanosterna black chest
mertensi named for R. Mertens
Mesoclemmys middle tortoise
mexicana from Mexico
miaria stained
microphyes small in stature
minor smaller, lesser
mobilensis from Mobile, Alabama
Morenia pertaining to sluggishness
mouhotii named for H. Mouhot
muhlenbergii named for G. H. E. Muhlenberg
murrayi named for L. T. Murray
mutica curtailed, cut off, unarmed, smooth
mydas wet
nanus dwarfed, small
nasuta, nasutus large nose
natalensis from Natal, Africa
nebulosa dark, clouded
nelsoni (Pseudemys) named for G. Nelson
nelsoni (Terrapene) named for E. W. Nelson
niger black
nigra black
nigricans black
nigrinoda black-knobbed
nigrita black
Nilssonia presumably referring to the proper name
  Nilsson
nogueyi named for G. Noguey
Notochelys back turtle
novaeguineae from New Guinea
oaxacae from Oaxaca, Mexico
oblonga oblong
Ocadia a word without meaning, coined by Gray
ocellata with eyes
oculifera eye-bearing
odoratum smelly
olivacea olive-colored
orbicularis circular
Orlitia a word without meaning, coined by Gray
ornata ornate
osceola from Osceola, Florida
ouachitensis from the Ouachita River, Louisiana
Palea wattles
pallida, pale
pallidipectoris pale breast
pallidipes pale foot
pani named for Pan Jionghua
pardalis like a leopard
parietalis pertaining to the parietal scale
parkeri (Chelodina) named for F. Parker
parkeri (Melanochelys) named for H. W. Parker
peersi named for V. S. Peers
peguense from Pegu, Tenasserim, Burma
Pelochelys mud turtle
Pelodiscus mud, disc
Pelomedusa mud dweller
peltifer shield-bearer
Peltocephalus shield head
Pelusios mud
peninsularis pertaining to a peninsula
perixantha yellow rim
petersi named for W. Peters
phantastica a product of fantasy
phayrei named for Capt. Phayre

Phrynops toad face
picta painted
pileata capped, covered with a cap
planicauda flat tail
Platemys flat tortoise
platycephala flat head
platynota flat back
Platysternon flat breast
Podocnemis foot leg
polyphemus many-voiced
postinguinale posteriorly placed inguinal scute
Psammobates sand walker
Pseudemydura false *Emydura*
Pseudemys false turtle
pseudogeographica falsely maplike
pulcherrima beautiful
pulchra beautiful
punctata spotted, punctuated
punctularia spotted, punctuated
Pyxidea boxlike
Pyxis box
quadriocellata bearing four ocelli
radiata rayed
radiolata rayed
Rafetus from the Raft River, India
reevesii named for J. R. Reeves, Jr.
reticularia netlike
Rheodytes current or stream diver
Rhinoclemmys nose turtle
rhizophorarum bearing roots
rhodesianus from Rhodesia (Zimbabwe)
rivulata stream
robertmertensi named for R. Mertens
rogerbarbouri named for R. W. Barbour
rossignonii named for J. Rossignon
rubida reddish color
rubriventris red-bellied
rufipes red foot
rugosa wrinkled, creased
sabinensis from the Sabine River, Louisiana
Sacalia scaled
salvinii named for O. Salvin
schlegelii named for H. Schlegel
scorpioides scorpionlike
scripta written
scutata armed with scales
senegalensis from Senegal, Africa
seriei Indian peoples
serpentina snakelike
sextuberculata bearing six tubercles
seychellensis from the Seychelles Islands, Indian Ocean
shiui named for O. Shiu
siebenrocki named for F. K. Siebenrock
Siebenrockiella named for F. K. Siebenrock
signata, signatus marked
silvatica forest dweller
sinensis from China
sinuatus bent, curved
smithi named for A. Smith
sonoriense from Sonora, Mexico
spekii named for J. H. Speke
spengleri named for L. Spengler
spinifera bearing spines
spinosa spiny
spixii named for J. B. de von Spix
spooneri named for C. S. Spooner
Staurotypus cross-shaped

steindachneri   named for F. Steindachner
stejnegeri   named for L. Stejneger
subglobosa   rounded beneath
subniger   black bottom
subplana   dim beneath
subrubrum, subrufa   red underside
subtrijuga   almost three-keeled
sulcata   furrowed, grooved
suwanniensis   from the Suwannee River, Florida
swinhoei   named for R. Swinhoe
sylhetensis   from Sylhet, Kahasia Hills, Bangladesh
tarascense   named after Tarascas Indian tribe, Michoacan,
  Mexico
taylori   named for E. H. Taylor
tcheponensis   from Tchepone, (=Sapone), southern Laos
tecta   roofed
tectifera   roofed
temminckii   named for C. J. Temminck
tentoria, tentorius   extended
tequesta   named after Tequesta Indian tribe
terrapen   turtle
Terrapene   turtle
terrapin   turtle
terrestris   of the land
Testudo   tortoise
texana   from Texas
thermalis   pertaining to heat
thurjii   thurgy, an Indian colloquial name for *Hardella
  thurjii*
tornieri   named for G. Tornier
Trachemys   rough tortoise
travancorica   from Travancore, India

tricarinata   three-keeled
trifasciata   three bands
trijuga   three-keeled
trimeni   named for R. Trimen
Trionyx   three-clawed one
triporcatus   triple ridged
tristernalis   three breast plates
triunguis   three-clawed
trivittata   three-striped
troostii   named for G. Troost
tuberculatus   knobbed
tuberosus   full of humps
umbrina   dark
unifilis   single thread
upembae   from Upemba National Park, Shaba Province,
  Zaire
vandenburghi   named for J. Van Denburgh
vanderhaegei   named for M. Vanderhaege
venusta   elegant, charming
verroxii   named for J. B. E. Verreaux
versa   turn, change
vicina   near
vogeli   named for Z. Vogel
vogli   named for C. Vogl
wermuthi   named for H. Wermuth
williamsi   named for E. E. Williams
yaquia   from the Yaqui drainage, Sonora, Mexico
yniphora   bearing, united (a fused gular scute)
yucatana   from Yucatán, Mexico
yunnanensis   from Yunnan Province, China
zarudnyi   named for N. Zarudny
zuliae   from Edo. Zulia, Venezuela

Index